Paul Stewart and Chris Bingham (Eds.)

Electrical Power and Energy Systems for Transportation Applications

MDPI

This book is a reprint of the Special Issue that appeared in the online, open access journal, *Energies* (ISSN 1996-1073) from 2014–2016 (available at: http://www.mdpi.com/journal/energies/special_issues/Electrical_Power_Energy_Systems).

Guest Editors
Paul Stewart
Institute for Innovation in Sustainable Engineering
UK

Chris Bingham
School of Engineering, University of Lincoln
UK

Editorial Office
MDPI AG
St. Alban-Anlage 66
Basel, Switzerland

Publisher
Shu-Kun Lin

Senior Assistant Editor
Guoping (Terry) Zhang

1. Edition 2017

MDPI • Basel • Beijing • Wuhan • Barcelona • Belgrade

ISBN 978-3-03842-242-6 (Hbk)
ISBN 978-3-03842-207-5 (PDF)

Table of Contents

List of Contributors

Jan Becker Institute for Power Electronics and Electrical Drives (ISEA), RWTH Aachen University, 52066 Aachen, Germany; Aachen Research Alliance, JARA-Energy, 52425 Jülich, Germany.

Joan Bergas-Jané Centre d'Innovació Tecnològica en Convertidors Estàtics i Accionaments (CITCEA-UPC), Departament d'Enginyeria Elèctrica, Universitat Politècnica de Catalunya, ETS d'Enginyeria Industrial de Barcelona, 08028 Barcelona, Spain.

Chris Bingham School of Engineering, University of Lincoln, Brayford Pool, Lincoln LN6 7TS, UK.

Wanli Cai State Key Laboratory of Advanced Electromagnetic Engineering and Technology, School of Electrical & Electronic Engineering, Huazhong University of Science and Technology, Wuhan 430074, China.

Joo-Ho Choi School of Aerospace and Mechanical Engineering, Korea Aerospace University, Goyang 412-791, Korea.

Kil To Chong Department of Electronics & Information Engineering, Chonbuk National University, Jeonju 567, Korea.

Thierry Coosemans Electrotechnical Engineering and Energy Technology, MOBI Research Group, Vrije Universiteit Brussel, 1050 Brussels, Belgium.

Cedric De Cauwer Electrotechnical Engineering and Energy Technology, MOBI Research Group, Vrije Universiteit Brussel, 1050 Brussels, Belgium.

Jong Deuk Lee Department of Electronics & Information Engineering, Chonbuk National University, Jeonju 567, Korea.

Johan Driesen Faculty of Engineering, Department of Electrical Engineering, Division Electrical Energy & Computer Architectures, KU Leuven, 3001 Leuven, Belgium.

Feng Gao The Key Laboratory of Smart Grid of Ministry of Education, Tianjin University, Tianjin 300072, China.

Rundong Ge Electrical and Electronic Engineering Institute, Mailbox 435, North China Electric Power University, Changping District, Beijing 102206, China.

Frederik Geth Faculty of Engineering, Department of Electrical Engineering, Division Electrical Energy & Computer Architectures, KU Leuven, 3001 Leuven, Belgium.

Shovon Goutam Mobility, Logistic and Automotive Technology Research Center (MOBI), Department of Electrical Engineering and Energy Technology (ETEC), Vrije Universiteit Brussel, Brussel 1050, Belgium.

Chenglin Gu State Key Laboratory of Advanced Electromagnetic Engineering and Technology, School of Electrical & Electronic Engineering, Huazhong University of Science and Technology, Wuhan 430074, China.

Takashi Hikihara Department of Electrical Engineering, Kyoto University, Katsura, Nishikyo, Kyoto 615-8510, Japan.

Ende Hu The State Key Laboratory of Power Transmission Equipment & System Security and New Technology, College of Electrical Engineering, Chongqing University, Chongqing 400044, China.

Xiaodong Hu State Key Laboratory of Advanced Electromagnetic Engineering and Technology, School of Electrical & Electronic Engineering, Huazhong University of Science and Technology, Wuhan 430074, China.

Zhaoguang Hu Department of Economy and Energy Supply & Demand Research, State Grid Energy Research Institute, Beijing 102200, China.

Yin Hua National Engineering Laboratory for the Automotive Electronic Control Technology, Shanghai Jiao Tong University, Shanghai 200240, China,

Taek Won Jeong Department of Electronics & Information Engineering, Chonbuk National University, Jeonju 567, Korea.

Jiuchun Jiang Collaborative Innovation Center of Electric Vehicles in Beijing, Haidian District, Beijing 100044, China; National Active Distribution Network Technology Research Center (NANTEC), Beijing Jiaotong University, Haidian District, Beijing 100044, China.

Yinxi Jin Department of Electrical Engineering, Harbin Institute of Technology, Harbin 150080, China.

Yoo Cheol Jung Department of Software Engineering, Chonbuk National University, Jeonju 567, Korea.

Yong Kang State Key Laboratory of Advanced Electromagnetic Engineering and Technology, School of Electrical and Electronic Engineering, Huazhong University of Science and Technology, Wuhan 430074, Hubei, China.

Alexandros Kordonis Department of Electrical Engineering, Kyoto University, Katsura, Nishikyo, Kyoto 615-8510, Japan.

Baoquan Kou Department of Electrical Engineering, Harbin Institute of Technology, Harbin 150080, China.

Jaewook Lee School of Aerospace and Mechanical Engineering, Korea Aerospace University, Goyang 412-791, Korea.

Niels Leemput Faculty of Engineering, Department of Electrical Engineering, Division Electrical Energy & Computer Architectures, KU Leuven, 3001 Leuven, Belgium.

Botong Li The Key Laboratory of Smart Grid of Ministry of Education, Tianjin University, Tianjin 300072, China.

Gengyin Li State Key Laboratory of Alternate Electrical Power System with Renewable Energy Sources, North China Electric Power University, Beijing 102206, China.

Huiyong Li Electrical and Electronic Engineering Institute, Mailbox 435, North China Electric Power University, Changping District, Beijing 102206, China.

Mian Li University of Michigan-Shanghai Jiao Tong University Joint Institute, Shanghai Jiao Tong University, Shanghai 200240, China; National Engineering Laboratory for the Automotive Electronic Control Technology, Shanghai Jiao Tong University, Shanghai 200240, China.

Minying Li Guangdong Zhicheng Champion Co., Dongguan 523718, Guangdong, China.

Xiaoyu Li School of Electrical Engineering and Automation, Harbin Institute of Technology, Harbin 150001, China.

Ruijin Liao The State Key Laboratory of Power Transmission Equipment & System Security and New Technology, College of Electrical Engineering, Chongqing University, Chongqing 400044, China.

Jinxiang Liu School of Mechanical Engineering, Beijing Institute of Technology, Beijing 100081, China.

Wenying Liu Electrical and Electronic Engineering Institute, Mailbox 435, North China Electric Power University, Changping District, Beijing 102206, China.

Fabrice Locment Sorbonne University, Université de Technologie de Compiègne, EA 7284 AVENUES, Centre Pierre Guillaumat CS 60319, Compiègne 60203 Cedex, France.

Bo Long School of Mechanical, Electronic, and Industrial Engineering, University of Electronic Science and Technology of China, Chengdu 611731, China.

Rengui Lu School of Electrical Engineering and Automation, Harbin Institute of Technology, Harbin 150001, China.

Quancheng Lv Electrical and Electronic Engineering Institute, Mailbox 435, North China Electric Power University, Changping District, Beijing 102206, China.

Chengbin Ma University of Michigan-Shanghai Jiao Tong University Joint Institute, Shanghai Jiao Tong University, Shanghai 200240, China.

Fukang Ma School of Mechanical and Power Engineering, North University of China, Taiyuan 030051, China; School of Mechanical and Vehicle Engineering, Beijing Institute of Technology, Beijing 100081, China.

Xuerui Ma National Engineering Laboratory for Automotive Electronic Control Technology, Shanghai Jiao Tong University, Shanghai 200240, China.

Shihong Miao State Key Laboratory of Advanced Electromagnetic Engineering and Technology, Huazhong University of Science and Technology, Wuhan 430074, China.

Daichi Nishihara Department of Electrical Engineering, Kyoto University, Katsura, Nishikyo, Kyoto 615-8510, Japan.

Liyong Niu Collaborative Innovation Center of Electric Vehicles in Beijing, Haidian District, Beijing 100044, China; National Active Distribution Network Technology Research Center (NANTEC), Beijing Jiaotong University, Haidian District, Beijing 100044, China.

Pol Olivella-Rosell Centre of Technological Innovation in Static Converters and Drives, Department of Electrical Engineering, College of Industrial Engineering of Barcelona, Universitat Politècnica de Catalunya-BarcelonaTech, 187-08036 Barcelona, Spain/

Noshin Omar Mobility, Logistic and Automotive Technology Research Center (MOBI), Department of Electrical Engineering and Energy Technology (ETEC), Vrije Universiteit Brussel, Brussel 1050, Belgium.

Xiaohong Ran State Key Laboratory of Advanced Electromagnetic Engineering and Technology, Huazhong University of Science and Technology, Wuhan 430074, China.

Matthias Rogge Juelich Aachen Research Alliance, JARA-Energy, 52425 Juelich, Germany; Electrochemical Energy Conversion and Storage Systems Group, Institute for Power Electronics and Electrical Drives (ISEA), RWTH Aachen University, 52066 Aachen, Germany.

Susanne Rothgang Aachen Research Alliance, JARA-Energy, 52425 Jülich, Germany; Institute for Power Electronics and Electrical Drives (ISEA), RWTH Aachen University, 52066 Aachen, Germany.

Dirk Uwe Sauer Juelich Aachen Research Alliance, JARA-Energy, 52425 Juelich, Germany; Institute for Power Generation and Storage Systems (PGS), Institute for Power Electronics and Electrical Drives (ISEA), RWTH Aachen University, Aachen, Germany.

Manuela Sechilariu Sorbonne University, Université de Technologie de Compiègne, EA 7284 AVENUES, Centre Pierre Guillaumat CS 60319, Compiègne 60203 Cedex, France.

Li-Wei Shi Aero-Power Science Technology Center, Nanjing University of Aeronautics and Astronautics, Nanjing 210016, China.

Kai Song School of Electrical Engineering and Automation, Harbin Institute of Technology, Harbin 150001, China.

Paul Stewart Institute for Innovation in Sustainable Engineering, Lonsdale House, Derby DE1 3EE, UK.

Andreas Sumper Centre of Technological Innovation in Static Converters and Drives, Department of Electrical Engineering, College of Industrial Engineering of Barcelona, Universitat Politècnica de Catalunya-BarcelonaTech, 187-08036 Barcelona, Spain.

Wenpeng Sun The Key Laboratory of Smart Grid of Ministry of Education, Tianjin University, Tianjin 300072, China.

Woosuk Sung Research and Development Division, Hyundai Motor Company, Hwaseong 445-706, Korea.

Ryo Takahashi Department of Electrical Engineering, Kyoto University, Katsura, Nishikyo, Kyoto 615-8510, Japan.

Jean-Marc Timmermans Mobility, Logistic and Automotive Technology Research Center (MOBI), Department of Electrical Engineering and Energy Technology (ETEC), Vrije Universiteit Brussel, Brussel 1050, Belgium.

Peter Van den Bossche Department of Industrial Engineering (INDI); Mobility, Logistic and Automotive Technology Research Center (MOBI), Department of Electrical Engineering and Energy Technology (ETEC), Vrije Universiteit Brussel, Brussel 1050, Belgium.

Joeri Van Mierlo Mobility, Logistic and Automotive Technology Research Center (MOBI), Department of Electrical Engineering and Energy Technology (ETEC), Vrije Universiteit Brussel, Brussel 1050, Belgium.

Juan Van Roy Faculty of Engineering, Department of Electrical Engineering, Division Electrical Energy & Computer Architectures, KU Leuven, 3001 Leuven, Belgium.

Roberto Villafafila-Robles Centre d'Innovació Tecnològica en Convertidors Estàtics i Accionaments (CITCEA-UPC), Departament d'Enginyeria Elèctrica, Universitat Politècnica de Catalunya, EU d'Enginyeria Tècnica Industrial de Barcelona, 187-08036 Barcelona, Spain.

Guo Wei School of Electrical Engineering and Automation, Harbin Institute of Technology, Harbin 150001, China.

Sebastian Wollny Electrochemical Energy Conversion and Storage Systems Group, Institute for Power Electronics and Electrical Drives (ISEA), RWTH Aachen University, 52066 Aachen, Germany.

Yingjie Wu National Engineering Laboratory for Automotive Electronic Control Technology, Shanghai Jiao Tong University, Shanghai 200240, China; State Key Laboratory of Advanced Electromagnetic Engineering and Technology, Huazhong University of Science and Technology, Wuhan 430074, China.

Min Xu National Engineering Laboratory for the Automotive Electronic Control Technology, Shanghai Jiao Tong University, Shanghai 200240, China.

Shimin Xue The Key Laboratory of Smart Grid of Ministry of Education, Tianjin University, Tianjin 300072, China.

Jin Yang School of Engineering and Applied Science, Aston University, Birmingham B4 7ET, UK.

Lijun Yang The State Key Laboratory of Power Transmission Equipment & System Security and New Technology, College of Electrical Engineering, Chongqing University, Chongqing 400044, China.

Yuqing Yang Collaborative Innovation Center of Electric Vehicles in Beijing; National Active Distribution Network Technology Research Center (NANTEC), Beijing Jiaotong University, Haidian District, Beijing 100044, China.

Chengliang Yin National Engineering Laboratory for Automotive Electronic Control Technology, Shanghai Jiao Tong University, Shanghai 200240, China.

Shifei Yuan National Engineering Laboratory for Automotive Electronic Control Technology, Shanghai Jiao Tong University, Shanghai 200240, China.

Yuan Yuan The Department of Material Engineering, Chongqing University, Chongqing 400044, China.

Fujun Zhang School of Mechanical and Vehicle Engineering, Beijing Institute of Technology, Beijing 100081, China.

He Zhang Department of Electrical Engineering, Harbin Institute of Technology, Harbin 150080, China.

Lu Zhang Department of Electrical Engineering, Harbin Institute of Technology, Harbin 150080, China.

Shuanlu Zhang School of Mechanical Engineering, Beijing Institute of Technology, Beijing 100081, China.

Weige Zhang Collaborative Innovation Center of Electric Vehicles in Beijing, Haidian District, Beijing 100044, China; National Active Distribution Network Technology Research Center (NANTEC), Beijing Jiaotong University, Haidian District, Beijing 100044, China.

Yan Zhang School of Mechanical and Vehicle Engineering, Beijing Institute of Technology, Beijing 100081, China.

Yu Zhang State Key Laboratory of Advanced Electromagnetic Engineering and Technology, School of Electrical and Electronic Engineering, Huazhong University of Science and Technology, Wuhan 430074, Hubei, China.

Changlu Zhao School of Mechanical and Vehicle Engineering, Beijing Institute of Technology, Beijing 100081, China.

Chen Zhao University of Michigan-Shanghai Jiao Tong University Joint Institute, Shanghai Jiao Tong University, Shanghai 200240, China.

Zhenfeng Zhao School of Mechanical and Vehicle Engineering, Beijing Institute of Technology, Beijing 100081, China.

Yanan Zheng Department of Economy and Energy Supply & Demand Research, State Grid Energy Research Institute, Beijing 102200, China.

Bo Zhou Aero-Power Science Technology Center, Nanjing University of Aeronautics and Astronautics, Nanjing 210016, China.

Ming Zhou State Key Laboratory of Alternate Electrical Power System with Renewable Energy Sources, North China Electric Power University, Beijing 102206, China.

Zhengxing Zuo School of Mechanical Engineering, Beijing Institute of Technology, Beijing 100081, China.

About the Guest Editors

Paul Stewart has been appointed as Research Chair in Energy and Environment in the Institute for Innovation in Sustainable Engineering (IISE) at the University of Derby. He will join IISE to lead and develop an industry focused research and enterprise group in fundamental and applied Research and Development (R&D) activity across the energy, automotive and aerospace sectors, growing IISE research, industrial innovation energy futures and legislation. He will also establish the structure for university-wide multidisciplinary collaboration in energy research and innovation.

Chris Bingham received the B.Eng. degree in Electronic Systems and Control Engineering, from Sheffield City Polytechnic, in 1989, and the M.Sc. (Eng.) degree in Control Systems Engineering, at the University of Sheffield, in 1990. He was awarded a Ph.D. by Cranfield University, U.K., in 1994, for research on control systems to accommodate non-linear dynamic effects in aerospace flight–surface actuators. From 1994 to 2010, Dr Bingham held academic positions at the University of Sheffield, as a researcher and Senior Lecturer. He now holds the position of Professor of Energy Conversion in the School of Engineering at the University of Lincoln, UK.

Preface to "Electrical Power and Energy Systems for Transportation Applications"

Dear colleagues, greetings from the Special Issue Editors.

This book contains the successful invited submissions to a Special Issue of Energies Journal on the subject area of "Electrical Power and Energy Systems for Transportation Applications".

Electrical power and energy systems are at the forefront of application developments in, for example, more-electric and all-electric aircraft, electric and hybrid road vehicles and marine propulsion applications. The associated hardware, technologies and control methods are crucial to achieving critical global targets in energy efficiency, low-carbon and low-emissions operations. The greatest challenges occur when we combine new technologies at large-scale and often complex system levels.

Topics of interest for the call included, but were not limited to:

- Novel Electrical Power Systems architectures and technologies;
- Energy vectors, integration with renewables, power and energy dense machines, converters and energy storage;
- Air, land and sea vehicles; electrical propulsion and actuation for land, sea and air vehicles;
- Electrical Machines, Drives, Systems and Applications—AC and DC machines and drives;
- Multiscale systems modeling; remote monitoring and diagnosis;
- Power Electronic Systems—Converters and emerging technologies;
- Modeling simulation and control, reliability and fault tolerance, safety critical operation;
- Electrical Power Generation Systems—Modeling and simulation of electrical power systems;
- Load management; power quality; distribution reliability; distributed and islanded power systems, sensor networks, communication and control;
- Electrical Power Systems Modeling and Control—Modeling and control methodologies and applications;
- Intelligent systems; optimization and advanced heuristics; adaptive systems; robust control.

Response to our call was excellent, with the following statistics:

- Submission: (101);
- Publication: (25);
- Rejection: (76);
- Article Type: Review Article (0); Research Article (25);

Authors' geographical distribution (published papers):

- China (16)
- Belgium (3)
- Spain (2)
- Korea (2)
- Germany (2)

Published submissions inform the broad spectrum of technologies interfacing energy with transport and fall into four general areas of Renewables and Transmission, Generators, Batteries, and Electric Vehicles as exemplified in the following diagram.

We found the task of editing and selecting papers for this collection to both stimulating and rewarding. We would also like to thank the staff and reviewers of Energies for their effort and input.

Yours…..

Prof. Paul Stewart
Prof. Chris Bingham
Guest Editors

Paul Stewart and Chris Bingham
Guest Editors

PID Controller Design for UPS Three-Phase Inverters Considering Magnetic Coupling

Yu Zhang, Minying Li and Yong Kang

Abstract: In three-phase inverters used in uninterruptible power supplies (UPSs), three-limb inductors and three-limb transformers are commonly used in consideration of cost and size. However, magnetic coupling exists between the three phases of the inverter, which can result in complex models. When instantaneous feedback control strategies are introduced to achieve high quality output waveforms, the transient analysis of the closed-loop inverters becomes difficult. In this paper, the phenomenon of magnetic coupling in three-phase inverters due to three-limb inductors and three-limb transformers is analyzed. A decoupled dynamic model is derived based on the instantaneous symmetrical components transformation, which comprises three decoupled equivalent circuits of instantaneous symmetrical components. Analyses based on this model indicate that magnetic coupling may have a significant impact on the performance of three-phase inverters under unbalanced load conditions and transient responses. For three-phase inverters in UPSs with Proportional-Integral-Differential (PID) closed-loop control strategies, the interactive influence between instantaneous closed-loop regulation and magnetic coupling is researched. Finally, a method of reliability analysis and PID controller design for inverters with magnetic coupling is derived. Simulation and experiment results validate the model and conclusions.

Reprinted from *Energies*. Cite as: Zhang, Y.; Li, M.; Kang, Y. PID Controller Design for UPS Three-Phase Inverters Considering Magnetic Coupling. *Energies* **2014**, 7, 8036–8055.

1. Introduction

In consideration of cost and size, the three-phase transformers and three-phase inductors of three-phase inverters used in uninterruptible power supplies (UPS) normally have magnet cores with a three-limb structure, also known as a core-type structure [1–3]. Hence, coupling exists between the three phases of the inverter because the main flux of each limb must pass through the other two limbs. In addition, a pulse width modulation (PWM) inverter in a UPS must introduce various instantaneous feedback control strategies, such as PID, multiple feedback controls, *etc.* to adapt for diverse, unpredictable nonlinear loads [4,5]. However, in the design of instantaneous control strategies, the appropriate dynamic models become complex when magnetic coupling is considered. Typically this coupling has been ignored, in other words, a three-limb transformer is treated as the combination of three single-phase transformers and a three-limb inductor is treated as the combination of three single-phase inductors [2,3]. For many practical problems, this approximation is acceptable; however, its applicability is not fully understood.

While magnetic coupling is not considered for three-phase inverters, dynamic models in the d-q rotation frame or in the α-β stationary frame are simpler than in the a-b-c frame. Two sets of regulators, for positive and negative sequence components, can be introduced for unbalanced load conditions [6–11]. Unfortunately, even ignoring magnetic coupling, transient analysis is complex due to strong coupling between axes [12–14].

In three-phase inverters, an unbalanced condition would occur when unbalanced loads or unbalanced bridge voltages are present. Such conditions can also appear during transient responses such as a sudden change of load, though even in steady state analyses based on symmetrical components (SC), it has been concluded that ignoring magnetic coupling in three-limb transformers and three-limb inductors is not correct under unbalanced conditions [15]. During transient responses, magnetic coupling cannot be ignored either.

SCs have been used for analyzing three-limb transformers and three-limb inductors under unbalanced conditions for many years, especially for fault situations of power systems [15–17]. Normally, periodic three-phase variables in each circle are assumed. They are analyzed by their fundamental and harmonic SCs separately [18]. For three-limb transformers, a duality derived model [19] and an equivalent circuit model in the a-b-c stationary frame have been proposed [20]. However, such models based on SCs in the frequency domain are more suitable for analyzing slow transient responses in a power system. In inverters where transient responses are fast, the three-phase variables can no longer be treated as periodic waves. Hence they cannot be expressed by SCs.

For inverters with a three-limb transformer and three-limb inductor, to obtain optimal instantaneous closed-loop control, magnetic coupling should be considered in the dynamic models so that transient performance can be properly considered. To the authors' knowledge, transient models in the design of instantaneous closed-loop controllers while considering the magnetic coupling has been seldom reported in literature.

A transformation known as the instantaneous symmetrical components transformation (ISCT) has been proposed that can be used in the analysis of transient responses in AC motors [21,22]. Its transformation matrix is similar to the symmetrical components transformation (SCT), but is performed on three-phase instantaneous variables. The transformation results in instantaneous symmetrical components (ISC), which have also been used for extracting instantaneous values or disturbances in three-phase power systems [9,23–25]. Because ISC-based dynamic models for three-limb transformers and three-limb inductors are simpler in formulation [26,27], it may also be a useful tool when magnetic coupling is taken into account in transient responses of inverters. However, it has been seldom reported in the literature.

In this paper, magnetic coupling is studied using ISCs, and we proceed as follows: First, in Section 2, the instantaneous symmetrical component transformation is reviewed. Next, in Section 3, a transient model for a three-phase PWM inverter is derived based on ISCs which considers magnetic coupling due to a three-limb inductor and three-limb transformer. Based on this model, in Section 4, the transient performance of three-phase inverters is evaluated, especially considering the influence of magnetic coupling. Furthermore, for inverters with PID closed-loop control strategies, the influence between instantaneous closed-loop control and magnetic coupling is analyzed. Then, a performance analysis and controller design for a three-phase inverter with magnetic coupling is derived. Finally, in Section 5, the results of the simulations are presented and validated with experiments and Section 6 concludes the paper.

2. The Instantaneous Symmetrical Components Transformation

Main text paragraph. In this section, the instantaneous symmetrical components transformation will be reviewed. The instantaneous symmetrical components transformation (ISCT), introduced by Lyon, W.V. [21,28], is expressed as:

$$
\begin{bmatrix} x^+(t) \\ x^-(t) \\ x^0(t) \end{bmatrix} = \frac{1}{\sqrt{3}} \begin{bmatrix} 1 & \alpha & \alpha^2 \\ 1 & \alpha^2 & \alpha \\ 1 & 1 & 1 \end{bmatrix} \cdot \begin{bmatrix} x_a(t) \\ x_b(t) \\ x_c(t) \end{bmatrix} \tag{1}
$$

where $\alpha = e^{j2\pi/3}$, the x_a, x_b, x_c are three-phase instantaneous variables, and x^+, x^-, x^0 are the resulting ISCs, including the instantaneous positive-sequence component (IPSC), the instantaneous negative-sequence component (INSC) and the instantaneous zero-sequence component (IZSC), respectively. The inverse transformation is expressed as:

$$
\begin{bmatrix} x_a(t) \\ x_b(t) \\ x_c(t) \end{bmatrix} = \frac{1}{\sqrt{3}} \begin{bmatrix} 1 & 1 & 1 \\ \alpha^2 & \alpha & 1 \\ \alpha & \alpha^2 & 1 \end{bmatrix} \cdot \begin{bmatrix} x^+(t) \\ x^-(t) \\ x^0(t) \end{bmatrix} \tag{2}
$$

According to Equation (1), the IZSC is a real variable and the IPSC and INSC are complex variables and are complex conjugates.

Typically, for normal three-phase, three-bridge inverters shown in Figure 1a, no current flows in the IZSC. In contrast, we consider the three-phase full-bridge inverter shown in Figure 1b, because the IZSC current output from the inverter

3

bridges can flow in the three-limb inductor and three-limb transformer. Hence, the IZSC will influence the inverter and should be included in the dynamic model.

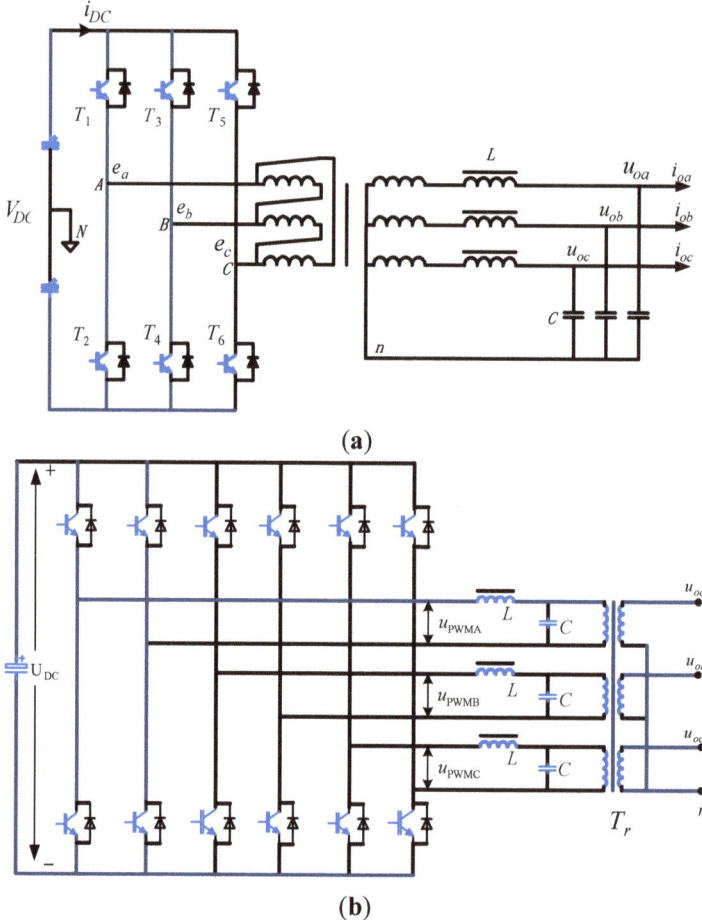

Figure 1. (**a**) Three-phase three bridge inverter; (**b**) Three-phase full bridge inverter.

3. ISC Model for Three-Phase Inverters

In this section, an ISC model for a three-phase inverter is presented, including a model for a three-limb transformer, a model for a three-limb inductor, a model for a three-phase load, and finally a complete model combining the three.

3.1. ISC Model for Three-Limb Transformer

The defining variables of the transformer are shown in Figure 2, where u_A, u_B, u_C and i_A, i_B, i_C are three-phase voltages and currents in the primary windings and

u_a, u_b, u_c and i_a, i_b, i_c are three-phase voltages and currents in the secondary windings. In addition, the number of turns of the primary and secondary windings are given by N_1 and N_2, respectively, φ_A, φ_B, φ_C are the flux linkages in the three primary windings, and φ_a, φ_b, φ_c are the flux linkages in the three secondary windings.

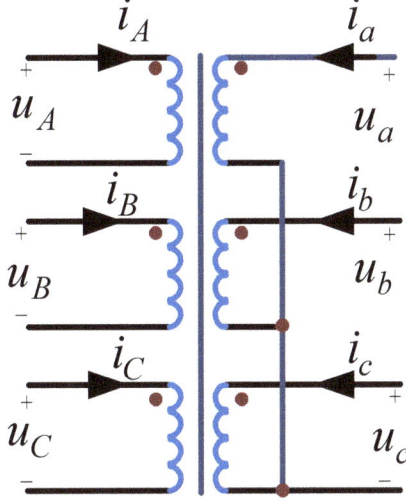

Figure 2. Variables of the six windings in a three-limb transformer.

In each winding, ignoring any nonlinearity, the flux linkage is determined by the currents linearly as:

$$
\begin{bmatrix} \varphi_A(t) \\ \varphi_B(t) \\ \varphi_C(t) \\ \varphi_a(t) \\ \varphi_b(t) \\ \varphi_c(t) \end{bmatrix} = \begin{bmatrix} L_{AA} & M_{AB} & M_{AC} & M_{Aa} & M_{Ab} & M_{Ac} \\ M_{BA} & L_{BB} & M_{BC} & M_{Ba} & M_{Bb} & M_{Bc} \\ M_{CA} & M_{CB} & L_{CC} & M_{Ca} & M_{Cb} & M_{Cc} \\ M_{aA} & M_{aB} & M_{aC} & L_{aa} & M_{ab} & M_{ac} \\ M_{bA} & M_{bB} & M_{bC} & M_{ba} & L_{bb} & M_{bc} \\ M_{cA} & M_{cB} & M_{cC} & M_{ca} & M_{cb} & L_{cc} \end{bmatrix} \times \begin{bmatrix} i_A(t) \\ i_B(t) \\ i_C(t) \\ i_a(t) \\ i_b(t) \\ i_c(t) \end{bmatrix} \tag{3}
$$

and the voltage is determined by the flux linkages and currents [29], as:

$$
\begin{bmatrix} u_A(t) \\ u_B(t) \\ u_C(t) \\ u_a(t) \\ u_b(t) \\ u_c(t) \end{bmatrix} = \begin{bmatrix} r_1 & 0 & 0 & 0 & 0 & 0 \\ 0 & r_1 & 0 & 0 & 0 & 0 \\ 0 & 0 & r_1 & 0 & 0 & 0 \\ 0 & 0 & 0 & r_2 & 0 & 0 \\ 0 & 0 & 0 & 0 & r_2 & 0 \\ 0 & 0 & 0 & 0 & 0 & r_2 \end{bmatrix} \cdot \begin{bmatrix} i_A(t) \\ i_B(t) \\ i_C(t) \\ i_a(t) \\ i_b(t) \\ i_c(t) \end{bmatrix} + \frac{d}{dt} \begin{bmatrix} \varphi_A(t) \\ \varphi_B(t) \\ \varphi_C(t) \\ \varphi_a(t) \\ \varphi_b(t) \\ \varphi_c(t) \end{bmatrix} \tag{4}
$$

5

In Equation (3), L_{xx} (with $x = A, B, C, a, b,$ or c) is the self-inductance of winding x and M_{xy} is the mutual inductance between winding x and winding y. In Equation (4), r_1 is the resistance of the primary windings and r_2 is the resistance of the secondary windings.

Now, assuming that the three-phase windings are symmetrical and that the three limbs of the core have the same permeance λ, then L_{xx} and M_{xy} can be expressed as [29]:

$$L_{AA} = L_{BB} = L_{CC} = L_{1m} + L_{1l} \tag{5}$$

$$L_{aa} = L_{bb} = L_{cc} = L_{2m} + L_{2l} \tag{6}$$

$$M_{AB} = M_{BA} = M_{BC} = M_{CB} = M_{AC} = M_{CA} = -\frac{1}{2}L_{1m} \tag{7}$$

$$M_{ab} = M_{ba} = M_{bc} = M_{cb} = M_{ac} = M_{ca} = -\frac{1}{2}L_{2m} \tag{8}$$

$$M_{Aa} = M_{aA} = M_{Bb} = M_{bB} = M_{Cc} = M_{cC} = M_{12} \tag{9}$$

$$M_{Ab} = M_{Ac} = M_{Ba} = M_{Bc} = M_{Ca} = M_{Cb} = M_{aB} = M_{aC} = M_{bA} = M_{bC} = M_{cA} = M_{cB} = -\tfrac{1}{2}M_{12} \tag{10}$$

In the above equations, L_{1l} is leakage inductance of each primary winding and L_{2l} is leakage inductance of each secondary winding. In addition, L_{1m} and L_{2m} are the inductances due to the main fluxes that flow through the limbs of the primary and secondary windings respectively, and M_{12} is mutual inductance between the primary and secondary winding due to the main fluxes in the limbs, which are given as:

$$L_{1m} = N_1 \cdot (i_A \cdot N_1) \cdot \lambda / i_A = \lambda N_1^2 \tag{11}$$

$$L_{2m} = N_2 \cdot (i_a \cdot N_2) \cdot \lambda / i_a = \lambda N_2^2 \tag{12}$$

$$M_{12} = N_1 \cdot (i_a \cdot N_2) \cdot \lambda / i_a = \lambda N_1 N_2 \tag{13}$$

All parameters of the secondary windings can be given in terms of those on the primary side, so that:

$$L'_{2m} = L_{2m} \cdot (N_1/N_2)^2 = M'_{12} = M_{12} \cdot (N_1/N_2) = L_{1m} \tag{14}$$

and L_{2l} and r_2 of the secondary windings become:

$$L'_{2l} = L_{2l} \cdot (N_1/N_2)^2 \tag{15}$$

$$r'_2 = r_2 \cdot (N_1/N_2)^2 \tag{16}$$

6

For convenience, the prime symbol is omitted in the remainder of this paper. Given the above, Equation (3) becomes:

$$
\begin{bmatrix}
\varphi_A(t) \\
\varphi_B(t) \\
\varphi_C(t) \\
\varphi_a(t) \\
\varphi_b(t) \\
\varphi_c(t)
\end{bmatrix}
=
\begin{bmatrix}
L_{1m}+L_{1l} & -\frac{1}{2}L_{1m} & -\frac{1}{2}L_{1m} & L_{1m} & -\frac{1}{2}L_{1m} & -\frac{1}{2}L_{1m} \\
-\frac{1}{2}L_{1m} & L_{1m}+L_{1l} & -\frac{1}{2}L_{1m} & -\frac{1}{2}L_{1m} & L_{1m} & -\frac{1}{2}L_{1m} \\
-\frac{1}{2}L_{1m} & -\frac{1}{2}L_{1m} & L_{1m}+L_{1l} & -\frac{1}{2}L_{1m} & -\frac{1}{2}L_{1m} & L_{1m} \\
L_{1m} & -\frac{1}{2}L_{1m} & -\frac{1}{2}L_{1m} & L_{1m}+L_{2l} & -\frac{1}{2}L_{1m} & -\frac{1}{2}L_{1m} \\
-\frac{1}{2}L_{1m} & L_{1m} & -\frac{1}{2}L_{1m} & -\frac{1}{2}L_{1m} & L_{1m}+L_{2l} & -\frac{1}{2}L_{1m} \\
-\frac{1}{2}L_{1m} & -\frac{1}{2}L_{1m} & L_{1m} & -\frac{1}{2}L_{1m} & -\frac{1}{2}L_{1m} & L_{1m}+L_{2l}
\end{bmatrix}
\times
\begin{bmatrix}
i_A(t) \\
i_B(t) \\
i_C(t) \\
i_a(t) \\
i_b(t) \\
i_c(t)
\end{bmatrix}
\quad (17)
$$

Next, using ISCT, the decoupled ISC equations can be derived from Equations (4) and (17), and are given as:

$$
\begin{cases}
\varphi_1^+(t) = L_1 \cdot i_1^+(t) + L_m \cdot i_2^+(t) \\
\varphi_1^-(t) = L_1 \cdot i_1^-(t) + L_m \cdot i_2^-(t) \\
\varphi_1^0(t) = L_{1l} \cdot i_1^0(t) \\
\varphi_2^+(t) = L_2 \cdot i_2^+(t) + L_m \cdot i_1^+(t) \\
\varphi_2^-(t) = L_2 \cdot i_2^-(t) + L_m \cdot i_1^-(t) \\
\varphi_2^0(t) = L_{2l} \cdot i_2^0(t)
\end{cases}
\quad (18)
$$

$$
\begin{cases}
u_1^+(t) = r_1 \cdot i_1^+(t) + L_1 \cdot \frac{d}{dt}[i_1^+(t)] + L_m \cdot \frac{d}{dt}[i_2^+(t)] \\
u_1^-(t) = r_1 \cdot i_1^-(t) + L_1 \cdot \frac{d}{dt}[i_1^-(t)] + L_m \cdot \frac{d}{dt}[i_2^-(t)] \\
u_1^0(t) = r_1 \cdot i_1^0(t) + L_{1l} \cdot \frac{d}{dt}[i_1^0(t)] \\
u_2^+(t) = r_2 \cdot i_2^+(t) + L_2 \cdot \frac{d}{dt}[i_2^+(t)] + L_m \cdot \frac{d}{dt}[i_1^+(t)] \\
u_2^-(t) = r_2 \cdot i_2^-(t) + L_2 \cdot \frac{d}{dt}[i_2^-(t)] + L_m \cdot \frac{d}{dt}[i_1^-(t)] \\
u_2^0(t) = r_2 \cdot i_2^0(t) + L_{2l} \cdot \frac{d}{dt}[i_2^0(t)]
\end{cases}
\quad (19)
$$

where $L_m = \frac{3}{2}L_{1m}$, $L_1 = L_{1l} + L_m$ and $L_2 = L_{2l} + L_m$. Also, the φ_1^+, φ_1^-, φ_1^0 and φ_2^+, φ_2^-, φ_2^0 are ISCs of φ_A, φ_B, φ_C and φ_a, φ_b, φ_c, the u_1^+, u_1^-, u_1^0 and u_2^+, u_2^-, u_2^0 are ISCs of μ_A, μ_B, μ_C and u_a, u_b, u_c and the i_1^+, i_1^-, i_1^0 and i_2^+, i_2^-, i_2^0 are ISCs of i_A, i_B, i_C and i_a, i_b, i_c. Here, a subscript "1" of an ISC denotes the ISC in the primary windings, and a subscript "2" denotes the ISC in the secondary windings. Using Equation (19), ISC equivalent circuits can be derived and are shown in Figure 3. It is observed that the equivalent circuits of the IPSC and INSC have the same form, which is different from that of IZSC, as indicated by the dash line in Figure 3c.

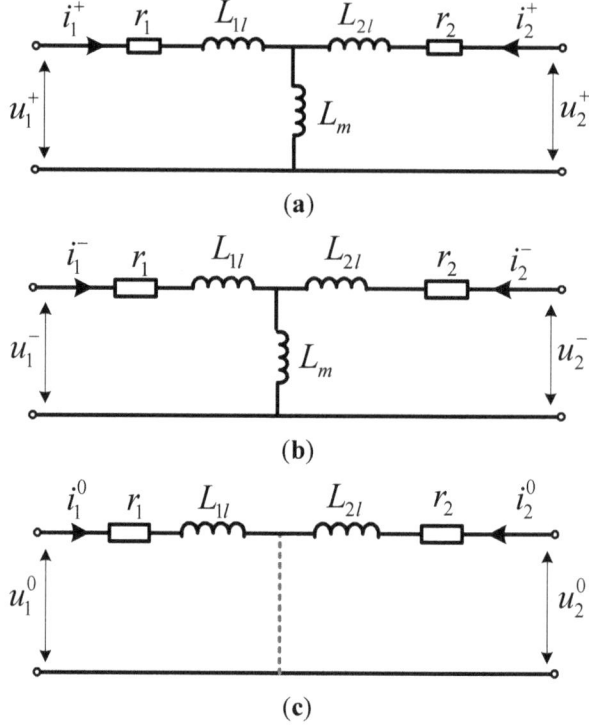

Figure 3. The ISC equivalent circuits for the three-limb transformer: **(a)** IPSC; **(b)** INSC; **(c)** IZSC.

Selecting i_1^+, i_2^+ as state variables, from Equation (19), IPSC state equations are given as:

$$\frac{d}{dt}\begin{bmatrix} i_1^+ \\ i_2^+ \end{bmatrix} = \begin{bmatrix} -\frac{r_1 L_2}{K_m L_m} & \frac{r_2}{K_m} \\ \frac{r_1}{K_m} & -\frac{r_2 L_1}{K_m L_m} \end{bmatrix} \cdot \begin{bmatrix} i_1^+ \\ i_2^+ \end{bmatrix} + \begin{bmatrix} \frac{L_2}{K_m L_m} & -\frac{1}{K_m} \\ -\frac{1}{K_m} & \frac{L_1}{K_m L_m} \end{bmatrix} \cdot \begin{bmatrix} u_1^+ \\ u_2^+ \end{bmatrix} \quad (20)$$

where $K_m = \frac{L_1 \cdot L_2}{L_m} - L_m$. The INSC has the same state equation form as the IPSC. Finally, the IZSC state equations are given as:

$$\frac{d}{dt}\begin{bmatrix} i_1^0 \\ i_2^0 \end{bmatrix} = \begin{bmatrix} -\frac{r_1}{L_{1l}} & 0 \\ 0 & -\frac{r_2}{L_{2l}} \end{bmatrix} \cdot \begin{bmatrix} i_1^0 \\ i_2^0 \end{bmatrix} + \begin{bmatrix} \frac{1}{L_{1l}} & 0 \\ 0 & \frac{1}{L_{2l}} \end{bmatrix} \cdot \begin{bmatrix} u_1^0 \\ u_2^0 \end{bmatrix} \quad (21)$$

3.2. ISC Model for Three-Limb Inductor

The three-limb inductor has the same magnet core as the three-limb transformer, but with only three windings. Variables r_L and L_l are the resistance and leakage inductance for each winding. Their flux linkages are defined as φ_{LA}, φ_{LB}, φ_{LC}, and their voltages and currents are u_{LA}, u_{LB}, u_{LC} and i_{LA}, i_{LB}, i_{LC}, respectively, with all variables related through:

$$\begin{bmatrix} \varphi_{LA}(t) \\ \varphi_{LB}(t) \\ \varphi_{LC}(t) \end{bmatrix} = \begin{bmatrix} L & -\frac{1}{2}L_m & -\frac{1}{2}L_m \\ -\frac{1}{2}L_m & L & -\frac{1}{2}L_m \\ -\frac{1}{2}L_m & -\frac{1}{2}L_m & L \end{bmatrix} \cdot \begin{bmatrix} i_{LA}(t) \\ i_{LB}(t) \\ i_{LC}(t) \end{bmatrix} \tag{22}$$

where $L = L_l + L_m$. Next, ISC equations are derived from Equation (22), as:

$$\begin{cases} \varphi_L^+(t) = L \cdot i_L^+(t) \\ \varphi_L^-(t) = L \cdot i_L^-(t) \\ \varphi_L^0(t) = L_l \cdot i_L^0(t) \end{cases} \tag{23}$$

where φ_L^+, φ_L^-, φ_L^0 are the ISCs of φ_{LA}, φ_{LB}, φ_{LC}, i_L^+, i_L^-, i_L^0 are the ISCs of i_{LA}, i_{LB}, i_{LC}. The equations of three-phase voltages in the inductors can also be simplified using Equation (4). The ISCs of u_{LA}, u_{LB}, u_{LC} are u_L^+, u_L^-, u_L^0, which can be expressed by:

$$\begin{cases} u_L^+(t) = r_L \cdot i_L^+(t) + L \cdot \frac{d}{dt}[i_L^+(t)] \\ u_L^-(t) = r_L \cdot i_L^-(t) + L \cdot \frac{d}{dt}[i_L^-(t)] \\ u_L^0(t) = r_L \cdot i_L^0(t) + L_l \cdot \frac{d}{dt}[i_L^0(t)] \end{cases} \tag{24}$$

are given as the equivalent circuits of the ISCs as shown in Figure 4.

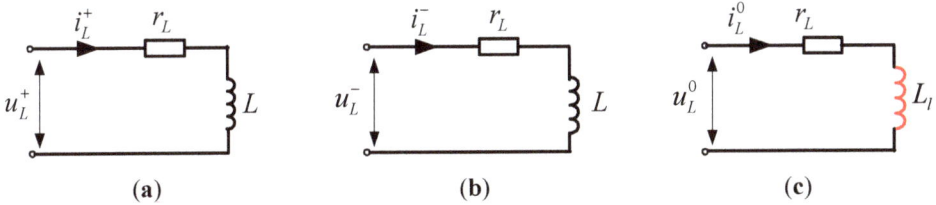

Figure 4. The ISC equivalent circuits for the three-limb inductor: (**a**) IPSC; (**b**) INSC; (**c**) IZSC.

It can be observed from Figure 4 that the equivalent circuit of the IPSC has the same form as that of the INSC, but different from that of the IZSC. Importantly, the impedance seen by i_L^0 is $r_L + j\omega L_l$, not $r_L + j\omega L$.

Finally, if i_L^+ and i_L^0 are selected as state variables, from Equation (24), IPSC and IZSC state equations are derived as:

$$\frac{d}{dt}[i_L^+] = -\frac{r_L}{L} \cdot [i_L^+] + \frac{1}{L} \cdot [u_L^+] \tag{25}$$

9

$$\frac{d}{dt}[i_L^0] = -\frac{r_L}{L_l} \cdot [i_L^0] + \frac{1}{L_l} \cdot [u_L^0] \tag{26}$$

and the INSC has the same state equation form as that of the IPSC.

3.3. ISC Model for Three-Phase Load

For a three-phase load with voltages u_{za}, u_{zb}, u_{zc} and currents i_{za}, i_{zb}, i_{zc}, in the Laplace domain, they are related by:

$$\begin{bmatrix} u_{za}(s) \\ u_{zb}(s) \\ u_{zc}(s) \end{bmatrix} = \begin{bmatrix} z_a & 0 & 0 \\ 0 & z_b & 0 \\ 0 & 0 & z_c \end{bmatrix} \cdot \begin{bmatrix} i_{za}(s) \\ i_{zb}(s) \\ i_{zc}(s) \end{bmatrix} \tag{27}$$

where $z_a = r_a + sL_a$, $z_b = r_b + sL_b$ and $z_c = r_c + sL_c$. Applying the ISCT to Equation (27), the ISC equation is given as:

$$\begin{bmatrix} u_z^+(s) \\ u_z^-(s) \\ u_z^0(s) \end{bmatrix} = \frac{1}{3} \begin{bmatrix} z^0 & z^- & z^+ \\ z^- & z^0 & z^- \\ z^+ & z^- & z^0 \end{bmatrix} \cdot \begin{bmatrix} i_z^+(s) \\ i_z^-(s) \\ i_z^0(s) \end{bmatrix} \tag{28}$$

where $\begin{bmatrix} z^+ \\ z^- \\ z^0 \end{bmatrix} = \begin{bmatrix} 1 & \alpha & \alpha^2 \\ 1 & \alpha^2 & \alpha \\ 1 & 1 & 1 \end{bmatrix} \cdot \begin{bmatrix} z_a \\ z_b \\ z_c \end{bmatrix}$.

For unbalanced loads, $z_a \neq z_b \neq z_c$, thus z^+ and z^- are not zero, while for balanced loads, $z_a = z_b = z_c = r_z + sL_z$, and Equation (28) becomes:

$$\begin{bmatrix} u_z^+(s) \\ u_z^-(s) \\ u_z^0(s) \end{bmatrix} = \begin{bmatrix} r_z + sL_z & 0 & 0 \\ 0 & r_z + sL_z & 0 \\ 0 & 0 & r_z + sL_z \end{bmatrix} \cdot \begin{bmatrix} i_z^+(s) \\ i_z^-(s) \\ i_z^0(s) \end{bmatrix} \tag{29}$$

3.4. ISC Model for Whole Inverter

For three-phase inverters as shown in Figure 1, combining the ISC equivalent circuits of the three-limb inductor and three-limb transformer, the ISC equivalent circuits can be given as shown in Figure 5, where u_{PWM}^+, u_{PWM}^-, u_{PWM}^0 are the ISCs of u_{PWMA}, u_{PWMB}, u_{PWMC}.

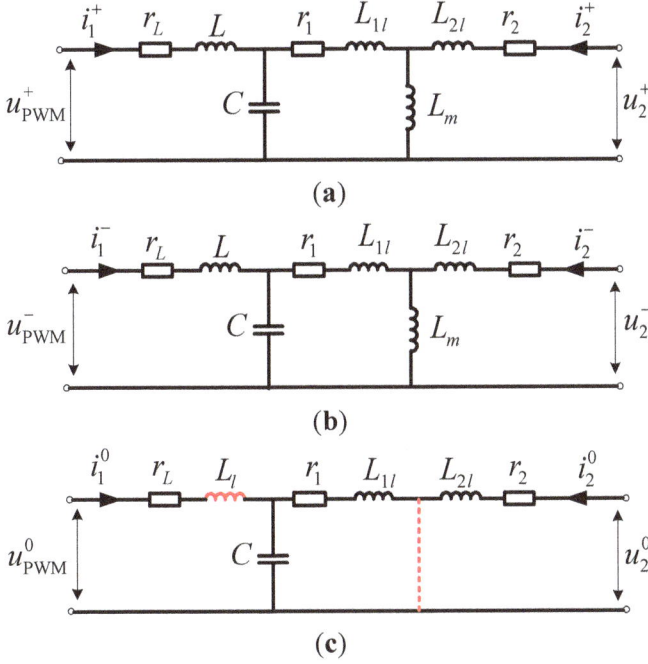

Figure 5. ISC equivalent circuits for three-phase inverters: (**a**) IPSC; (**b**) INSC; (**c**) IZSC.

It should be noted that, a short circuit exists in the IZSC equivalent circuit as indicated by the dashed, red line in Figure 5c, and the inductance of the inductor highlighted in red is L_l, not L. So, this inverter has different characteristics from three-phase inverters with three single-phase inductors and three single-phase transformers, wherein the IPSC, INSC and IZSC all have the same equation.

It is also important to note that coupling does not exist between the IPSC, INSC and IZSC equivalent circuits. Therefore, the equivalent circuits in Figure 5 give an easy way to analyze the transient responses of three-phase inverters with magnetic coupling. In this paper, the transient model for a three-phase inverter with magnetic coupling is expressed by these three ISC equivalent circuits.

4. Transient Characteristics of Three-Phase Inverters Considering Magnetic Coupling

In this section, the transient behavior of three-phase inverters will be considered including magnetic coupling. To begin, in Figure 5c, the impedance of capacitor C

is given as $Z_C = \frac{1}{sC}$. If we define $Z_{Ll} = r_L + sL_l$ and $Z_{1l} = r_1 + sL_{l1}$, the parallel combination of Z_C and Z_{1l} results in:

$$Z_{para} = Z_{1l} \parallel Z_C = \frac{Z_{1l}Z_C}{Z_{1l} + Z_C} = \frac{Z_{1l}}{Z_{1l} \cdot sC + 1} \tag{30}$$

Note that at low frequencies, $|sC| \approx 0$, thus $Z_{para}; Z_{1l}$. The current i_1^0 is given as:

$$i_1^0(s) = \frac{u_{\text{PWM}}^0(s)}{Z_{Ll} + Z_{1l}} = \frac{u_{\text{PWM}}^0(s)}{(r_L + r_1) + s(L_{Ll} + L_{1l})} \tag{31}$$

In three-wire three-bridge inverters, i_1^0 does not exist; however, for inverters such as those in Figure 1b, current i_1^0 is an inherent, serious problem that should be evaluated. At low frequencies, the r_L, r_1, ωL_l, ωL_{1l} are small, and therefore a small voltage u_{PWM}^0 can result in a large current i_1^0, which may lead to over-current in power switches and shutdown of inverters.

For inverters in UPSs, it is necessary to introduce various instantaneous control strategies, such as PID and multiple feedback controls, to adapt for diverse nonlinear loads [4,5]. Among various instantaneous control strategies, the PID control strategy has a simple form and high robustness and is thus analyzed in this paper. Figure 6 gives a diagram of a three-phase inverter with three PID controllers in which the voltages of the three-phase capacitor, u_{CA}, u_{CB}, u_{CC}, are sampled. In Figure 6, $\mathbf{T_r}$ is the three-limb transformer and $\mathbf{L}$ is the three-limb inductor. A PID controller $R(s)$ is used for each phase, which has the form:

$$R(s) = K_P + \frac{K_I}{s} + K_D s \tag{32}$$

Based on the ISC equivalent circuits in Figure 5, the IPSC equivalent circuit of Figure 6 is shown in Figure 7, where u_m^+ is the IPSC of the modulation signals u_{mA}, u_{mB}, u_{mC}; u_C^+ is the IPSC of the capacitor voltages u_{CA}, u_{CB}, u_{CC}; u_{ref}^+ is the IPSC of the reference voltages u_{refA}, u_{refB}, u_{refC}; u_o^+ is the IPSC of the load voltages u_{oa}, u_{ob}, u_{oc}; and i_o^+ is the IPSC of the load currents i_{oa}, i_{ob}, i_{oc}. For inverters in UPSs, currents i_{oa}, i_{ob}, i_{oc} and their ISCs are treated as disturbances in the model because the diverse sets of loads for UPS are unpredictable.

12

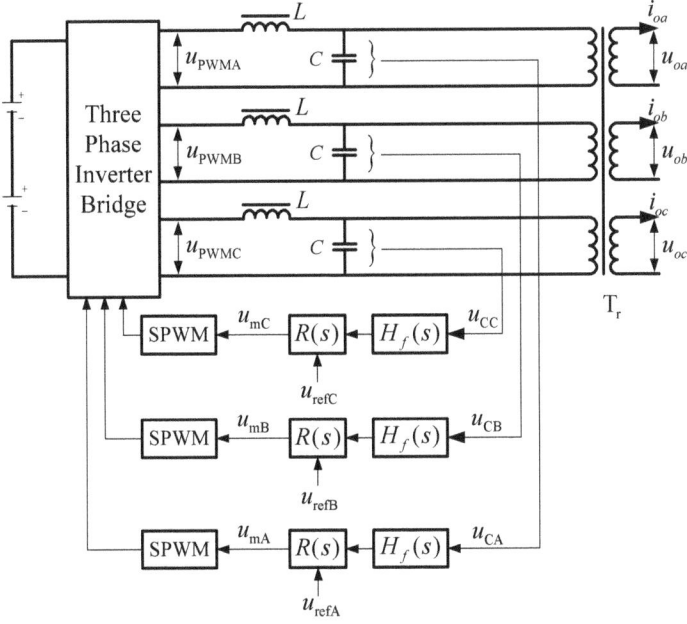

Figure 6. A three-phase inverter with PID closed-loop controllers.

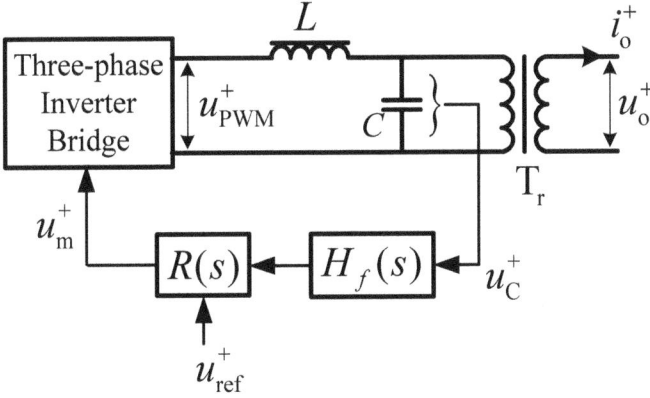

Figure 7. IPSC equivalent circuit for three-phase inverters with a PID controller.

Treating the transistors (see Figure 1), typically IGBTs, as ideal switches, when the DC bus voltage is constant and the switching frequency is high enough, the inverter bridges will amply low frequency modulation signals linearly and can be modeled as an amplifier with gain K_b [4,5], given by

$$u_{PWM}^+(s) = K_b \cdot u_m^+(s) \tag{33}$$

13

Neglecting r_1, r_2 and L_m, the transfer function block diagram of the IPSC derived from Figure 7 is shown in Figure 8. The inductance L_{leak} is defined as:

$$L_{leak} = L_{1l} + L_{2l} \tag{34}$$

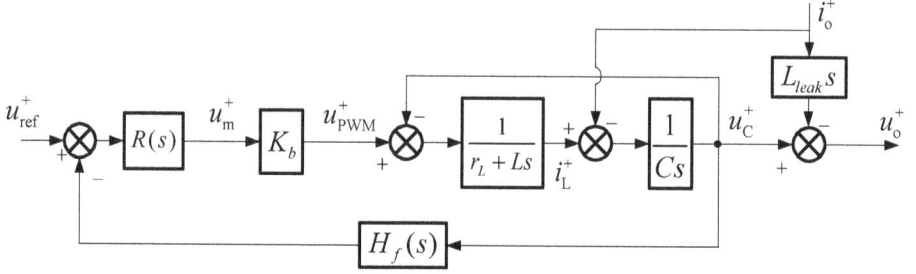

Figure 8. IPSC transfer function diagram for three-phase inverters with instantaneous feedback control.

Hence, from Figure 8, the IPSC transfer function is derived as:

$$u_o^+(s) = \frac{K_b R(s)}{P_{close}(s)} u_{ref}^+(s) - \frac{L(s)}{P_{close}(s)} i_o^+(s) - L_{leak}s \cdot i_o^+(s) \tag{35}$$

where $P_{close}(s) = LCs^2 + r_L Cs + K_b H_f(s)R(s) + 1$ and $L(s) = Ls + r_L$.

Since the INSC is always the complex conjugate of the IPSC, its equivalent circuit and the transfer function all have the same forms as those of IPSC. The IZSC equivalent circuit is shown in Figure 9.

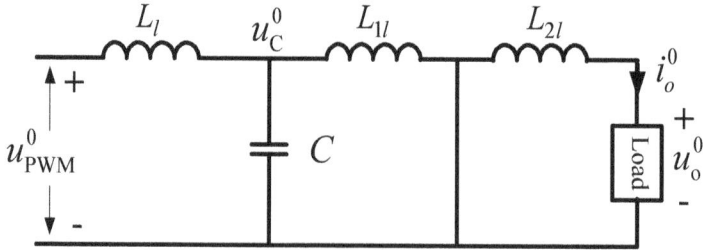

Figure 9. IZSC equivalent circuit for three-phase inverters with instantaneous feedback control.

From Figure 9, the IZSC of the load voltages can be expressed as:

$$u_o^0(s) = -Z_{inv}^0(s) \cdot i_o^0(s) \tag{36}$$

14

where the impedance Z_{inv}^0 is:

$$Z_{inv}^0(s) = L_{2l} \cdot s \tag{37}$$

Next, Equation (2) can be used to solve for the three-phase voltages of the load, as:

$$u_{oa}(s) = \frac{K_h R(s)}{P_{close}(s)}[u_{refA}(s) - \frac{u_{ref}^0(s)}{\sqrt{3}}] - \frac{L(s)}{P_{close}(s)}[i_{oa}(s) - \frac{i_o^0(s)}{\sqrt{3}}] - L_{leak}s \cdot i_{oa}(s) + L_{1l}s \cdot \frac{i_o^0(s)}{\sqrt{3}} \tag{38}$$

$$u_{ob}(s) = \frac{K_h R(s)}{P_{close}(s)}[u_{refB}(s) - \frac{u_{ref}^0(s)}{\sqrt{3}}] - \frac{L(s)}{P_{close}(s)}[i_{ob}(s) - \frac{i_o^0(s)}{\sqrt{3}}] - L_{leak}s \cdot i_{ob}(s) + L_{1l}s \cdot \frac{i_o^0(s)}{\sqrt{3}} \tag{39}$$

$$u_{oc}(s) = \frac{K_h R(s)}{P_{close}(s)}[u_{refC}(s) - \frac{u_{ref}^0(s)}{\sqrt{3}}] - \frac{L(s)}{P_{close}(s)}[i_{oc}(s) - \frac{i_o^0(s)}{\sqrt{3}}] - L_{leak}s \cdot i_{oc}(s) + L_{1l}s \cdot \frac{i_o^0(s)}{\sqrt{3}} \tag{40}$$

The corresponding block diagram for Equations (38)–(40) is shown in Figure 10. It can be observed that the IZSC voltage u_{ref}^0 and current i_o^0 are added as two disturbances, which influence the three-phase inverter in unbalanced conditions.

In the design of the PID controller, stability and high performance should be guaranteed while magnetic coupling exists. In a three-phase inverter, since the three-phase variables can be described by ISCs and its dynamic model can be expressed by ISC equivalent circuits, the analysis of the three-phase variables can be decomposed into the analysis of the IPSC, INSC and IZSC. In Figure 5, the IPSC and INSC equivalent circuits are the same, and they are also the same as a single-phase inverter when magnetic coupling is neglected so that the three-phase inverter is seen as three single-phase inverters. So, at the beginning of this analysis, the three-limb transformer and three-limb inductor can be treated as three, separated single-phase transformers and single-phase inductors, and the equivalent circuits of the IPSC and INSC can be obtained naturally. Hence the analysis results for a single-phase inverter are also valid for the IPSC and INSC equivalent circuit. In contrast, the IZSC equivalent circuit is different from that of the IPSC and INSC, so it should be evaluated separately; essentially, the effect of u_{ref}^0 and i_o^0 must be evaluated. Based on this conclusion, the design procedure for the instantaneous controller should include two steps:

1. The magnetic coupling is first neglected, and the three-phase inverter is modeled as three single-phase inverters so that the controller design can ensure the dynamic performance of the IPSC.
2. The voltage u_{ref}^0 and current i_o^0 are then evaluated to guarantee the performance of IZSC.

For the case shown in Figure 6, $R(s)$ can be designed according to the block diagram shown in Figure 8. Next, according to Figure 9, u_{PWM}^0 should be kept zero, otherwise compensation is needed. In practice, however, keeping $u_{ref}^0 = 0$ is sufficient. The reference u_{ref}^0 will appear in parallel inverter systems, and the independent regulation of three phase voltages for restraining circulating current in each phase will result in considerable voltage u_{PWM}^0, giving rise to serious problems. In addition, the 3rd order IZSC harmonics due to dead zones in the drive circuits of the power switches contribute to u_{PWM}^0 [30].

Unbalanced load currents contain i_o^0, which contribute u_o^0 through $Z_{2l}=r_2 + L_{2l}s$ according to Figure 9. This is indeed an advantage of a three-limb transformer when compared with three single-phase transformers, because u_o^0 is not significant due to small Z_{2l}. However, on the other hand, u_o^0 can't be regulated by instantaneous feedback controls because voltage u_{PWM}^0 is bypassed by the three-limb transformer.

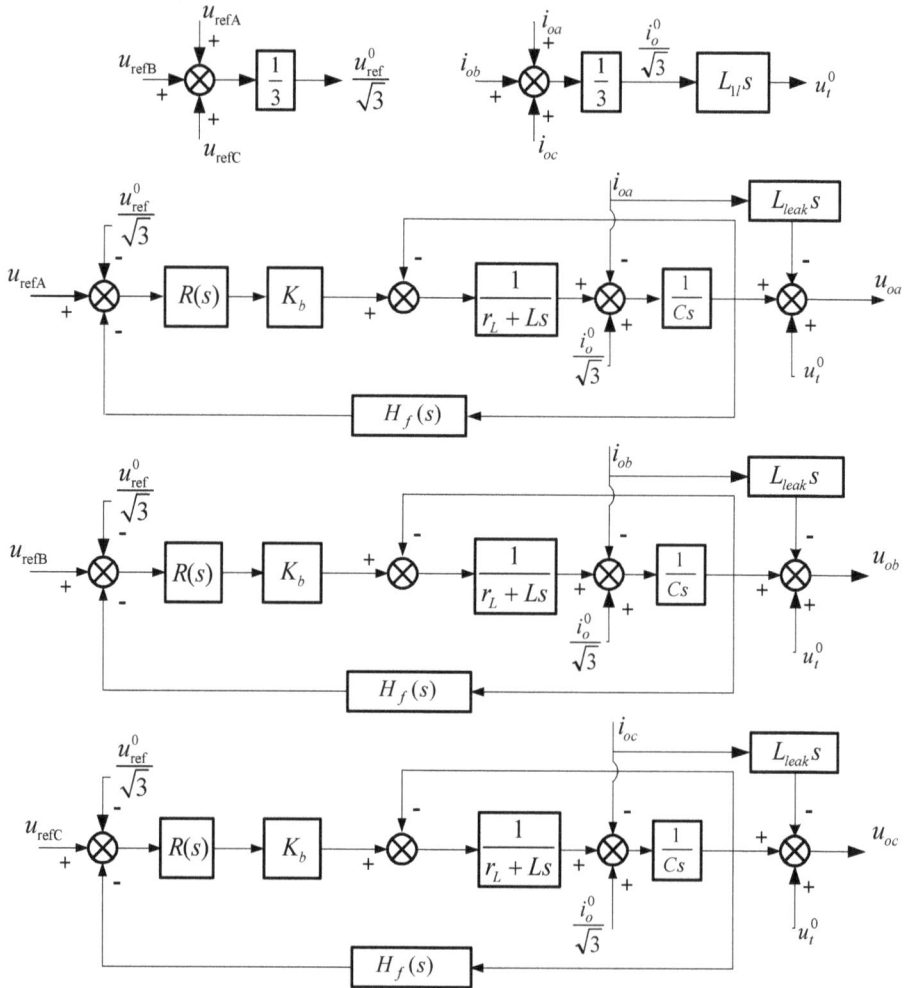

Figure 10. Block diagram for a three-phase inverter with PID control considering magnetic coupling.

5. Simulation and Experiment Results

A prototype of 50 kVA three-phase inverter was built for the following experiments, which had the structure given in Figure 1b. The three-limb transformer T_r is used as the output transformer and the three-limb inductor L is used as a filter. A PID controller is used for $R(s)$. The parameters of the inverter and its controller are shown in Table 1.

Table 1. Parameters of Inverter and Controller.

L	C	r	L_{1l} (per unit)	L_{2l} (per unit)	K_b	H_f (s)	$R(s)$
0.6 mH	140 uF	0.12 Ω	2.5%	2.5%	53.8	0.019	$9.4 + \frac{2.06 \times 10^4}{s} + 1.85 \times 10^{-3}s$

The bode diagram of IPSC is shown in Figure 11. It can be seen that the PID controller designed for IPSC behaves good performance. Then the INSC has the same performance. However, according to Equation (36), the IZSC cannot be regulated by PID controller due to magnetic couples.

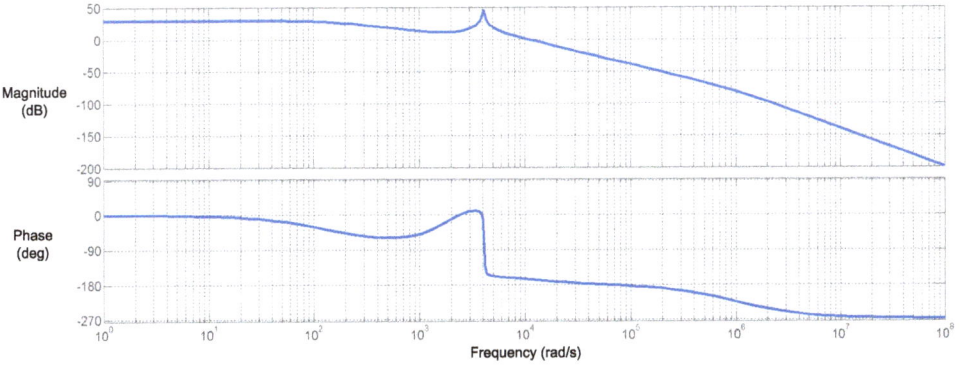

Figure 11. Bode diagram of IPSC.

A simulation to analyze the effect of the IZSC is shown in Figure 12a. When $u^0_{PWM}(s) = 0$, the primary side current i_A in the transformer is negligible and is not shown. The phase-shift between u_{PWMA} and u_{PWMB} is changed from 120° to 122.3°. The simulation shows that the current i_A is 20A, even without load. The distortion of i_A is due to dead-times of the power switches. In the experiment, i_A is measured by an oscilloscope and is shown in Figure 12b. The experimental result is the similar to that of the simulation shown in Figure 12a.

The proposed model in Figure 10 is also validated with a simulation incorporating an unbalanced load step-up which is realized by a step up of the resistive load during phase "a". When the magnetic couples are neglected, the

17

waveform of the load voltages is shown in Figure 13a. It can be seen that, only phase "a" is influenced during step-up process. Next, the magnetic couples are considered and the waveform of the load voltages is shown in Figure 13b. During single-phase load step-up, due to magnetic coupling, all three phases are influenced by the IZSCs. Also, due to the superposition of i_o^0, which has the same waveform as i_a, the voltage magnitudes of phases "b" and "c" are changed by i_o^0. They have different values as indicated by the dashed line labeled as "Line1" in Figure 13b. The largest voltage magnitude appears in phase "c" and the lowest magnitude voltage appears in the phase "a".

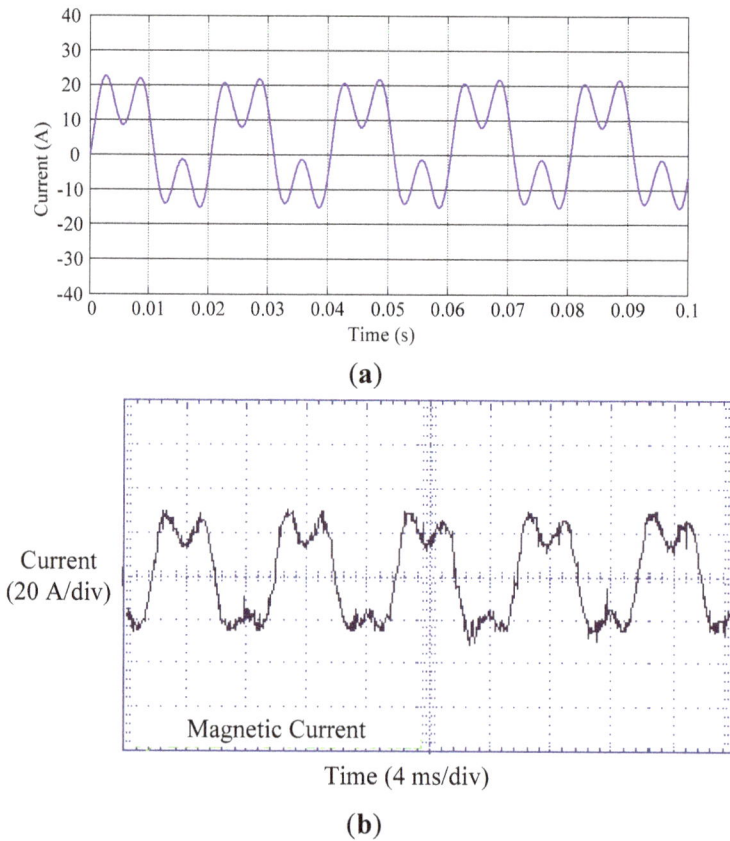

(a)

(b)

Figure 12. Primary current of three-limb transformer in unbalanced condition: (a) Simulation; (b) Experiment.

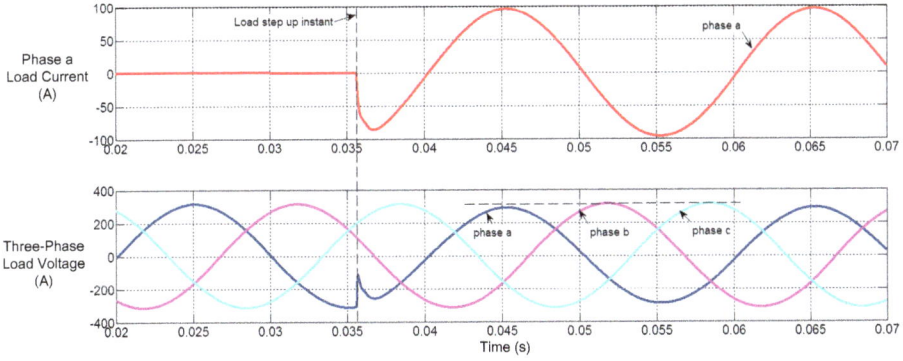

(a)

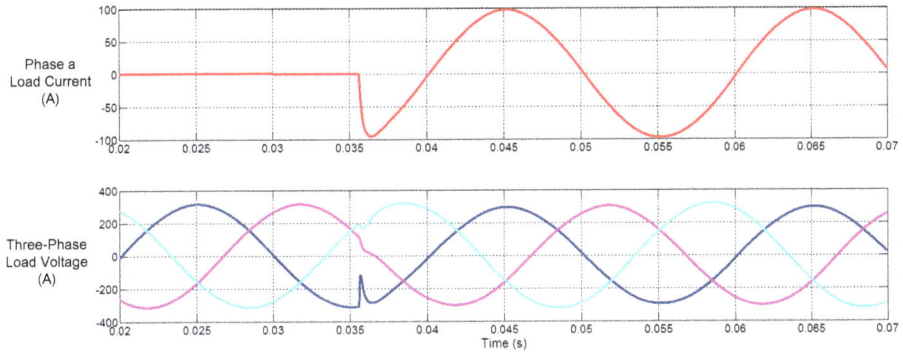

(b)

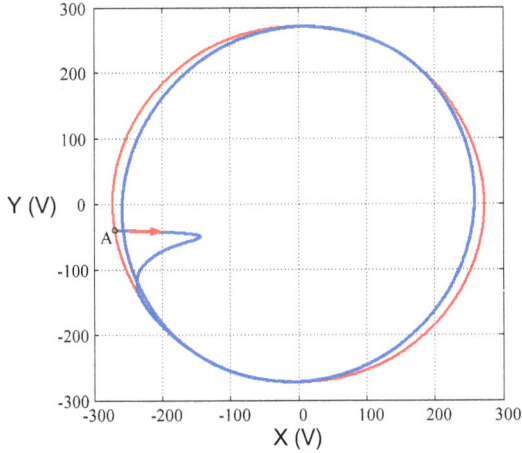

(c)

Figure 13. Simulation waveforms of the three phase load voltage (single-phase load step up): (**a**) No magnetic couples; (**b**) With magnetic couples; (**c**) IPSC locus of load voltage with magnetic couples.

19

The corresponding IPSC locus is shown in Figure 13c. The dot labeled as "A" denotes the moment of load step-up. Before load step-up, the locus is a circle as indicated by the thin, red line. However, after step-up, the locus changes to an ellipse due to the unbalance three-phase load voltages as indicated by the thick line.

A corresponding experiment was performed also by a step-up of resistive load in phase "a", and the corresponding waveforms are shown in Figure 14. The waveform is similar to that of simulation, except for small oscillations in phases "b" and "c", which are caused by the parasitic inductances and capacitance in the inverter. Unfortunately, these effects are difficult to include in the simulation.

In this example, simulation and experiment show that the influence of magnetic couples on the inverter system is acceptable even though the IZSCs are not compensated, because the incurred IZSC is small due to small Z_{inv}^0 and $u_{ref}^0 = 0$. However, for many applications, such as parallel inverter system, u_{ref}^0 or Z_{inv}^0 are large, then the compensations of IZSCs should be considered.

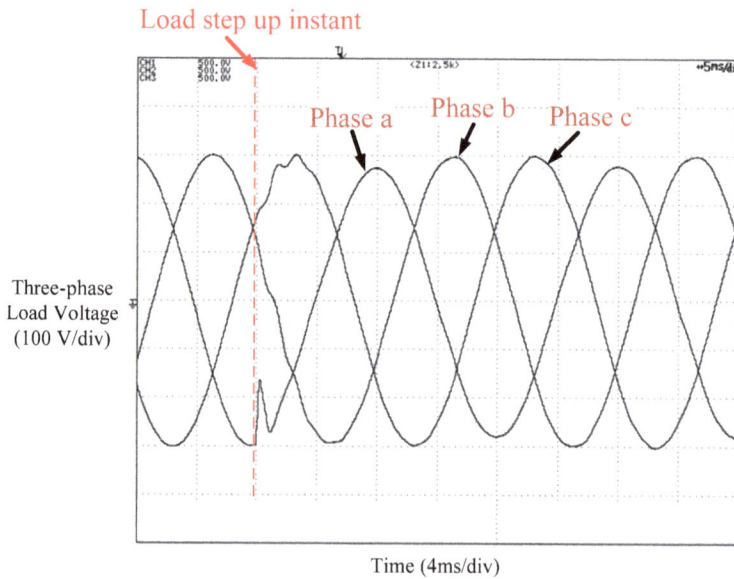

Figure 14. Experimental waveforms of three-phase load voltages (step up of single-phase resistive load)

6. Conclusions

For three-phase inverters used in UPSs, three-limb transformers and three-limb inductors are commonly used in their construction, which can bring about magnetic coupling. Here, we derived a dynamic model based on the instantaneous symmetrical components transformation, which considers this magnetic coupling phenomenon. The model includes IPSC, INSC and IZSC equivalent circuits. There is no coupling between these circuits even with magnetic coupling, and hence the model is simple to solve and is suitable for analyzing the transient behavior of three-phase inverters with magnetic coupling. Based on this model, analysis indicated that magnetic coupling might have a significant impact on the dynamic performance of three-phase inverters under unbalanced conditions. For inverters with instantaneous PID closed-loop control strategies, the influence between closed-loop control and magnetic coupling was studied and a corresponding method for performance analysis and controller design for three-phase inverters considering magnetic coupling was derived. Finally, simulation and experimental results validated the model and conclusions.

Acknowledgments: This work is supported by the Project of National Natural Science Foundation of China (50877032) and the Guangdong Zhicheng Champion Co., Dongguan, Guangdong, China.

Author Contributions: Yu Zhang completed the research work and drafted the article. Minying Li and Yong Kang provided the funding and evaluated the work.

Conflicts of Interest: The authors declare no conflict of interest.

References

1.   Harlow, J.H. Transformers. In *The Electric Power Engineering Handbook*; Grigsby, L.L., Ed.; CRC Press: Boca Raton, FL, USA, 2001.
2.   Lin, C.E.; Yeh, J.C.; Huang, C.L.; Cheng, C.L. Transient model and simulation in three-phase three-limb transformers. *IEEE Trans. Power Deliv.* **1995**, *2*, 896–905.
3.   Dolinar, M.; Dolinar, D.; Stumberger, G.; Polajzer, B.; Ritonja, J. A three-phase core-type transformer iron core model with included magnetic cross saturation. *IEEE Trans. Magn.* **1995**, *10*, 2849–2851.
4.   Zhang, Y.; Yu, M.; Liu, F.; Kang, Y. Instantaneous current sharing control strategy for parallel operation of UPS modules using virtual impedance. *IEEE Trans. Power Electron.* **2013**, *1*, 432–440.
5.   Abdel-Rahim, N.M.; Quaicoe, J.E. Analysis and design of a multiple feedback loop control strategy for single-phase voltage-source UPS inverters. *IEEE Trans. Power Electron.* **1996**, *4*, 532–541.
6.   Suh, Y.; Lipo, T.A. Control scheme in hybrid synchronous stationary frame for PWM AC/DC converter under generalized unbalanced operating conditions. *IEEE Trans. Ind. Appl.* **2006**, *3*, 825–835.
7.   Yazdani, A.; Iravani, R. A unified dynamic model and control for the voltage-sourced converter under unbalanced grid conditions. *IEEE Trans. Power Deliv.* **2006**, *3*, 1620–1629.
8.   Song, H.S.; Nam, K. Dual current control scheme for PWM converter under unbalanced input voltage conditions. *IEEE Trans. Ind. Electron.* **1999**, *5*, 953–959.
9.   Song, H.S.; Joo, I.W.; Nam, K. Source voltage sensorless estimation scheme for PWM rectifiers under unbalanced conditions. *IEEE Trans. Ind. Electron.* **2003**, *6*, 1238–1245.

10. Suh, Y.; Lipo, T.A. Modeling and analysis of instantaneous active and reactive power for PWM AC/DC converter under generalized unbalanced network. *IEEE Trans. Power Deliv.* **2006**, *3*, 1530–1540.

11. Zmood, D.N.; Holmes, D.G. Stationary frame current regulation of PWM inverters with zero steady-state error. *IEEE Trans. Power Electron.* **2003**, *3*, 814–822.

12. Mao, H.; Boroyevich, D.; Lee, F.C. Novel reduced-order small-signal model of a three-phase PWM rectifier and its application in control design and system analysis. *IEEE Trans. Power Electron.* **1998**, *3*, 511–521.

13. Kazmierkowski, M.P.; Malesani, L. Current control techniques for three-phase voltage-source PWM converters: A survey. *IEEE Trans. Ind. Electron.* **1998**, *5*, 691–703.

14. Hsu, P.; Behnke, M. A three-phase synchronous frame controller for unbalanced load. In Proceedings of the IEEE Power Electronics Specialist Conference, Fukuoka, Japan, 17–22 May 1998; pp. 1369–1374.

15. Kothari, D.P.; Nagrath, I.J. *Power System Engineering*, 2nd ed.; McGraw-Hill: New York, NY, USA, 2008.

16. Lotfi-fard, S.; Faiz, J.; Iravani, R. Improved overcurrent protection using symmetrical components. *IEEE Trans. Power Deliv.* **2007**, *2*, 843–850.

17. Gandelli, A.; Leva, S.; Morando, A.P. Topological considerations on the symmetrical components transformation. *IEEE Trans. Circuits Syst. Fundam. Theory Appl.* **2000**, *8*, 1202–1211.

18. Rioual, P.; Pouliquen, H.; Louis, J.P. Regulation of a PWM rectifier in the unbalanced network state using a generalized model. *IEEE Trans. Power Electron.* **1996**, *3*, 495–502.

19. Chen, X.; Venkata, S.S. A three-phase three-winding core-type transformer model for low-frequency transient studies. *IEEE Trans. Power Deliv.* **1997**, *2*, 775–782.

20. Suonan, J.; Xu, L.; Jiao, Z. New equivalent circuit of three-phase three-limb transformer based on magnetic circuit characteristics. In Proceedings of the International Conference on Advanced Power System Automation and Protection (APAP2011), Beijing, China, 16–20 October 2011; pp. 1678–1683.

21. Paap, G.C. Symmetrical components in the time domain and their application to power network calculations. *IEEE Trans. Power Syst.* **2000**, *2*, 522–528.

22. Ghosh, A.; Joshi, A. A new approach to load balancing and power factor correction in power distribution system. *IEEE Trans. Power Deliv.* **2000**, *1*, 417–422.

23. Iravani, M.R.; Karimi-Ghartemani, M. Online estimation of steady state and instantaneous symmetrical components. *IEEE Proc. Gener. Transm. Distrib.* **2003**, *5*, 616–622.

24. Suma, J.; Mishra, M.K. Instantaneous symmetrical component theory based algorithm for characterization of three phase distorted and unbalanced voltage sags. In In Proceedings of the IEEE International Conference on Industrial Technology (ICIT 2013), Cape Town, South Africa, 25–28 February 2013; pp. 845–850.

25. Tummuru, N.R.; Mishra, M.K.; Srinivas, S. Multifunctional VSC controlled microgrid using instantaneous symmetrical components theory. *IEEE Trans. Sustain. Energy* **2014**, *1*, 313–322.

26. Zhang, Y.; Peng, L.; Duan, S.; Kang, Y. The dynamic model of three-phase inverters with magnetic couples. In Proceedings of the 33rd Annual Conference of IEEE Industrial Electronics Society (IECON2007), Taipei, Taiwan, 5–8 November 2007; pp. 380–385.

27. Rao, U.K.; Mishra, M.K.; Ghosh, A. Control strategies for load compensation using instantaneous symmetrical component theory under different supply voltages. *IEEE Trans. Power Deliv.* **2008**, *4*, 2310–2317.

28. Lyon, W.V. *Applications of the Method of Symmetrical Components*; McGraw-Hill: New York, NY, USA, 1937.

29. Mohan, N.; Undeland, T.M.; Robbins, W.P. *Power Electronics—Converters, Applications, and Design*, 3rd ed.; John Wiley & Sons Inc.: Hoboken, NJ, USA, 2003.

30. Zhang, Y.; Chen, X.; Kang, Y.; Chen, J. The restrain of the dead time effects in parallel inverters. In Proceedings of the IEEE International Electric Machines and Drives Conference (IEMDC'05), San Antonio, TX, USA, 15 May 2005; pp. 797–802.

Energy Management of a Hybrid AC–DC Micro-Grid Based on a Battery Testing System

Bo Long, Taek Won Jeong, Jong Deuk Lee, Yoo Cheol Jung and Kil To Chong

Abstract: Energy Recovery Battery Testing Systems (ERBTS) plays an important role in battery manufacture. The conventional ERBTS configuration contains a fundamental transformer, and a bidirectional Direct Current (DC)–DC and Alternating Current (AC)–DC converter. All ERBTS are connected in parallel, thus constituting a special and complicated AC micro-grid system. Aiming at addressing their low energy recovery efficiency, complex grid-connected control algorithm issues for islanded detection, and complicated power circuit topology issues, a hierarchical DC-link voltage hybrid AC–DC micro-grid that contains composite energy storing devices is proposed. Moreover, an energy management optimal scheme for the proposed scheme is put forward. The system configuration of the proposed scheme is described in detail. Compared to the conventional scheme, the proposed scheme has the merits of simplified power circuit topology, no need for phase synchronous control, and much higher energy recovery efficiency and reliability. The validity and effectiveness of the proposed technique is verified through numerous experimental results.

Reprinted from *Energies*. Cite as: Long, B.; Jeong, T.W.; Lee, J.D.; Jung, Y.C.; Chong, K.T. Energy Management of a Hybrid AC–DC Micro-Grid Based on a Battery Testing System. *Energies* **2015**, *8*, 1181–1194.

1. Introduction

With the recent rapid development of electric vehicle technology, many countries and enterprises have invested heavily in developing electric vehicles (EVs). During the last year the total number of battery manufacturing companies in China reached 700–800. The development of power batteries is the key issue for the massive industrialization of electric vehicles. Battery performance greatly influences the driving mileage and reliability of EVs, hence, the characteristics of the produced batteries must be carefully evaluated. The common method of battery performance analysis is the charge/discharge test experiment. In order to improve the accuracy of test currents, most battery companies use the power-grid to supply the energy for charging tests. When performing the discharging test, to minimize the current ripple, the power transistors are forced to operate in the linear region instead of switching state to get high accuracy current test performance data, with a large power resistance adopted as the load; This discharged energy is totally wasted. According to the statistics from Chinese government, this power consumption can reach up to 736 gigawatt (GW). The corresponding economic loss of the wasted energy could reach up to 1.18 billion dollars.

To solve the above problems, Energy Recovery Power Battery Test Systems (ERPBTS) have been developed by some companies such as Aeroviroment and Bitrode Limited in the USA and Arbin and Digatron in Germany. Generally speaking, a qualified ERBTS should satisfy the following requirements:

(1) High-voltage and large-current output capability (such as 450 V/500 A, 225 kW).
(2) Full adjustable test voltage and current. Time response of adjustable current that ranges from zero to the command value should be kept within 15 ms, and load current should be held steady while performing internal resistance tests.
(3) High power factor on the grid side with low total harmonic distortion (THD).
(4) Reversible energy converting ability so that the discharged energy can be recovered to the power-grid with high efficiency.

Based on the aforementioned requirements, we have conducted ample research work. In [1], parameter design and power flow control strategy of an energy recovery battery testing system were presented. Considering that harmonic distortion usually occurs for non-linear loads sharing the same AC-grid, a multi-functional ERPBTS is designed in [2], in which the proposed scheme can simultaneously realize energy recovery and harmonic compensation with fast time response. Moreover, modeling and direct power control (DPC) of ERPBTS using an optimal switching table are discussed in [3,4], respectively.

If many battery testing systems (BTSs) are simultaneously used by a battery company, massive amounts of discharging energy can be recycled. However, we found that these instruments have not been practically adopted due to the following reasons.

1. Most enterprises focus on power circuit topology optimization and sophisticated control strategy for BTS. For massive grid-connected instruments in an AC-micro-grid, very few companies consider the energy saving issues as a whole.
2. In the utility, since the maximum output power of a single ERBTS can reach 250 kW, the associated instantaneous power input, when interfacing with the AC-grid, might cause an obvious rise and fall of the voltage at the point of common coupling (PCC). Moreover, if the test instrument is not well controlled, it might cause serious harmonic and voltage distortion problems on the nearby devices sharing the same grid. Because of these unresolved issues, the electric power department usually inhibits the interfacing of such high power ERPBTS to the AC-grid directly.

3. The conventional BTS has much lower energy recovery efficiency due to the complex composite power circuit topology which is usually composed by a bidirectional DC-DC converter and a bidirectional AC-DC converter with a fundamental frequency isolation transformer as the interface. Moreover, the conventional ERPBTS is very expensive. For a 200 kW testing system, the selling price can reach about 240,000 USD.

In recent years, due to its obvious advantages, the DC micro-grid system has received extensive attention in electric power systems. The DC micro-grid system is especially suitable for applications with renewable energy sources, such as photovoltaic (PV) energy, wind power, tidal energy, and geothermal energy. Compared with the conventional structure, the DC micro-grid system can achieve much higher efficiency and simpler power circuit topology due to the elimination of extra AC–DC and DC–AC conversion stages, since many renewable DGs, energy storage systems (ESSs) [5], and an increasing number of loads directly utilize DC power. Additionally, the DC system offers improved controllability because the complex phase synchronization, and the reactive power compensation control problems intrinsic to the AC grid can be ignored [6,7]. Furthermore, the DC micro-grid can be fully decoupled from the utility grid by an interface converter, enabling the seamless transition between grid-connected and islanded operation modes [8–10].

For the aforementioned reasons, the DC micro-grid is receiving increased attention, especially for small-scale commercial and residential applications [11]. For AC and DC micro-grids, energy management control is one of the crucial points for micro-grid operation. The control objective is coordinating the distributed micro-grid terminals and to provide a simple, reliable, stable, and cost-effective power supply for both local customers and the utility. Much research work have been done to develop a standardized hierarchical power management framework [12–14].

Considering the merits and demerits of the AC and DC micro-grid system, we found that very few papers describe the application of hybrid AC-DC micro-grid architecture in ERPBTS. Hence, this paper proposes a hybrid AC-DC micro-grid system based on ERPBTS, and the proposed scheme resolves the above drawbacks of the conventional scheme. In the proposed system, the converter for battery testing is a current-feed DC-DC converter, and all the testing instruments are connected to a DC-bus, to form a DC-micro-grid system [15,16]. In this way, the recovered energy can be reused in other local loads, which increase the system efficiency and simplify the control strategy for a DC-micro-grid.

This paper is arranged as follows: in Section 2, we explain the system configuration and operation principles of conventional ERPBTS based on an AC micro-grid. In Section 3, the system configuration of a hybrid AC-DC micro-grid for ERPBTS is illustrated. The topology and energy flow chart comparisons chart between the conventional AC and proposed hybrid AC-DC micro-grids are elaborated. In Section 4, an optimal energy management control scheme and the stability of the proposed system are analyzed. The corresponding experimental results are also presented. Finally, in Section 5, the main contributions of this paper are summarized and suggested future work is also presented.

2. System Configuration of the Conventional ERPBTS

In this section, we first illustrate the system configuration of the conventional scheme. Then, control strategies for an AC micro-grid are presented, and drawbacks of the direct AC-micro-grid are also shown.

Figure 1 demonstrates the block diagram of the massive AC-grid-connected ERPBTS, in which the red line represents the main power grid. The black line symbolizes the local micro-grid (AC 380 V), dividing the testing area into two sections. The high-power testing instruments are connected to the high AC-voltage (380 VAC) side, and relatively low-power test instruments are interconnected on the single-phase (220 VAC) side. Each section contains a bidirectional AC–DC–DC converter and the series-connected battery packs. The configurations show that all the testing instruments are connected in parallel to form an AC-micro-grid. Reactive power, power quality, reliability and frequency control are the main issues in this system.

In this way, the battery testing instrument fulfills the battery testing and energy recovery requirements. The discharged energy produced during the testing procedure is recycled to the other loads sharing the AC-grid.

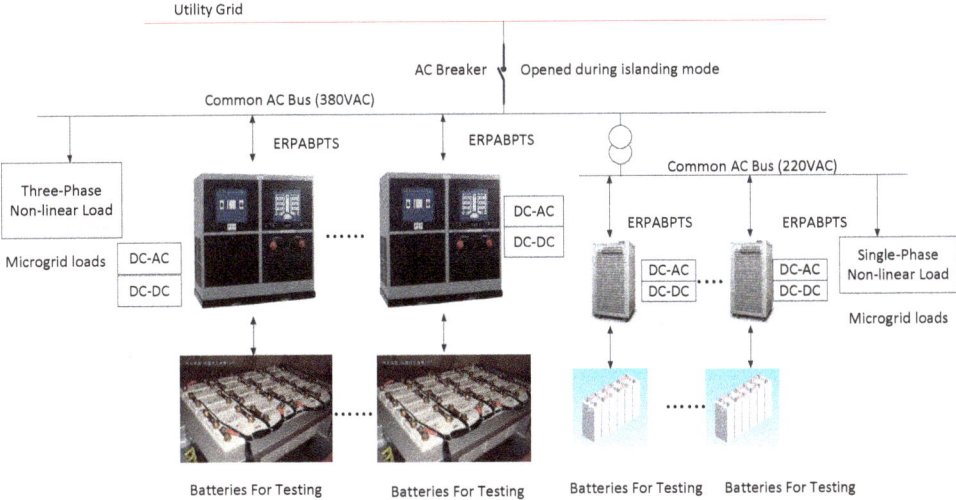

Figure 1. System configuration of conventional AC micro-grid based on ERPBTS.

Figure 1 demonstrates the configuration of the AC micro-grid systems, which contains the AC 380 and AC 220 V systems. The popular ways of voltage/current reference signals generation for power electronics converter in an AC-micro-grid are summarized below.

(1) The first solution uses a master-slave controller, in which a master controller is used to communicate with all sources using a fast communication technique. All micro-sources (grid-connected converters) inform the master controller of the current status of their local variables (such as voltage/current, frequency, *etc.*). Based on that information, the master controller computes and transmits the reference voltage/current and frequency values for all the sources. The requirement of a fast communication technique increases the cost of the whole system. Moreover, the reliability of the system is reduced due to the presence of an additional communication circuit and the master controller [17].

(2) The second method uses a droop controller and measures only the local variables of a source to calculate the reference/current signal of the power converter. This decentralized control scheme does not need a fast communication system between grid-connected converters for operation and has the merits of high reliability, low cost, and easy scalability. Because of these advantages, droop controller is extensively adopted in AC micro-grids.

Compared with the AC-micro-grid, the DC micro-grid has many advantages. First, it does not require synchronous or frequency tracking control and so is more reliable, controllable, and suitable for distributed generation system (DGs). Secondly, in DC micro-grid systems, compared with a cascaded power converter, the topology of the one-stage power converter is much simpler, resulting in much higher energy recovery efficiency. Thirdly, the load voltage is not influenced by voltage regulation, voltage dropdown, or unbalanced loads. Finally, the DC-voltage in transmission is not influenced by the impulse current; thus, there is no need to consider the line reactive power component, and the line power loss can be ignored. Therefore, combining a DC micro-grid with an AC micro-grid to form a hybrid AC–DC micro-grid can solve both the efficiency and energy balance problems for ERPBTS.

3. A Hybrid AC–DC Micro-Grid System for ERPBTS

In this section, a hybrid AC–DC micro-grid battery testing system is proposed. System configuration, operation principle, and merits of the proposed scheme are elaborated.

3.1. System Configuration

Figure 2 illustrates the layout of the hybrid AC–DC micro-grid based on ERPBTS. The configuration is established based on the assumption that all the battery factories are connected to the utility grid and share a common micro-grid. Seen from the outside, each test factory interfaced with the main grid can be regarded as a module. Each battery test factory is regarded as the local microgrid, interfaced with the main grid via an AC-breaker. Hence, seen from the grid-side, all the battery factories form the AC-microgrid.

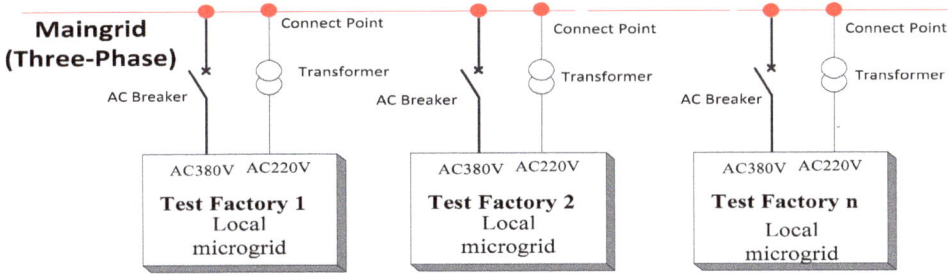

Figure 2. The hybrid AC–DC micro-grid layout based on ERPBTS.

The internal configuration of a representative factory is shown in Figure 3, showing that the testing instruments forms a DC micro-grid containing two-level DC-link voltage (high and low DC-link voltage). The large power testing instruments are connected on the high voltage side; the relatively small power output testing instruments are connected at the low voltage side. Moreover, a DC–DC converter (blue color in Figure 3) is inserted between high and low DC-links for bidirectional energy transmission.

In order to support the stability of the DC-link voltage, an ultracapacitor-battery composite power supply system is adopted in the proposed system, and the composite energy storing system (ESS) can satisfy the stable and peak power needs. In addition, since the half-bridge DC–DC converter in this system usually operates as a current-source converter, the operation condition of the DC micro-grid can be dominated by ESS for power balance.

A bidirectional DC–AC converter is inserted between the main grid and the DC-micro-grid in case the ESS and battery testing instrument reserves are not adequate to meet the demands of the local loads. In such circumstances, the deficit can be compensated by importing the energy from the utility; On the other hand, excess energy reserve of the ESS and battery testing instruments leads to unbalance in the DC-link power, and the additional energy can be exported to the utility. Generally speaking, the DC–AC converter functions as a utility interface converter to maintain the power balance within the DC-micro-grid.

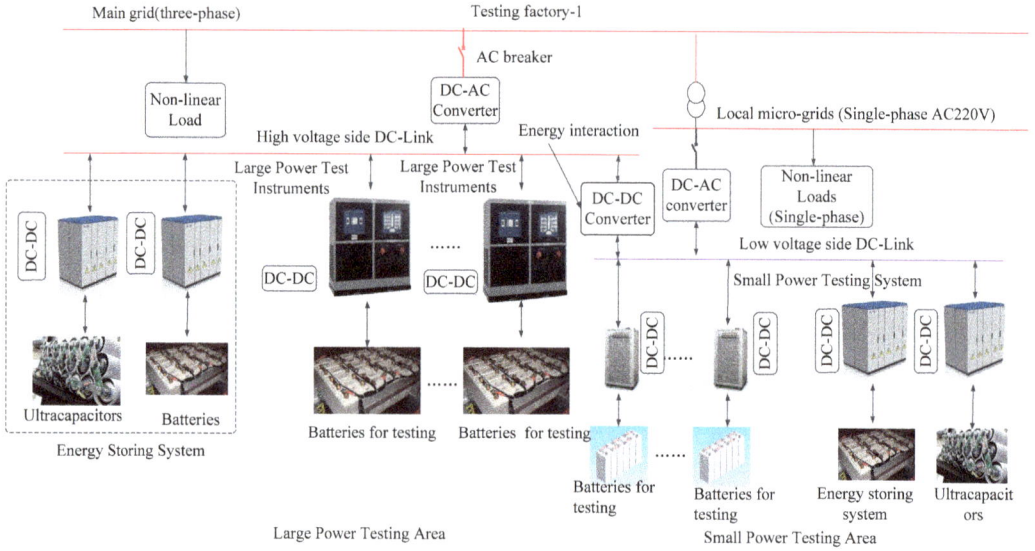

Figure 3. Architecture of the multiple converters with hierarchal DC-link voltage configuration in the micro-grid.

3.2. Topology Comparison

Figure 4 illustrates the operation principle comparisons between the conventional and proposed schemes used for battery testing instruments. The comparison results show that the proposed scheme has at least three distinct advantages:

(1) In the conventional AC micro-grid system based on ERPBTS, the internal power converter topology is composed of DC–DC and DC–AC converters connected in series. Thus, the two systems are strongly coupled; a decoupler control algorithm is needed if no energy storing system is interfaced with the DC-link. However, the power converter in the proposed scheme is composed of a single DC–DC converter, resulting in considerably high accuracy and energy recovery efficiency. Moreover, since the topology of the converter is greatly simplified, the amount of software operations is considerably reduced, and reliability of the testing instrument is maximally enhanced.

(2) In conventional ERBTS, the inverter is controlled to synchronize with the power-grid on phase angle, frequency, and amplitude. In addition, in order to eliminate the total harmonic distortion (THD) of the grid-connected inverter, a complicated current filter (such inductance-capacitance-inductance (LCL) type filter) is usually implemented, requiring complex closed-loop control algorithms. In the proposed scheme, only the DC-link voltage should be maintained stable and constant, resulting in a much simpler structure of the proposed micro-grid.

(3)   In the proposed scheme, the energy interactions between different testing instruments can be completed in a single stage (DC micro-grid), However, the conventional scheme needs at least two stages. Hence, the proposed scheme is much simpler, and the total expense of the proposed scheme can be greatly reduced.

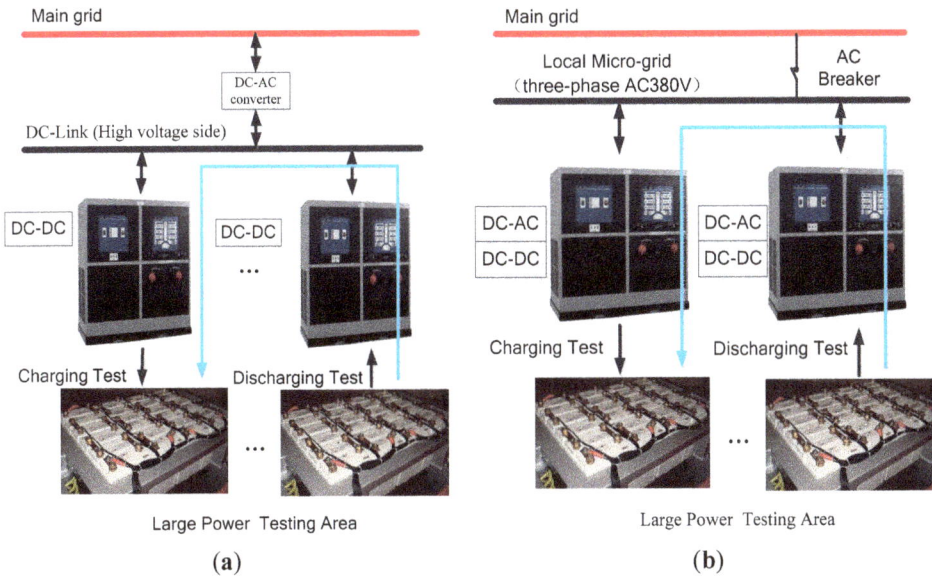

Figure 4. Energy chart between the conventional and proposed architectures. (**a**) Proposed architecture; and (**b**) Conventional architecture.

4. Energy Management Strategy of a DC Micro-grid Based on ERPBTS and an Energy Storage System

4.1. Energy Optimal Control

For an AC micro-grid system, the crucial points for its operation are intelligent control and management. The control objective is to coordinate the distributed micro-grid terminals in order to mitigate the power intermittency and uncertainty. To provide a stable, reliable, and economic power supply for both local customers and the utility, a series of control strategies have been published for AC micro-grids. Yet, the control strategies for AC-micro-grid cannot be simply migrated to DC micro-grids due to the aforementioned special features of the DC system.

Like an AC-grid, the control methods for a DC micro-grid system can also be categorized into two classes of centralized and decentralized control structures. For centralized control, all the micro-source converters sharing the same DC-bus are controlled by a central energy controller through a communication link (such as CAN-bus). A data center (server) is needed for acquiring the information of voltage, current, and the node address of each converter. This solution has the drawbacks of limited nodes on power converters, difficulty expanding the nodes, need for an additional central controller, and a required hardware circuit for fast communication.

Compared with the centralized solution, a decentralized control structure is proposed in which all the distributed terminals operate independently, and the control decisions are made based on local information [18]. The common droop control methods used in DC micro-grids for power management are Voltage/Current (V/I) and Voltage/Power (V/P) control. The core issue for V/I and V/P control methods is using the deviation of DC-bus voltage for autonomous power sharing among different power sources [19]. The limitation of the V/I and V/P schemes is that all the terminals sharing the same DC bus have to rigidly follow the preset droop curves without a flexible mode transition mechanism, especially when a voltage change occurs in the DC micro-grid [20].

From [5], when the DC-link voltage is regulated by different power sources, the operation modes for the proposed system are: (a) Utility-dominating mode; (b) Energy storage system-dominating mode; (c) Battery testing instruments-dominating mode. Hence, a seamless transition from one dominating mode to another according to the DC-bus voltage deviation is the key issue. Power variables of different terminals can be defined as:

P_{ESS_CH}, P_{ESS_DSC}—The maximum charging and discharging power values, these values are determined by the status of charge (SOC) in the energy storage system; if the SOC of batteries and the ultra-capacitor is zero, this means that the ESS is fully discharged, $P_{ESS\_DSC} = 0$.

V_{dcbus}—DC-bus voltage;

P_{load}—Local load power consumption, including the linear and non-linear loads sharing the same DC-bus;

V_{upper}, V_{lower}—Threshold DC-link voltages for the operation modes.

P_{BTS_CHG}, P_{BTS_DSC}—The total charging power input and discharging power output of ERPBTS.

The characteristics of each dominating mode determined by the DC-link voltage range are listed in Table 1, which shows that the power characteristics of each terminal determines the rise and fall of the DC-link voltage. In mode I, when the DC-link voltage falls below the lower limit, the utility will dominate the DC-link voltage. In mode II, when the DC-link voltage falls between the upper and lower voltage limits, the battery-ultracapacitor energy storage system will dominate the DC-link voltage. In mode III, when there is redundant power output to charge the capacitors and give rise of the DC-link voltage, the ERPBTS will dominate the DC-bus.

Table 1. Energy optimal scheme for hybrid AC-DC microgrid.

Dominating Mode	Power Characteristics	Voltage Range	DC-link Voltage Regulation
Utility (Mode I)	$P_{load} + P_{BTS\_CHG} + P_{ESS\_CH} > P_{ESS\_DSC} + P_{BTS\_DSC}$	$V_{dcbus} < V_{lower}$	Utility Units
ESS (Mode II)	$P_{ESS\_CH} < P_{load} + P_{BTS\_CHG} - P_{BTS\_DSC} < P_{ESS\_DSC}$	$V_{lower} < V_{dcbus} < V_{upper}$	ESS Units
ERBTS (Mode III)	$P_{BTS\_DSC} > P_{load} + P_{BTS\_CHG} + P_{ESS\_CH} - P_{ESS\_DSG}$	$V_{dcbus} > V_{upper}$	ERPBTS Units

4.2. Experimental Results

In order to verify the stable behavior of the proposed configuration and droop-controlled strategy, a scaled-down laboratory prototype of a three-node, three-branch DC-micro-grid is developed. The component specifications used in experiment are illustrated in Figure 5. The schematic diagram of this system in which each micro-source converter is a DC–DC converter is illustrated in Figure 2.

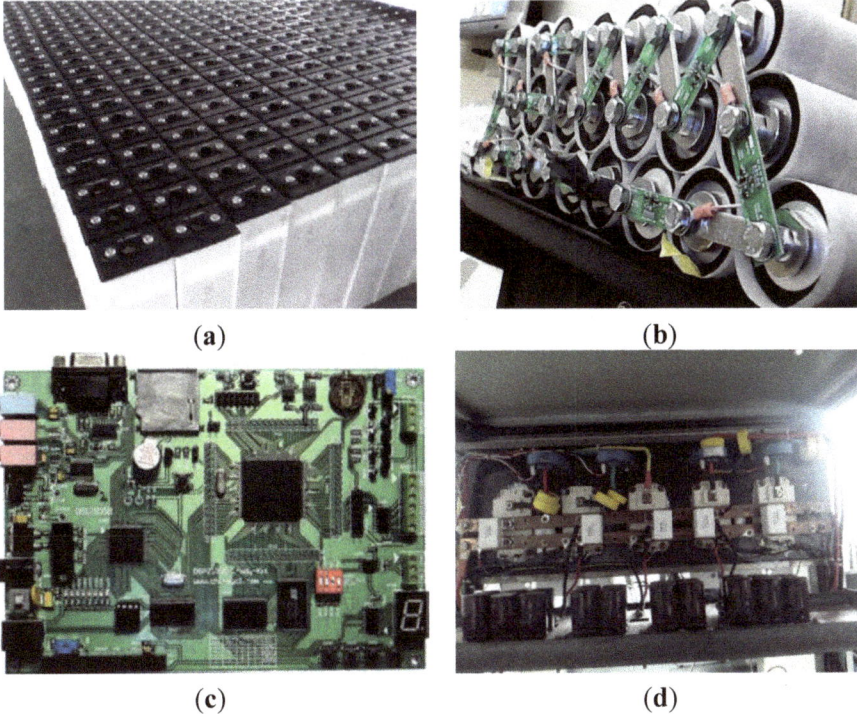

(a)

(b)

(c)

(d)

Figure 5. Hardware setup of the proposed scheme. (**a**) Batteries for testing; (**b**) Ultra-capacitors for energy storing system (ESS); (**c**) DSP control board; (**d**) AC–DC converter for utility dominating operation.

Specifications of the bidirectional DC–DC and AC–DC converters are given in Tables 1 and 2 in the Appendix A. The core controller (CPU) chosen for implementation of voltage and current droop control algorithm is the TMS320F28335 from Texas Instruments (Dallas, TX, USA).

The experimental hardware setup for the proposed hybrid AC–DC micro-grid contains four parts (refer to Table 2): (1) The utility grid that contains a three-phase voltage source converter (VSC) with isolation transformer; (2) Two bidirectional DC–DC converters for the ESS system; (3) Two bidirectional DC–DC converters for the battery testing systems, one for charging, another for discharging; and (4) Two power resistors serves as the loads.

Table 2. List of the all the terminals used in experiments.

Energy Source	Converter Type	Rated Power	Quantity
Utility Grid	Three-phase voltage source inverter with isolation transformer	30 kW	1
Ultracapacitor-Battery Energy storing System (ESS)	Bidirectional DC–DC converter for batteries + Bidirectional DC–DC converter for ultra-capacitors	5.0 kW	2
ERPBTS	Bidirectional DC–DC converter	15 kW	2
Loads	Power resistors	3 kW	2

For safety considerations, the conventional 380 V three-phase AC distribution system is reduced to a 75 V three-phase AC-grid using a step-down transformer; the DC-bus voltage of the tested DC micro-grid system is set between 200 and 250 V. The corresponding voltage of each operation mode is allocated within this range. The thresholds of the DC-bus voltage range are listed in Table 3.

Table 3. DC-link voltage range definition for each operation mode.

Operation Mode	(Mode I)	(Mode II)	(Mode III)
DC-bus Voltage	<200	200~250 V	>250 V

Table 3 gives the operation mode definition for each operation mode, totally, there are three operation modes.

Table 1 shows that, when the DC-bus voltage falls below the lower limit V_{lower}, the DC-bus will be dominated by the utility grid via a voltage source converter (VSC) that operates as a PWM rectifier. Figure 6a illustrates the waveforms of active and reactive current components when the DC micro-grid is interfacing with the AC micro-grid using unit power factor control (PFC) method. Figure 6b shows the phase voltage and phase current of the VSC rectifier when there is a deficit in power needs in a DC micro-grid. By dynamically controlling the reactive current component, the DC-bus voltage is maintained stable (in mode I).

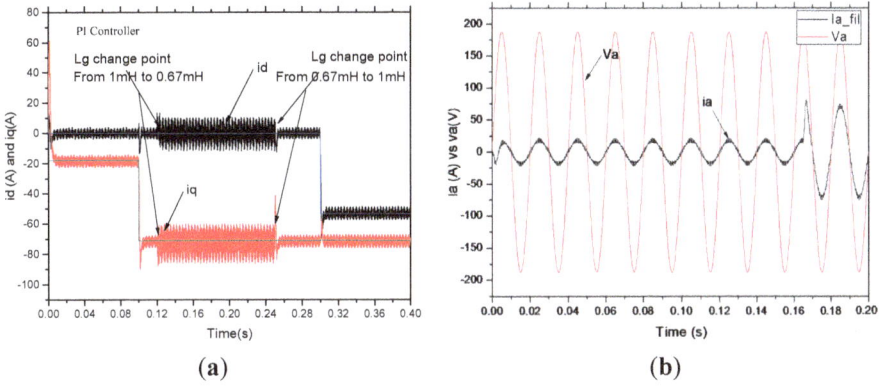

(a)   (b)

Figure 6. (**a**) Waveforms of the active and reactive current components when the DC micro-grid is interfacing with the AC micro-grid. (**b**) Waveforms of the phase voltage and current at the point of common coupling.

Figure 7a illustrates the dominating mode transition process of the three-phase converter (from mode II to mode I) when the DC micro-grid initially operates in ESS storage-dominating mode for charging testing. When a power deficit occurs, the domination mode will automatically change to mode I, and the utility will compensate the additional power needs. Figure 7b demonstrates the dominating mode transition process of the three-phase VSC converter (from grid connected mode to islanded mode).

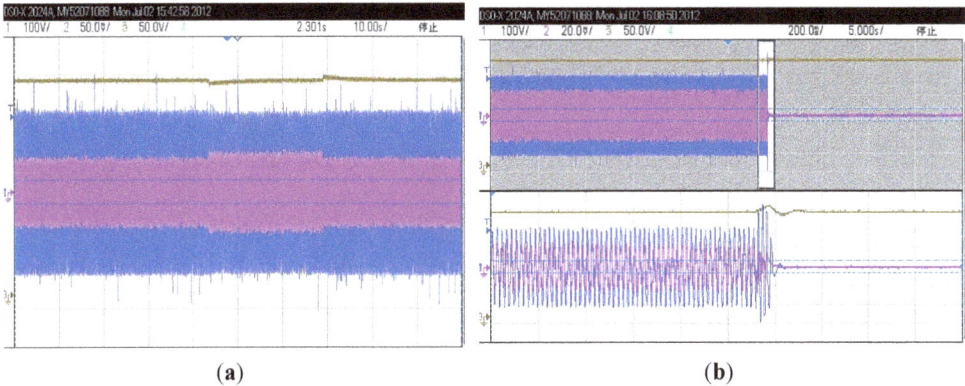

(a)   (b)

Figure 7. (**a**) Waveform of the converter switching from grid-connected mode to rectifier mode; (**b**) Waveform of the converter switching from grid-connected mode to islanded operation mode.

Figure 8a shows the transition process of the utility-dominating mode switching from islanded mode to grid connected mode. The battery charging current in ERPBTS is illustrated in Figure 8b, showing that the dominating mode of dc-link voltage in dc-micro-grid changes from utility-dominating mode to ERPBTS-dominating mode. In this case, the DC-bus voltage V_{dcbus} is greater than the upper limit of reference voltage V_{upper} ($V_{dcbus} > V_{upper}$), implying sufficient power output in the DC-link. This additional energy will automatically charge the capacitors and increase the DC-bus voltage, ultimately resulting in a high voltage breakdown of the power transistors. Therefore, in order to release the additional energy, the operation mode of ERPBTS should be changed from discharging mode to charging mode. Figure 8b shows the waveform of mode transition when the battery test current changes from +20 A (discharging current) to −20 A (charging current). The experimental results show that the proposed scheme is feasible and applicable.

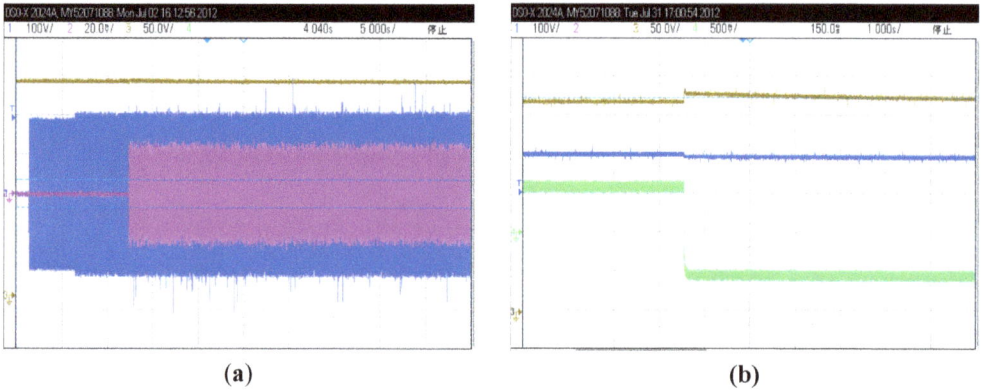

(a)   (b)

Figure 8. (**a**) Waveform of the AC-DC converter switching from islanded mode to grid-connected mode; (**b**) Waveform of the battery charging current.

5. Conclusions

This paper describes a novel system configuration of hybrid AC–DC micro-grid system for massive grid-connected energy recovery battery testing systems. Energy flow comparisons between the proposed and conventional schemes are illustrated. The comparison results show that the proposed scheme has the merits of simplified power circuit topology, high energy recovery efficiency and high current/voltage test resolution. Based on the analysis results, an optimal energy management decentralized control strategy was suggested. Detailed implementation based on the laboratory AC–DC hybrid micro-grid test system verifies the performance of the proposed strategy.

Acknowledgments: This work was supported by the Fundamental Research Funds for the Central Universities of China (NO. ZYGX2012J095), China Postdoctoral Science Foundation Funded Project (2013M542266), State Key Laboratory of Control and Simulation of Power System Generation Equipment (SKLD14KM03), Tsinghua University, and the National Research Foundation of Korea (NRF) funded by the government of South Korea (MEST) (No. 2013009458 and No. 2013068127). The authors would like to thank all the reviewers for their advice and suggestions on improving this paper.

Author Contributions: Bo Long conceived and developed the ideas behind the present research and proposed the hybrid AC–DC micro-grid system for ERBTS. Bo Long, Won Jeong and Chong Deuk Lee executed the hardware setup and software implementation, literature review, and manuscript preparation. Final review, including final manuscript corrections, was performed by Kil To Chong and Bo Long.

Appendix A.

Table 1. Power transistor specifications of the three-phase VSC converter.

Components	Part name/Manufacturer	Rating values
IGBT	SKM400GB128D/SEMIKRON	1200 V–400 A
Fast DIODE	SKKD75F12/SEMIKRON	1200 V–75 A
Capacitor	YDK	450 V–5000 μF

Table 2. Specifications of the proposed ERPBTS.

Elements	Parameters	Values
Power Battery Pack	Battery terminal voltage V_{bat}	240 V
	Battery type	Li-ion
Utility Grid	Grid voltage (line to line voltage) V_g	380 V
	Line frequency f_n	50 Hz
	Equivalent inductance at grid side L_g	1 mH
L Type Filter	Converter side inductor L	1 mH
DC–DC Converter	Nominal power	175 kW
	Inductor for boost chopping L	4 mH
	Switching frequency f_s	5000 Hz
	Dead time t_d	2 us
DC–AC Converter	Nominal power P_e	175 kW
	Two series DC-link capacitor C_{dc}	16,000 uF
	Initial DC-link capacitor voltage V_{c0}	500 V
	DC-link voltage reference V_{dc_ref}	900 V
	IGBT switching frequency f_{inv}	2000 Hz
	Dead time t_d	2 us

Conflicts of Interest: The authors declare no conflict of interest.

References

1. Bo, L.; Chong, K.T. Parameter design and power flow control of energy recovery power accumulator battery pack testing system. *J. Electr. Eng. Technol.* **2013**, *8*, 787–798.
2. Long, B.; Ryu, J.H.; Lim, S.T.; Chong, K.T. Design and control of a multi-functional energy recovery power accumulator battery pack testing system for electric vehicles. *Energies* **2014**, *7*, 1376–1392.
3. Long, B.; Kil, T.C. Modeling and direct power control of energy recovery power battery testing system under charging mode—A new approach. *Int. Rev. Electr. Eng.* **2012**, *7*, 5993–6004.
4. Long, B.; Ryu, J.H.; Chong, K.T. Optimal switching table-based sliding mode control of an energy recovery Li-Ion power accumulator battery pack testing system. *Energies* **2013**, *6*, 5200–5218.
5. Gu, Y.; Xiang, X.; Li, W.; He, X. Mode-Adaptive decentralized control for renewable DC micro-grid with enhanced reliability and flexibility. *IEEE Trans. Power Electron.* **2014**, *29*, 5072–5080.
6. Eghtedarpour, N.; Farjah, E. Power control and management in a hybrid AC–DC micro-grid. *IEEE Trans. Smart Grid* **2014**, *5*, 1494–1505.
7. Loh, P.C.; Ding, L.; Yi Kang, C.; Blaabjerg, F. Autonomous operation of hybrid micro-grid with AC and DC subgrids. *IEEE Trans. Power Electron.* **2013**, *28*, 2214–2223.
8. Blaabjerg, F.; Teodorescu, R.; Liserre, M.; Timbus, A.V. Overview of control and grid synchronization for distributed power generation systems. *IEEE Trans. Ind. Electron.* **2006**, *53*, 1398–1409.
9. Rocabert, J.; Luna, A.; Blaabjerg, F.; Rodriguez, P. Control of Power Converters in AC Micro-grids. *IEEE Trans. Power Electr.* **2012**, *27*, 4734–4749.
10. He, J.W.; Li, Y.W.; Munir, M.S. A Flexible Harmonic Control approach through voltage-controlled DG-grid interfacing converters. *IEEE Trans. Ind. Electron.* **2012**, *59*, 444–455.
11. Kakigano, H.; Miura, Y.; Ise, T. Low-voltage bipolar-type DC micro-grid for super high quality distribution. *IEEE Trans. Power Electr.* **2010**, *25*, 3066–3075.
12. Guerrero, J.M.; Loh, P.C.; Lee, T.L.; Chandorkar, M. Advanced control architectures for intelligent micro-grids-part II: Power quality, energy storage, and AC–DC micro-grids. *IEEE Trans. Ind. Electron.* **2013**, *60*, 1263–1270.
13. Guerrero, J.M.; Chandorkar, M.; Lee, T.L.; Loh, P.C. Advanced Control Architectures for intelligent micro-grids-part I: Decentralized and hierarchical control. *IEEE Trans. Ind. Electron.* **2013**, *60*, 1254–1262.
14. Kim, J.; Guerrero, J.M.; Rodriguez, P.; Teodorescu, R.; Nam, K. Mode adaptive droop control with virtual output impedances for an inverter-based flexible AC micro-grid. *IEEE Trans. Power Electr.* **2011**, *26*, 689–701.
15. Sanchez, S.; Molinas, M.; Degano, M.; Zanchetta, P. Stability evaluation of a DC micro-grid and future interconnection to an AC system. *Renew. Energy* **2014**, *62*, 649–656.

16. Hemmati, M.; Amjady, N.; Ehsan, M. System modeling and optimization for islanded micro-grid using multi-cross learning-based chaotic differential evolution algorithm. *Int. J. Electr. Power* **2014**, *56*, 349–360.

17. Anand, S.; Fernandes, B.G. Reduced-Order Model and Stability Analysis of low-voltage DC micro-grid. *IEEE Trans. Ind. Electron.* **2013**, *60*, 5040–5049.

18. Ito, Y.; Zhongqing, Y.; Akagi, H. DC micro-grid based distribution power generation system. In Proceedings of the 4th International Power Electronics and Motion Control Conference, IPEMC 2004, Xi'an, China, 14–16 August 2004; Volume 1743, pp. 1740–1745.

19. Lie, X.; Dong, C. Control and operation of a DC micro-grid with variable generation and energy storage. *IEEE Trans. Power Deliv.* **2011**, *26*, 2513–2522.

20. Rodriguez, M.; Stahl, G.; Corradini, L.; Maksimovic, D. Smart DC Power management system based on software-configurable power modules. *IEEE Trans. Power Electron.* **2013**, *28*, 1571–1586.

Coordinated Charging Strategy for Electric Taxis in Temporal and Spatial Scale

Yuqing Yang, Weige Zhang, Liyong Niu and Jiuchun Jiang

Abstract: Currently, electric taxis have been deployed in many cities of China. However, the charging unbalance in both temporal and spatial scale has become a rising problem, which leads to low charging efficiency or charging congestion in different stations or time periods. This paper presents a multi-objective coordinated charging strategy for electric taxis in the temporal and spatial scale. That is, the objectives are maximizing the utilization efficiency of charging facilities, minimizing the load unbalance of the regional power system and minimizing the customers' cost. Besides, the basic configuration of a charging station and operation rules of electric taxis would be the constraints. To tackle this multi-objective optimizing problems, a fuzzy mathematical method has been utilized to transfer the multi-objective optimization to a single optimization issue, and furthermore, the Improved Particle Swarm Optimization (IPSO) Algorithm has been used to solve the optimization problem. Moreover, simulation cases are carried out, Case 1 is the original charging procedure, and Cases 2 and 3 are the temporal and spatial scale optimized separately, followed with Case 4, the combined coordinated charging. The simulation shows the significant improvement in charging facilities efficiency and users' benefits, as well as the better dispatching of electric taxis' charging loads.

Reprinted from *Energies*. Cite as: Yang, Y.; Zhang, W.; Niu, L.; Jiang, J. Coordinated Charging Strategy for Electric Taxis in Temporal and Spatial Scale. *Energies* **2015**, *8*, 1256–1272.

1. Introduction

In recent years, Electric Vehicle (EV) technologies have developed rapidly with the high attention from governments all over the world. In China, the government document, Energy Saving and New Energy Vehicles Industry Development Planning (2011–2020), put forward that the number of EVs should reach 5 million in 2020, and according to the research report from the Ministry of Industry and Information Technology, the figure is predicted to be 60 million in 2030 [1]. At the same time, with the development of the EV industry, related charging facilities must be built to meet the anticipated significantly increased charging demand. There are also many local plans for the construction of charging stations, chargers and intelligent charging service system, such as in Beijing, Shenzhen.

As pioneers, electric buses and electric taxis are first to be demonstrated for utilization of EVs. In Shenzhen, an EV charging network has been set up with electric buses, electric taxis, charging stations and related charging services.

In this paper, only electric taxis are considered. After running for a while, some issues have arisen with the operation of electric taxis. According to investigation of operational data from Shenzhen, taxi drivers work two shifts, one for day-time, the other for night. The shift-swapping time and position are not strict, but are usually around 5:00/17:00 and somewhere near the drivers' places. In terms of the hidden rules that electric taxis should start with full State of Charge (SOC), two charging peaks occur before the work shifts. Moreover, charging twice is not enough to operate for the whole day, and another two charges are needed during the operation periods. All of these facts result in four load peaks in the temporal scale. At the same time, unbalanced distribution in the spatial scale also arises for the reason that most drivers prefer the charging station to be near their places.

If the charging load distribution of EVs is unbalanced in either the temporal or spatial scale, the utilization efficiency of charging facilities will be much lower. It may also trigger some relative load unbalance problems when EV load penetration gets higher, for example, more charging costs for drivers and more feeder losses in regional power systems.

Some researchers have proposed some charging strategies to solve similar electric taxi operation problems. In [2], with some investigation of EV taxi data from Shenzhen, the authors understood that the status (e.g., operational patterns, driver income and charging behaviors) of EV taxis can provide invaluable information to policy makers and studied the patterns from two aspects: operational behaviors and charging behaviors, but a rescheduling strategy was not proposed. In [3–6], the research group mainly focused on maximizing the profit to reach optimal charging for electric taxis, minimizing their charging cost in face of time-varying electricity prices and some pricing schemes for electric taxis to track the load profile, whose scope is mainly for cost or benefit optimization from a temporal perspective without consideration of the spatial scope. Besides, other work [7] proposed a facility optimization model to minimize the life circle cost (LCC) of charging/swapping facilities, the time value of electric taxis under the constraints of queuing model and the price spread between oil and electricity. A new dispatching policy also presented in [8] with consideration of the taxi demand, the remaining power of electrical taxis, and the availability of battery charging/switching stations in order to reduce the waiting time for power recharging and thus increase the workable hours for taxi drivers.

Most of the works above have a relatively narrow vision for electric taxis. In this paper, we summarize the problems of electric taxi operation to be on a temporal and spatial scale. Moreover, the issue of coordinated charging for electric taxis at both scales was modeled by using a multi-objective approach based on three separate objectives: maximizing the utilization efficiency of charging facilities, minimizing the load unbalance co-operated with the regional power system and minimizing the customers' cost. Then the problem was solved by a fuzzy mathematical method and Improved PSO algorithm.

In summary, the originality of this paper, as well as the differences between the proposed strategy and the EV optimization in the literature, is: (i) in this paper, with the investigation of electric taxis operated in Shenzhen, most of the settings are assumed based on real operation data. The optimization formulation proposed is oriented to handle the practical problems arising from real electric taxi operation; (ii) in the previous research, electric taxi optimization was formulated for electricity cost, or for the charging facilities, or for the customer demands, but those studies only considered this issue from specific perspective. This paper proposes an integrated strategy for electric taxis' coordinated charging, considering the charging facilities, charging load influence related to the regional power system and customers' composite cost; and (iii) most previous research only considers the electric taxi optimization problem from a time perspective. This paper explored the electric taxi charging problem with different visions, with a temporal, spatial and combined scope, so the new strategy involves the solution of a multi-objective and multi-scale constrained optimization model for coordinated charging.

Furthermore, we focused mainly on coordinated charging for electric taxis. However, the proposed procedure can also be extended easily to other types of EV users, e.g., commuting EVs charging at stations. The remainder of this paper is organized as follows: Section 2 formulates our coordinated charging model and shows the strategy proposed for solving the mentioned charging problem; Section 3 presents the mathematical methodology to solve the proposed model, fuzzy method and Improved PSO algorithm. Our simulations for different cases are presented in Section 4. Section 5 summarizes the conclusions.

2. Coordinated Charging Strategy

In this paper, the proposed strategy is oriented to tackle the unbalance problem in the temporal and spatial domain. Figure 1 shows the relationship and communication between electric taxis and charging stations. Electric taxis would send the real time data to a charging station through 3G/4G, including geographical information, battery management information and operation information. The charging station would send the command of the time and station to charge for each electric taxi after the optimization procedure.

The main purpose of the coordinated charging for electric taxis is to determine the optimal time and the optimal station to charge. More preciously, for coordinated charging in the temporal scale, the objective is to seek the best time period to charge; and for coordinated charging in the spatial scale, the objective is to find the best station to charge. The coordinated charging variable, the time period and station to charge, would be optimized under minimizing or maximizing one or more objective functions while satisfying the several equality and inequality constraints. Its mathematical model can established as:

$$\text{Min} f(x)$$
$$\text{s.t.} g(x) = 0, h(x) \leqslant 0 \tag{1}$$

where, f is the objective function to be optimized; g and h are the equality and inequality constrains, respectively; x is the vector of time or station selection variable.

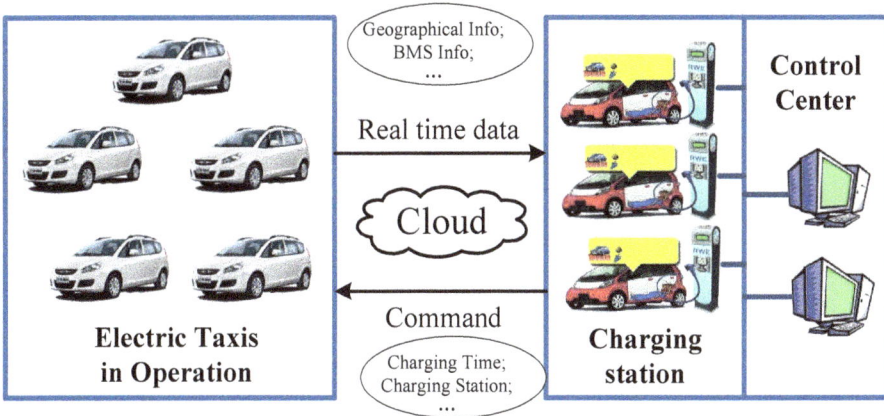

Figure 1. Framework of charging station and electric taxis.

In this paper, three optimization objectives are put forward for electric taxis' coordinated charging, including maximizing the utilization efficiency of charging facilities, minimizing the load unbalance in the regional power system and minimizing the customers' cost. Their mathematical models and constraints are described as follows.

2.1. Objective Functions

2.1.1. Maximizing the Utilization Efficiency of Charging Facilities

In terms of the operation characteristics of electric taxis, the unbalanced utilization of charging equipment leads to most electric taxis turning to charge at a certain period of time or at a certain station, which reduces the temporal and spatial utilization rate of the charging equipment. From the perspective of charging station operators, in order to improve the utilization efficiency of charging infrastructure in temporal and spatial scale, a scheme should be proposed to average the temporal and spatial characteristic of charging behaviors, which is shown as below:

$$\text{Time:} F_{1,\text{time}} = \min \sum_{t=1}^{T} \left(\sum_{m=1}^{M} x_{m,t} P_m - \frac{1}{T} \sum_{t=1}^{T} \sum_{m=1}^{M_t} x_{m,t} P_m \right)^2 \tag{2}$$

$$\text{Space}: F_{1,\text{space}} = \min \sum_{n=1}^{N} \left(\sum_{m=1}^{M} x_{m,n} P_m - \frac{C_n}{\sum_{j=1}^{N} C_n} \sum_{m=1}^{M} x_{m,n} P_m \right)^2 \tag{3}$$

$t = 1, 2, ..., T; m = 1, 2, ..., M; n = 1, 2, ..., N;$

where, M and N is the number of electric taxis and stations in this region, respectively; T is the simulation duration; $x_{m,t}$ and $x_{m,n}$ are the selection variable for temporal and spatial optimization. Besides, P_m is the charging power of electric taxi m; and C_n is the number of chargers in charging station n.

2.1.2. Minimizing the Load Unbalance in the Regional Power System

The unbalanced charging load will also lead to an uneven distribution in the regional power system in both the temporal and spatial scale. The corresponding bad side impact is higher power losses that are one of the most important issues for power system operation. By this means, minimizing power losses is also considered to be an objective, shown as below:

$$\text{Time}: F_{2,\text{time}} = \min \sum_{t=1}^{T} (P_{\text{loss},t}^2 + Q_{\text{loss},t}^2) \tag{4}$$

$$\text{Space}: F_{2,\text{space}} = \min \sum_{n_{brh}=1}^{N_{brh}} (P_{\text{loss},n_{brh}}^2 + Q_{\text{loss},n_{brh}}^2) \tag{5}$$

where, P_{loss} and Q_{loss} are the active and reactive power loss and t and n_{brh} are the time point and branch number, respectively.

However, to obtain P_{loss} and Q_{loss}, the power flow calculation should be involved, which would severely delay the calculation speed for optimization. Additionally, Q_{loss} is usually neglected after the power factor correction. To simplify the model, based on the conclusions shown in [9,10] that the power losses are approximately linearly related to load variance in a radial power system, we extended load variance from the temporal scale to the spatial scale, and use this value to estimate power losses in the regional power system:

$$\text{Time} : F'_{2,\text{time}} = \min \sum_{t=1}^{T} (P_t - \overline{P}_T)^2 \tag{6}$$

$$\text{Space} : F'_{2,\text{space}} = \min \sum_{n_{\text{bus}}=1}^{N_{\text{bus}}} (P_{n_{\text{bus}}} - \overline{P}_{N_{\text{bus}}})^2 \tag{7}$$

where, P_t and $P_{n_{\text{bus}}}$ are the active load of regional power system in period t; active load of n_{bus}, respectively. Besides, $\overline{P}_T$ and $\overline{P}_{N_{\text{bus}}}$ are the average active power in the corresponding scale.

2.1.3. Minimizing the Customers' Cost

Electric taxis are operated for making profits, however, the charging cost is one of the basic costs for electric taxis, just like the oil expense for fuel vehicles. Besides, the charging procedure, which needs at least two hours with 1/2C charging rate from 0% to 100% Battery State of Charge (SOC), is not like refilling the oil. In this way, the time cost (the charging time and the time waiting to charge) and travel cost (travel to charging station) are also involved into the cost function, which is depicted below,

$$F_3 = \min C_{\text{time}} + C_{\text{travel}} + C_{\text{utility}} \tag{8}$$

where, C_{time}, C_{travel} and C_{utility} are the cost of time, travel and the electricity bills, respectively (Unit:RMB). This equation could be used in the three scales above. For the details:

$$C_{\text{time}} = \alpha \cdot \sum (T_c + T_w) \tag{9}$$

$$C_{\text{travel}} = \beta \cdot \sum L_{\text{travel}} \tag{10}$$

$$C_{\text{utility}} = \sum_{t=1}^{T} c_t (\sum_{n=1}^{N} c_n \sum_{m=1}^{M} x_{m,n,t} P_m \Delta t) \tag{11}$$

In the equation above, T_c and T_w represent the time to charging and to wait for charging (Unit: h); L_{travel} means the travel distance to the station for charging (Unit: km); α, β are the price coefficients of time and travelling; c_t, c_n are the service price of charging station in each period and each station (unit: RMB/kWh).

Besides, the EV battery cost (e.g., degradation and replacement) is not considered in this paper, because the optimizing formulation proposed is to operate the electric taxi charging issue for both temporal and spatial scale in the period duration of one day. The EV battery degradation and replacement cost in one day

45

is negligible, compared to the operation cost for the whole day, so to simplify this paper, the battery degradation cost has been neglected.

2.2. Constraints

2.2.1. Constraints with Electric Taxi Operation

(1) In the strategy proposed, the electric taxis have limited charging times for one simulation duration. For example, we consider 24 h to be one simulation duration, which also serves as two shifts for taxi operation, and four times is the charging limitation:

$$\sum_{t=1}^{T} x_{m,t} = 4 \tag{12}$$

(2) Besides, the electric taxi drivers who are willing to charge have N options to choose the station, but only one could be the final decision:

$$\sum_{n=1}^{N} x_{m,n} = 1 \tag{13}$$

(3) Next, to make sure enough electricity for the electric taxi reaches the assigned charging station, the travel distance during each charging interval should be less than the driving mileage (DM) of the electric taxi. In Equation (14), the $L_{operation}$ is the electric taxis' operation distance between two adjacent chargings is:

$$L_{operation} \leqslant DM \tag{14}$$

2.2.2. Electric Taxi Battery Constraints

(1) For the Charging Power, the charging power for a single charger is also constrained by the capability of the charging facilities. The limitation is shown as follows:

$$P_{c-min} \leqslant P_c \leqslant P_{c-max} \tag{15}$$

where, P_{c-min} and P_{c-max} are the bottom and upper power limit of the charger.

(2) For SOC of battery, the SOC of battery also has some constraints:

$$SOC_{min} \leqslant SOC_{m,t} \leqslant SOC_{max} \tag{16}$$

where, SOC_{min} is the bottom power limit of SOC; SOC_{max} is the upper power limit of SOC; $SOC_{m,t}$ is the SOC of vehicle m in period t.

(3) Next, as the experiment data indicates, lithium battery charging is usually done in the Constant Current Constant Voltage mode [11,12]. Shown in Figure 2, the

five charging profiles are overlapped no matter what the value of the original SOC is. In this way, we can draw the conclusion that the entire charging pattern is fixed as the initiation of 0%, and any charging power profile with a certain initial SOC could be decided by the entire charging pattern and its value of initial SOC [13].

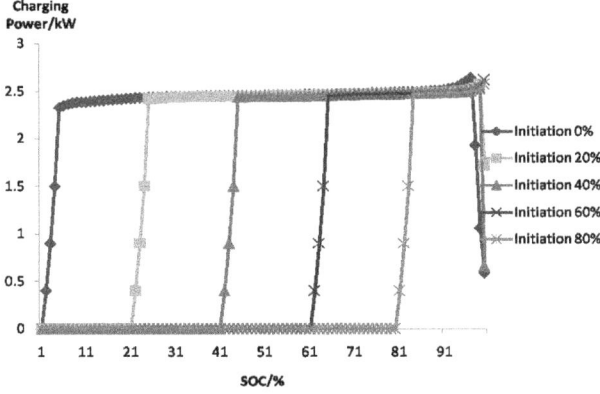

Figure 2. The charging pattern of various SOC at the beginning of charging.

3. Methodology

3.1. Fuzzy Mathematical Method for Multi-Objective Problems

In this paper, the problem for coordinated charging turns out to be a multi-objective optimization problem. Usually, the weighting method and the fuzzy mathematics method are utilized to handle each objective weight of a multi-objective optimization problem. Compared to subjective setting or several tests for weighting methods to get proper weight coefficients, the fuzzy mathematics method uses a membership function to perform objective function fuzzification to form the fuzzy multi-objective function. There is no need to use weighting coefficients to tackle multi-objective problems in the fuzzy mathematics method, and the solution is obtained objectively [14].

For objectively considering the weight of each optimization objective, the linear membership function is used to make the objective functions fuzzification to form the fuzzy multi-objective function. The linear membership function can be described as:

$$\mu_x(x) = \begin{cases} 0 & f_i(x) \le c_{i,\min} \\ \dfrac{f_i(x) - c_{i,\min}}{c_{i,\max} - c_{i,\min}} & c_{i,\min} < f_i(x) < c_{i,\max} \quad i = 1,2,3,4 \\ 1 & f_i(x) \ge c_{i,\max} \end{cases} \tag{17}$$

47

where, $f_i(x)$ is ith objective function of fuzzy multi-objective problem; $\mu_i(x)$ is the membership function of $f_i(x)$; m is the number of objective functions; $c_{i,min}$, $c_{i,max}$ are the upper and lower limit values of $f_i(x)$, respectively; $c_{i,min}$ is the optimal value obtained by a single objective function; and $c_{i,max}$ is the initial value of each objective function, the details of which are summarized in Table 1. The membership function curves given by Equation (17) are shown in Figure 3.

Table 1. Case setting of charging period.

Time/h	1	2	3	4	5	6	7	8
Waiting coefficient	0.1	0.5	1	0.5	0.1	0	0	0.1
Charging price	0.5	0.5	0.5	0.5	0.5	0.5	0.5	1
Nominated load profile	0.76	0.71	0.67	0.66	0.65	0.66	0.69	0.72
Time/h	9	10	11	12	13	14	15	16
Waiting coefficient	0.5	0.8	0.5	0.1	0.1	0.5	1	0.5
Charging price	1	1	2	2	2	2	2	1
Nominated load profile	0.81	0.89	0.93	0.94	0.91	0.87	0.85	0.85
Time/h	17	18	19	20	21	22	23	24
Waiting coefficient	0.1	0	0	0.1	0.5	0.8	0.5	0.1
Charging price	1	1	2	2	2	1	1	0.5
Nominated load profile	0.89	0.99	1.00	0.98	0.97	0.93	0.86	0.80

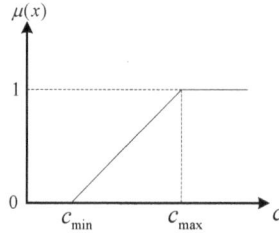

Figure 3. Membership function of sub-objective.

3.2. Improved Self-Adaptive PSO Algorithm with Detection Function

The coordinated charging problem in the strategy of Section 2 is considered a complex multi-constraints, nonlinear optimization problem. Compared to many classical algorithms, such as linear programming, quadratic programming, gradient descending method and other numerical algorithms, heuristic algorithms are novel algorithms for solving optimization problems and are much easier to implement and extend. The typical heuristic methods include genetic algorithms (GA), particle swarm optimization (PSO), differential evolution (DE), artificial immune algorithm, and artificial bee colony (ABC) algorithm. These algorithms are based on multi-point stochastic searching, and they can effectively solve optimization problems in different situations. Their global convergence capacity is better than that of the classical algorithms.

The PSO algorithm possesses superior performance in its implementation and a good trade-off between exploration and exploitation ability. It was introduced by Kennedy and Eberhart in 1995 [15], and originates from the simulation of birds' behaviour and fish behaviour. PSO is an algorithm with a simple structure, simple parameter setting and fast convergence speed, which has been widely applied in function optimization, mathematical modelling, system control, and some other areas [16].

In basic PSO algorithms, w, c_1 and c_2 are fixed values. For the search accuracy and search speed, many improved PSO methods were proposed to modify the parameters, especially the inertia weight, so that the search space can be changed steadily from the global to the local. In this paper, the improved inertia weight is shown in Equation (20). The algorithm may adjust w dynamically via Equation (20), so that it can optimize dynamically by taking both global search and local search into account during changing. The improved PSO is shown as follows:

$$v_{id}^{k+1} = w(k)v_{id}^k + c_1r_1(p_{id} - z_{id}^k) + c_2r_2(p_{gd} - z_{id}^k) \tag{18}$$

$$z_{id}^{k+1} = z_{id}^k + v_{id}^{k+1} \tag{19}$$

$$w(k) = w_{start} - (w_{start} - w_{end})(\frac{k}{T})^2 \tag{20}$$

where, w_{start} and w_{end} represent the initial value and the final value of w, respectively; T is the maximum number of evolutionary generations; k is the current number of the evolutionary generation.

3.3. The Mathematical Procedure

In this section, we present the mathematical procedure for the coordinated charging strategy proposed based on multi-objective and multi-scale optimization. During this procedure, the operation of the charging process has been simulated, and besides, there should be a monitor system collecting the operation data for the optimization platform. After the optimization calculation, the command from the control center would be sent to execute the elements which decide when and where to charge the electric taxi. All of these are performed in the aggregator, the brain of this system. In details, this is expressed as follows:

STEP 1: During operation, the electric taxi information is collected at the control center through the monitoring system, including the ID of vehicles, the geographical data, the SOC condition, etc. For simulation, all of these data would be generated by a stochastic process.

STEP 2: The optimization process:

- Case 2: For temporally-coordinated charging: the operation scheduling of electric taxis for one simulation duration is collected, which is the original operation. Then, the scheduling would be optimized through the proposed multi-objective strategy. The new electric taxi scheduling would be better in facility utilization, charging load profile and operation cost.
- Case 3: For spatially-coordinated charging: this optimization is quite similar to the temporal model, with the difference that the variable is the choice of charging station in a certain period rather than charging period for a long duration.
- Case 4: For combination coordinated charging: When the temporally-coordinated charging is running, once the charging period dispatching for one simulation interval is accomplished, then the spatial coordinated charging procedure would be called for optimal charging station selection.

STEP 3: When the temporal or spatial coordinated charging is called, the Improved PSO algorithm for the corresponding fitness function, where the fuzzy mathematical method has turned multi-objective optimization into single objective optimization, would be executed. The optimization procedure-based improved PSO is shown in Figure 4.

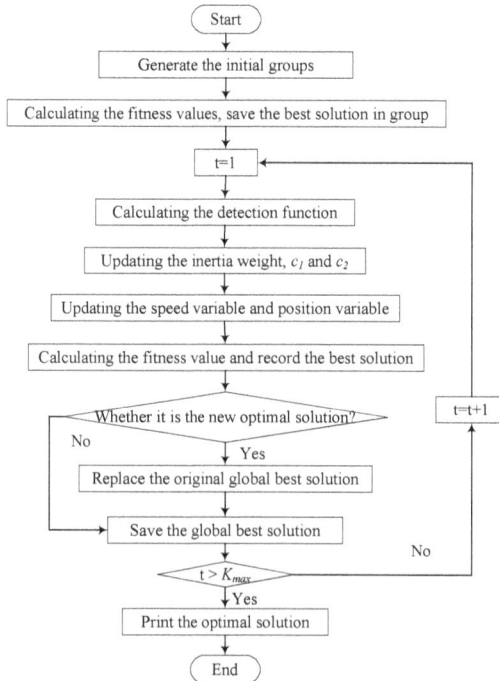

Figure 4. The flow chart of the improved PSO algorithm.

4. Simulations and Results

4.1. Case Setting

In this section, some operation data are generated based on the original operation statistics of electric taxis in Shenzhen. Assuming that:

(1) There are N charging station and M electric taxis running in the region with a charging demand distribution in four time intervals, according to the statistical results in [17]. In this case, $N = 8$; $M = 100$.

(2) The value of selection variables, $x_{m,t}$ and $x_{m,n}$, either 1 or 0, $x_{m,t} = 1$ means the electric taxi m would start charging at period t, $x_{m,t} = 0$ means not; and $x_{m,n} = 1$ means the electric taxi m would choose to charge at station n, $x_{m,n} = 0$ also means not.

(3) Assume that the distance between the position of the electric taxis and their assigned charging station shows a normal distribution $N\sim(10, 2)$ (unit: km). Besides, we neglect the bad traffic conditions, such as the traffic congestion, and suppose that electric taxi m is running with the speed of v_m. The speeds of electric taxis are also in a normal distribution $N\sim(40, 5)$ (unit: km/h).

(4) In Equations (9) and (10), α, β are the price coefficients of extra time and travel for EV operation, respectively. According to the investigation, we assume that the wages and the working hours of electric taxi drivers during a working shift would be 300 yuan and 12 h, respectively. So 25 (unit: RMB/h) is considered to be the value of α for time coefficient in the simulation cases. Besides, with the speed value of each taxi, v_m, the coefficient of travel could be transferred from the coefficient of time. By this mean, β is considered to be α/v_m (unit: yuan/km).

(5) The charging rate for electric taxis used in this simulation is 0.5 C, and it is constant for all the charging facilities, which also comes from the investigation for the Shenzhen case.

(6) The time shifts and the time starting to charge distribution are assumed as in Figure 5 below.

(7) The SOC of vehicles when arriving at the station are assumed to be in a normal distribution, with the mean value of 50% and standard deviation of 10% shown in Figure 6. Besides, the charging duration would be derived through the initial SOC of electric taxis.

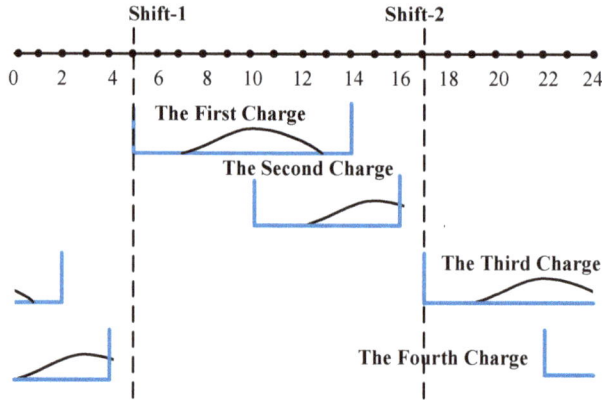

Figure 5. The starting time to charge distribution.

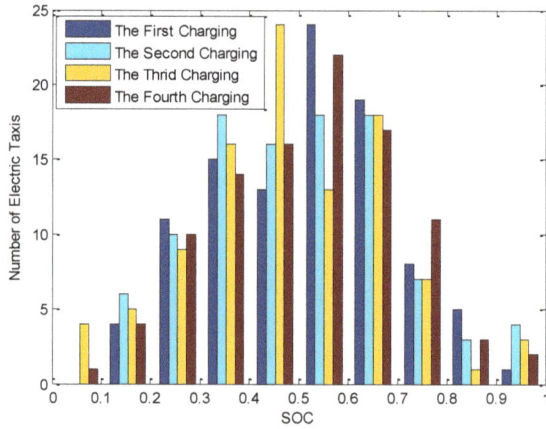

Figure 6. *SOC* distribution when electric taxis return.

(8)   The waiting coefficient of different time period stations are assumed in terms of the statistical results, and the charging price of temporal scale is assumed based on the peak-valley price, which are shown in Tables 1 and 2.

Table 2. Case setting of charging stations.

No.	1	2	3	4	5	6	7	8
Number of chargers	6	8	4	10	4	8	10	6
Waiting coefficient	0.05	0.20	0.20	0.05	0.05	0.20	0.20	0.05
Charging price	1	1	1	1	1	1	1	1

(9) Next, the regional power system and charging station geographical information are shown in Figure 7, which is a modified IEEE 33 bus system with charging station loads. The load in each bus and the fluctuation are emulated in Table 2.

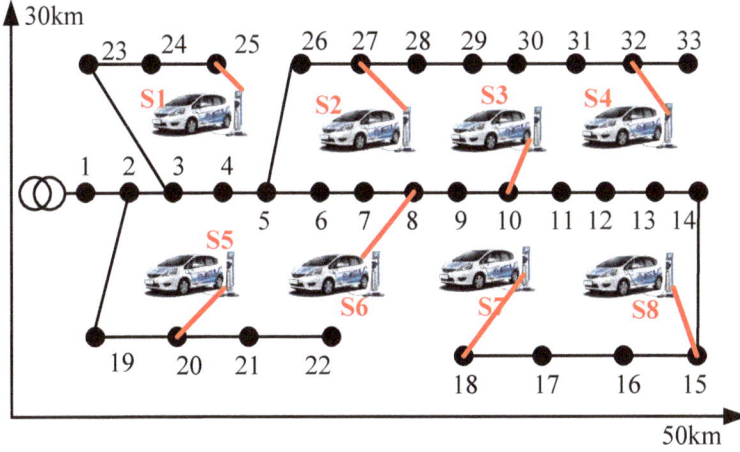

Figure 7. Case setting for stations.

(10) In the simulation cases, the parameters of the Improved PSO Algorithm, c1, c2, ω_{start} and ω_{end}, have been assigned to be 1.49, 1.49, 0.9 and 0.4, respectively.

(11) The upper and lower limitations $c_{i,min}$, $c_{i,max}$ of the fuzzy mathematic method are shown in Table 3.

Table 3. Upper and lower sub-objective limits.

Sub-Objective	Objective 1	Objective 2	Objective 3
Original value, $c_{i,max}$ (Temporal)	4.7683×10^8	4.8068×10^8	1.1824×10^6
Optimal value, $c_{i,min}$ (Temporal)	2.0288×10^7	2.7157×10^7	9.2963×10^5
Original value, $c_{i,max}$ (Spatial)	9.8929×10^8	1.2391×10^8	2.2636×10^4
Optimal value, $c_{i,min}$ (Spatial)	4.0234×10^8	3.5364×10^3	2.1236×10^4

4.2. Results

4.2.1. The Effectiveness of the Improved PSO Algorithm

In this paper, we utilized the Improved PSO Algorithm to solve a multi-constraints, nonlinear optimization problem. In Figure 8, the fitness optimization process of the Improved PSO and the original PSO are compared. Compared to the original PSO algorithm, we can see that the Improved PSO with the changing inertia weight, still has the better search ability and better optimization results.

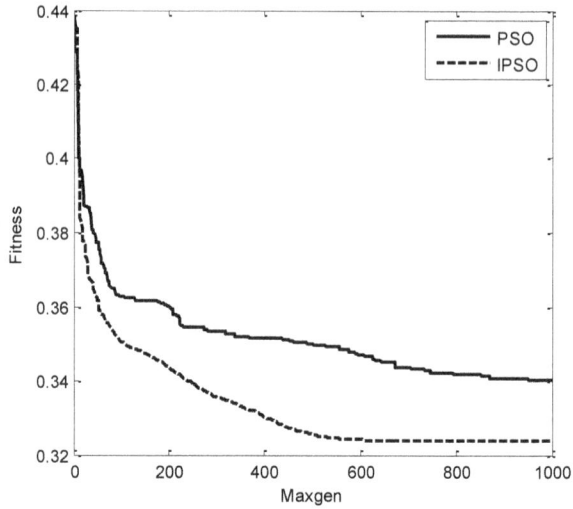

Figure 8. Fitness comparison of original and improved PSO algorithms.

4.2.2. Results of Cases 1, 2 and 4

With the charging strategy in temporal scale, the charging power dispatches are compared in Figure 9, that is, the original charging load profile, generated based on the operation characteristics shown in Figure 5, and the temporal and combined coordinated charging profiles obtained from the corresponding strategy. Table 4 displays the sub-objective values and multi-objective values in the three different cases.

In the Figure above and Table below, it is shown that the charging load curve has a better deployment with a much better choice of charging time, for both spatial and combined coordinated charging. The charging peak has declined from approximately 2000 kW to 1000 kW and 1500 kW, respectively.

Table 4. Sub-objective value of temporal coordinated charging strategy.

Objective	Objective 1	Objective 2	Objective 3	Multi-Objective
Original charging	4.7683×10^8	4.8068×10^8	1.1824×10^6	3.0000
Temporal coordinated charging	2.2597×10^7	3.1242×10^7	1.1490×10^6	0.8819
Combined coordinated charging	6.5921×10^7	7.1438×10^7	1.1118×10^6	0.9183

Additionally, the Table depicts the details of the optimization process, with an apparent decrease in all of the three objectives, especially for Objective 1 and 2, as well as the multi-objective. It is noticed that the combined coordinated charging has a modest optimizing effect, compared to separated spatial coordinated charging, with the reason that the combined dispatching considered only one time interval with a limited vision for global optimization.

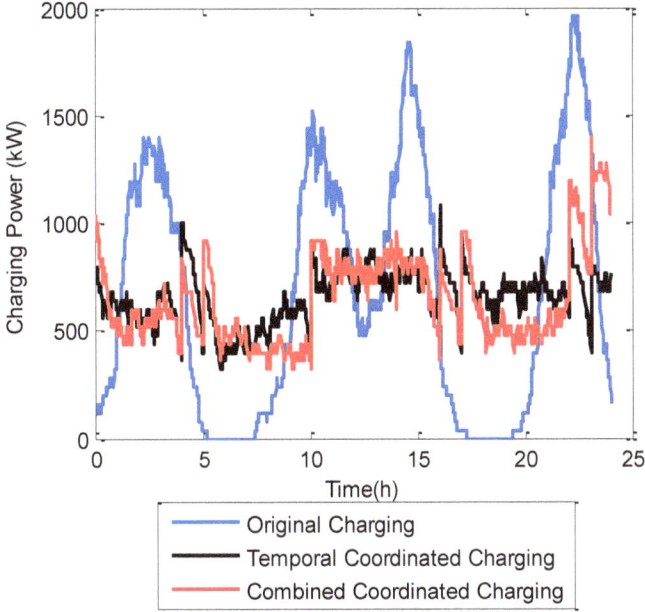

Figure 9. The comparison of original, temporal and combined coordinated charging load.

Besides, we could see the effectiveness of the temporal and combined coordinated charging strategy in a modified IEEE 33 bus test system. In Table 5, the Cases 2 and 4 both show the improvement in power loss and voltage deviation, and Case 2 is even better, which conforms to the conclusion of Table 4.

Table 5. Effects in modified IEEE 33-bus system with coordinated charging strategy.

Case	Case 1	Case 2	Case 4
Power loss (kWh)	7081.4	6818.9	6824.6
Voltage deviation (kV^2)	2.5318×10^4	2.4427×10^4	2.4439×10^4

4.2.3. Results of Cases 1, 3 and 4

With the charging strategy in spatial scale, the charging selection for stations are compared in Figure 10. That is, the original charging, generated based on the positions of eight charging stations shown in Figure 5, with the assumption that the geographical information is the only factor that taxi drivers consider, and the spatial and combined coordinated charging results obtained from the corresponding strategy. Table 6, displays the sub-objective values and multi-objective values in three different cases.

Table 6. Sub-objective value of spatial coordinated charging strategy.

Objective	Objective 1	Objective 2	Objective 3	Multi-Objective
Original charging	9.8929×10^8	1.2391×10^7	2.2636×10^4	3.0000
Spatial coordinated charging	4.9374×10^8	6.4124×10^5	2.1961×10^4	0.6787
Combination coordinated charging	5.0361×10^8	1.8533×10^6	2.2036×10^4	0.7589

In Figure 10, the original charging scenario is supposed to result in charging congestion and charging delay at related stations, and the dispatching is much better after applying the new dispatching rules for station selection. The sub-objective values have been reduced, especially for Objective 2, which verifies the effectiveness of the spatial and combined coordinated charging strategy. The combined coordinated charging still has the same problem, showing restrained improvement compared to separate optimization.

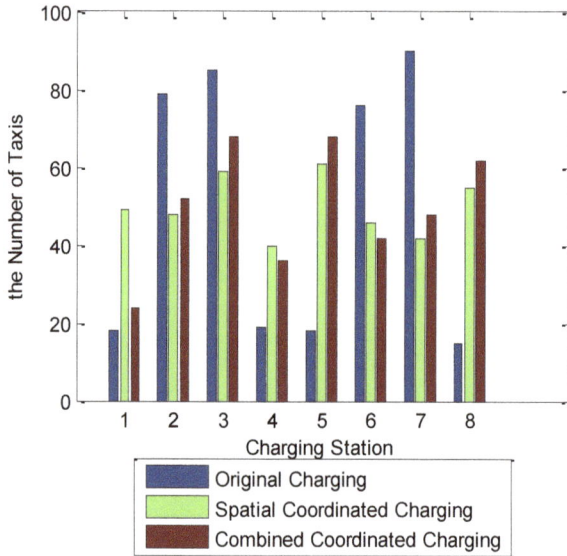

Figure 10. The comparison of original, spatial and combined coordinated charging distribution.

5. Conclusions

In this paper, a new strategy has been proposed for electric taxi coordinated charging, subject to the charging facilities, charging load influence in the regional power system and customers' composite cost; additionally, this work also applies different visions to evaluate this problem, in the temporal, spatial and combined scales, and the key of this problem is to intelligently identify the appropriate charging periods and stations to charge. The new strategy involves the solution of a constrained multi-objective and multi-scale optimization model for coordinated charging. Moreover, the simulations could prove that this strategy shows significant effectiveness for coordinated charging in the temporal and spatial scale.

Acknowledgments: This work was supported by National Key Technology R&D Program (2013BAA01B03). Besides, we also want to thank the investigation support from Potevio New Energy.

Author Contributions: The initial design and implementation of the coordinated charging framework and the solution based on the fuzzy method and Improved PSO was done by Yuqing Yang. She was also responsible for the simulation implementation for case study. Besides, this work was also performed under the advisement and regular feedback from Weige Zhang, who also revised the manuscript critically. Moreover, Liyong Niu and Jiuchun Jiang also gave some useful suggestions for this work and helped revise the manuscript.

Conflicts of Interest: The authors declare no conflict of interest.

References

1. Zhang, W.; Wu, B.; Li, W.; Lai, X. Discussion on development trend of battery electric vehicles in China and its energy supply mode. *Power Syst. Technol.* **2009**, *33*, 1–5.
2. Tian, Z.; Wang, Y.; Tian, C.; Zhang, F.; Tu, L.; Xu, C. Understanding operational and charging patterns of electric vehicles using GPS records. In Proceedings of the 2014 IEEE 17th International Conference on Intelligent Transportation System (ITSC), Qingdao, China, 8–11 October 2014.
3. Sun, L.; Yang, J.; Yang, Z. Optimal charging strategy of plug-in electric taxi. In Proceedings of the 10th IEEE International Conference on Control and Automation (ICCA), Hangzhou, China, 12–14 June 2013.
4. Yang, Z.; Sun, L.; Ke, M.; Shi, Z.; Chen, J. Optimal charging strategy for plug-in electric taxi with time-varying profits. *IEEE Trans. Smart Grid* **2014**, *5*, 2787–2797.
5. Yang, J.; Yang, Z. Pricing scheme for aggregate load scheduling of plug-in electric taxi fleet. In Proceedings of the 33rd Chinese Control Conference, Nanjing, China, 28–30 June 2014.
6. Yang, Z.; Sun, L.; Chen, J.; Yang, Q.; Chen, X.; Xing, K. Profit maximization for plug-in electric taxi with uncertain future electricity prices. *IEEE Trans. Power Syst.* **2014**, *29*, 3058–3068.
7. Jing, Z.; Fang, L.; Lin, S.; Shao, W. Modelling for electric taxi load and optimization model for charging and swapping facilities for electric taxi. *ITEC Asia Pac.* 2014.
8. Lu, J.; Yeh, M.; Hsu, Y.C.; Yang, S.; Gan, C.; Chen, M. Operation electric taxi fleets: A new dispatching strategy with charging plans. In Proceedings of the 2012 IEEE International Electric Vehicle Conference (IEVC), Greenville, SC, USA, 4–8 March 2012.
9. Eric, S.; Mohammad, H.; MacPherson, M.J. Coordinated charging of plug-in hybrid electric vehicles to minimize distribution system losses. *IEEE Trans. Smart Grid.* **2011**, *2*, 198–205.

10. You, H.; Jiang, X. Research on relationship of load variance characteristics and electric distribution line energy loss. *Power Demand Side Manag.* **2008**, *10*, 13–15.

11. Wen, J. *Studies of Lithium-Ion Power Battery Optimization Charging Theory for Pure Electric Vehicle*; Bejing Jiaotong University: Beijing, China, 2011.

12. Wen, F. *Study on Basic Issues of the Li-ion Battery Pack Management Technology for Pure Electric Vehicles*; Beijing Jiaotong University: Beijing, China, 2010.

13. Yang, Y.; Jiang, J.; Bao, Y.; Zhang, W.; Huang, M.; Su, S. Dynamic coordinated charging strategy and positive effects in regional power system. In Proceedings of the 11th IEEE International Conference on Control and Automation (ICCA), Taichung, Taiwan, 18–20 June 2014.

14. He, X.; Wang, W. Fuzzy Multi-objective opimal power flow based on modified artificial bee colony algorithm. *Math. Probel. Eng.* **2014**, *2014*.

15. Kennedy, J.; Eberhart, R. Particle swarm optimization. In Proceedings of the IEEE International Conference on Neutral Networks, Perth, WA, USA, 27 November–1 December 1995.

16. Zhang, Y.C.; Xiong, X.; Zhang, Q.D. An improved Self-adaptive PSO algorithm with detection function for multimodal function optimization problem. *Math. Probel. Eng.* **2013**, *2013*.

17. Li, Y.; Huang, M.; Zhang, W. An estimation method for daily charging load of electric taxis. *Autom. Electr. Power Syst.* **2014**, *38*, 55–60.

MV and LV Residential Grid Impact of Combined Slow and Fast Charging of Electric Vehicles

Niels Leemput, Frederik Geth, Juan Van Roy, Pol Olivella-Rosell, Johan Driesen and Andreas Sumper

Abstract: This article investigates the combined low voltage (LV) and medium voltage (MV) residential grid impact for slow and fast electric vehicle (EV) charging, for an increasing local penetration rate and for different residential slow charging strategies. A realistic case study for a Flemish urban distribution grid is used, for which three residential slow charging strategies are modeled: uncoordinated charging, residential off-peak charging, and EV-based peak shaving. For each slow charging strategy, the EV hosting capacity is determined, with and without the possibility of fast charging, while keeping the grid within its operating limits. The results show that the distribution grid impact is much less sensitive to the presence of fast charging compared to the slow charging strategy. EV-based peak shaving results in the lowest grid impact, allowing for the highest EV hosting capacity. Residential off-peak charging has the highest grid impact, due the load synchronization effect that occurs, resulting in the lowest EV hosting capacity. Therefore, the EV users should be incentivized to charge their EVs in a more grid-friendly manner when the local EV penetration rate becomes significant, as this increases the EV hosting capacity much more than the presence of fast charging decreases it.

Reprinted from *Energies*. Cite as: Leemput, N.; Geth, F.; Van Roy, J.; Olivella-Rosell, P.; Driesen, J.; Sumper, A. MV and LV Residential Grid Impact of Combined Slow and Fast Charging of Electric Vehicles. *Energies* **2015**, *8*, 1760–1783.

1. Introduction

Electric vehicles (EVs) are available on the market and their sales are growing. A primary difference with conventional vehicles is the way of refueling. EVs can charge at every location that offers a compatible electrical socket, of which the most common ones are at home and the workplace [1]. Due to the typically long standstill times at these locations, and the low average daily driven distances [2], a low charging power at these locations is sufficient to fulfill the majority of the mobility needs, thereby keeping the charging infrastructure investments low.

Typically, mode 2 or mode 3 charging, as defined in the IEC 61851-1 standard [3], are used to charge EVs at their standstill locations. Mode 2 charging makes use of a standard domestic socket, whereas mode 3 makes use of a dedicated grid connection.

In Europe, single-phase charging is commonly rated at 10 A for mode 2 and 16 A for mode 3 [4]. As 90% of the daily EV energy consumption is below 18 kWh [2], the daily charging time for mode 2 and mode 3, respectively, remains below 9 and 6 h, 90% of time.

For occasional long-distance trips, fast charging is a necessary addition to slow charging, given the typical driving range of EVs nowadays. The implementation of fast charging networks that cover large contiguous regions make EVs a viable alternative for conventional vehicles [5,6]. The power rating for fast chargers vary from 50 kW [1] up to 120 kW [6], which allows recharging EVs within an acceptable time span, typically within half an hour. Multistandard fast chargers provide compatibility with the different types of fast charging standards that are used nowadays [1].

EV charging will impact the power system, e.g., peak power demand, load profile and voltage magnitude deviations [7]. For slow charging, the residential low voltage (LV) grid impact may be significant, due to the simultaneity between the residential power peak and the plugging in of EVs when arriving at home, which starts the charging process for uncoordinated charging. Therefore, an extensive amount of research is conducted on coordinated slow charging strategies [8].

For fast charging, two typical infrastructure configurations are used. The first one is a network of single- or dual-outlet fast chargers that allow reaching the next fast charger. This is a configuration for the initial rollout of fast charging infrastructure [9]. The fast chargers are connected to the local LV grid, and therefore, the grid impact will strongly depend on the local situation, e.g., being connected to the distribution transformer through a separate feeder or through an existing feeder that contains other loads.

In the second configuration, fast charging stations with multiple fast chargers (>4) are located next to busy traffic arteries. These stations resemble conventional highway refueling stations in their setup, and are typically commercially operated [1]. Because the total power rating of such stations is in the order of magnitude of a typical European residential LV grid (>200 kVA), they are connected to the medium voltage (MV) grid through a dedicated transformer.

1.1. Literature Overview

1.1.1. Slow Charging

Different large-scale coordination objectives have been considered for mode 2 and mode 3 EV charging, e.g., integrating a higher share of intermittent renewable energy in the power system [10–12], congestion management [13], frequency regulation [14], portfolio imbalance reduction [15], charging cost minimization [16–18]. Large-scale coordinated charging shows to be effective to obtain such objectives. However, fast charging is not taken into account in their coordination strategies. Because the large-scale strategies require a substantial high EV penetration rate for the benefits to be noticeable [19], widespread coordinated charging is not expected soon. However, EV clusters may occur soon, which significantly impact the distribution grid.

EV charging can be controlled locally to smoothen the power profile of the distribution system [7], e.g., distribution grid peak shaving as explicit objective [20–22], or as a constraint in cost-minimizing coordination strategies [23,24]. Distribution grid peak shaving can be chosen as an optimization objective, to mitigate local load peaks in the distribution grid. This reduces the simultaneity of household and EV power demand, which positively impacts voltage deviations [20] and grid losses [21]. Voltage deviation reduction can also be considered as a coordination objective during times of excessive deviations [7].

A wide range of programming techniques are used to implement optimized charging strategies, e.g., linear programming [24,25], sequential quadratic programming [7,26], dynamic programming [7,14], convex quadratic programming [21], and heuristic programming [11,27]. The choice of the programming technique is a trade-off between optimality, completeness, accuracy, robustness, and execution time. Because no optimization model will be used here to represent control decision, but rather a set of rules, a further analysis of the different techniques is out of scope.

All of the above mentioned charging optimization strategies require communication between the grid and the EVs. Therefore, such strategies are ineffective in case of malicious or absent communication. To overcome this problem, local parameters can be used for a rule-based control of the EV charging process, opposed to the abovementioned optimization strategies. For example, the EV charging power rating can be reduced based on the required charging energy, which can be calculated by the EV itself based on the time until the next departure, which can be delivered as an input by the EV user. This concept is hereafter referred to as EV-based peak shaving. In previous work, this rule-based charging strategy has been investigated for one LV distribution grid [28]. In the considered grid, EV-based peak shaving allowed for a higher EV hosting capacity compared to other strategies. Fast charging was not included there.

1.1.2. Fast Charging

Several aspects of fast charging are investigated in the literature: economic assessments, infrastructure usage optimization, MV grid impact assessments, and the power electronic converter design. An economic analysis on fast charging infrastructure is performed for Germany [29] and China [30]. For both countries, it is concluded that fast charging infrastructure is hardly profitable with the present energy pricing and battery costs, and a high EV penetration rate is required for a profitable exploitation [29]. Both [29] and [30] do not use a time-based fast charging scenario in their modeling, as this is not the scope of an economic analysis.

Traffic modeling is used to determine the fast charging demand in space and time in [31]. The optimization of EV charging scheduling for highway fast chargers is discussed in [32] and [33]. The time-based occupation of the fast charging stations, and the variation of the EV battery state of charge at the start of the highway trip are based on a mathematical distribution, not on mobility behavior or slow charging behavior.

The MV grid impact of fast charging stations is discussed in [34–36]. A static worst-case load scenario is modeled in [34], assuming all fast chargers are used simultaneously at their rated power. Time-based vehicle arrival pattern of conventional refueling stations are used in [35], and time-based road occupation profiles are used in [36], to model the demand for fast charging. No slow charging behavior is taken into account, and in [35] and [36] it is assumed that the EVs continuously draw the rated fast charging power. Three-phase balanced loads are assumed in the MV grid, thereby not taking into account phase unbalance.

The electric and power electronic design of fast chargers is discussed in [37] and [38]. As these components need to be sized for the peak load, a worst-case scenario for the fast charging demand is used. The design of a fast charging station with local energy storage and local photovoltaic power production is discussed in [39]. The fast charging demand is based upon mobility behavior and takes into account residential slow charging. Load flow is not assessed, because the scope is on the sizing of the fast charging station and the power flow between its components.

1.2. Scope

To the best of the authors' knowledge, neither the combination of slow and fast charging, nor the combined MV and LV grid impact assessment is investigated in the literature. Both slow and fast charging behavior are modeled here, because they influence each other, and therefore, their interaction influences the grid impact. Grid impact is assessed for MV and LV, as both network levels influence each other.

A realistic Flemish distribution grid topology and a scenario realistic for the Flemish situation, are used here. Realistic single-phase Flemish household LV load profiles and measured photovoltaic (PV) power production profiles are used. The EV charging behavior is based upon Flemish mobility behavior and passenger vehicle fleet composition.

Three residential slow charging strategies are compared in terms of their EV hosting capacity, with and without the presence of fast charging. Fast charging is modeled to supplement slow charging at home and the workplace, when the battery capacity is insufficient to fulfill the mobility requirements otherwise. The fast charging infrastructure is assumed to consist out of charging stations with multiple fast chargers, sized to fulfill the fast charging demands of the area it covers. The fast charging stations are connected to the MV grid through a dedicated transformer.

2. Materials and Methods

2.1. Distribution Grid

Three feeders are used to model three residential LV grid topologies, as illustrated in Figure 1 and with the parameters summarized in Table 1. These feeders are provided by the Flemish distribution system operators (DSOs) and are identified as (semi-) urban topologies, with a TT (Terra-Terra) grounding arrangement [40]. Cable parameters are taken from the design specifications of the cable standard NBN C33-322 [41]. All household loads h are assumed to have a single-phase grid connection between one of the three phases ($p \in \{a, b, c\}$) and the common neutral conductor (n), through feeder branching from the main one. The houses are connected alternatingly to each of the three phases. The rated neutral-to-phase voltage U^{nom} is 230 V.

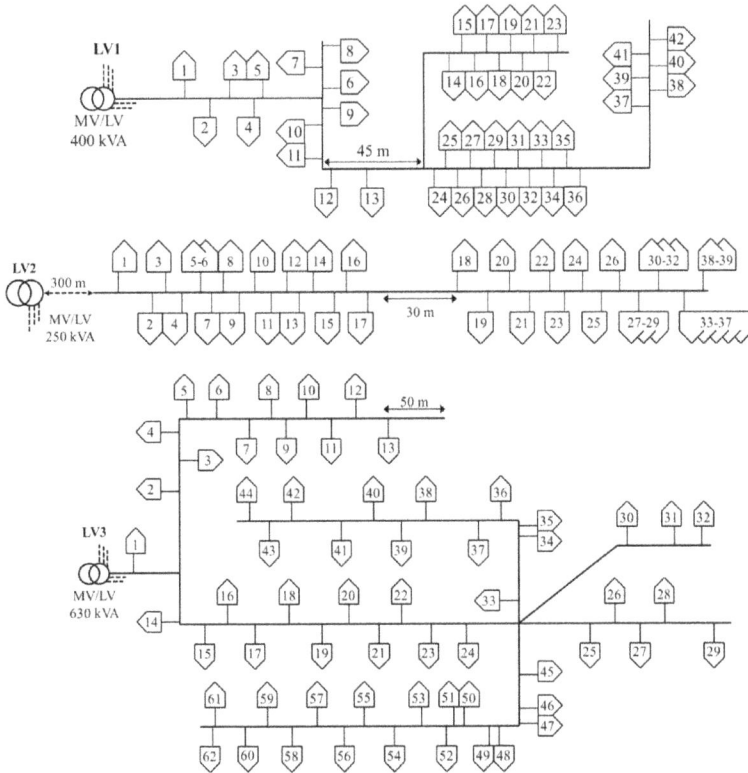

Figure 1. LV grid topologies.

Table 1. LV grid parameters.

Grid		LV1	LV2	LV3
Cable type	Primary	Al-4 × 95 mm²	Al-4 × 150 mm²	
	Secondary	Cu-4 × 16 mm² (35 mm² at 2th last node LV2)		
I_{MAX} [A]	Primary	245	315	
	Secondary	120 (175)		
Z [Ω/km]	Primary	$0.320 + 0.078i$	$0.206 + 0.078i$	
	Secondary	$1.15 + 0.083i$ ($0.524 + 0.081i$)		
# nodes/feeder		42	29	62
# loads/feeder		42	39	62
Parallel feeders	# Added feeders	5	3	5
	Length [m]	250	250	300
Total # LV loads		252	156	372
Z transformer [Ω]		$0.008 + 0.029i$	$0.013 + 0.045i$	$0.004 + 0.020i$

The three feeders are connected to three different MV/LV transformers (Delta-Wye+N), which have three commonly occurring power ratings for Flanders: 250, 400, and 630 kVA. The transformer impedances are derived from [42], and their taps are set at 1 pu. A higher transformer tap could be selected to increase the lowest occurring voltages, which would reduce the impact of the EV charging load on the voltage deviations. However, this would also increase the highest occurring voltages, as no on-load tap changers are used at MV/LV transformers in Flanders. This can cause overvoltage events, at moments of high residential PV production, as residential PV power production reduces the operational margins [43]. Therefore, the taps are chosen at 1 pu here, as the goal is not to find the optimal tap settings, but to provide a realistic scenario for a Flemish distribution grid with PV power generation. Different tap settings would obviously influence the results, but the qualitative results of the comparative analysis of the different charging strategies would remain the same, as the tap settings remain fixed during this analysis.

Because multiple LV feeders are connected to each MV/LV transformer, a number of simplified parallel feeders, as summarized in Table 2, are added to each of the three grid topologies, as in [29]. An aggregated load, with an equivalent number of households as for each of the three detailed feeders, is added to the end of each parallel feeder. The three LV grid topologies are each used three times, together with the fast charging station (FCS), to connect to a MV feeder with 9 nodes, as illustrated in Figure 2. As a result, a total of 2340 household loads are connected to the MV feeder. The rated MV line-to-line voltage is 11 kV, and the distance between each node is 600 m. This is a realistic urban MV feeder topology for Flanders [44], where MV grids are operated in an open-ring topology with few or absent laterals [45]. The MV cable is a three-core armored aluminum conductor, i.e., Al 11 kV-3 × 95 mm², with an impedance of $0.411 + 0.105i$ Ω/km, and a current rating of 200 A [46].

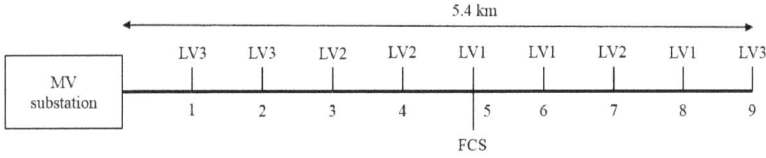

Figure 2. MV grid topology to which the LV grids and the fast charge station (FCS) are connected.

Table 2. LV feeder locations of the randomly assigned PV installations.

MV Node	Houses with a PV Installation
1	1, 17, 21, 24, 42, 46, 50
2	6, 15, 20, 27, 43, 57, 58
3	10, 16, 17, 23
4	10, 17, 20, 25
5	1, 6, 17, 21, 37
6	3, 26, 28, 30, 42
7	9, 21, 25, 31
8	2, 8, 22, 27, 33
9	6, 7, 18, 23, 37, 49, 58

The end-user distribution grid voltages should stay within the operation limits of the EN50160 standard [47], which are defined as follows:

$$0.9 \cdot U^{\text{nom}} \leqslant \left| U_{p,i,k} \right| \leqslant 1.1 \cdot U^{\text{nom}},$$

for >95% of all 10 min intervals of any week

(1)

$$0.85 \cdot U^{\text{nom}} \leqslant \left| U_{p,i,k} \right| \leqslant 1.1 \cdot U^{\text{nom}},$$

(2)

$$F_{i,k}^{\text{VU}} \leqslant 2\%, \text{ for } > 95\% \text{ of all 10 min intervals of any week}$$

(3)

U^{nom} is the nominal voltage magnitude, *i.e.*, 230 V. $\left| U_{p,i,k} \right|$ and $F_{i,k}^{\text{VU}}$ are, respectively, the voltage rms magnitude and the voltage unbalance factor at phase p, grid node i, and time step k.

These voltage magnitude and unbalance constraints, together with the above mentioned feeder current constraints, will determine how much EV charging load the distribution grid can accept, additional to the residential load and the PV power generation. As a result, these constraints will determine the EV hosting capacity. Opposed to the current and voltage constraints, the transformer power rating is not a hard real-time constraint. Exceeding the transformer power rating accelerates its ageing, which reduces the technical lifetime of the transformer. If this technical lifetime goes below the economic lifetime, the transformer will need to be replaced

65

earlier than initially expected. This is a cost that should be taken into account, but it does not require immediate action.

2.2. Residential Load and PV Generation

A residential electric load profile generator is used to create a statistically representative set H_o of single-phase Flemish household electric power consumption profiles h, with a 15 min time resolution, as presented in [48]. Variations between week and weekend days, as well as seasonal variations, are taken into account in the profiles. Reactive power consumption is neglected, because this is not generated by the profile generator. This is because residential grid-connected PV power generation amounts to 10% of the residential power consumption in Flanders [49]. Therefore, a PV installation is randomly assigned to 10% of the houses. The random assignment of the PV installation locations on the detailed feeders, as summarized in Table 2, is done to take into account the locational sensitivity of their grid impact.

The PV power profiles are based upon full-year measurements on an installation at the KU Leuven, with a 15 min time resolution. Therefore, seasonal variations in PV power generation are taken into account. The profiles are scaled to match the annual energy generation to the annual energy consumption at the selected household, taking into account the single-phase regulatory inverter power rating limit of 5 kVA in Flanders [50]. A power factor (PF) of 1 is assumed for the inverter [51].

2.3. EV Charging Load

A mobility profile generator is used to create a representative Flemish set ε_0 of EV charging profiles e (1 vehicle per household = 100% EV penetration rate) with a 1 min time resolution T_s. The model is explained in detail in [29], and consists of three sub-models: (1) a mobility behavior model; (2) a battery energy storage model; and (3) a charge behavior model.

The mobility profile of each EV is created with the mobility simulation tool that is discussed in [52]. This tool uses statistical data on Flemish transportation behavior, to create realistic driving patterns. For each vehicle, it is known when it is driving, standing still, and also where it is standing still: at home, at the workplace, or at other locations. In this model, the difference in mobility behavior during weekdays and weekend days is included. Seasonal variations are not taken into account, because for Flanders, no useful data is available on such variations.

The diversity of vehicle types and vehicle fuels within the fleet, and the resulting variations in specific power consumption and yearly driven distances are taken into account. The fleet consists out of three vehicle types: subcompact, midsize, and large vehicles. The vehicle types are modeled, based upon two vehicles within that category. The type-specific parameters are summarized in Table 3 for the modeled vehicles. Other parameters remain identical for all the vehicle type, e.g., the tire rolling resistance factor (0.01) and the auxiliary power consumption (500 W). A detailed explanation, including all assumptions, parameters, and drive cycles, can be found in [52]. The resulting fleet consists out 945 subcompact (40.4%), 1124 midsize (48%) and 271 large vehicles (11.6%).

Table 3. Summary of the parameters for the vehicle types [52].

	Vehicle model	Mass [kg]	Frontal Surface [m^2]	C_X
Subcompact	Mistubishi *i*-MiEV	1100	2.37	0.33
	Smart Electric Drive	975	2.40	0.35
Midsize	Nissan Leaf	1521	2.70	0.28
	Chevrolet Volt	1800	2.55	0.28
Large	Toyota RAV4 EV	2000	3.10	0.33
	eRUF Cayenne	2670	3.30	0.36

The resulting specific power consumption for each vehicle category is, respectively, 0.181, 0.213 and 0.302 kWh/km. The differences are significant, due to the differences in vehicle weight and air resistance. Given the average vehicle speed in Flanders of 42 km/h, the average battery discharging power P_k^d for each vehicle category is, respectively, 7.6, 8.9, and 12.7 kW, during driving.

The battery energy storage model, as discussed in detail in [53], is used to calculate the evolution of the battery energy content E_k and the state of charge SOC_k during each one-minute time step of the one-week simulation as specified in Section 2.5($k \in \{1, 2, \ldots, 10\,080\}$):

$$E_k = E_{k-1} - \delta_k^{sd} + \left(\eta^{ch} T_s P_k^{ch} - T_s P_k^d / \eta^d \right) \tag{4}$$

$$SOC_k = 100 \cdot E_k / E^{nom} \tag{5}$$

The self-discharge δ_k^{sd} is equal to 3% of the rated battery capacity E^{nom} per month, which is 20, 30, and 40 kWh for the three vehicle types, respectively. P_k^{ch} and P_k^d are the charging and discharging power during each time step k. Both Equations (4) and (5) are valid for all time steps of the simulation, as they are both applicable during driving and during standstill. Obviously, an EV cannot be simultaneously charging and discharging ($P_k^{ch} \cdot P_k^d = 0$), as an EV cannot be grid connected and driving at the same time. Therefore, during driving $P_k^{ch} = 0$, and during standstill $P_k^d = 0$. Furthermore, when an EV is not grid connected during standstill or when the battery is already fully charged, then also $P_k^{ch} = 0$. The charging and discharging efficiencies η^{ch} and η^d are 90.25%, which includes the electrochemical (95%) and the power electronic (95%) conversion efficiencies. The battery is never discharged below 20% SOC, to extend the battery cycle life [54].

67

The EVs are grid-connected at home and at the workplace, whenever they are standing still for more than 15 min. For shorter standstill times, it is assumed that the EV user will not make the effort to connect the EV to the charging infrastructure, because of the limited SOC increase during such a short time span. Mode 3 charging is assumed at these locations [3], with the commonly occurring single-phase charging power rating P^{ch} of 3.3 kW, which results in a current of 16 A at 90% of U_{nom} in continental Europe [4]. The end-of-charge power limit P^{EOC} (SOC_k), as illustrated in Figure 3, is based upon measurements performed on real vehicles, as discussed in [54]. When the SOC exceeds 92%, P^{EOC} (SOC_k) will limit the effective charging to below P^{ch}.

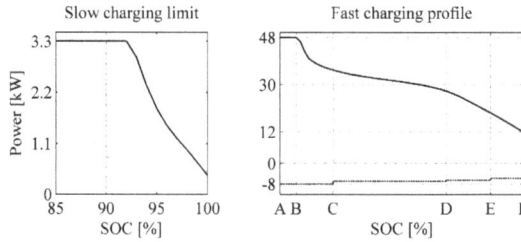

Figure 3. (**left**) charging power limit for slow charging; and (**right**) active (solid line) and reactive (dashed line) power profile for fast charging.

The fast charging demand is modeled as discussed in [40]: the EVs will interrupt their trip if the SOC goes below the threshold value $SOC(th)$, which is chosen uniformly between 20% and 30% for each trip: $SOC(th) \in \{20\% \ldots 30\%\}$. This is to take into account that an EV will not be exactly located at a fast charging station when the SOC reaches the 20% limit, as discussed above. When fast charging occurs, it is assumed that the battery will be fast charged up to 80% SOC. The fast charging power profile P^{FC} (SOC_k), as illustrated in Figure 3, is based upon measurements that are discussed in [55]. Point A and point F coincide with $SOC(th)$ and point 80%, respectively. As a result, depending on $SOC(th)$, points B to E occur at a different SOC values. The calculation of these points is summarized in Table 4. As can be seen, the fast chargers inject reactive power Q^{FC} into the grid, which are included in the simulations. The reactive power injection reduces in 4 discrete steps, as a function of the active fast charging power that is drawn. This behavior is implemented in the fast charger by the manufacturer as a way to reduce the voltage drop induced by the fast charger, *i.e.*, phase leading behavior.

2.4. Slow Charging Cases

Slow charging occurs in all cases *c*, which are summarized in Table 5 and are compared to the case without EVs, *i.e.*, case 0. Each case is simulated (a) without and (b) with fast charging. The EVs are modeled as extended range electric vehicles, in which the combustion engine is only engaged when the battery is depleted. Therefore, even if the battery is depleted in the cases without fast charging, all mobility requirements are met. Three slow charging strategies are modeled, which represent three typical charging objectives: high comfort, low charging cost, and low grid impact. Further assessments and more advanced charging strategies could be considered, but this is out of scope for this article.

Table 4. Calculation of the SOC values for the fast charging profile.

Point	SOC Value
A	$SOC(th)$
B	$SOC\,(th) + ((SOC(F) - SOC(th)) \cdot (1/15))$
C	$SOC\,(th) + ((SOC(F) - SOC(th)) \cdot (1/5))$
D	$SOC\,(th) + ((SOC(F) - SOC(th)) \cdot (2/3))$
E	$SOC\,(th) + ((SOC(F) - SOC(th)) \cdot (5/6))$
F	80%

Table 5. Summary of the slow charging cases.

Case c	Description
0	No EVs
1a	Uncoordinated charging without fast charging
1b	Uncoordinated charging with fast charging
2a	Off-peak charging without fast charging
2b	Off-peak charging with fast charging
3a	EV-based peak shaving without fast charging
3b	EV-based peak shaving with fast charging

For $c = 1$, the objective is to maximize the user's comfort. Each EV immediately starts charging when it is grid-connected at home and at the workplace. The charging power set point for case 1 $P_{k,1}^{set}$ is equal to the rated Mode 3 charging power P^{ch}, i.e., 3.3 kW:

$$P_{k,1}^{set} = P^{ch} \tag{6}$$

For $c = 2$, the objective is to decrease the residential charging cost. Therefore, the EVs only charge at home during the off-peak tariff period, when the end user electricity price is about 30% lower than during the on-peak tariff period in Flanders [56]. This tariff occurs when the time of day T_k is between 10 pm and 7 am:

$$P_{k,2}^{set} = \begin{cases} P^{ch}, & 10\ pm \leqslant T_k \leqslant 7\ am \\ 0, & 7\ am < T_k < 10\ pm \end{cases} \tag{7}$$

At the workplace, the EVs will still be able to charge between 7 am and 10 pm, because the charging cost at the workplace might be significantly different from the one at home. This rule-based strategy only makes use of the time of day (T_k), which is known by the onboard computer.

For $c = 3$, the objective is to reduce the local grid impact with a strategy that can be implemented without interaction between the end user and the grid. EV-based peak shaving, a rule-based strategy as explained in detail in [29], is used: the nominal set point $P_{k,3}^{set,nom}$ is the minimum power rating required to get the battery charged in the time span until the next departure T^{dep}, taking into account the upper limit of P^{ch} and the end-of-charge power limitation $P^{EOC} (SOC_k)$:

$$P_{k,3}^{set} = f\left(E^{nom}, E_k, T^{dep}, P^{ch}, P^{EOC} (SOC_k)\right) \tag{8}$$

As a result, the battery charged to the same level at the departure time as with case 1. This charging strategy makes use of one user input: the next departure time. The other parameters are already available in the onboard computer.

For all cases, the end of charge behavior limits the effective charging power $P_{k,c}^{ch}$ at SOC values above 92%:

$$P_{k,c}^{ch} = \begin{cases} P_{k,j}^{set}, & SOC_k \leqslant 92\% \\ P^{EOC} (SOC_k), & SOC_k > 92\% \end{cases} \tag{9}$$

Neither one of the three charging cases makes use of an optimization strategy. For $c = 1$, the charging behavior is directly determined by the mobility behavior. For $c = 2/3$, the charging behavior is controlled by a rule-based strategy that makes use of onboard parameters, and an additional user input for $c = 2$.

2.5. Simulation Approach

The simulations are conducted for a one-week scenario, resulting in a number of time steps $n_t = 10{,}080$, i.e., $k \in \{1, 2, \ldots, n_t\}$. The one-week scenario is conducted for a week that is representative for the first quarter of the year, because this is the period of the year when high grid loads occur in Flanders. Therefore, the grid constraints will be exceeded the soonest when additional load, due to EV charging, is added. The simulations are conducted for an EV penetration rate that increases in steps of 10% until 100%, or until the load flow offers no feasible solution. The LV feeder locations of the vehicles that are added for each step are summarized in Table 6. As can be seen, the location is diversified, to take into account the locational sensitivity of the results.

For each 1 min time step T_s, the evolution of the SOC of the EVs is calculated as formulated in Equations (4) and (5). The resulting residential EV charging profile is added to the residential load and generation profile at the respective LV grid nodes. The residential load and generation profiles are kept constant for their 15 min period. The aggregated fast charging load is added to the 5th node on the MV grid through a separate MV/LV transformer, as illustrated in Figure 2. The workplace charging profiles are not added, as it is assumed that there are no workplace charging locations in this residential grid.

Table 6. Location (house number) of the EVs on the LV feeders.

EV Group	MV Grid Node								
	1	2	3	4	5	6	7	8	9
10%	4, 7, 13, 16, 27, 37, 54	10, 17, 31, 33, 44, 58, 60	4, 25, 27, 37	4, 7, 8, 30	10,11, 13, 24, 50	6, 11, 14, 28, 38	22, 25, 28, 38	2, 11, 18, 33	3, 6, 8, 10, 20, 22, 23
20%	10, 15, 41, 44, 46, 57	14, 32, 41, 43, 46, 53	3, 7, 14, 22	6, 17, 18, 35	8, 25, 27, 42	1, 3, 19, 30	7, 17, 34, 37	5, 13, 19, 42	7, 27, 32, 40, 52, 55
30%	6, 11, 12, 34, 35, 51	16, 24, 27, 35, 45, 49	16, 19, 33, 39	3, 5, 28, 38	12, 15, 38, 39	7, 12, 32, 33	1, 14, 30, 32	24, 26, 30, 32	17, 33, 41, 43, 46, 62
40%	18, 21, 30, 40, 43, 61	13, 37, 42, 51, 56, 62	2, 11, 21, 23	15, 20, 25, 29	4, 6, 7, 41	5, 8, 9, 39	6, 13, 19, 29	7, 20, 36, 40	21, 24, 43, 49, 53, 60
50%	20, 36, 39, 52, 53, 56	3, 8, 28, 34, 52, 55	1, 15, 17, 31	1, 13, 19, 32	9, 16, 18, 34	2, 16, 18, 29	3, 8, 21, 26	15, 35, 39, 41	2, 9, 14, 26, 37, 56
60%	2, 19, 23, 24, 33, 38, 60	5, 9, 20, 26, 29, 50, 61	10, 24, 28, 35	21, 27, 31, 33	14, 22, 30, 33, 37	10, 15, 23, 27, 41	24, 27, 31, 39	4, 8, 9, 25, 29	13, 18, 35, 47, 48, 54, 61
70%	3, 8, 17, 28, 42, 48	18, 21, 36, 38, 47, 59	9, 26, 29, 34	11, 14, 16, 26	2, 19, 24, 35	20, 21, 35, 40	2, 12, 16, 33	3, 17, 23, 38	11, 15, 34, 42, 57, 59
80%	5, 14, 22, 25, 47, 55	2, 6, 7, 12, 48, 57	8, 24, 28, 37	2, 10, 36, 39	3, 28, 31, 32	4, 22, 26, 36	5, 18, 20, 35	6, 10, 27, 34	4, 12, 16, 29, 45, 58
90%	1, 31, 32, 45, 50, 59	1, 4, 19, 22, 23, 25	5, 6, 12, 20	12, 22, 24, 37	1, 16, 21, 29	13, 25, 37, 42	4, 11, 23, 36	12, 16, 28, 37	1, 19, 25, 28, 36, 39
100%	9, 26, 29, 49, 58, 62	11, 15, 30, 39, 40, 54	13, 18, 30	9, 23, 26	5, 20, 26, 36	17, 24, 31, 34	9, 10, 15	1, 21, 22, 31	5, 30, 31, 38, 50, 51

To obtain grid impact results as discussed in Section 3.3, Section 3.4, Section 3.5, a three-phase unbalanced load flow algorithm is implemented in MATLAB. Unbalanced loads are taken into account, as well as the resulting LV neutral conductor voltage drop. The backward-forward sweep technique is used, because of the radial layout of the grid [57]. The problem is converged when the worst time step voltage error eV is below 0.1 V. All loads are modeled as constant power loads with unity power factor, except for the fast chargers, which will inject a reactive power $P^{FC}(SOC_k)$ during fast charging, as illustrated in Figure 3.

For each charging case, the EV hosting capacity is calculated, which is the highest EV penetration rate that can be achieved without exceeding the feeder current constraints and the grid voltage constraints, as discussed in Section 2.1.

3. Results and Discussion

3.1. User Impact

To assess the dependency of the EVs on their range extender for the cases without fast charging, the utility function F^U is defined as the ratio of the electrically driven distances to the total driven distances [58]. Obviously, F^U is 100% for each EV when fast charging is possible, given that the fast charging stations have a sufficient geographical spreading. For all cases without fast charging, the maximum F^U is 100%, i.e., there is at least one vehicle in the fleet of 2340 EVs that will drive purely electrically during the simulated week. For $c = 1a/3a$, the minimal and mean F^U are 84.2% and 96.7%, respectively. The values are identical for both cases, because they provide the same SOC by the next departure time. For $c = 2a$, it has a marginally lower minimal and mean F^U, namely 82.6% and 96.2%, respectively. The difference compared to $c = 1a/3a$ is very small, because the charging time is usually well below the standstill time within the off-peak time span.

The total required electric driving energy for the EVs is 170,636 kWh, for the fleet of 2340 EVs during the one-week simulation period that is created here. This account for the energy being delivered by the EV batteries, by slow and fast charging for the b-cases, and by slow charging and the range extender for the a-cases. The share of charging energy for the different locations is summarized in Table 7. The majority of EV charging occurs at home, while workplace charging only accounts for 5.6% to 6.7% of the charging energy, due to the low average distance between home and the workplace. For $c = 2$, the slightly higher share in workplace charging energy is due to the limitation of the charging time at home. Therefore, sometimes the EV is not fully charged when leaving home to go to work. As a result, more charging energy is delivered at the workplace.

Table 7. Share of charging energy for the different locations.

Case	1/3a	1/3b	2a	2b
Home (slow)	94.0	83.3	93.3	81.1
Work (slow)	6.0	5.6	6.7	6.0
Fast	/	11.1	/	12.9

Fast charging delivers 11.1% up to 12.9% of the charging energy, depending on the slow charging case. The presence of fast charging mainly reduces the share of residential charging energy, as can be seen when comparing the shares of charging energy at home of the a-cases with the b-cases. The share of fast charging energy strongly depends on the assumptions being made. In theory, fast charging could be used to charge the batteries exactly up to the SOC that is required to reach the next charging destination, instead of charging up to 80% by default. Also, if the fast charging infrastructure would be perfectly spread out, each EV would reach a fast charge station when their battery SOC reaches the 20% threshold value. Under these theoretical assumptions, the share of fast charging energy would equal the share of energy being delivered by the range extenders for the cases without fast charging, i.e., 3.3% for $c = 1b/3b$, and 3.8% for $c = 2b$. Furthermore, larger battery capacities will also decrease the need for fast charging. For example, a doubling of the battery capacities decreases the share of fast charging energy to 2.3% for $c = 1b/3b$, and 3.2% for $c = 2b$.

3.2. Charging Behavior

As a detailed illustration of the difference in charging behavior between the different charging cases, the charging profile for a single EV is shown in Figure 4 for a 17 h timespan, for $c = 1b/2b/3b$. Also, the evolution of the SOC is shown. In this example, the fast charging profiles are identical for the three cases, but the residential charging profiles are different. The latter is similar in shape for $c = 1b$ and $c = 2b$, but shifted in time. For $c = 1b$, the charging process at home starts immediately when the EV arrives there (around 21.15 h), while for $c = 2b$, the charging starts at 22:00, when the off-peak tariff period starts. For $c = 3b$, the charging also starts immediately when arriving at home, but with charging power significantly reduced, because in this example, there is a long standstill time at home before the next departure (more than 15 h until the next departure). As a result, the SOC rises more slowly for $c = 3b$, compared to $c = 1b/2b$. This clearly illustrates how the slow charging profiles are significantly affected by the applied charging strategy.

To illustrate the aggregated grid impact of the EV charging behavior in detail, the profiles for a 40% EV penetration rate (936 EVs) are illustrated in Figure 5, for a 12 h timespan. As can be seen, the residential grid impact differs significantly for the three cases. The highest residential charging peak occurs for $c = 2b$, which results in the highest MV feeder load. This is due to the synchronization effect that occurs at the start of the off-peak tariff period, which does not occur for $c = 1b$, because the distribution of the arrival at home of EVs. The lowest peak power occurs for $c = 3b$, resulting in the lowest MV feeder load of the three cases. This is due to the combination of the spread on the EV arrival times at home, the SOC when arriving at home, and the time until the next departure time. Therefore, there is a spread on the charging time and charging power, which significantly reduces the peak load, as can be seen in Figure 5.

The fast charging profiles are identical for $c = 1b$ and $c = 3b$, as the SOC for both cases is identical at the time of departure for each trip. As a consequence, their SOC profiles will be identical during driving, so there is an identical need for fast charging for both cases. The fast charging profile is different for $c = 2b$, but it has similar peak values, as the need for fast charging is only slightly higher, which can be seen when comparing the fast charge energy share of the different cases in Table 7. The fast charge peak load is significantly lower than the residential slow charging peak, because it only accounts for a small share of the charging actions. Therefore, the grid impact of the slow charging strategy is more significant than the presence of fast charging.

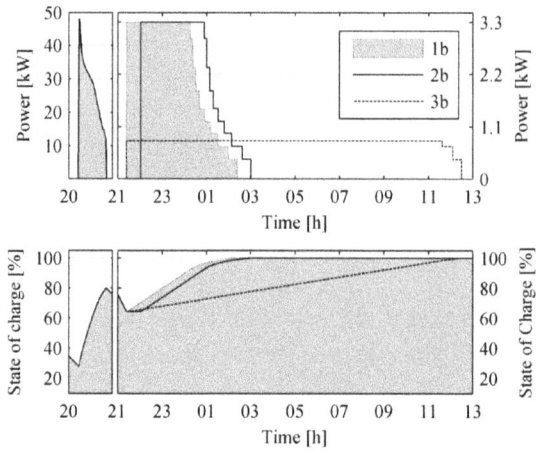

Figure 4. Fast (**left**) and slow (**right**) charging profiles (**top**), and the evolution of the state of charge (**bottom**).

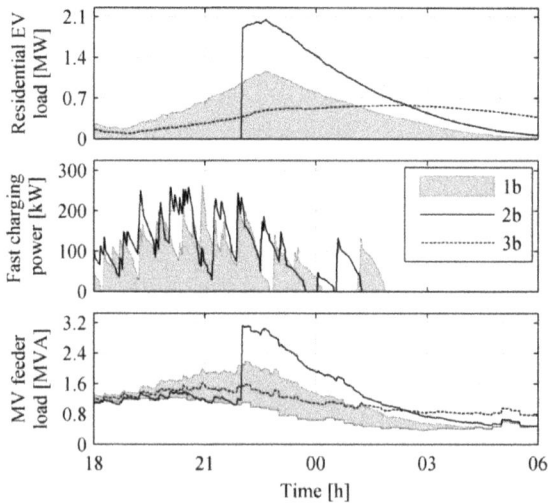

Figure 5. Load impact for a 40% EV penetration rate. The white surface in the bottom figure represents the non-EV residential load.

3.3. EV Hosting Capacity

To illustrate the impact of the EV load on the grid voltages in detail, the per-phase nodal voltages are illustrated in Figure 6 for a one-hour time period between 19 h and 20 h, for $c = 1b$, with a 40% EV penetration rate. The per-phase MV feeder load and the resulting per-phase voltages at the 9th node of the MV feeder are shown. Also, the per-phase LV grid load at the 9th MV node are illustrated, as well as the nodal voltages at the LV node the farthest away from the MV/LV transformer are illustrated, *i.e.*, node 62 of feeder topology LV3 that is connected to the 9th MV node. In this way, the per-phase loads and voltages at the MV and LV level are illustrated in detail for a one-hour timespan when uncoordinated EV charging occurs.

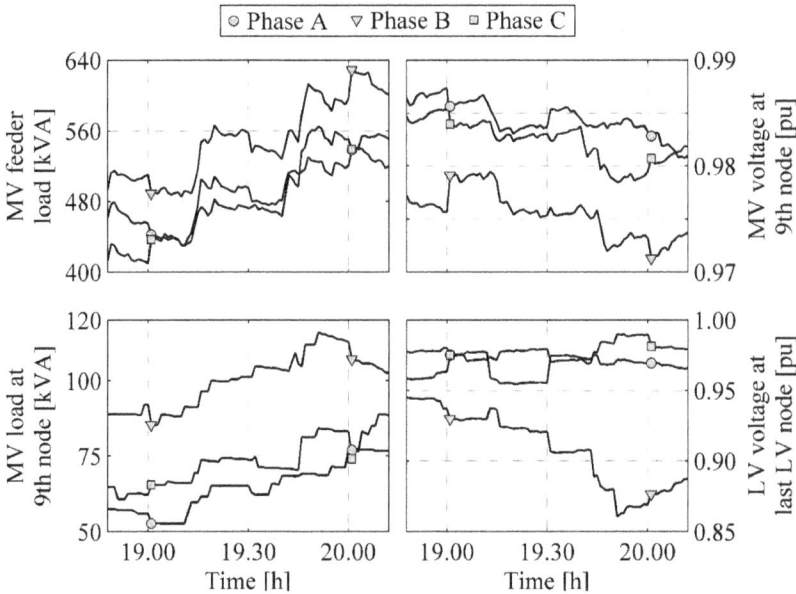

Figure 6. Per-phase load profiles (**left**) and voltage magnitudes (**right**), at the last MV node (**top**) and the last LV node (**bottom**), for case 1b with a 40% EV penetration rate.

As can be seen, there are notable differences in load of the phases. This is due to the fact that the households have a single-phase grid connection, which a common type of residential grid connection in Flanders. As a result, there is unbalance in the instantaneous grid load, due to the differences in household power consumption profiles. The modeled EV chargers are also single-phase, causing unbalance as well. It is seen that even on the aggregated MV level, this unbalance in the per-phase loads is not negligible. Also, the PV installation are not evenly spread over the three phases, resulting in an unbalanced PV power injection. The fast chargers each have a three-phase grid connection, therefore they act as a balanced grid load.

The unbalanced loads result in noticeable differences in the nodal phase voltages. The difference in the per-phase voltage magnitudes exceeds 0.01 pu at the 9th node of the MV feeder, and they exceed 0.05 pu at the 62th node of the LV feeder. This illustrates the added value of performing an unbalanced load flow (as will be discussed below), otherwise the voltage deviations are underestimated when assuming a balanced situation at the MV/LV interface. As a result, the EV hosting capacity would be overestimated.

For each charging case c, the EV hosting capacity is calculated as the maximum EV penetration rate R_{MAX}^{EV} that can be accepted, while not exceeding the grid voltage constraints and the feeder current constraints, as discussed in Section 2.1. The resulting EV hosting capacity for the charging cases are summarized in Table 8. For each case considered here, the voltage magnitude constraints, *i.e.*, Equations (1) and (2), are the most stringent ones. As can be seen, the cases without fast charging result in an equal or higher R_{MAX}^{EV} than the cases with fast charging. However, the differences between the cases with and without fast charging are limited to only 10 percentage points, which is equal to only 1 step of increase in the EV penetration rate. This is due to the limited share of fast charging in the charging behavior of the EVs, while residential charging accounts for the vast majority of charging actions.

Table 8. EV hosting capacity [%].

Case	1a	1b	2a	2b	3a	3b
R_{MAX}^{EV}	80	70	50	40	100	100

R_{MAX}^{EV} is more influenced by the slow charging strategies than whether or not fast charging occurs, as differences up to 60 percentage points occur between the different slow charging cases. These results show that the additional grid impact due to the presence of fast charging can easily be compensated for by the implementation of a more grid-friendly residential charging strategy. For instance, by incentivizing EV users to start charging immediately when arriving at home ($c = 1a/1b$) instead of waiting until the off-peak tariff period starts ($c = 2a/2b$), a significantly higher R_{MAX}^{EV}, even if fast charging is possible. This is due to the synchronization effect of EV charging at the start of the off-peak tariff period for $c = 2a/2b$, which creates a peak in the power demand. This power peak results in a voltage drop that negates the beneficial effect of shifting the charging load away from the time of the residential peak demand. An even higher R_{MAX}^{EV} is possible when EV-based peak shaving is applied ($c = 3a/3b$), as it allows for a 100% EV penetration rate for the scenario discussed here, whether or not fast charging occurs. Therefore, when the local EV penetration rate increases to significant levels, the EV users should be incentivized to charge their EVs in a more grid-friendly manner, as this can allow for a substantially higher EV hosting capacity with the same grid infrastructure.

3.4. Fast Charging Requirements

The fast charging power peak and the one percent peak (OPP) value are illustrated in Figure 7 for EV penetration rates up to R_{MAX}^{EV}. The OPP is the value above which only 1% of the values are situated. For $c = 1b$ and $c = 3b$, the values are identical for fast charging, but the values for penetration rates above 70% are only applicable for $c = 3b$, as this is the R_{MAX}^{EV} for $c = 1b$. For $c = 2b$, the values are only marginally different from those of the other two cases, as the fast charging requirements very similar. The OPP for fast charging power is significantly smaller than the peak value, which is related to the fact that the fast charging power profile for each EV, as illustrated in Figure 3, only draws its maximal power for a short period and it quickly declines as the SOC increases. Therefore, the MV/LV transformer that connects the fast charging station to the MV grid, might be sized for a power rating smaller than the expected peak value, as this value only occurs for a very short time.

The peak value and the OPP value for the occupation of fast chargers is also illustrated in Figure 7. As can be seen, 19 fast chargers are needed to cover the requirements at all times. However, for 99% of time, 16 fast chargers are sufficient. Therefore, it might be more efficient to allow for a limited waiting time during these peak situations, and thereby avoiding the investment in additional chargers that are only useful for 1% of time. Furthermore, this would reduce the fast charging power peak, which would lower the grid impact of the fast charging infrastructure. These results are in line with [40], where the number of fast chargers is reduced with 40%, which resulted in waiting times below 10 min for 99.7% of time.

For an increasing EV penetration rate, *i.e.*, from 10% to 100% for $c = 3b$, the ratio of fast chargers to EVs lowers significantly: from 3% to 0.9% for the peak value, and from 1.7% to 0.7% for the OPP. This means that there is a need for one fast charger for each 111–142 EVs, when there is a sufficiently high amount of EVs on the roads. This number can be compared to the ratio of fuel stations per car in Europe, which varies from 1/1500 (Luxembourg) to 1/3500 (Germany) stations per vehicle [59]. Thus, for a fast charging infrastructure with the same spatial spreading as the current fueling infrastructure and a 100% EV penetration rate, the fast charging infrastructure would consist out of stations with on average 11–25 fast chargers to cover the OPP, or with on average 14–32 fast chargers to cover the peak value.

An increase in EV battery capacity would decrease the ratio of fast chargers to EVs significantly, because the share of fast charging increases significantly when the EV battery increases, as discussed in Section 3.1. For the cases discussed here, a doubling of the EV battery capacities would reduce the peak value and the OPP value for the ratio to 0.4% and 0.3%, for a 100% EV penetration rate. The resulting fast charging infrastructure would consist out of stations with on average 5–11 fast chargers to cover the OPP, or on average 6–14 fast chargers to cover the peak value. These results show that the fast charging infrastructure needs to be extensive, even though fast charging only account for a minority of the charging actions. This is due to the fact that a fast charger fuels a vehicle more slowly than a fuel pump.

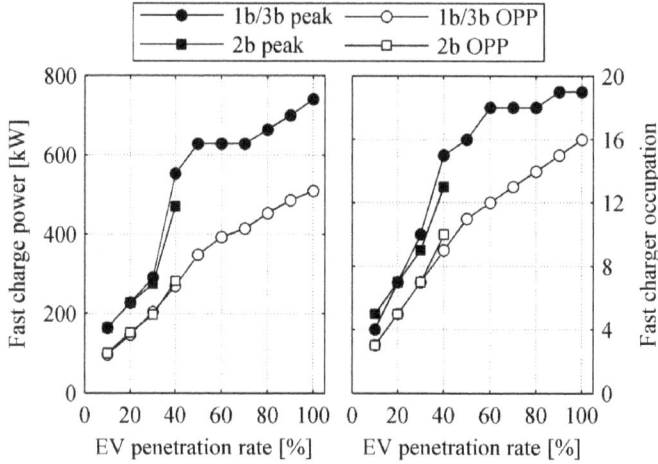

Figure 7. Peak values and one percent peak (OPP) for the fast charging load (**left**); and fast charger occupation (**right**).

3.5. Peak Load

The MV distribution feeder peak load, and the highest MV/LV residential grid transformer load are illustrated in Figure 8. As can be seen, the presence of fast charging reduces the LV residential peak load, but only slightly, as the presence of fast charging only accounts for a limited amount of the charging energy. Conversely, the presence of fast charging increases the peak load on the MV feeder, as this is where the fast charging infrastructure is connected to the grid. These opposing trends show how the presence of fast charging substitutes a fraction of the home charging actions. However, the difference in grid impact between the three slow charging cases is substantially larger than the difference between the situation with or without fast charging. Therefore, the increased grid impact of fast charging can be mitigated in an effective manner by implementing a grid-friendly slow charging strategy.

For $c = 1a/1b$ and $c = 2a/2b$, the LV residential peak loads are approximately the double at R_{MAX}^{EV}, compared to the case without EVs. The peak load is above 100% for $c = 1a$ and $c = 2a/2b$, at the higher EV penetration rates. This means that at least one of the 9 residential grid MV/LV transformers is loaded above its rated power. It must be assessed if the increase in transformer ageing is acceptable, or if a transformer upgrade is required. For $c = 3a/3b$, the peak load increases to a smaller degree, because the EVs on average charge at a significantly lower effective power rating. As a result, the transformer peak load stays well below the rated power for $c = 3a/3b$. The initial drop in the MV feeder load is due an initial reduction in load unbalance, when the low amount of EV charging load reduces the load unbalance on the grid.

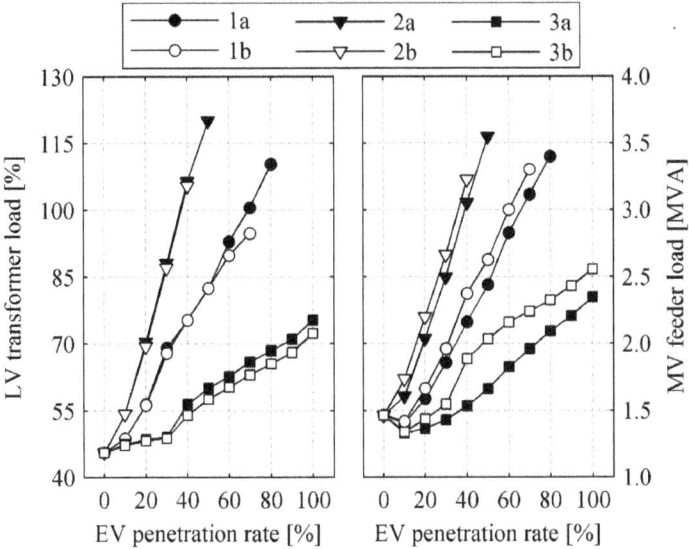

Figure 8. Highest MV/LV transformer peak load (**left**) and MV feeder peak load (**right**).

4. Conclusions

The distribution grid impact of EV charging is expected to be far more sensitive to the applied residential slow charging strategy, than to the presence or absence of fast charging. For a given residential slow charging strategy, the presence of fast charging decreases the EV hosting capacity with 10 percentage points or less. A more grid-friendly residential slow charging strategy increases the EV hosting capacity much more than the presence of fast charging decreases it. This is because EV charging at home accounts for the vast majority of EV charging, due to the long standstill times there. Fast charging only accounts for a low share of the EV charging energy, which even decreases further if EV battery capacities increase. As a result, fast charging stations have a limited impact on the medium voltage level, because the load increase is limited, compared to the load increase caused by residential slow charging. However, despite the limited share in the charging energy, fast charging infrastructure is indispensable to allow the vehicle fleet to drive close to purely electrically.

The choice of the slow charging strategy at home strongly influences the EV hosting capacity, which varies from 40% up to 100%. The highest hosting capacity occurs when EV-based peak shaving is implemented, both with and without fast charging, as for most of the time the effective charging power will be significantly lower than the rated power. This is due to the typical long standstill times at home and the typical low daily driven distances. Therefore, EV-based peak shaving significantly reduces the distribution grid impact, compared to the other two strategies. The lowest EV hosting capacity occurs for off-peak residential charging, due to the synchronization of the residential charging actions at the start of the off-peak period. This synchronization of EV load creates a peak that is higher than for the other cases. Therefore, distribution grid operators should consider adapting the peak/off-peak tariff scheme for households with an EV, when a significant local EV penetration rate occurs. By incentivizing more grid-friendly EV charging strategies, a higher EV penetration rate can be accepted without the need to invest in grid infrastructure.

Fast charging stations are fundamentally different from refueling stations for conventional vehicles. The latter are responsible for all of the refueling needs, while the former are only responsible for a fraction of the charging needs. However, because the charging actions take much more time than for conventional refueling, the infrastructure will still be substantial. Furthermore, the fast chargers have a three-phase power electronic interface that is permanently connected to the grid. Therefore, this widespread infrastructure of permanently grid-connected power electronic interfaces offers possibilities to supply grid services, even without the presence of stationary energy storage, e.g., active filtering, reactive power support, and grid balancing. Such services might be useful, as voltage deviations are observed at the MV grid level, due to the load situation of the grid, but also due to load unbalance.

Acknowledgments: Niels Leemput has a Ph.D. grant of the Institute for the Promotion of Innovation through Science and Technology in Flanders (IWT-Vlaanderen). Niels Leemput and Juan Van Roy are enrolled in the Ph.D. School of EIT-KIC InnoEnergy. KU Leuven is a partner of EnergyVille, Dennenstraat 7, 3600 Genk, Belgium.

Author Contributions: N.L., P.O.-R. and A.S. conceived and designed the experiments; N.L. performed the simulations; N.L., P.O.-R. and A.S. analyzed the data; N.L., F.G., J.V.R. and J.D. contributed materials and analysis tools; N.L. wrote the paper.

Conflicts of Interest: The authors declare no conflict of interest.

References

1. Bruninga, R.; Sorensen, J.A.T. Charging EVs efficiently now while waiting for the smart grid. In Proceedings of the IEEE Green Technologies Conferences, Denver, CO, USA, 4–5 April 2013; pp. 1–7.
2. Leemput, N.; van Roy, J.; Geth, F.; Driesen, J.; de Breucker, S. Grid and fleet impact mapping of EV charge opportunities. In *Data Science and Simulation in Transportation Research*; Janssens, D., Ed.; IGI Global: Hershey, PA, USA, 2013; pp. 364–390.
3. *Electric Vehicle Conductive Charging System—Part 1: General Requirements*; IEC Standard 61851-1; International Electrotechnical Commission: Geneva, Switzerland, 2010.
4. Schneider Electric. Connection System on the Recharging Spot: A Key Element for Electric Vehicles. Available online: http://www.evplugalliance.org/en/doc/ (accessed on 8 December 2014).
5. Fast Charging along the Highway. Available online: http://www.fastned.nl/en (accessed on 8 December 2014).
6. Supercharger. Available online: http://www.teslamotors.com/supercharger (accessed on 8 December 2014).
7. Clement-Nyns, K.; Haesen, E.; Driesen, J. The impact of charging plug-in hybrid electric vehicles on a residential distribution grid. *IEEE Trans. Power Syst.* **2010**, *25*, 371–380.
8. Leemput, N.; van Roy, J.; Geth, F.; Tant, P.; Claessens, B.; Driesen, J. Comparative analysis of coordination strategies for electric vehicles. In Proceedings of the 2011 IEEE PES ISGT Europe, Manchester, UK, 5–7 December 2011; pp. 1–8.
9. Electric Car Charging: Electric Vehicle Charging Points in Ireland. Available online: http://www.esb.ie/electric-cars/electric-car-charging.jsp (accessed on 8 December 2014).
10. Kempton, W.; Tomić, J. Vehicle-to-grid power implementation: From stabilizing the grid to supporting large-scale renewable energy. *J. Power Sources* **2005**, *144*, 280–294.
11. Saber, A.Y.; Venayagamoorthy, G.K. Plug-in vehicles and renewable energy sources for cost and emission reductions. *IEEE Trans. Ind. Electron.* **2011**, *4*, 1229–1238.
12. Gallus, M.D.; la Fauci, R.; Andersson, G. Investigating PHEV wind balancing capabilities using heuristics and model predictive control. In Proceedings of the 2010 IEEE PES General Meeting, Minneapolis, MN, USA, 25–29 July 2010; pp. 1–8.
13. Tran-Quoc, T.; Braun, M.; Marti, J.; Kieny, C.; Hadjsaid, N.; Bacha, S. Using control capabilities of DER to participate in distribution system operation. In Proceedings of the 2007 IEEE PowerTech, Lausanne, Switzerland, 1–5 July 2007; pp. 561–566.
14. Sekyung, H.; Soohee, H.; Sezaki, K. Development of an optimal vehicle-to-grid aggregator for frequency regulation. *IEEE Trans. Smart Grid* **2010**, *1*, 65–72.
15. Vandael, S.; de Craemer, K.; Boucké, N.; Holvoet, T.; Deconinck, G. Decentralized coordination of plug-in hybrid vehicles for imbalance reduction in a Smart Grid. In Proceedings of the 10th International Conferences on Autonomous Agents and Multiagent System, Taipei, Taiwan, 2–6 May 2011; pp. 803–810.
16. Acha, S.; Green, T.C.; Shah, N. Optimal charging strategies of electric vehicles in the UK power market. In Proceedings of the 2011 IEEE PES ISGT, Anaheim, PA, USA, 17–19 January 2011; pp. 1–8.

17. Rotering, N.; Ilic, M. Optimal charge control of plug-in hybrid electric vehicles in deregulated electricity markets. *IEEE Trans. Power Syst.* **2010**, *26*, 1021–1029.

18. Doostizadeh, M.; Khanabadi, M.; Esmaeilian, A.; Mohseninezhad, M. Optimal energy management of a retailer with smart metering and plug-in hybrid electric vehicle. In Proceedings of the 10th International Conferences on Environment and Electrical Engineering, Rome, Italy, 8–11 May 2011; pp. 1–5.

19. Lyon, T.P.; Michelin, M.; Jongejan, A.; Leahy, T. Is "smart charging" policy for electric vehicles worthwhile? *Energy Policy* **2012**, *41*, 259–268.

20. Leemput, N.; Geth, F.; Claessens, B.; van Roy, J.; Ponnette, R.; Driesen, J. A case study of coordinated electric vehicle charging for peak shaving on a low voltage grid. In Proceedings of the 2012 IEEE PES ISGT Europe, Berlin, Germany, 14–17 October 2012; pp. 1–7.

21. Sortomme, E.; Hindi, M.M.; MacPherson, S.D.J.; Venkata, S.S. Coordinated charging of plug-in hybrid electric vehicles to minimize distribution system losses. *IEEE Trans. Smart Grid* **2011**, *2*, 198–205.

22. Shao, S.; Zhang, T.; Pipattanasomporn, M.; Rahman, S. Impact of TOU rates on distribution load shapes in a smart grid with PHEV penetration. In Proceedings of the 2010 IEEE PES Transmission and Distribution Conference and Exposition, New Orleans, LA, USA, 19–22 April 2010; pp. 1–6.

23. Vandael, S.; Claessens, B.; Hommelberg, M.; Holvoet, T.; Deconinck, G. A scalable three-step approach for demand side management of plug-in hybrid vehicles. *IEEE Trans. Smart Grid* **2013**, *4*, 720–728.

24. De Craemer, K.; Vandael, S.; Claessens, B.; Deconinck, G. An event-driven dual coordination mechanism for demand side management of PHEVs. *IEEE Trans. Smart Grid* **2014**, *5*, 751–760.

25. Richardson, P.; Flynn, D.; Keane, A. Local *versus* centralized charging strategies for electric vehicles in low voltage distribution systems. *IEEE Trans. Smart Grid* **2012**, *3*, 1020–1028.

26. Hajimiragha, A.; Cañizares, C.A.; Fowler, M.W.; Elkamel, A. Optimal transition to plug-in hybrid electric vehicles in Ontario, Canada, considering the electricity-grid limitations. *IEEE Trans. Ind. Electron.* **2010**, *57*, 690–701.

27. Schneider, K.P.; Gerkensmeyer, C.E.; Kintner-Meyer, M.C.W.; Fletcher, R. Impact assessment of plug-in hybrid vehicles on pacific northwest distribution systems. In Proceedings of the 2008 IEEE PES General Meeting, Pittsburg, PA, USA, 20–24 July 2008; pp. 1–6.

28. Leemput, N.; Geth, F.; van Roy, J.; Delnooz, A.; Büscher, J.; Driesen, J. Impact of electric vehicle on-board single-phase charging strategies on a Flemish residential grid. *IEEE Trans. Smart Grid* **2014**, *5*, 1815–1822.

29. Schroeder, A.; Traber, T. The economics of fast charging infrastructure for electric vehicles. *Energy Policy* **2012**, *43*, 136–144.

30. Li, Z.; Ouyang, M. The pricing of charging for electric vehicles in China-Dilemma and solution. *Energy* **2011**, *36*, 5765–5778.

31. Bae, S.; Kwasinski, A. Spatial and temporal model of electric vehicle charging demand. *IEEE Trans. Smart Grid* **2012**, *3*, 394–403.

32. Yang, S.N.; Cheng, W.S.; Hsu, Y.C.; Gan, C.H.; Lin, Y.B. Charge scheduling of electric vehicles in highways. *Math. Comput. Model.* **2013**, *57*, 2873–2882.

33. Coninckx, C.; Claes, R.; Vandael, S.; Leemput, N.; Holvoet, T.; Deconinck, G. Anticipatory coordination of electric vehicle allocation to fast charging infrastructure. In Proceedings of the 12th International Conferences on Practical Applications of Agents and Multi-Agent Systems, Salamanca, Spain, 4–6 June 2014; pp. 1–12.

34. Etezadi-Amoli, M.; Choma, K.; Stefani, J. Rapid-charge electric-vehicle stations. *IEEE Trans. Power Deliv.* **2010**, *25*, 1883–1887.

35. Yunus, K.; Zelaya-De La Parra, H.; Reza, M. Distribution grid impact of plug-in electric vehicles charging at fast charging stations using stochastic charging model. In Proceedings of the 2011 European Conferences Power Electronics and Application, Birmingham, UK, 30 August–1 September 2011; pp. 1937–1947.

36. Mauri, G.; Valsecchi, A. Fast charging stations for electric vehicle: The impact on the MV distribution grids of the Milan metropolitan area. In Proceedings of the IEEE Internationa Energy Conferences Exhibition, Florence, Italy, 9–12 September 2012; pp. 1055–1059.

37. Arancibia, A.; Strunz, K. Modeling of an electric vehicle charging station for fast DC charging. In Proceedings of the 2012 IEEE International Electric Vehicle Conferences, Greenville, CA, USA, 4–8 March 2012; pp. 1–6.

38. Aggeler, D.; Canales, F.; Zelaya-De La Parra, H.; Coccia, A.; Butcher, N.; Appeldoorn, O. Ultra-fast DC-charge infrastructures for EV-mobility and future smart grids. In Proceedings of the 2010 IEEE PES ISGT Europe, Gothenburg, Sweden, 11–13 October 2010; pp. 1–8.

39. Machiels, N.; Leemput, N.; Geth, F.; van Roy, J.; Buscher, J.; Driesen, J. Design criteria for electric vehicle fast charge infrastructure based on Flemish mobility behavior. *IEEE Trans. Smart Grid* **2014**, *5*, 320–327.

40. Lacroix, B.; Calvas, R. *Earthing Systems Worldwide and Evolutions*; Technical paper ECT173; Schneider Electric: Rueil-Malmaison, France, 1995.

41. *Cables for Underground Construction, Full Synthetic Insulation and Fortified Cloak (Type 1 kV)*; NBN Standard C 33-322; Bureau for Standardization: Brussels, Belgium, 1975.

42. Belmans, R.; Deconinck, G.; Driesen, J. Deel IV: Transformatoren. In *Elektrische Energie*; Deel 1; ACCO: Leuven, Belgium, 2010; pp. 269–270.

43. Gonzalez, C.; Weckx, S.; Efkarpidis, N.; Vingerhoets, P.; de Rybel, T.; Driesen, J. Constrained PV penetration level in LV distribution networks based on the voltage operational margin. In Proceedings of CIRED 22, Stockholm, Sweden, 10–13 June 2013; pp. 1–4.

44. Lehtonen, M. Fault management in electrical distribution systems. In Proceedings of the CIRED 1999, Nice, France, 1–4 June 1999.

45. Gonzalez-de Miguel, C.; de Rybel, T.; Driesen, J. Enhancing reliability in medium voltage distribution networks with directional fault passage indicators without voltage sensors. *Acta Electrotech.* **2013**, *54*, 73–80.

46. Nexans, S.A. 6–36 kV Medium Voltage Underground Power Cables: XLPE Insulated Cables. Available online: http://www.nexans.be/Belgium/2013/Medium%20Voltage% 20Underground%20Power%20Cables%20Catalogue.pdf (accessed on 8 December 2014).

47. *Voltage Characteristics of Electricity Supplied by Public Electricity Networks*; CENELEC Standard EN 50160; European Committee for Electrotechnical Standardization (CENELEC): Brussels, Belgium, 2010.

48. Labeeuw, W.; Deconinck, G. Residential electrical load model based on mixture model clustering and Markov models. *IEEE Trans. Ind. Inform.* **2013**, *9*, 1561–1569.

49. De Cijfers. Available online: http://www.ode.be/zonnestroom/de-cijfers (accessed on 8 December 2014).

50. *Specific Technical Instructions for Decentralized Production Installations Working in Parallel with the Distribution Grid*; Synergrid Standard C10/11 (Rev. 12); Synergrid: Brussels, Belgium, 2009.

51. SMA Solar Technology AG. Sunny Boy 240. Available online: http://files.sma.de/dl/ 18925/SB240-DEN1442web.pdf (accessed on 3 February 2015).

52. Van Roy, J.; Leemput, N.; de Breucker, S.; Geth, F.; Tant, P.; Driesen, J. An availability analysis and energy consumption model for a Flemish fleet of electric vehicles. In Proceedings of the European Electric Vehicle Congress, Brussels, Belgium, 26 May 2011; pp. 1–12.

53. Tant, J.; Geth, F.; Six, D.; Driesen, J. Multiobjective battery storage to improve PV integration in residential distribution grids. *IEEE Trans. Sustain. Energy* **2013**, *4*, 182–191.

54. Dogger, J.; Roossien, B.; Nieuwenhout, F. Characterization of Li-ion batteries for intelligent management of distributed grid-connected storage. *IEEE Trans. Energy Convers.* **2011**, *26*, 256–263.

55. Seljeseth, H.; Taxt, H.; Solvang, T. Measurements of network impact from electric vehicles during slow and fast charging. In Proceedings of the CIRED 2013, Stockholm, Sweden, 10–13 June 2013; pp. 1–4.

56. VREG. Available online: http://www.vreg.be/ (accessed on 8 December 2014).

57. Cheng, C.; Shirmohammadi, D. A three-phase power flow method for real-time distribution system analysis. *IEEE Trans. Power Syst.* **1995**, *10*, 671–679.

58. Duvall, M.; Knipping, E.; Alexendar, M.; Tonachel, L.; Clarc, C. *Environmental Assessment of Plug-in Hybrid Electric Vehicles, Volume 1: Nationwide Greenhouse Gas Emissions*; Technical Report 1015325; Electric Power Research Institute: Palo Alto, CA, USA, 2007.

59. Evolutie van het Aantal Tankstations. Available online: http://www.petrolfed.be/ nl/petroleumindustrie/economie/evolutie-van-het-aantal-tankstations (accessed on 8 December 2014).

Research on a Small Signal Stability Region Boundary Model of the Interconnected Power System with Large-Scale Wind Power

Wenying Liu, Rundong Ge, Quancheng Lv, Huiyong Li and Jiangbei Ge

Abstract: For the interconnected power system with large-scale wind power, the problem of the small signal stability has become the bottleneck of restricting the sending-out of wind power as well as the security and stability of the whole power system. Around this issue, this paper establishes a small signal stability region boundary model of the interconnected power system with large-scale wind power based on catastrophe theory, providing a new method for analyzing the small signal stability. Firstly, we analyzed the typical characteristics and the mathematic model of the interconnected power system with wind power and pointed out that conventional methods can't directly identify the topological properties of small signal stability region boundaries. For this problem, adopting catastrophe theory, we established a small signal stability region boundary model of the interconnected power system with large-scale wind power in two-dimensional power injection space and extended it to multiple dimensions to obtain the boundary model in multidimensional power injection space. Thirdly, we analyzed qualitatively the topological property's changes of the small signal stability region boundary caused by large-scale wind power integration. Finally, we built simulation models by DIgSILENT/PowerFactory software and the final simulation results verified the correctness and effectiveness of the proposed model.

Reprinted from *Energies*. Cite as: Liu, W.; Ge, R.; Lv, Q.; Li, H.; Ge, J. Research on a Small Signal Stability Region Boundary Model of the Interconnected Power System with Large-Scale Wind Power. *Energies* **2015**, *8*, 2312–2336.

1. Introduction

With the facts of fossil energy exhaustion and global ecology environment deterioration, wind power generation [1–3], as the most mature and effective renewable energy technology, is developing rapidly all over the world. In particular, China plans to construct nine ten-million kilowatt wind power bases [4,5] and parts of these bases have been put into operation. After these large-scale wind farms were put into operation, the large-scale wind power and traditional power system were connected to each other, sending out wind-thermal-bundled [6,7] power, which has become the typical operation mode of the power system. However, the current structure of the wind power transmission grid is still not strong enough to satisfy the scale of wind power integration. Thus, small signal stability has become the bottleneck restricting the sending-out of large-scale wind power as well as the transmission capacity of the power grid. To maintain the stability margin of the power grid, wind farms may even need to cease power production frequently. Thus, it is necessary to research the small signal stability of power grids with large-scale wind power integration.

Some [8–10] have studied the small signal stability of power systems with wind power integration so far. Generally speaking, however, most of this research is limited to a certain grid and typical working conditions, which is not a universal situation and may even yield conflicting results. The study of the small signal stability region boundary has broader perspectives and shows all the critical operating points of the small signal stability, which provides a more scientific and reasonable basis for the safety monitoring, defense and control of power systems.

In theory, the small signal stability region boundary [11,12] consists of saddle node bifurcation (SNB), Hopf bifurcation (HB) and singularity induced bifurcation (SIB). Among them, HB [13,14], which physically represents a pair of conjugate eigenvalues crossing the imaginary axis, is related to the system oscillatory instability. This paper mainly focuses on stability region boundary consisting of Hopf bifurcations.

At present, some papers also report research on the small signal stability region boundary. Seydel [15], Roose *et al.* [16] and Hiskens [17] studied the methods for obtaining the small signal stability region boundary based on a direct method and a continuation method, respectively. Sun and Yu [18] proposed fitting a boundary consisting of Hopf bifurcations using hyper-planes. Based on an implicit function, Yang *et al.* [19] presented a method to obtain the small signal stability region boundary through polynomial approximation. Jia *et al.* [20,21], and Li *et al.* [22] studied the influences on the small signal stability region boundary of exciter voltage limits, time delays and saturated links, respectively. Generally, these works were mainly focused on quickly finding the small signal stability region boundary of the traditional power system and the influences of various factors on the small signal stability region boundary, without paying attention to the boundary of the interconnected power system with large-scale wind power. Besides, most of the research methods in these papers are just massive calculations and iterations based on the electromagnetic model of the power system, unable to directly provide enough information to reveal the topological properties of the small signal stability region boundary on the whole. In fact, for the interconnected power system with large-scale wind power, the small signal stability region boundary is still determined by the eigenvalue. From the perspective of the eigenvalues' changes, under certain conditions, a pair of complex eigenvalues may turn into real eigenvalues or from the opposite direction [23,24]. This phenomenon that the eigenvalues' change under critical conditions essentially shows obvious catastrophe indications and it's a kind of typical catastrophe phenomenon. Catastrophe theory [25–27] is often used in systems with obvious catastrophe indications. A significant advantage of catastrophe theory is that it can establish a functional relationship between control variables and state variables without contacting any special inner mechanism to avoid solving differential equations of the whole power system. At present, catastrophe theory has been widely applied in road traffic [28], biology [29], geotechnical engineering [30], and sociology [31], but seldom in power systems. Sallam and Dineley [32], and Wvong and Mihiring [33,34] studied the dynamic stability and transient stability of power systems based on catastrophe theory. Mahmoud [35] used catastrophe theory to analyze the voltage stability of a distribution network and proposed some new voltage indexes. Yusheng *et al.* [36] set up a cusp catastrophe model for over-heating faults of oil-immersed transformers and further proposed a remaining life prediction method for this kind of transformer. However, catastrophe theory has no application in small signal stability or other related wind power research as of yet. Hence, this paper establishes the small signal stability region boundary model of an interconnected power system with large-scale wind power based on catastrophe theory and the catastrophe indications of the eigenvalues.

The paper is organized as follows: the first section, as a preface, expounded the current situation of the research and problems studied in this paper. The second section analyzed typical characteristics and mathematical models of the interconnected power system with large-scale wind power. The third section discussed the disadvantages of traditional analysis methods of small signal stability region boundary. Section 4 introduced the basic principles of catastrophe theory. Section 5 analyzed the catastrophe indications of eigenvalues to apply catastrophe theory to the analysis of eigenvalues. Section 6 established the small signal stability region boundary model of interconnected power system with large-scale wind power in two-dimensional power injection space and extend the 2D model to multi dimensions. In Section 7, simulations were conducted to verify the model. In the end, we summarized our work of this paper.

2. An Interconnected Power System with Large-Scale Wind Power and Its Model

2.1. Interconnected Power System with Large-Scale Wind Power and Its Typical Characteristics

Due to that wind power resources in China are centralized and far from the load centers, wind power development always uses the manner of centralized development and long-distance transportation. For now, China is planning to construct nine ten-million kilowatts wind power bases in Gansu, Xinjiang, Hebei *etc.*, which are connected to the traditional power system and together form the interconnected power system with large-scale wind power. These systems show some visible characteristics as follows: the capacity of wind power is much larger than that of the local loads which makes it hard for it to be consumed on site; wind farms cover large areas and the power of each wind farm is transmitted into a higher voltage power grid after collection at power collection points through transmission lines; wind power needs long distance lines to be sent out; wind power bases are always located together at the end of the power grid and the outward power delivery structure is relatively weak, which results in serious stability problems, *etc.* Figure 1 shows the Jiuquan wind power base and its outward power transmission channels.

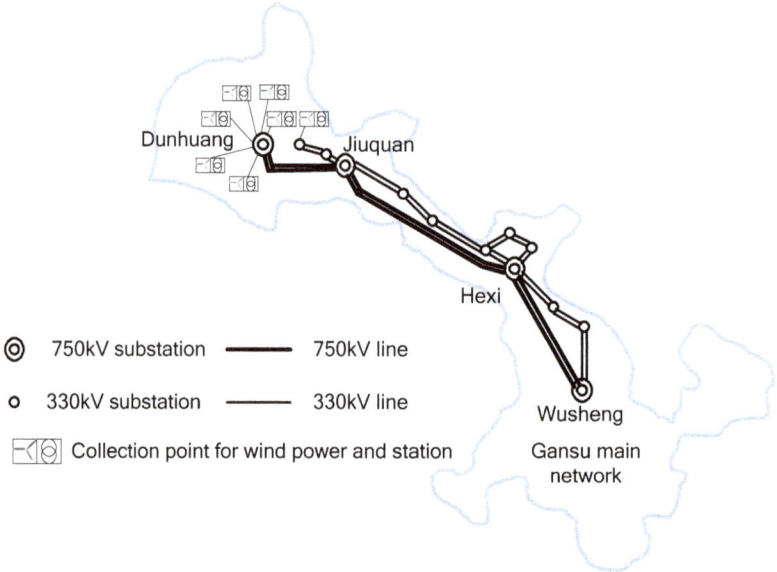

Figure 1. Outward power transmission channels of Jiuquan wind power base.

2.2. Equivalence of Large-Scale Wind Power Base

In the interconnected power system with large-scale wind power, a wind power base is composed of a large number of wind farms, covering a large area. Thus, wind power from the same area is always collected at certain HV bus power collection points and then sent out at a higher voltage class through a transformer. For example, the first-stage project of Jiuquan wind power base, mainly located in Yumen and Guazhou, forms nine power collection points, including seven 330 kV booster stations and two 330 kV substations. The wind power is then sent out through a 750 kV line.

The wind farms of the same power collection point include large numbers of wind turbines, so it is impossible to establish different models for each wind turbine. On the other hand, the power output of these wind turbines show similar change trends [7] and the wind turbines have similar states when large amounts of wind is available. Therefore, it's feasible to establish an equivalent model to simulate the operation of a wind farm group. That is to say, regard the power collection point as a unit and replace the wind farms of the same power collection point by an equivalent wind turbine, which specifically comes down to two steps: firstly, replace m wind turbines of the same wind farm by one equivalent wind turbine adopting the capacity weighted method [37]. The calculation of the parameters of the equivalent wind turbine is shown in Equation (1):

$$
\begin{cases}
S_{eq} = \sum_{i=1}^{m} S_i, \quad P_{eq} = \sum_{i=1}^{m} P_i, \quad Q_{eq} = \sum_{i=1}^{m} Q_i \\[2mm]
x_{m-eq} = \sum_{i=1}^{m} \frac{S_i}{S_{eq}} x_m, \quad x_{s-eq} = \sum_{i=1}^{m} \frac{S_i}{S_{eq}} x_s, \quad x_{r-eq} = \sum_{i=1}^{m} \frac{S_i}{S_{eq}} x_r \\[2mm]
r_{s-eq} = \sum_{i=1}^{m} \frac{S_i}{S_{eq}} r_s, \quad r_{r-eq} = \sum_{i=1}^{m} \frac{S_i}{S_{eq}} r_r \\[2mm]
H_{eq} = \sum_{i=1}^{m} \frac{S_i}{S_{eq}} H_i, \quad K_{eq} = \sum_{i=1}^{m} \frac{S_i}{S_{eq}} K_i, \quad D_{eq} = \sum_{i=1}^{m} \frac{S_i}{S_{eq}} D_i
\end{cases}
\tag{1}
$$

In Equation (1), subscript m is the total number of wind turbines in a wind farm. Subscript i is the number of wind turbines. Subscript eq means equivalent parameters. x_m is the magnetizing reactance, x_s and r_s represent the reactance and resistance of the stator circuit respectively, x_r and r_r represent the reactance and resistance of the rotor circuit respectively, the units of which are Ω. S is the capacity of the wind turbine in MVA. P is the active output of the wind turbine in MW. Q is the reactive output in Mvar. H is the inertia time constant in s. K represents the shafting stiffness coefficient in N·m/rad. D is the shafting damping coefficient in s.

After the wind farm is made equivalent to one wind turbine, each power collection point will connect several equivalent wind farms. A typical sending-out model of wind power at the same power collection point is shown in Figure 2.

In Figure 2, $P_i + jQ_i$ is the output power of wind farm i. The PCC bus of each wind farm is connected to the collection bus through the transmission line whose impedance is $R_i + jX_i$, and the collection point is connected to the high voltage access point through the transmission line whose impedance is $R + jX$ and the transformer.

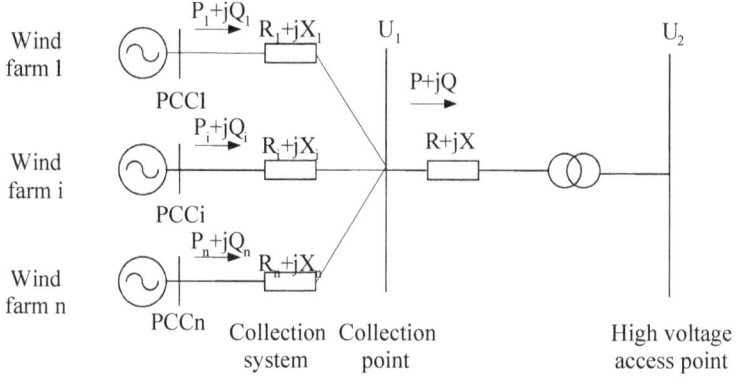

Figure 2. Typical sending-out model of wind power.

For the typical system with wind power sending-out in Figure 2, with the same method, several equivalent wind turbines of the same power collection point were replaced by another equivalent wind turbine. For the collection system, an equivalent power loss method was adopted, assuming that different wind farms have the same voltage U. The voltage of the equivalent wind farm is also U and $R_\Sigma + jX_\Sigma$ represents the impedance of the equivalent collection system, thus:

$$\sum_{i=1}^{n} \frac{(P_i^2 + Q_i^2)}{U^2}(R_i + jX_i) = \frac{\sum_{i=1}^{n}(P_i^2 + Q_i^2)}{U^2}(R_\Sigma + jX_\Sigma) \tag{2}$$

Then we obtain:

$$\begin{cases} R_\Sigma = \dfrac{\sum\limits_{i=1}^{n}(P_i^2 + Q_i^2)R_i}{\sum\limits_{i=1}^{n}(P_i^2 + Q_i^2)} \\[4mm] X_\Sigma = \dfrac{\sum\limits_{i=1}^{n}(P_i^2 + Q_i^2)X_i}{\sum\limits_{i=1}^{n}(P_i^2 + Q_i^2)} \end{cases} \tag{3}$$

So far, we have replaced the wind farm group of the same power collection point with an equivalent wind turbine. Because a wind power base always consists of limited power collection points, we replace the base with a group of wind turbines. Now, the typical structure of interconnected power system with large-scale wind power is as shown in Figure 3.

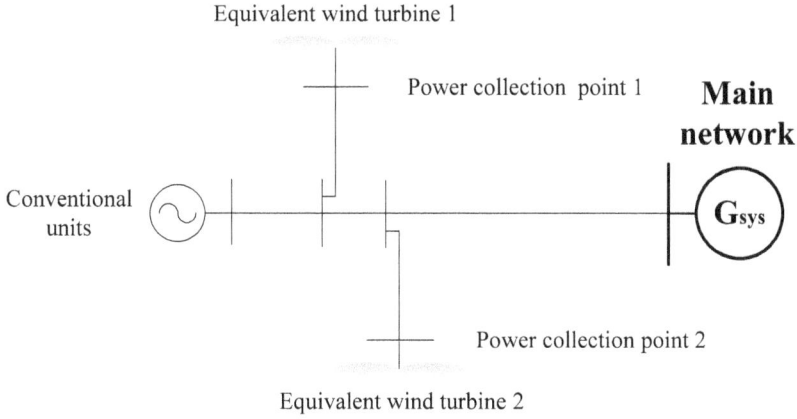

Equivalent wind turbine 1

Power collection point 1

Main network

Conventional units

Gsys

Power collection point 2

Equivalent wind turbine 2

Figure 3. Typical structure of interconnected power system with large-scale wind power.

2.3. Model of Interconnected Power System with Large-Scale Wind Power

For the interconnected power system with large-scale wind power, the dynamic elements including the wind turbine can be expressed as a set of differential algebraic equations [38,39]:

$$\begin{cases} \dot{x}_i = f_i(x_i, u_i) \\ i_i = g_i(x_i, u_i) \end{cases} \tag{4}$$

In Equation (4), x_i is state variable of each dynamic element; u_i is the terminal voltages of each dynamic element; i_i is the current injected into the network.

Collecting all of these equations of dynamic elements we obtain a set of differential algebraic equations of an interconnected power system with large-scale wind power:

$$\begin{cases} \dot{x} = f(x, u) \\ i = g(x, u) \end{cases} \tag{5}$$

where x is the state variable vector of the power system; u is the vector of the bus terminal voltage; i is the current vector injected into the bus; f and g are nonlinear differential equations composing of f_i, g_i.

In Equation (5), the current i and voltage u follow the basic principle of the circuit:

$$i = Y_n u \tag{6}$$

where Y_n is the network admittance matrix. Equations (5) and (6) together form the model of an interconnected power system with large-scale wind power.

91

3. Traditional Analysis Method of Small Signal Stability Region Boundary and Its Disadvantage

In the traditional analysis method, we establish a state matrix to acquire the small signal stability region boundary based on the model of the power system. Concrete steps are as follows:

Equation (5) can be linearized as Equation (7) at the equilibrium point:

$$\begin{cases} \Delta\dot{x} = A_D\Delta x + B_D\Delta u \\ \Delta i = C_D\Delta x + Y_D\Delta u \end{cases} \tag{7}$$

where $A_D = \partial f/\partial x$, $B_D = \partial f/\partial u$, $C_D = \partial g/\partial x$, $Y_D = \partial g/\partial u$; Δx is the micro-increment of x; Δu is the micro-increment of u and Δi is the micro-increment of i.

From Equation (6) we obtain:

$$\Delta i = Y_n\Delta u \tag{8}$$

Substituting Equation (8) into Equation (7) we can reach Equation (9) after rearranging:

$$\Delta\dot{x} = [A_D + B_D(Y_N - Y_D)^{-1}C_D]\Delta x = A\Delta x \tag{9}$$

The eigenvalues of the system state matrix A will decide whether the system reaches the small signal stability region boundary under some critical conditions: (1) SNB occurs when a pair of real eigenvalues cross the imaginary axis; (2) HB occurs when a pair of complex conjugate eigenvalues cross the imaginary axis; (3) SIB occurs when $(Y_N - Y_D)$ is singular. These three kinds of bifurcation form the small signal stability region boundary $\partial\Omega_{sssr}$:

$$\partial\Omega_{sssr} = \{SNBs\} \cup \{HBs\} \cup \{SIBs\} \tag{10}$$

where SNBs is the point set of SNB, HBs is the point set of HB, SIBs is the point set of SIB.

The key to analyzing the small signal stability region boundary is figuring out when the parameters of the power system reach a critical condition. Conventional methods acquire the boundary by continuous adjustment of the system parameters and calculating the eigenvalues of the system state matrix. Because of the huge order and complex coupling properties of the power system model, conventional methods are trapped in complicated calculations and lack enough information to directly reveal the topological properties of the small signal stability region boundary on the whole.

In contrast, based on catastrophe theory, we can directly establish a function between control variables and state variables without contacting any special inner mechanism, thus avoiding solving the differential equations of the whole system and making it very applicable to the analysis of systems with an unknown or difficult to obtain inner mechanism. Nevertheless, catastrophe theory itself has inherent drawbacks. On one hand, in a practical complex system, it's difficult to establish its potential function through derivation. On the other hand, catastrophe theory is mainly applied to systems with no more than five control variables. However, this does not affect the application of catastrophe theory in this paper. Thus we adopt catastrophe theory to discuss the model of the small signal stability region boundary of interconnected power systems with large-scale wind power.

4. Catastrophe Theory

Catastrophe theory was first proposed in 1972 by Thom, a French mathematician. This theory held that system dynamics can be derived through a smooth potential and critical points can be classified according to potential function [27].

The form of the system potential function is as follows:

$$f: V(X \times C) \tag{11}$$

where X is the state variable vector $(x_1, x_2, \ldots, x_n)$; C is the control variable vector $(c_1, c_2, \ldots, c_r)$ which represents r independent control variables. The equilibrium surface M, which is a subset of $R^n \times R^r$ space, represents the system's stable operating points, defined as Equation (12):

$$V\prime(x) = 0 \tag{12}$$

where superscript "$\prime$" represents the derivation of x for V. The definition of singularity set S is a subset of M consisting of all degenerate critical points of V, namely:

$$V''(x) = 0 \tag{13}$$

Mapping S to the control space C, then we obtain the bifurcation set B which is the control parameter set when catastrophe happens to the system's operating state. The bifurcation set B can be obtained by combing Equations (12) and (13) to eliminate the state variable vector X. Appendix A shows a typical catastrophe mechanism explaining the principle of catastrophe theory.

Thom's proof [25,36] indicated that the property of the potential function was not decided by state variables but the number of control variables. Generally, when the number was no more than four, there were only seven different kinds of catastrophe. Table 1 shows different potential functions of these seven basic kinds of catastrophe, where x, y are state variables and v, μ, w, t are control variables.

Table 1. Elementary catastrophe models and their potential functions.

Catastrophe model	Potential functions
Fold	$V(x) = x^3 + vx$
Cusp	$V(x) = x^4 + \mu x^2 + vx$
Swallowtail	$V(x) = x^5 + \mu x^3 + vx^2 + wx$
Butterfly	$V(x) = x^6 + tx^4 + \mu x^3 + vx^2 + wx$
Elliptic	$V(x, y) = x^3 - xy^2 + w(x^2 + y^2) + \mu x + vy$
Hyperbolic	$V(x, y) = x^3 + y^3 + wxy + \mu x + vy$
Parabolic	$V(x, y) = y^4 + x^2 y + wx^2 + ty^2 + \mu x + vy$

5. Eigenvalue Catastrophe Indications

Catastrophe theory is often used in systems with obvious catastrophe indications. The catastrophe indications [26,27,36] are certain obvious characteristics of a catastrophe system including catastrophe, multi-modes and divergence, *etc.* Among them catastrophe is the most basic and obvious catastrophe indication, indicating some sudden change of the system state; multi-modes means that the number of the system states may be more than one; divergence shows the instability of the path perturbation of control parameters. Usually, a small perturbation of control parameters usually only causes small changes of the state variables except the perturbation near the degenerate critical point, which may causes huge changes in the final value of state variables.

Small signal stability is decided by the eigenvalues of the system. From the perspective of the change of eigenvalues, under some certain conditions, a pair of complex eigenvalues may turn into real eigenvalues or from the opposite direction, which is a typical kind of catastrophe phenomenon remarkably similar to changes of the equations' roots in Appendix A. This process also shows multi-mode characteristics. In addition, a small perturbation of parameters near the critical point would cause huge changes in the properties of the eigenvalues, thus showing the typical characteristics of divergence. Therefore, a change of eigenvalues indicates obvious catastrophe indications and shows that the use of catastrophe theory is both feasible and reasonable.

There are two kinds of application of catastrophe theory to specific problems, which are the analysis method and the empirical method. The analysis method is deducing where the equilibrium position of dynamical system is, while the empirical method establishes a system catastrophe model according to system external characteristics with mathematical descriptions of the unknown bifurcation set. It builds equilibrium surface equations of the system by data fitting or qualitative fitting [26,27] and chooses an appropriate catastrophe model according to the number of control variables and state variables.

6. Small Signal Stability Region Boundary Model of the Interconnected Power System with Large-Scale Wind Power

Hopf bifurcation is related to the system oscillatory instability and this paper mainly focuses on the stability region boundary consisting of Hopf bifurcations. Here, we established the small signal stability region boundary model of the interconnected power system with large-scale wind power in two-dimensional power injection space and extend the 2D model to multiple dimensions to obtain the small signal stability region boundary model in multidimensional power injection space.

6.1. Small Signal Stability Region Boundary Model in Two-Dimensional Power Injection Space

In this paper, we assumed power sources 1 and 2 were any two power sources, which can influence the dominant oscillation mode and their type were wind power or conventional power units. We choose a large unit which is far from and unrelated to the dominant oscillation mode as the system balancing machine. Keeping the structure and parameters of the system unchanged and at the same time keeping the node power injected unchanged except for the balancing machine, then active power injected by source 1 and 2 forms a two-dimensional small signal stability region boundary.

Essentially, this boundary is decided by the system eigenvalues. Because the analysis method involves complex nonlinear differential equations, this paper adopts the empirical method to establish the catastrophe model. We choose P_1, the active power of source 1 and P_2, the active power of source 2 as control variables and the eigenvalues of the dominant oscillation mode as state variable x. There are two control variables and one state variable in total, corresponding to the cusp catastrophe model in Table 1.

In catastrophe theory, the potential function is expanded according to a Taylor series abandoning high-order terms and simplifies the model by differentiable homeomorphism. The simplest model is called a canonical form [25] such as the potential functions in Table 1. The differentiable homeomorphism of the cusp catastrophe model only involves linear transformation [25], so the differences between each parameter and the canonical form of the cusp catastrophe model are no more than a coefficient and a constant, that is:

$$V(x) = a(x - x_0)^4 + b(P_1 + c)(x - x_0)^2 + (P_2 + d)(x - x_0) \qquad (14)$$

where a, x_0, b, c, d are unknown constants. The formulation of equilibrium surface is as follows:

$$V'(x) = 4a(x - x_0)^3 + b(P_1 + c)(x - x_0) + P_2 + d = 0 \qquad (15)$$

Hopf bifurcation is the operating point of the equilibrium surface and the state variable in accord to the Hopf bifurcation point is $x = 0 \pm jn$. Substitute $x = 0 \pm jn$ into Equation (15) and we obtain:

$$\begin{cases} -4ax_0^3 + 12an^2x_0 - (4an^2 - 12ax_0^2)x_0 + P_2 + d = 0 \\ b(P_1 + c) + 6ax_0^2 - 2an^2 = 0 \end{cases} \tag{16}$$

After arranging Equation (16):

$$\begin{cases} P_1 = \frac{2a}{b}n^2 - \frac{1}{b}(bc+6ax_0^2) \\ P_2 = -8ax_0n^2 - 8ax_0^3 - d \end{cases} \tag{17}$$

By eliminating variable n we get:

$$e_{12}P_1 + e_{21}P_2 = f_{12} \tag{18}$$

where:

$$\begin{cases} e_{12} = 4x_0b \\ e_{21} = 1 \\ f_{12} = -(d + 4bcx_0 + 32ax_0^3) \end{cases} \tag{19}$$

Here e_{12} and f_{12} are both constants. According to Equation (18), P_1 and P_2 show a linear change relationship on the small signal stability region boundary. Equation (18) is the small signal stability region boundary model of the interconnected power system with large-scale wind power in two-dimensional power injection space. We can determine the small signal stability region boundary according to e_{12} and f_{12} acquired by data fitting.

It needs to be indicated that Hopf bifurcations is not the bifurcation set B of catastrophe theory. The bifurcation set B corresponds to catastrophe points of the catastrophe model. However, the catastrophe point of the catastrophe theory in fact represents a sudden jump of the properties of eigenvalues, including complex eigenvalues turning into real eigenvalues or real eigenvalues turning into complex eigenvalues while Hopf bifurcation represents eigenvalues crossing the imaginary axis. The comparison sketch between Hopf bifurcation and catastrophe point is shown in Figure 4.

This difference should not discourage the application of catastrophe theory. The potential function of catastrophe theory is expanded according to a Taylor series and abandons unnecessary high-order terms, which is a standard technique in catastrophe theory. Catastrophe theory provides a rigorous proof [25] and truncation to make this approach has a reliable foundation. Thus, the potential function obtained in this way still has enough accuracy for all the operating points of the system. The equilibrium surface equation consists of the derived functions is also widely valid and applicable. That is to say, the equilibrium surface equations are still reasonable for Hopf bifurcation.

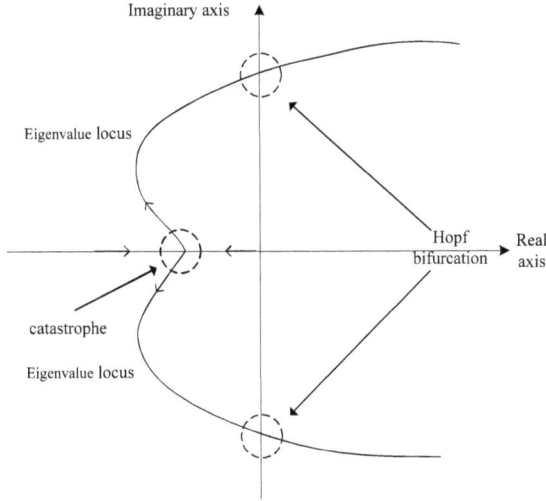

Figure 4. The comparison sketch between Hopf bifurcation and catastrophe point.

6.2. Small Signal Stability Region Boundary Model in Multidimensional Power Injection Space

Based on the two-dimensional boundary model, this section deduced the small signal stability region boundary model in multidimensional power injection space. Assume power source $1, 2, \ldots, n$, which can influence the dominant oscillation mode and their type are wind power or conventional power units. These n power sources can form these two-dimensional boundaries as Equation (20) shows:

$$
\left\{
\begin{array}{l}
e_{12}P_1 + e_{21}P_2 = f_{12}\{P_3, P_4, \cdots P_n \text{ constant}\} \\
e_{13}P_1 + e_{31}P_3 = f_{13}\{P_2, P_4, \cdots P_n \text{ constant}\} \\
\cdots\cdots\cdots\cdots\cdots \\
e_{1n}P_1 + e_{n1}P_n = f_{1n}\{P_2, P_3, \cdots P_{n-1} \text{ constant}\}
\end{array}
\right.
\tag{20}
$$

Arrange Equation (20) and make the coefficient of P_1 to be 1and we get:

$$
\left\{
\begin{array}{l}
P_1 + \frac{e_{21}}{e_{12}}P_2 = \frac{f_{12}}{e_{12}}\{P_3, P_4, \cdots P_n \text{ constant}\} \\
P_1 + \frac{e_{31}}{e_{13}}P_3 = \frac{f_{13}}{e_{13}}\{P_2, P_4, \cdots P_n \text{ constant}\} \\
\cdots\cdots\cdots\cdots\cdots \\
P_1 + \frac{e_{n1}}{e_{1n}}P_n = \frac{f_{1n}}{e_{1n}}\{P_2, P_3, \cdots P_{n-1} \text{ constant}\}
\end{array}
\right.
\tag{21}
$$

97

When their outputs change synchronously, these n sources form an n-dimensional boundary as Equation (22) shows:

$$k_1 P_1 + k_2 P_2 + \cdots + k_i P_i \cdots + k_n P_n = r \tag{22}$$

where $k_1, \cdots, k_i, \cdots, k_n$, and r can be constants as well as functions. Then make the coefficient of P_1 to be 1 and we get:

$$P_1 + \frac{k_2}{k_1} P_2 + \cdots + \frac{k_i}{k_1} P_i + \cdots + \frac{k_n}{k_1} P_n = \frac{r}{k_1} \tag{23}$$

For P_1 and P_i ($i \in \{2, \cdots, n\}$), keep the other power injection in Equation (23) fixed to obtain Equation (24):

$$P_1 + \frac{k_i}{k_1} P_i = \frac{r}{k_1} - \frac{k_2}{k_1} P_{20} - \cdots - \frac{k_{i-1}}{k_1} P_{(i-1)0} - \frac{k_{i+1}}{k_1} P_{(i+1)0} - \cdots - \frac{k_n}{k_1} P_{n0} \tag{24}$$

Subscript 0 represents fixed value. According to Equation (21), the relationship between P_1 and P_i is linear, with a slope of e_{i1}/e_{1i}. To make Equation (24) satisfy the relationship in Equation (21), we kept $k_i/k_1 = e_{i1}/e_{1i}$.

Because i represents any number of $\{2, \cdots, n\}$, we obtain $k_2/k_1 = e_{21}/e_{12}, \cdots, k_n/k_1 = e_{n1}/e_{1n}$. All of the coefficients above are constants. For Equation (24), r/k_1 must be constant, or else the right side of Equation (24) will be variable and fail to satisfy the linear relationship in Equation (21).

Therefore, coefficients in Equation (22) are determined:

$$P_1 + e_2 P_2 + \cdots + e_i P_i \cdots + e_n P_n = f \tag{25}$$

where, $e_2 = e_{21}/e_{12}, \cdots, e_n = e_{n1}/e_{1n}$. $e_2, \cdots, e_n; f$ are all constants.

Equation (25) is the small signal stability region boundary model in multidimensional power injection space. We can determine the small signal stability region boundary according to the coefficients of variables acquired by data fitting.

Sun and Yu [18] proposed a method to fit the small signal stability region boundary using a hyper-plane but lacked theories to support their arguments. However, the research in this paper provides theoretical support for their proposal.

In the analysis above, this paper never limits the type of power source, so large-scale wind power injection doesn't change the form of the small signal stability region boundary model but only changes the dimensions of the hyper-plane and the values of the parameters. When wind power is injected into system instead of common units, both the number of power sources and the dimensions of the hyper-plane will remain unchanged. However, values of the parameters will change because the difference between properties of wind power and common units is very big. Wind power injection into a certain node in the power system makes the number of power sources increase, leading to changes of the parameters as well as the dimension of hyper-plane.

7. Simulation Verification

To verify the validity of the small signal stability region boundary model proposed in this paper, we tested two examples below.

7.1. Example 1

We built the simulation system shown in Figure 5 using DIgSILENT/ PowerFactory software, where bus D is an infinite bus, buses A, B, C are respectively connected to normal generators G_1 with excitation system, DFIG wind farm 1 and DFIG wind farm 2. The capacity of wind farm 1 and wind farm 2 are 210 MVA and 300 MVA, respectively. Both of these wind farms consist of several sets of DFIG with the same parameters and operation conditions in parallel. In the analysis, each wind farm is equivalent to a wind turbine. We chose the 2 MW wind turbine built into the DIgSILENT/PowerFactory software as the wind turbine model and Hansen *et al.* [39] introduced the model in detail. Appendix A shows the parameters of each element in the system(Table 1 shows the parameters of one single DFIG; Table 2 shows the parameters of generator G_1; Table 3 shows the parameters of the excitation system equipped on G_1; Appendix A shows the parameters of the transmission lines.).

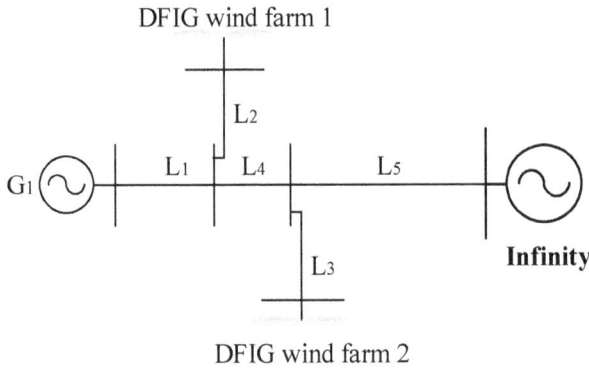

Figure 5. The wire map of the power system.

7.1.1. Test of the Two-Dimensional Boundary Model

Assume that P_1 is the active power of G_1, P_2 is the active power of wind farm 1 and P_3 is the active power of wind farm 2. Every power source operates with unity power factor. Keep $P_3 = 200$ MW and adjust the values of P_1 and P_2. When

Hopf bifurcation occurs, take a sample of the data. Table 4 of Appendix A shows the sampling points.

Substitute the data in Table 4 into Equation (18) and obtain $e_{12} = 1.3616$, $f_{12} = 413.1675$, that is:

$$1.3616P_1 + P_2 = 413.1675 \tag{26}$$

Figure 6 shows the comparison between sampling points and the fitting boundary, where P_1 is the abscissa and P_2 is the ordinate. According to Figure 6, they have good consistency. In order to accurately express the error between the sampling points and the fitting boundary, define the error σ as follows:

$$\sigma = \frac{d}{\sqrt{P_1^2 + P_2^2}} \tag{27}$$

where d is the distance from sampling points to the fitting boundary. After calculation, we obtained that the average error was 0.319% and the biggest was 0.645%, which had good accuracy and showed that the model was reasonable and feasible. Detailed errors are shown in Table 5 of Appendix A.

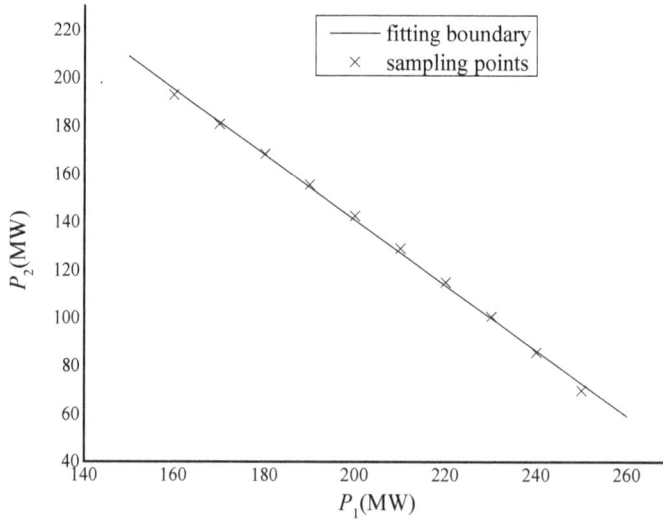

Figure 6. The two-dimensional stability region boundary contrast figure.

7.1.2. Verification of the Multidimensional Boundary Model

Every power source operates with unity power factor and we adjusted P_1, P_2 and P_3. When Hopf bifurcation occurs, take a sample of the data. Table 6 in Appendix A shows the sampling points.

Substitute the data in Table 6 into Equation (25) and obtain $e_2 = 0.7339$, $e_3 = 0.5871$, $f = 421.2902$, that is:

$$P_1 + 0.7339P_2 + 0.5871P_3 = 421.2902 \tag{28}$$

Figure 7 shows the contrast between sampling points and the fitting plane, where P_1, P_2, P_3 are three axes. According to Figure 7, they have good consistency.

In order to accurately express the error between sampling points and the fitting boundary, define the error σ as follows:

$$\sigma = \frac{d}{\sqrt{P_1^2 + P_2^2 + P_3^2}} \tag{29}$$

where d is the distance from sampling points to the fitting plane. After calculation, we obtained that the average error was 0.088% and the biggest was 0.18%, which had good accuracy and showed that the model was reasonable and feasible. Detailed errors are shown in Table 7 of Appendix A.

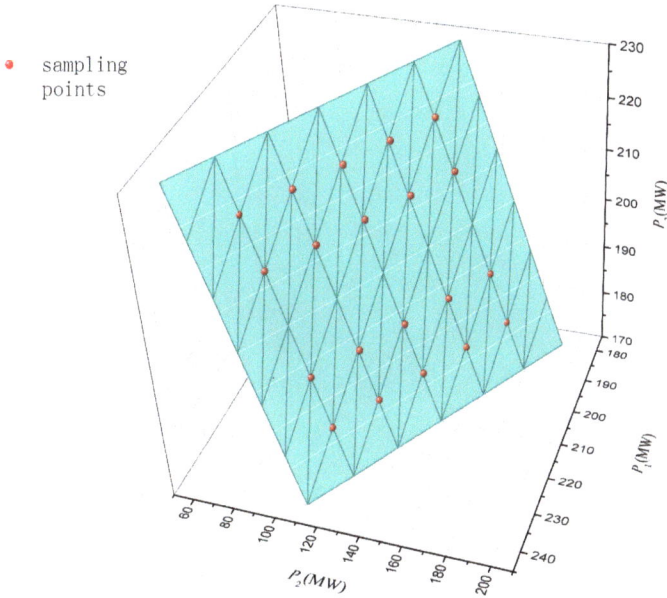

Figure 7. The three-dimension stability boundary contrast figure.

To make further analysis of the boundary, assume $P_3 = 200$ MW and turn Equation (28) into:

$$1.3626P_1 + P_2 = 414.0485 \tag{30}$$

This is the boundary equation degenerating into two dimensions. The difference between Equations (26) and (30) is very small, showing that the theory is reasonable and feasible.

7.1.3. Time-Domain Simulation Verification

To further test the accuracy of the boundary, we choose two operating points near the boundary: point 1 (210, 139, 180) and point 2 (210, 150, 180) and observe the stability of the system affected under a small disturbance by adopting time-domain simulation. Operating point 1 was in the boundary while point 2 was out of the boundary. The small disturbance was set as the mechanical torque's increasing 0.01 p.u of generator G_1 at 1 s, then restoring to the original level at 2 s. The simulation time was set at 50 s. Under the above disturbance, the power angle swing curves of generator G_1 at operating point 1 and point 2 were shown in Figure 8.

As can be seen from Figure 8, the swing amplitude of the angle at operation point 2 becomes larger and larger and the system would finally lose stability; while at operation point 1, the swing amplitude of the angle showed a decreasing trend and finally was stable. Thus, the time-domain simulation results also verified the validity of our theory on the stability boundary.

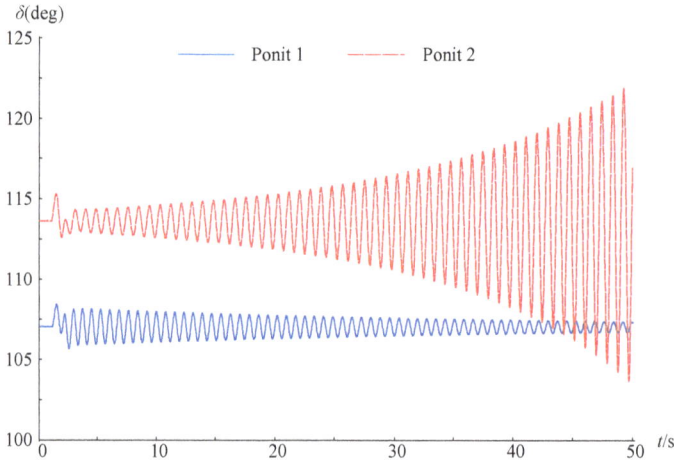

Figure 8. The angle swing curves of generator G1 at operation point 1 and point 2.

7.2. Example 2

The second simulation test uses the improved three generators and nine buses system. The length of each line in the IEEE three generators and nine buses system [40] is extended to 220% of the original value. Active power of load A (Figure 9) is set at 200 MW, load B at 250 MW, and load C at 200 MW. The remaining values stay unchanged. The DFIG wind farm 1 and wind farm 2 are connected to BUS7 and BUS9, respectively, through transformers. The system with the wind farm connected to it is shown in Figure 9.

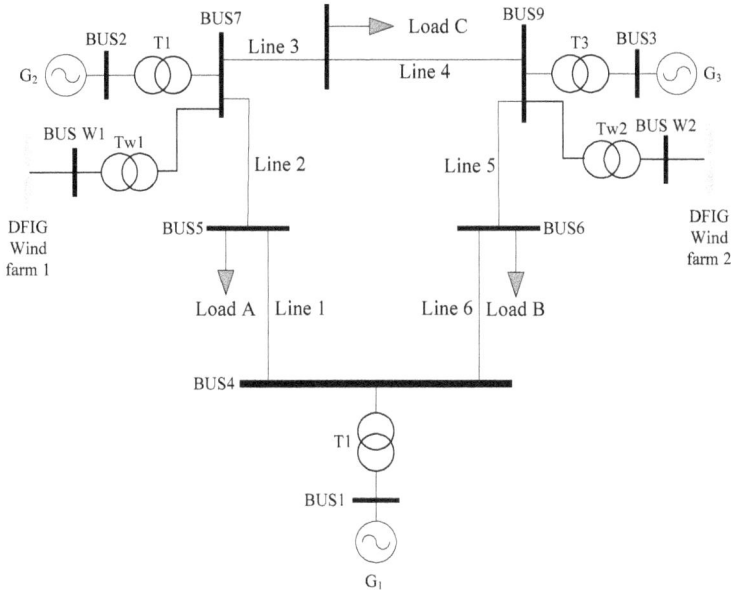

Figure 9. The improved model diagram of the three generators and nine buses system.

New equipment in this system include DFIG wind farm 1, DFIG wind farms 2, transformer Tw1, transformer Tw2, bus BUS W1, and bus BUS W2. Among them, DFIG wind farm 1 and 2 are both equivalent wind farms with a capacity of 150 MVA. Each single wind turbine still uses the 2 MW wind turbine model in the DIgSILENT/PowerFactory in the equivalence. The parameters of the transformers Tw1 and Tw2 are shown in Table 2. The system was modeled by DIgSILENT/PowerFactory software.

Table 2. New transformer parameters.

Transformer	Rated voltage (kV)	Rated power (MVA)	Short-Circuit voltage U_k (%)	No-Load current I_0 (%)	Connection type
Tw1, Tw2	20/0.69	166.65	5	3	YN/yn0

There are two oscillation modes in the improved system. The frequency of one mode is around 1.15 HZ and the other is around 0.2 HZ. The 0.2 HZ mode

103

has a strong damping in any case and the corresponding eigenvalues are far away from HB. Thus the paper does not consider this situation and mainly focuses on the 1.15 HZ mode.

Mark the active output of conventional unit G_2 as P_1, the active output of DFGI wind farm 1 as P_2, the active output of conventional unit G_3 as P_3 and the active output of DFGI wind farm 2 as P_4. The new wind farms adopt the unity power factor operation mode. Set G_1 as the balance unit and G_2, G_3 as PV nodes. Adjust the value of P_1, P_2, P_3, P_4 and sample data when HB occurs. Table 8 in Appendix A shows the sampling points.

Substitute the data in Table 8 into Equation (25) and obtain $e_2 = 1.0542$, $e_3 = 0.8195$, $e_4 = 0.878$, $f = 489.563$. That is

$$P_1 + 1.0542P_2 + 0.8195P_3 + 0.878P_4 = 489.563 \tag{31}$$

Similarly, define the error as follows:

$$\sigma = \frac{d}{\sqrt{P_1^2 + P_2^2 + P_3^2 + P_4^2}} \tag{32}$$

In Equation (32), d is the distance between the sampling points and the plane. The average value of the error was calculated to be 0.0384% and the maximum error was 0.0843%, which was of high accuracy. Detailed errors are shown in Table 9 of Appendix A. The results of time domain simulation were similar to Figure 8 and it was unnecessary to go into detail. The simulation above also verified the multi-dimensional boundary model.

From the above analysis, the stability region boundary provides a critical stable operating range of the system. The shorter the distance between current operating point and stability region boundary is, the more likely it is the system will lose stability. Thus, when the operation point is close to the stability region boundary, we need to adjust the output of each power supply timely and make the system away from the boundary, in order to avoid oscillation and instability of the system.

8. Conclusions

This paper provides a new method of analysis of the small signal stability region boundary for an interconnected power system with large-scale wind power. Compared with conventional analysis methods, based on the external characteristic modeling method of the running track, catastrophe theory directly established a function between control variables and state variables without contacting any special inner mechanism. Applying catastrophe theory, this paper established a small signal stability region boundary model consisting of Hopf bifurcations and drawn the following conclusions:

(1)   The small signal stability region boundary model in two-dimensional power injection space is a straight line. When the other injected power doesn't change, the power from the two power sources influencing the dominant oscillation mode shows a linear relation between them.

(2)   The small signal stability region boundary model in multidimensional power injection space is a hyper-plane. When the other injected power doesn't change, the power from sources influencing the dominant oscillation mode forms a hyper-plane.

(3)   Compared with the conventional system, large-scale wind power integration doesn't change the form of the small signal stability region boundary model but only the dimensions of the hyper-plane and the values of the parameters.

This paper developed a deeper understanding of the small signal stability region boundary of the interconnected power system with large-scale wind power and also presented new ideas and methods for its study.

Acknowledgments: This work is supported by the project "Research of the Impact Mechanism on Self-organized Critical State by Large-scale Cluster Grid-connected Wind Power and Identification", funded by the National Science Foundation of China (51377053), and the project "The Research of Key Techniques of Loss Reduction in Power System Based on Load-Net-Source Coordinated Control", funded by Major Project of State Grid Corporation of China (52272214002).

Author Contributions: Wenying Liu contributed to design, directed, and made suggestions for the research. Rundong Ge contributed to analysis, writing, and interpretation of the research. Quancheng Lv and Huiyong Li made some suggestions and translated this article from Chinese to English. Jiangbei Ge also made suggestions for the research. All co-authors contributed to the writing of the final research article.

Appendix A. Appendices

Appendix A: A typical catastrophe mechanism explaining the basic principle of catastrophe theory, (Figure 1)

The process of this mechanism is very simple. Firstly, pick two nearly identical rubber bands and cut from thin cardboard a disk whose diameter is one unit. Push a drawing pin through the disc at a point Q near the circumference with the point of this pin upwards. Mount the disc on a suitable base by pushing a second pin through the center, O. Loop the two rubber bands over the pin at Q, and use a third drawing pin to fasten the other end of one of the rubber bands to a point R on the base, two units from O. The remaining end, P, is left free.

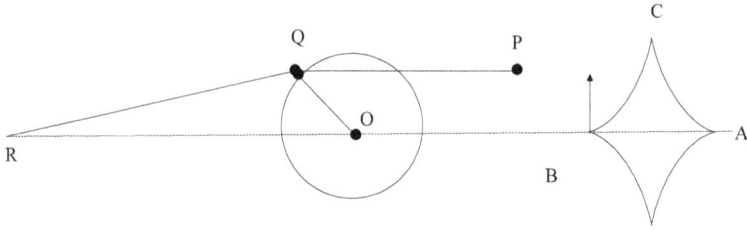

Figure 1. A typical catastrophe mechanism.

We operate the machine by moving P slowly in the plane of the machine. If we experiment for some time, we discover some curious features. The most obvious of these is that while the machine almost always responds smoothly to small changes in the position of P, is occasionally jumps suddenly. If we mark on the base the positions of P at which these jumps occur, we find that they form the perimeter of a curved diamond.

Saunders [25] made a detailed analysis of this sudden jump and turned the potential function of the mechanism into the following form:

$$V(x) = x^4 + \mu x^2 + vx \tag{A1}$$

That was the simplest model of the cusp catastrophe and the equilibrium surface was:

$$V'(x) = 4x^3 + 2ux + v = 0 \tag{A2}$$

Equation (A2) is a cubic equation, which has three real roots or just one real root. The literature [25] reveals that the mathematical essence of the sudden jump is changes in the real roots' number, namely a pair of complex roots turning into real roots or real roots turning into complex roots. The number of real roots was decided by the discriminant:

$$\Delta = 8u^3 + 27v^2 \tag{A3}$$

When $\Delta < 0$, there were three real roots; when $\Delta > 0$, there was only one real root, while $\Delta = 0$ was the critical condition of the changes of the number of real roots. It was easy to verify that $V'(x) = \Delta = 0$ was totally equivalent to $V'(x) = V''(x) = 0$, which was the bifurcation set B of the catastrophe system. Figure 2 shows the Equilibrium surface and bifurcation set of the cusp catastrophe model.

The equilibrium surface was divided into an upper lobe, middle lobe, and lower lobe. The bifurcation set B was the projection of two creases of the equilibrium surface on the u–v surface. If control variables changed to the bifurcation set B, there must be sudden jumps of equilibrium points (upper lobe jumped to lower lobe or lower lobe jumped to upper lobe). Obviously, the outline in Figure 1 is composed of the bifurcation set B in Figure 2.

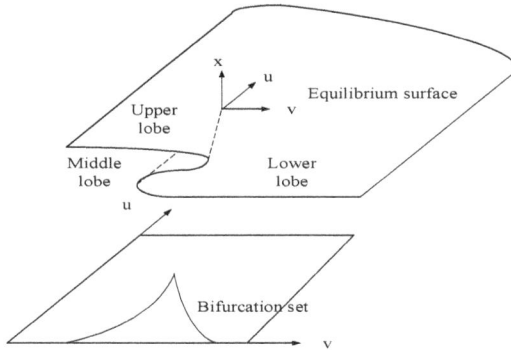

Figure 2. The Equilibrium surface and the bifurcation set of the cusp catastrophe model.

Appendix B

Appendix B1: DFIG Parameters

Table 1. DFIG parameters.

P_n (MW)	U_s (V)	R_s (p.u)	X_s (p.u)	X_m (p.u)	R_r (p.u)	X_r (p.u)	H_w (s)	H_g (s)	K
2	690	0.01	0.1	3.5	0.01	0.1	4.02	0.47	80.27

Appendix B2: Generator Parameters

Table 2. Generator parameters.

Generator	Capacity (WVA)	Voltage (KV)	x_d (p.u)	x'_d (p.u)	x_q (p.u)	x'_q (p.u)	T'_{d0} (s)	T'_{q0} (s)
G_1	300	18	1.72	1.66	0.23	0.378	0.8	0.12

The voltage and power reference values are this machine's rated voltage and capacity.

Appendix B3: Excitation Parameters

Details of the excitation system model refer to the A type exciter in Appendix A of reference [40]. The specific parameters are as follows:

Table 3. Excitation parameters.

τ_R	τ_{A1}	K_A	τ_{A2}	V_{Rmax}	V_{Rmin}	τ_E	K_E	K_F	τ_F	$E1$	$Se1$	$E2$	$Se2$
0.001	0.05	400	0.01	3	−3	0.95	−0.17	0.04	1	3.66	0.03	4.89	0.1

Appendix B4: Line Parameters

Line type is LGJ-400, L_1 of 60 km in length, L_2 of 30 km in length, L_3 of 30 km in length, L_4 of 30 km in length and L_5 of 100 km in length.

Appendix C

Table 4. The two-dimension boundary sampling points of small-signal stability region.

P_1 (MW)	P_2 (MW)	Real part	Imaginary part
160	192.95	−0.00002	7.371781
170	180.84	0.00001	7.42214
180	168.34	−0.00004	7.473076
190	155.49	0.00001	7.518541
200	142.43	−0.00003	7.544407
210	128.98	0	7.568171
220	115.07	0	7.593426
230	100.64	−0.00004	7.62154
240	85.64	0.00001	7.65214
250	69.94	0.00002	7.689427

Table 5. The errors of the two-dimension sampling points.

P_1 (MW)	P_2 (MW)	Error (%)
160	192.95	0.557664
170	180.84	0.204026
180	168.34	0.062567
190	155.49	0.247486
200	142.43	0.381506
210	128.98	0.419963
220	115.07	0.346774
230	100.64	0.151014
240	85.64	0.172708
250	69.94	0.644714

Table 6. The three-dimension boundary sampling points of small-signal stability region.

P_1 (MW)	P_2 (MW)	P_3 (MW)	Real part	Imaginary part
190	170.76	180	0.0001	7.536991
200	157.61	180	0	7.585456
210	144.3	180	−0.00002	7.610067
220	130.58	180	−0.00002	7.63315
230	116.39	180	−0.00001	7.656929
190	163.21	190	0	7.52991
200	150.12	190	−0.00002	7.565746
210	136.74	190	−0.00004	7.590114
220	122.94	190	0.00001	7.613221
230	108.64	190	0.00001	7.638675
190	147.67	210	0.00002	7.49535
200	134.54	210	0	7.520848
210	121	210	0.00003	7.545433
220	106.98	210	0.00003	7.572346
230	92.41	210	−0.00002	7.603393
190	139.64	220	0.00002	7.470371
200	126.43	220	0	7.496321

Table 7. The errors of the three-dimension sampling points.

P_1 (MW)	P_2 (MW)	P_3 (MW)	Error (%)
190	170.76	180	0.067958
200	157.61	180	0.013502
210	144.3	180	0.067639
220	130.58	180	0.051357
230	116.39	180	0.044867
190	163.21	190	0.008951
200	150.12	190	0.100204
210	136.74	190	0.141877
220	122.94	190	0.111855
230	108.64	190	0.002365
190	147.67	210	0.085745
200	134.54	210	0.16861
210	121	210	0.182398
220	106.98	210	0.116044
230	92.41	210	0.040263
190	139.64	220	0.079899
200	126.43	220	0.148584
210	112.78	220	0.144009
220	98.64	220	0.058873
230	83.92	220	0.119396

Appendix D

Table 8. The four-dimension boundary sampling points of small-signal stability region.

P_1 (MW)	P_2 (MW)	P_3 (MW)	P_4 (MW)	Real part	Imaginary part
164.25	130	90	130	0.00002	6.182942
162	130.11	93	130	−0.00005	6.135284
160	132	95	128.23	−0.00002	6.130719
155	135	92	133.33	−0.00009	6.108222
155	135	89	136.1	−0.00006	6.11538
150	139.94	90	135	−0.00012	6.123751
153.94	140	90	130	0.00003	6.195451
156.63	130	100	130	0	6.015732
155.96	130	90	140	0	6.026069
151	135	108	123.01	0.00001	5.995063
154.24	140	95	125	0.00001	6.186703
142.92	145	103	125	0.00005	6.069223
143.43	150	90	130	0.00003	6.213094
147	135	100	134.51	−0.00008	5.94512
152.053	130	105	130	0	5.92228
136	148.31	80	150	0.00013	6.061767
130.98	150	105	130	−0.00007	5.96777

Table 9. The errors of the four-dimension sampling points.

P_1 (MW)	P_2 (MW)	P_3 (MW)	P_4 (MW)	Error (%)
164.25	130	90	130	0.084283
162	130.11	93	130	0.019028
160	132	95	128.23	0.003579
155	135	92	133.33	0.033593
155	135	89	136.1	0.028496
150	139.94	90	135	0.040916
153.94	140	90	130	0.03755
156.63	130	100	130	0.031047
155.96	130	90	140	0.01475
151	135	108	123.01	0.042525
154.24	140	95	125	0.036604
142.92	145	103	125	0.065856
143.43	150	90	130	0.031106
147	135	100	134.51	0.050112
152.053	130	105	130	0.066951
136	148.31	80	150	0.001088
130.98	150	105	130	0.064887

Conflicts of Interest: The authors declare no conflict of interest.

References

1. Baños, R.; Manzano-Agugliaro, F.; Montoya, F.G.; Gil, C.; Alcayde, A.; Gómez, J. Optimization methods applied to renewable and sustainable energy: A review. *Renew. Sustain. Energy Rev.* **2011**, *15*, 1753–1766.
2. Hernández-Escobedo, Q.; Saldaña-Flores, R.; Rodríguez-García, E.R.; Manzano-Agugliaro, F. Wind energy resource in Northern Mexico. *Renew. Sustain. Energy Rev.* **2014**, *32*, 890–914.
3. Montoya, F.G.; Manzano-Agugliaro, F.; López-Márquez, S.; Hernández-Escobedo, Q.; Gil, C. Wind turbine selection for wind farm layout using multi-objective evolutionary algorithms. *Expert Syst. Appl.* **2014**, *41*, 6585–6595.
4. Li, J.F. *2012 China Wind Power Outlook*; China Environmental Science Press: Beijing, China, 2012; pp. 28–30. (In Chinese)
5. Kang, J.J.; Yuan, J.H.; Hu, Z.G.; Xu, Y. Review on wind power development and relevant policies in China during the 11th Five-Year-Plan period. *Renew. Sustain. Energy Rev.* **2012**, *16*, 1907–1915.
6. Chen, D.Z.; Ma, S.Y.; Song, Y.T.; Wang, N.B.; Fang, P.Y.; Ren, Y. Research on transient stability under HVDC block fault in wind-thermal-bundled power base transmitted by AC/DC system. In Proceedings of the 2014 International Conference on Power System Technology, Chengdu, China, 20–22 October 2014.
7. Xiao, C.Y.; Wang, N.B.; Ding, K.; Zhi, J. System power regulation scheme for jiuquan wind power base. *Proc. CSEE* **2010**, *30*, 1–7. (In Chinese)
8. Slootweg, J.G.; Kling, W.L. The impact of large scale wind power generation on power system oscillations. *Electr. Power Syst. Res.* **2003**, *67*, 9–20.
9. Mendonca, A.; Peas Lopes, J.A. Impact of large scale wind power integration on small signal stability. In Proceedings of the 2005 International Conference on Future Power Systems, Amsterdam, The Netherlands, 16–18 November 2005.
10. Gautam, D.; Vittal, V.; Harbour, T. Impact of increased penetration of DFIG-based wind turbine generators on transient and small signal stability of power systems. *IEEE Trans. Power Syst.* **2009**, *24*, 1426–1434.
11. Kwatny, H.G.; Fischl, R.F.; Nwankpa, C. Local bifurcation in power systems: Theory, computation and application. *Proc. IEEE* **1995**, *83*, 1453–1483.
12. Venkatasubramanian, V.; Schattler, H.; Zaborszky, J. Dynamics of large constrained nonlinear systems—A taxonomy theory. *Proc. IEEE* **1995**, *83*, 1530–1560.
13. Canizares, C.A.; Mithulananthan, N.; Milano, F. Linear performance indices to predict oscillatory stability problems in power systems. *IEEE Trans. Power Syst.* **2004**, *19*, 1104–1114.
14. Abed, E.H.; Varaiya, P.P.; Milano, F. Nonlinear oscillations in power systems. *Int. J. Electr. Power Energy Syst.* **1984**, *6*, 37–43.
15. Seydel, R. *Practical Bifurcation and Stability Analysis*, 3rd ed.; Springer: London, UK, 2010; pp. 199–257.

16. Roose, D.; Hlavacek, V. A direct method for the computation of Hopf bifurcation point. *SLAM J. Appl. Math.* **1985**, *45*, 879–894.

17. Hiskens, I.A. Analysis tool for power systems-contending with nonlinearities. *Proc. IEEE* **1995**, *83*, 1573–1587.

18. Sun, Q.; Yu, Y.X. Hyper-plane approximation of boundary of small signal stability region and its application. *J. Tianjin Univ.* **2008**, *41*, 647–652. (In Chinese)

19. Yang, S.; Liu, F.; Zhang, D.; Mei, A.W. Polynomial approximation of the small-signal stability region boundaries and its credible region in high-dimensional parameter space. *Int. Trans. Electr. Energy Syst.* **2013**, *23*, 784–801.

20. Jia, H.J.; Yu, X.D.; Cao, X.D. Impact of the exciter voltage limit to small signal stability region of a tree-bus power system. *Int. J. Electr. Power Energy Syst.* **2011**, *33*, 1598–1607.

21. Jia, H.J.; Yu, X.D.; Yu, Y.X.; Wang, C.S. Power system small signal stability region with time delay. *Int. J. Electr. Power Energy Syst.* **2008**, *30*, 16–22.

22. Li, G.Q.; Li, J.; Gao, S. Power system small signal stability region with saturation element considered. *Autom. Electr. Power Syst.* **2010**, *34*, 14–17. (In Chinese)

23. Pulgar, P.; Hector, A. *Wind Farm Model for Power System Stability Analysis*; The University of Illinois at Urbana-Champaign: Champaign, IL, USA, 2010.

24. Ash, R.; Ash, G. Numerical computation of root loci using the newton-raphson technique. *IEEE Trans. Autom. Control* **1968**, *13*, 576–582.

25. Saunders, P.T. *An Introduction to Catastrophe Theory*, 1st ed.; Cambridge University Press: New York, NY, USA, 1980; pp. 1–60.

26. Zeeman, E.C. *Catastrophe Theory*; Springer: Berlin/Heidelberg, Germany, 1976; pp. 1–83.

27. Robert, G. *Catastrophe Theory for Scientists and Engineers*; Wiley: New York, NY, USA, 1993; pp. 1–92.

28. Papacharalampous, A.E.; Vlahogianni, E.I. Modeling microscopic freeway traffic using cusp catastrophe theory. *IEEE Intell. Transp. Syst. Mag.* **2014**, *6*, 6–16.

29. Jones, D.D.; Walters, C.J. Catastrophe theory and fisheries regulation. *J. Fish. Res. Board Can.* **1976**, *33*, 2829–2833.

30. Qin, S.; Jiao, J.J.; Wang, S. A cusp catastrophe model of instability of slip-buckling slope. *Rock Mech. Rock Eng.* **2001**, *34*, 119–134.

31. Helbing, D.; Farkas, I.; Vicsek, T. Simulating dynamical features of escape panic. *Nature* **2000**, *407*, 487–490. PubMed]

32. Sallam, A.A.; Dineley, J.L. Catastrophe theory as a tool for determining synchronous power system dynamic stability. *IEEE Trans. Power Appar. Syst.* **1983**, *102*, 622–630.

33. Wvong, M.D.; Mihirig, A.M. Catastrophe theory applied to transient stability assessment of power systems. *IEE Proc. Gener. Transm. Distrib.* **1986**, *133*, 314–318.

34. Mihirig, A.M.; Wvong, M.D. Transient stability analysis of multimachine power systems by catastrophe theory. *IEE Proc. Gener. Transm. Distrib.* **1989**, *136*, 254–258.

35. Mahmoud, G.A. Voltage stability analysis of radial distribution networks using catastrophe theory. *Gener. Transm. Distrib. IET* **2012**, *6*, 612–618.

36. Quan, Y.S.; Chen, J.; Li, W.; Zhang, X.D.; Peng, X.J.; Lao, G.Q. Methodology of forecasting the oil-immersed transformer over-hot fault tendency based on catastrophe theory. *Proc. CSEE* **2011**, *31*, 100–106. (In Chinese)

37. Zhou, H.Q.; Song, Z.P.; Wang, J.P.; Xue, Y. A review on dynamic equivalent methods for large scale wind farms. In Proceedings of the 2011 Asia-Pacific Power and Energy Engineering Conference (APPEEC 2011), Wuhan, China, 25–28 March 2011.

38. Kundur, P. *Power System Stability and Control*; McGraw-Hill: New York, NY, USA, 1994; pp. 700–705.

39. Hansen, A.D.; Jauch, C.; Sørensen, P.; Cutululis, N.; Jauch, C.; Blaabjerg, F. *Dynamic Wind Turbine Models in Power System Simulation Tool DIgSILENT*; The Risø National Laboratory: Roskilde, Denmark, 2003; pp. 1–82.

40. Anderson, P.M.; Fouad, A.A. *Power System Control and Stability*, 2nd ed.; Wiley: Southern Gate, UK, 2002; pp. 555–582.

The Three-Phase Power Router and Its Operation with Matrix Converter toward Smart-Grid Applications

Alexandros Kordonis, Ryo Takahashi, Daichi Nishihara and Takashi Hikihara

Abstract: A power router has been recently developed for both AC and DC applications that has the potential for smart-grid applications. This study focuses on three-phase power switching through the development of an experimental setup which consists of a three-phase direct AC/AC matrix converter with a power router attached to its output. Various experimental switching scenarios with the loads connected to different input sources were investigated. The crescent introduction of decentralized power generators throughout the power-grid obligates us to take measurements for a better distribution and management of the power. Power routers and matrix converters have great potential to succeed this goal with the help of power electronics devices. In this paper, a novel experimental three-phase power switching was achieved and the advantages of this operation are presented, such as on-demand and constant power supply at the desired loads.

Reprinted from *Energies*. Cite as: Kordonis, A.; Takahashi, R.; Nishihara, D.; Hikihara, T. The Three-Phase Power Router and Its Operation with Matrix Converter toward Smart-Grid Applications. *Energies* **2015**, *8*, 3034–3046.

1. Introduction

Distribution power-grids are undoubtedly one of the most complex systems humanity has ever created. Their structural vulnerability is a crucial factor [1] and the introduction of smart-grids and renewable energy sources has increased the difficulty of the analysis [2] since many parameters have to be taken into consideration in contrast with the straight-forward classic power supply. This increase in complexity requires new methods of power routing. In other words, it is obvious that power management is necessary in order to perform an effective power routing of this vast amount of newly incorporated renewable resources.

The initial stage of a power router has already been developed for DC applications [3] where the power is transferred on demand from the desired source to the desired load. This is now expanded to include AC applications [4,5] with circuit switching of two power lines and multiple routers in both parallel and series configurations, *i.e.*, an AC power routing network system can be created. The purpose of this project is to create a power network that can be readily adapted to the existing one, and can ultimately provide better regulation, distribution, and

114

transmission of power. Power packet dispatching systems [6] and intelligent power switches [7,8] have also been investigated for the purpose of improved power delivery. The feasibility of the power router requires further testing for practical applications; however, this study is aimed at demonstrating the feasibility of power router switching in three-phase systems. Such a system is investigated in this work because three-phase systems are commonly used in the majority of power-grids.

Another emerging technology for power systems is the matrix converter [9], which also has potential in applications with renewable resources and smart-grids [10–12] as it can convert energy like a classic transformer but also offers other advantages such as control of bidirectional power flow and better isolation [13,14]. Previous studies have already experimented with a matrix converter [15] and AC conversion [16], and the combination of both in a single apparatus is a natural extension of this research.

Matrix converters have the potential to be an essential part of the power electronic transformer (PET) [17,18] as a crucial means of coordinating between centralized and decentralized control, as also one of the "great challenges" of the power systems mentioned in [19]. The conventional way of transformer reaches the end of its era with the constant development of power electronics and the introduction of matrix converter and PET. Matrix converter has shown good efficiency results with the use of SiC devices in higher switching frequencies [15], fact that allows the replacement of the classic transformer by matrix converter topologies. Furthermore, a matrix converter with high frequency transformers has already shown its superiority *versus* the conventional ways as in [20,21]. PET has the possibility of achieving this challenge since it implements power electronics devices that can easily function under control. Other great advantages of PET include reduction of the overall size and cost [22] by implementing high frequency operation. It is important to mention, that in the final PET design other issues have to be taken into consideration such as voltage insulation, core size, flux densities, *etc.* [23].

This paper consists of four sections including this introduction. Section II of this paper describes the power router and matrix converter, and how the overall circuit is assembled. Section III describes the experimental and simulated results for various switching scenarios. The final section presents our conclusions.

2. Matrix Converter and Power Router

With the power control being increasingly decentralized, we consider the case of an area where power routers can be installed in every house or commercial facility, as shown in Figure 1. The utility-grid is linked to the smart-grid through the point of common coupling (PCC) with various loads and decentralized power generators (DPG) distributed throughout the latter smart-grid. The introduction of power routers, represented by the red blocks, enables the partitioning of the smart-grid into smaller segments, called micro-grids.

115

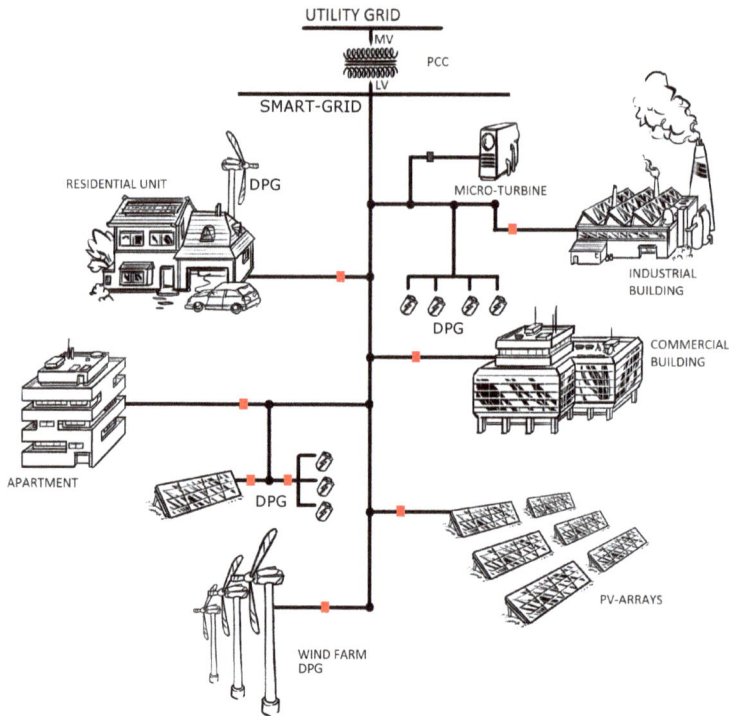

Figure 1. Smart-grid schematic. Power routers are distributed throughout the smart-grid (red blocks) partitioning it into smaller parts called micro-grids. Connection with the utility-grid is achieved at the point of common coupling (PCC). The conventional PCC (bulk transformer) will be replaced by a power electronic transformer (PET) which is smaller and more effective than the conventional one. The decentralized power generators (DPG) can be wind generators, solar cells, batteries, electric vehicles, *etc.*

The PCC is conventionally a bulk transformer that can transform the medium level voltage supply to low voltage supply. The introduction of PET [24,25] however leads to better regulation and control of the bidirectional power flow between the utility- and smart-grids. A possible PET assembly compromises two AC/AC matrix converters with a high frequency link wherein a compact electronic transformer is installed. Various PET topologies have been compared in previous literature [26] and the efficiency superiority of high frequency transformer *versus* the conventional low frequency one is investigated in [27].

Power routers, which can be distributed according to the load-demand, can perform switching to connect particular loads with the desired input sources. We assume the routers and matrix converters to have m inputs and n outputs, as shown in Figure 2. The matrix converter receives the input phases from the utility-grid and outputs them to the power routers installed on the smart-grid which may also receive power from DPGs. One or more routers together can form a micro-grid according to the localized control system needs. This study considers the most common case scenario of a three-phase system. Therefore, we will scale down our experiments to only three-phases. It can be understood from Figure 2 that the power routers can form a power network that will be consisted not only from power utilities but also from information ones. This can also be described as a next step towards the so-called Enernet [28,29], where the electrical power (*i.e.*, electrical energy) can coexist with the information (*i.e.*, internet, communications, *etc.*) [30]. Another important part is to distinguish the roles of matrix converter and power router. Matrix converters have the ability to replace the bulk transformers by offering active control of the point of common coupling. That is the main reason a matrix converter is used in this experimental work. On the other hand, the main focus is on power line switching with the aid of power routers.

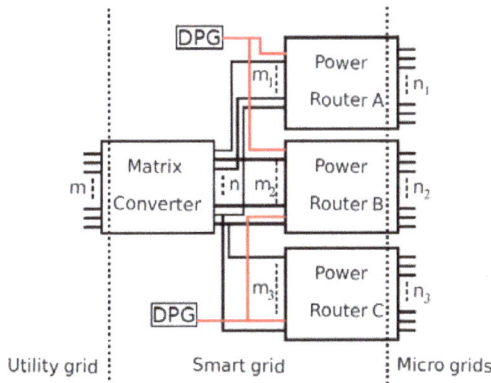

Figure 2. Generalized idea of matrix converter and power routers. All can have multiple input/output ports and various decentralized power generators (DPG) are distributed within the smart-grid.

The matrix converter has already been studied through both simulation and experiment from the authors [15]. This is called direct topology and its main difference from the indirect one is the lack of the DC link. Several advantages of the DC link have been proposed for better power quality such as in [31]. On the other hand, the authors' goal is mainly focused on a solid-state solution with no DC link for more compact final design. The behavior of the combined power router and matrix converter represents the next step towards the implementation of a system of PET with power routers. The system into consideration is shown in Figure 3.

117

A three-phase voltage source feeds nine bidirectional switches with power supplied through the input filter. Ideal switches can be considered for the simulations but in the experiment two power MOSFETs with their sources connected can form a bidirectional switch. After the output filter, the lines are distributed to the inputs of the power router. It is important to note that a power router has several inputs. Apart from the three-phase utility-grid sources, renewable energy power sources could also be connected with the power router. This is exactly where the advantage and the full potential of the power router lie on. It can accordingly distribute the supplies to the equivalent loads by ensuring constant power at certain loads. In addition, it can use DPGs to their maximum capacity by ensuring a not overloaded utility-grid as well as a novel and agile power delivery.

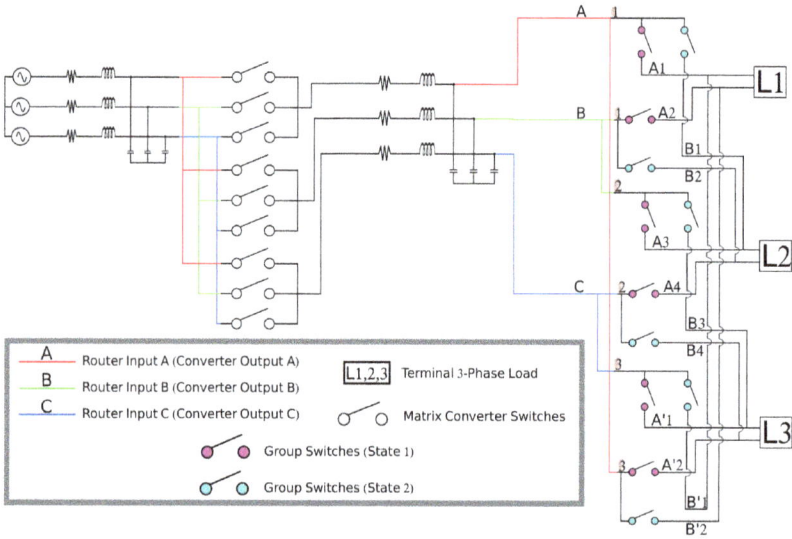

Figure 3. Simplified circuit topology of the direct matrix converter and the power router attached to its output. Both are three-phase input/output systems. The switch groups for state 1 and 2 of Table 1 are also obvious.

Table 1. State 1 is the steady state before switching. Each load is fed with a different phase-to-phase voltage after switching (State 2). AB, BC, and CA refer to the phase-to-phase voltages.

Load No.	State 1	State 2
1	AB	CA
2	BC	AB
3	CA	BC

118

The matrix converter utilizes the Venturini control modulation [32] described in [33]. This modulation creates an output voltage via the feedback of the input and target output voltages. As it was mentioned in the beginning, PET usually performs medium level voltage transformation into low level voltage (e.g., 6.6 kV to 100 V). In this paper, since the limitation of facilities at high voltage in laboratory, test level voltages are applied. As a result for the sake of equivalency in the experimental setup, the input voltage is set to 10 V and the voltage ratio of the matrix converter is set to 0.1 so that the output voltage of the converter is 1 V. Keeping the generality of discussion, let us consider the system base for both experiment and simulation results represented in per unit quantities. The LC filters are included in order to cut off any high frequency harmonics and are set as follows: L = 33 mH, C = 180 ⁻F, Load$_{1,2,3}$ = 15 Ω with the line resistances of a few Ohms for the sake of device protection. The switching frequency is set to 24.4 kHz and the pulses are digitally created utilizing a field programmable gate array (FPGA).

As it was mentioned before, smart power distribution, circuit switching, and constant power delivery to the desired loads are very challenging and important tasks. Ideally, all loads should have the capability of being fed with power from more than one input source. This basic switching scenario is also shown in Figure 3. The switches of the power router are separated into two groups. At first, the purple switches are on, and state 1 is set to the output. State 2 is set according to Table 1 during which the blue switches are on after the moment of switching. The loads are always fed with power but from different sources after the switching. The experimental prototype power router is shown in Figure 4. The line connections of the power cables are also schematically presented. Power router has four inputs and eight outputs and each input can be programmed to be outputted at one of its equivalent two outputs. Power line communication (PLC), which is the black box in the photograph, is used for controlling the power router through a personal computer. As Figure 4 shows, a power router input is considered to be a matrix converter output phase-to-phase voltage. When switching is performed such that a line is considered to fail, the equivalent phase-to-phase voltage is not reproduced at the output of the power router.

The power router specifications are a maximum operating voltage of 100–200 V$_{rms}$, continuous maximum power of 1 kW, and with Si-MOSFETs implemented as switches (450 V, 17 A). Increase of maximum power can be succeeded by implementing wide-bandgap SiC devices and such power routers are under development by our research group. Every ideal switch has to be implemented with the combination of two power MOSFETs. An insight figure of the power router is shown at Figure 5. How a single input phase can be transmitted to either of the two possible outputs is shown in the low part of the figure. Four bidirectional switches with their equivalent drive circuits are the main parts of the system. On the top, one of the bidirectional switches with its control circuit is graphically represented. It is obvious that the two MOSFETs have their sources connected so that only one drive circuit is sufficient. A photocoupler is the main responsible circuit for the bidirectional switch control. The combination of output ports among the different switch circuit units realises the circuit switching function of the power router. The design of the power router has scalability since the change of the number of switching circuit units does not affect the generality of the function. The following experimental results, although performed in test level voltages, can be scaled to higher voltage setup when the device ratings and power capacities are increased.

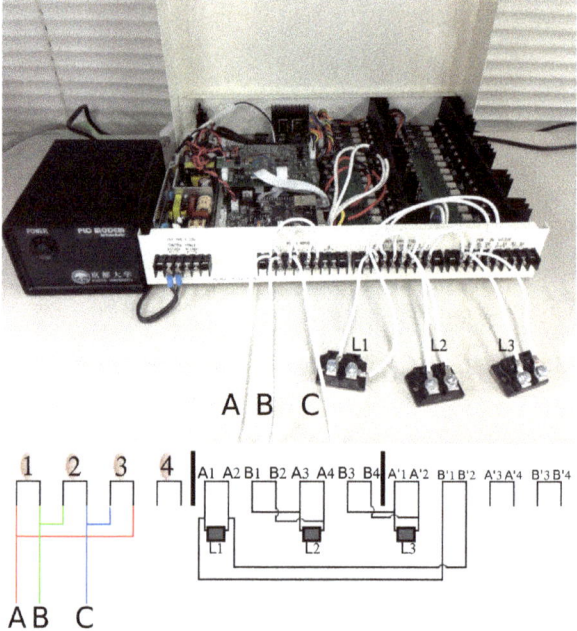

Figure 4. Experimental prototype power router and connection setup of input and output phases. This power router can be fed with up to four input sources and has eight outputs. For example, input 1 can be outputted at port A1-A2 or B1-B2 by the equivalent control. The last input of the power router is unconnected since three-phase switching is desired.

Although the isolation can be achieved in the PET topology, it can also be achieved in the topology of Figure 3 and especially of that of Figure 2. This ability is based on the power routers. The input of the system, *i.e.*, the matrix converter, can be isolated in case of an emergency if the equivalent power router disconnects from its loads. As a result, the operation of the input and the other power routers can remain safe.

120

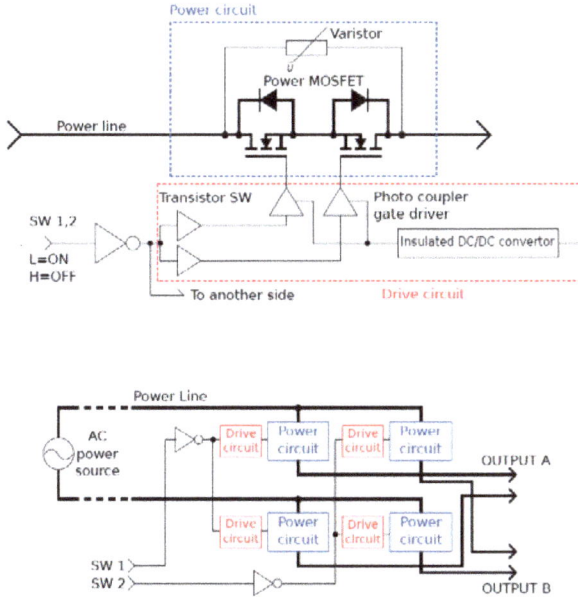

Figure 5. Power router switch layout and how a single input is transmitted to either of the two available outputs.

3. Results and Discussion

In the beginning, the scenario discussed in the previous section will be presented. It is a three-phase switching during which all the loads are connected to different inputs. In particular, Table 1 shows load 1 which is initially connected to phase AB, but after switching it is connected to phase CA. Load 2 switches from BC to AB, and load three switches from CA to BC. The experimental results are shown in Figure 6a. In the small circuit diagram, the colored lines indicate the line connections after switching and refer to the phase-to-phase voltages. The experimental result is also confirmed from the simulations in Figure 6b. If the practical system is considered (medium level voltage at 6.6 kV and low level voltage at 100 V) then 1 p.u. corresponds to 100 V. The simulations are generated from a Simulink Matlab model based on Figure 3.

Various switching scenarios were further tested. Instead of a phase-to-phase voltage from now on, we will just refer to phase A, B, and C for the sake of simplicity. For the case of a critical load in phase C (load 3), phases B and C are disconnected from the output representative of a system failure. In this case, switching between phases A and C should take place as shown in Figure 7a. The voltage has been maintained at output C as a result of the successful power routing. Verification of the experiment is shown in the simulation of Figure 7b. There is a small overvoltage in

121

the output of power router after the moment of switching, which is possibly due to a resonance between the power router and the output filter of the matrix converter. On the other hand, some line resistances have been installed in the experimental power router input for its protection and also for the suppression of such kind of undesired transients.

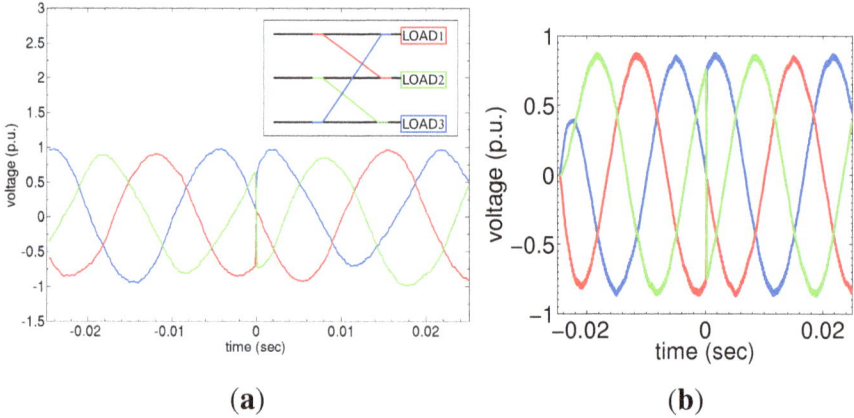

(a) (b)

Figure 6. (a) shows the experimental three-phase switching based on Table 1. The loads are fed with power but from different sources after power routing. (b) is the simulation verification. There is good correspondence between the two methods.

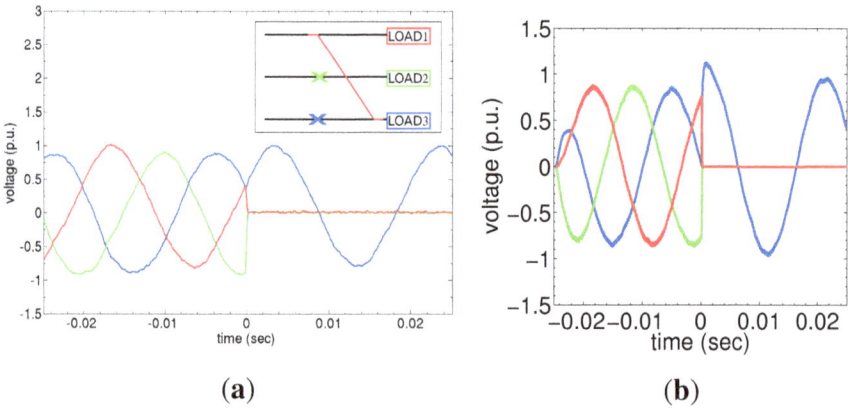

(a) (b)

Figure 7. (a) shows the experimental switching with one critical load and failure in 2 power router outputs. The critical load is successfully fed with power; (b) is the simulation verification. Small transients at the moment of switching caused by resonances between the power router and matrix converter output filter.

The final switching scenario is as shown in Figure 8a. In this case, phase B fails but load 2 is successfully fed with power due to switching with phase A. Phase C on the other hand feeds load 1 in phase A after power switching. In this case, load 3 has no power but the other two loads are successfully functioning. The transients are more obvious here, as shown in Figure 8b simulation results. Currents show similar behavior since we are experimenting with resistive loads.

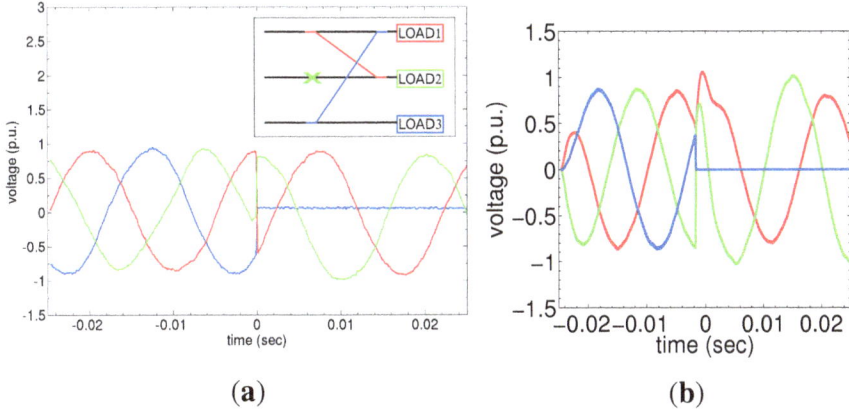

(a)   (b)

Figure 8. (a) shows the experimental power switching in a three-phase load while there is one failure in a power router output and two line switchings; (b) is the simulation verification. The transients become apparent at the moment of switching. Line resistances in the experimental power router inputs were implemented to ensure safe operation.

In general, power router realizes the switching between two input lines at the moment when the two powers of these lines are equal in order to achieve soft-switching and reduce the power losses. During the above experiments however, several switchings are simultaneously performed so that hard-switching was also implemented. The Simulink model implements hard-switching techniques in all cases.

Future studies should include the investigation of new control methods with faster and lower power loss switching, as well as power routing with both DC and AC sources. Investigation of unbalanced input sources is also of great importance as the majority of loads throughout the grid are unbalanced. The matrix converter creates an unbalanced output supply and power routing is performed as shown in Figure 9. There is also a characteristic delay between the switching due to the fact that the power router is unable to function properly for some msec due to the voltage imbalance (*i.e.*, power imbalance), but in the end it performs the necessary switchings. As it was mentioned before, power comparison decides the moment of switching. The unbalanced scenario creates unbalanced powers so as a result the comparison between them is a more complicated task for the control unit which is based on a more symmetrical scenario. We are currently working on such topics such as this delay reduction in order to reduce the switching losses. Load imbalance scenarios for power router are also investigated in [5]. In addition, the control modulation should cooperate with the power router so that matrix converter can create the desired voltage amplitude and phase correction according to the power need of each load. The main next topic is the inclusion of a PET in this system and the investigation of dynamics of the overall system. In detail, research remains to be done as far as the practical materials and engineering problems of PET are concerned including isolation, devices, and so on. These are the future problems as the history of conventional power apparatuses.

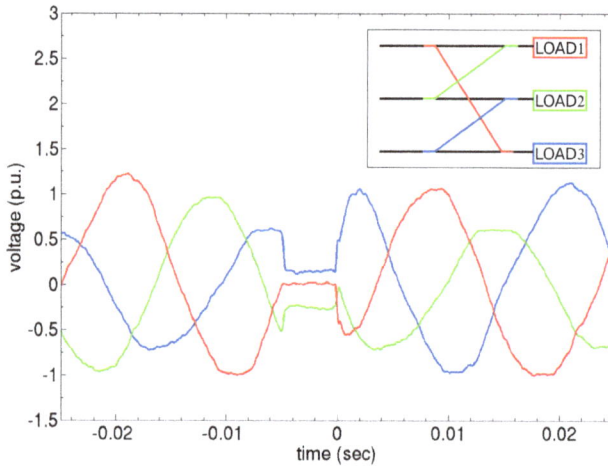

Figure 9. Experimental power routing for unbalanced sources. Power router can successfully make the switching but there is a characteristic delay between switching caused by this voltage imbalance.

4. Conclusions

This paper demonstrates a novel three-phase power routing. An experimental prototype power router was used with a direct AC/AC matrix converter. The overall system was tested under various switching scenarios and was successfully operated. Matrix converters and power routers will constitute the next generation of power-grids so that their investigation in a unified system is critical. The overall system successfully performed power routing but with some delay in the unbalanced sources scenario.

Acknowledgments: This work is partially supported by the Regional Innovation Cluster Program "Kyoto Environmental Nanotechnology Cluster", JST supported Super Cluster Project of Kyoto, Consultation of a High-Efficiency Energy Utilization System Realizing Clean and Low Environmental Load Society, and the Excellent Graduate Schools Program in Kyoto

University. Moreover this work was supported by Global Warming Mitigation Measure Technology and Demonstration Research Project: "Developing a Local-Exchange Energy System that Uses the Rail System to Create Heat and Power Source Independency for Existing Equipment" by the Japanese Ministry of Environment (fiscal years 2012–2014). The authors are grateful to the reviewers for their comments and careful reading of the manuscript. We would also like to acknowledge the assistance of ROHM Semiconductor Co. Ltd. for the prototype Silicon Carbide semiconductor providing. Alexandros Kordonis would also like to appreciate the help of Alexander Draude for his contribution of Figure 1.

Author Contributions: Alexandros Kordonis contributed to the simulation/experimental results and to the general organization of the article writing process. Ryo Takahashi and Daichi Nishihara contributed to the development of the power router and have the responsibilities on the programed operation. Takashi Hikihara has a responsibility on the project to develop PET and power router, including circuit design, based on the applications of SiC power devices. All of the authors contributed to the writing process of this research article.

Conflicts of Interest: The authors declare no conflict of interest.

References

1. Mei, S.; Cao, M.; Zhang, X. *Power Grid Complexity*; Springer: Berlin, Germany, 2011.

2. Hammons, T. Integrating renewable energy sources into european grids. *Int. J. Electr. Power Energy Syst.* **2008**, *30*, 462–475.

3. Takuno, T.; Koyama, M.; Hikihara, T. In-home power distribution systems by circuit switching and power packet dispatching. In Proceedings of the First IEEE International Conference Smart Grid Communications (SmartGridComm), Gaithersburg, MD, USA, 4–6 October 2010; pp. 427–430.

4. Takuno, T.; Kitamori, Y.; Takahashi, R.; Hikihara, T. Ac power routing system in home based on demand and supply utilizing distributed power sources. *Energies* **2011**, *4*, 717–726.

5. Takahashi, R.; Kitamori, Y.; Hikihara, T. Ac power local network with multiple power routers. *Energies* **2013**, *6*, 6293–6303.

6. Tashiro, K.; Takahashi, R.; Hikihara, T. Feasibility of power packet dispatching at in-home dc distribution network. In Proceedings of the IEEE Third International Conference Smart Grid Communications (SmartGridComm), Tainan, Taiwan, 5–8 November 2012; pp. 401–405.

7. He, M.M.; Reutzel, E.M.; Jiang, X.; Katz, R.H.; Sanders, S.R.; Culler, D.E.; Lutz, K. An architecture for local energy generation, distribution, and sharing. In Proceedings of the IEEE Conference Energy 2030, Atlanta, GA, USA, 17–18 November 2008; pp. 1–6.

8. Lu, G.; De, D.; Song, W.Z. Smartgridlab: A laboratory-based smart grid testbed. In Proceedings of the First IEEE International Conference Smart Grid Communications (SmartGridComm), Gaithersburg, MD, USA, 4–6 October 2010; pp. 143–148.

9. Szczesniak, P. *Three-Phase AC–AC Power Converters Based on Matrix Converter Topology*; Springer: London, UK, 2013.

10. Chang, Y.H. Design and analysis of power-cmos-gate-based switched-capacitor dc-dc converter with step-down and step-up modes. *Int. J. Circuit Theory Appl.* **2003**, *31*, 483–511.

11. Sumithira, T.; Kumar, A.N. An experimental investigation on off-grid solar photovoltaic power system using matrix converter. *J. Sci. Ind. Res.* **2014**, *73*, 124–128.

12. Cárdenas, R.; Peña, R.; Tobar, G.; Clare, J.; Wheeler, P.; Asher, G. Stability analysis of a wind energy conversion system based on a doubly fed induction generator fed by a matrix converter. *IEEE Trans. Ind. Electr.* **2009**, *56*, 4194–4206.

13. Shah, J.; Gupta, R.K.; Mohapatra, K.K.; Mohan, N. Power management with a dynamic power limit by a power electronic transformer for micro-grid. In Proceedings of the Power and Energy Society General Meeting, Minneapolis, MN, USA, 25–29 July 2010; pp. 1–5.

14. Mohapatra, K.K.; Gupta, R.; Thuta, S.; Somani, A.; Umarikar, A.; Basu, K.; Mohan, N. New research on ac–ac converters without intermediate storage and their applications in power-electronic transformers and ac drives. *IEEJ Trans. Electr. Electron. Eng.* **2009**, *4*, 591–601.

15. Kordonis, A.; Hikihara, T. Dynamic model of direct matrix converter and its experimental validation. *Int. J. Circuit Theory Appl.* **2015**, doi:10.1002/cta.2072.

16. Kordonis, A.; Takahashi, R.; Hikihara, T. Ac/Ac converter towards power routing systems in smart-grids: Advantage on operation by nonlinear dynamics. In Proceedings of the IEEE 2nd Global Conference Consumer Electronics (GCCE), Tolyo, Japan, 1–4 October 2013; pp. 158–159.

17. Manjrekar, M.D.; Kieferndorf, R.; Venkataramanan, G. Power electronic transformers for utility applications. In Proceedings of the Industry Applications Conference, Rome, Italy, 8–12 October 2000; Volume 4, pp. 2496–2502.

18. Ronan, E.R.; Sudhoff, S.D.; Glover, S.F.; Galloway, D.L. A power electronic-based distribution transformer. *IEEE Trans. Power Deliv.* **2002**, *17*, 537–543.

19. Amin, S.M.; Wollenberg, B.F. Toward a smart grid: Power delivery for the 21st Century. *IEEE Power Energy Mag.* **2005**, *3*, 34–41.

20. Garces, A.; Molinas, M. A study of efficiency in a reduced matrix converter for offshore wind farms. *IEEE Trans. Ind. Electron.* **2012**, *59*, 184–193.

21. Mohapatra, K.K.; Mohan, N. Matrix converter fed open-ended power electronic transformer for power system application. In Proceedings of the Power and Energy Society General Meeting—Conversion and Delivery of Electrical Energy in the 21st Century, Pittsburgh, PA, USA, 20–24 July 2008; pp. 1–6.

22. Shah, J.K. Dynamic Power Flow Control for a Smart Micro-grid by a Power Electronic Transformer. Ph.D. Thesis, University of Minnesota, Minneapolis, MN, USA, 2011.

23. Erickson, R.W.; Dragan, M. *Fundamentals of Power Electronics*; Springer: New York, NY, USA, 2011.

24. Iman-Eini, H.; Schanen, J.; Farhangi, S.; Barbaroux, J.; Keradec, J. A power electronic based transformer for feeding sensitive loads. In Proceedings of the Power Electronics Specialists Conference, PESC 2008 IEEE, Rhodes, Greece, 15–19 June 2008; pp. 2549–2555.

25. Basu, K.; Gupta, R.K.; Nath, S.; Castelino, G.F.; Mohapatra, K.K.; Mohan, N. Research in matrix-converter based three-phase power-electronic transformers. In Proceedings of the IEEE International Conference Power Electronics Conference (IPEC), Sapporo, Japan, 21–24 June 2010; pp. 2799–2803.

26. Sabahi, M.; Goharrizi, A.Y.; Hosseini, S.H.; Sharifian, M.B.B.; Gharehpetian, G.B. Flexible power electronic transformer. *IEEE Trans. Power Electron.* **2010**, *25*, 2159–2169.

27. Kang, M.; Enjeti, P.N.; Pitel, I.J. Analysis and design of electronic transformers for electric power distribution system. *IEEE Trans. Power Electron.* **1999**, *14*, 1133–1141.

28. Metcalfe. B. Presented at Singularity University, Moffett Field, CA, USA, July 2009.

29. Patterson, B.T. DC, come home: DC microgrids and the birth of the "Enernet". *IEEE Power Energy Mag.* **2012**, *10*, 60–69.

30. Ricciardi, S.; Santos-Boada, G.; Klinkowski, M.; Careglio, D.; Palmieri, F. *Towards Service Orchestration between Smart Grids and Telecom Networks*; Springer: Berlin, Germany, 2013.

31. Watson, A.J.; Dang, H.Q.S.; Wheeler, P.W.; Clare, J.C.; Mondal, G.; Rufer, A.R.; Kenzelmann, S.; De Novaes, Y. A novel multilevel converter structure integrated into power systems and its performance evaluation. In Proceedings of the 13th European Conference on Power Electronics and Applications, Barcelona, Spain, 8–10 September 2009; pp. 1–10.

32. Venturini, M.; Alesina, A. The generalised transformer-a new bidirectional sinusoidal waveform frequency converter with continuously adjustable input power factor. In Proceedings of the PESC'80, Power Electronics Specialists Conference, Atlanta, GA, USA, 16–20 June 1980; pp. 242–252.

33. Wheeler, P.W.; Rodriguez, J.; Clare, J.C.; Empringham, L.; Weinstein, A. Matrix converters: A technology review. *IEEE Trans. Ind. Electron.* **2002**, *49*, 276–288.

Estimation of State of Charge for Two Types of Lithium-Ion Batteries by Nonlinear Predictive Filter for Electric Vehicles

Yin Hua, Min Xu, Mian Li, Chengbin Ma and Chen Zhao

Abstract: Estimation of state of charge (SOC) is of great importance for lithium-ion (Li-ion) batteries used in electric vehicles. This paper presents a state of charge estimation method using nonlinear predictive filter (NPF) and evaluates the proposed method on the lithium-ion batteries with different chemistries. Contrary to most conventional filters which usually assume a zero mean white Gaussian process noise, the advantage of NPF is that the process noise in NPF is treated as an unknown model error and determined as a part of the solution without any prior assumption, and it can take any statistical distribution form, which improves the estimation accuracy. In consideration of the model accuracy and computational complexity, a first-order equivalent circuit model is applied to characterize the battery behavior. The experimental test is conducted on the $LiCoO_2$ and $LiFePO_4$ battery cells to validate the proposed method. The results show that the NPF method is able to accurately estimate the battery SOC and has good robust performance to the different initial states for both cells. Furthermore, the comparison study between NPF and well-established extended Kalman filter for battery SOC estimation indicates that the proposed NPF method has better estimation accuracy and converges faster.

Reprinted from *Energies*. Cite as: Hua, Y.; Xu, M.; Li, M.; Ma, C.; Zhao, C. Estimation of State of Charge for Two Types of Lithium-Ion Batteries by Nonlinear Predictive Filter for Electric Vehicles. *Energies* **2015**, *8*, 3556–3577.

1. Introduction

Global warming, the petroleum crisis, and legislation pushing for higher fuel economy and lower emissions, are leading to the development of electric vehicles (EVs) [1,2]. As the key component of any electric vehicle, the energy storage system attracts more and more attention. A variety of electrochemical energy storage devices are currently used in EV applications, such as lithium-ion (Li-ion) battery, nickel metal hydride (NiMH) battery, lead acid (LA) battery, and ultracapacitor (UC). Among them, Li-ion batteries are viewed as the most promising energy storage units for EVs, for its high energy density, high power density, low self-discharging rate, and long lifespan [3,4].

However, strict requirements should be satisfied when using Li-ion batteries, and a battery management system (BMS) is required to provide the functions

128

of monitoring, estimation, and protection to ensure the safe operations of Li-ion batteries. The state of charge (SOC), acting the similar role as the fuel meter for the internal combustion engine system, is the most important factor for batteries which should be accurately estimated by the BMS. The battery SOC indicates the residual capacity of the battery system and has significant importance in predicting the remaining driving range of EVs. Besides, accurate SOC estimation can also prevent the batteries from over-charging and over-discharging conditions and thus can extend the battery cycle life [5]. However, since the battery SOC cannot be directly measured and it is affected by many factors, such as current, temperature and battery age, estimation of the battery SOC is still a challenging problem that needs to be solved.

A number of SOC estimation methods have been proposed. Each method has its own advantages and limitations. Generally, these methods can be mainly classified into two kinds: (1) direct measurement based estimation and (2) model based estimation. The first kind of method directly uses the measurements from battery system to calculate the SOC, such as current integration method [6], open circuit voltage (OCV) based method [7]. The current integration method is easy to implement with low computation, but it suffers from the low estimation accuracy due to the accumulative errors caused by current sensor noises. In addition, it is also difficult to obtain the initial SOC when using the current integration method. Therefore, the open circuit voltage method is usually used complementarily with the current integration method to recalibrate the SOC and to provide the initial SOC. However, a long rest time of the tested battery is required to reach the open circuit voltage, which is usually unrealistic for real world applications.

In the second kind of method, the battery model is utilized when estimating the battery SOC. One of the model based methods for SOC estimation is based on the black-box battery models, such as neural networks (NN) [8], fuzzy logic (FL) [9], and support vector machine (SVM) [10]. Eddahech *et al.* [8] developed a recurrent neural network as a SOC predictor that takes into account operational conditions, the results show that the predictor allows very precise SOC estimation. Salkind *et al.* [9] utilized the fuzzy logic to estimate the battery SOC by using the training datasets obtained by impedance spectroscopy and coulomb counting techniques. Anton *et al.* introduced a support vector machine based SOC estimator for a high-capacity lithium iron manganese phosphate (LiFeMnPO$_4$) battery cell, using cell current, cell voltage, and cell temperature as independent variables. The results show that the SVM SOC estimator maintains a high level of accuracy. According to the literatures, the black-box model based methods can be quite accurate if sufficient experimental data is used to train the model. However, their performance highly depends on the quantity and quality of the training data set, a large amount of offline battery tests are necessary to obtain a good model which can be very time-consuming. Optimum state filtering method is another kind of model based method for battery SOC estimation. This method usually performs SOC estimation based on an equivalent circuit battery model [11,12]. Many different state filtering methods have been investigated, such as extended Kalman filter (EKF) [13–15], sigma point Kalman filter (SPKF) [16–18], adaptive extended Kalman filter (AEKF) [19], adaptive unscented Kalman filter (AUKF) [20], particle filter (PF) [21] and others [22–26]. Plett [13–17] established the EKF and UKF based SOC estimation methods using different orders of equivalent circuit battery models for simultaneous state and parameters estimation of LiPB packs. Both of them obtained very good results, and the methods were robust to different initial states. Han *et al.* adopt the AEKF method for SOC estimation by adaptively updating the process and measurement noise covariance which improved the estimation accuracy. Similarly, Hu *et al.* applied the AUKF for SOC estimation. All of the Kalman filter (KF) based methods achieve very good estimation performance because of online state error correction capability. However, there are some shortcomings for KF based methods. For instance, the statistic distribution of the process noise is assumed to be the zero mean white Gaussian process and prior knowledge of the noise covariance should be known before estimating the battery SOC. In fact, it is difficult to obtain the accurate information of the process noises for real world applications, and the filter performance will decrease or even diverge with inaccurate noise information. Besides, the assumption of the zero-mean Gaussian process noises usually cannot be met in practice which can decrease the estimation accuracy [27]. PF is another optimum state filtering methodology for SOC estimation [21]. It is able to represent any probability density function for the state by using Monte Carlo sampling methods which improves the estimation accuracy. However, the computational effort is high for PF due to the large amount of particles which makes it difficult to apply for real-world applications.

In this paper, the nonlinear predictive filter (NPF) is proposed to estimate the SOC of Li-ion batteries and the proposed method is evaluated on the battery cells with different chemistries. The NPF method was firstly proposed by Crassidis *et al.*, for spacecraft attitude estimation and obtained satisfactory results [28]. It is a nonlinear optimum state estimation method implemented on continuous-discrete time systems, which is particularly suitable for battery systems [29]. In the NPF method, the nonlinear dynamic system is treated as a preliminary model with a to-be-determined model error part, where the preliminary model describes the system dynamic and the model error mainly represents the system process noise. The significant advantage of NPF method is that the process noise is treated as an unknown model error determined as a part of the solution, and it is able to represent any distribution form. Compared to the aforementioned direct measurement based estimation method, the NPF method can provide more accurate results with better robust performance of initial values due to its capability of correcting the state error online. In comparison with black-box model based method, the large amount of training data sets is not required for NPF method which saves the computational effort. The KF based method usually assumes a zero mean white Gaussian process noise and requires the prior knowledge. In contrary, the process noise in NPF can be any type without any restricting pre-assumptions. Therefore, the NPF based SOC estimation method has better estimation performance and is more practical for real-world applications. Compared to the PF method, the NPF method has lower computation, since a large number of particles needs to be used and numerous matrix operations are required for PF method which greatly increase the requirements for hardware system. Additionally, the aforementioned literatures mainly perform their methods on one type of batteries; the robustness performance of different battery types is not discussed. In this study, the NPF method is performed on $LiCoO_2$ (LCO) and $LiFePO_4$ (LFP) battery cells, the experimental results show that the NPF based method can accurately estimate the battery SOC with good robustness to different initial values. Meanwhile, the estimation result of LFP battery suffers from a lower accuracy than that of LCO battery due to the characteristics of the flat open circuit voltage. Furthermore, the comparison study between NPF and EKF with the same experimental conditions indicates that the proposed NPF method has better estimation accuracy and faster convergence rate.

The rest of the paper is organized as follows. In Section 2, a first-order equivalent circuit battery model is introduced. The description for SOC estimation of the Li-ion battery cell using NPF is presented in Section 3. In Section 4, the estimation results are analyzed and discussed to verify the proposed method. Finally, the conclusion is provided in Section 5.

2. Battery Model

2.1. Model Structure

In order to apply NPF for SOC estimation, a suitable battery model is required to characterize the electrochemical properties of Li-ion batteries, including: ohmic resistance, charge transfer and diffusion. Different types of battery models have been proposed in the literature. Among them, the equivalent circuit models (ECMs) are the most commonly used ones for battery state estimation. The ECMs capture the battery input-output dynamics through electrical circuit elements, such as resistor, capacitor and voltage source, and can be easily used for model based estimation. Hu *et al.* introduced a comprehensive study for different types of ECMs, and their results indicate that the first-order ECM achieves an excellent compromise between accuracy and complexity [30]. Therefore, as shown in Figure 1, a first-order ECM composed of an open circuit voltage (OCV) source, a resistor, and an RC network, is used in this study. The resistor represents the electrical resistance of battery components with the accumulation and dissipation of charge in the electrical double-layer. The RC network describes the charge transfer and diffusion effect, and the voltage source indicates the battery's open circuit voltage which is a function of the battery SOC.

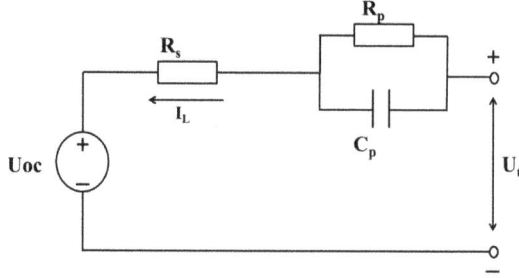

Figure 1. Schematic diagram of the first-order battery model.

The SOC definition for the Li-ion battery is formulated as:

$$SOC(t) = SOC(0) + \int_0^t \frac{\eta I_L(t)dt}{C_n} \tag{1}$$

where $SOC(0)$ is the initial SOC value, $SOC(t)$ is the battery SOC at time t, η is the Coulombic efficiency (in this paper η is assumed as 1), I_L is the input current (positive for charge, negative for discharge) and C_n is the nominal capacity.

Equations (2) and (3) describe the electrical behavior of the Li-ion battery:

$$\dot{U}_p(t) = -(R_pC_p)^{-1}U_p(t) + C_p^{-1}I_L(t) \tag{2}$$

$$U_t(t) = U_{oc}(SOC(t)) + U_p + I_L(t)R_s \tag{3}$$

where U_p is the polarization voltage, R_p and C_p are the polarization resistance and capacitance, respectively, U_t is the battery terminal voltage, and R_s is the ohmic resistance.

A spline function is employed to describe the relationship between the battery open circuit voltage U_{oc} and battery SOC, given by Equation (4):

$$U_{oc}(t) = Spline(SOC(t)) \tag{4}$$

2.2. Model Parameter Identification

2.2.1. Experimental Setup

In this work, two different types of battery cells, LiCoO$_2$ (LCO) and LiFePO$_4$ (LFP), are tested. The standard specifications of the tested cells are listed in Tables 1 and 2. The schematic diagram of the battery test bench is shown in Figure 2. It consists of a Takasago ZX-800LA electric power, a Kikusui PLZ150U electric load, a NI cDAQ-9174 data acquisition system, a host PC, and a thermal chamber. The ZX-800LA electric power can charge the battery cell with the maximum current of 80A at the maximum voltage of 84V, while the PLZ150U electric load is able to provide the maximum discharge current of 30A with the maximum voltage of 150V. The electric power and load are remotely controlled by the host PC to determine the charge/discharge power load of the tested battery cell. The data acquisition system with a sampling rate of 10 Hz is used to capture the current, voltage, and temperature of the tested battery and to transfer the obtained data to the host PC.

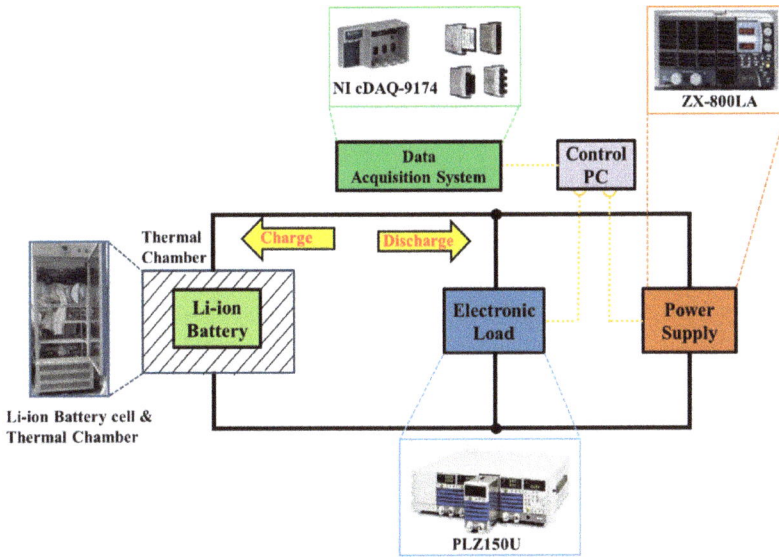

Figure 2. Schematic of battery test bench.

Table 1. Battery cell (LCO) specification.

Item	Specification
Cell Dimensions (mm)	Ø 18 × 69
Cell Weight (g)	48.2
Cell Capacity (nominal, Ah)	2.6
Cell Voltage (nominal, V)	3.7
Gravimetric Energy Density (nominal, Wh/kg)	180
Volumetric Energy Density (nominal, Wh/L)	464
Operating Temperature	$-20\,^{\circ}$C to $60\,^{\circ}$C

Table 2. Battery cell (LFP) specification.

Item	Specification
Cell Dimensions (mm)	Ø 32 × 113
Cell Weight (g)	205
Cell Capacity (nominal, Ah)	4.5
Cell Voltage (nominal, V)	3.3
Gravimetric Energy Density (nominal, Wh/kg)	71
Volumetric Energy Density (nominal, Wh/L)	161
Operating Temperature	−30 °C to 55 °C

2.2.2. Parameter Identification

For the first-order battery model, the values of the model parameters as well as the SOC-OCV relationship need to be identified. A series of battery tests, including: capacity test, pulse current test, and open circuit voltage test, are conducted to extract these parameters. The content of these tests are described as follows:

(1) Capacity test: The capacity test discharges the battery cell from the fully charged state (upper-limit voltage) to the fully discharged state (lower-limit voltage) with 0.5 C rate, and the cell capacity is referred as the total Ampere-hours drained out of the battery during the test. The cut-off voltages used during the test for LCO battery are V_{max} = 4.2 V, V_{min} = 2.8 V, and the cut-off voltages for LFP battery are V_{max} = 3.6 V, V_{min} = 2 V. The experimental results of the capacities for the tested LCO and LFP cells are 2.62 Ah and 4.29 Ah, respectively.

(2) Pulse current test: To identify the values of the electrical circuit elements in the first-order ECM, a pulse current test is conducted on the battery cells at 10% SOC intervals starting from 0.9 to 0.3. During the test, the environment temperature is controlled at 25 °C. The detailed test procedure can be found in [31]. In this study, the time period between two current pluses, when no current is applied, is used for parameters identification. The current and voltage profiles during this time period are shown in Figure 3. The ohmic resistance R_s can be expressed as:

$$R_s = \frac{U_s}{I_L} \tag{5}$$

where U_s is the instantaneous voltage response within one second, and I_L is the current before the rest time period. The polarization voltage U_t at time t can be expressed by:

$$U_t = U_p(1 - e^{-\frac{t}{R_p C_p}}), U_p = R_p I_L \tag{6}$$

134

where U_p is the maximum polarization voltage during the rest time period. The polarization resistance and capacitance can be identified by minimizing the difference between the model output and voltage measurement using the nonlinear least square method. The identification results of R_s, R_p, C_p for LCO and LFP cells are shown in Figure 4. It can be seen that the parameters vary at different battery SOC points. However, to reduce the complexity of the battery model, R_s, R_p, C_p at different SOC points are averaged to obtain the final model parameters and the results are shown in Table 3.

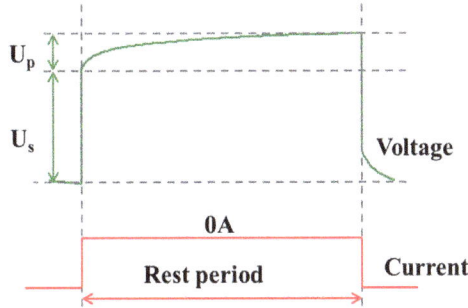

Figure 3. Voltage profile during the rest time period.

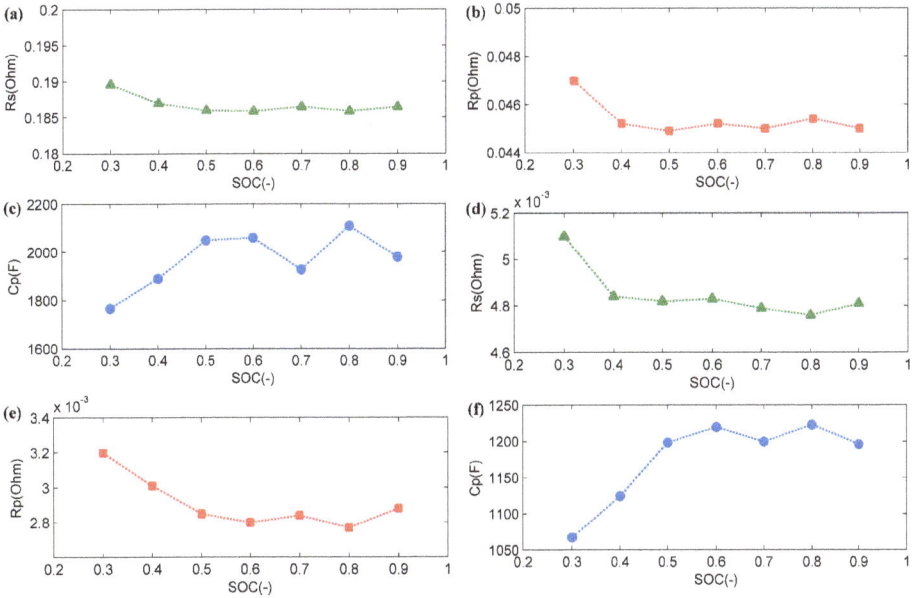

Figure 4. Parameters identification results: (**a**) R_s results of LiCoO$_2$ (LCO) cell; (**b**) R_p results of LCO cell; (**c**) C_p results of LCO cells; (**d**) R_s results of LiFePO$_4$ (LFP) cell; (**e**) R_p results of LFP cell; (**f**) C_p results of LFP cells.

135

(3) Open circuit voltage test: To calibrate the nonlinear SOC-OCV relationship, an open circuit voltage test is conducted as follows. The battery cell is discharged using 0.5 C constant current at 5% SOC interval from 100% SOC to 15% SOC. After each discharge period, the battery cell is rested for 3 hours to reach the close-to-equilibrium open-circuit potential for each SOC point. A similar procedure is conducted to get the SOC-OCV curve under the battery charge condition. Since the possible hysteresis voltage is neglected in this paper, the SOC-OCV relationship for the battery model is defined as the average of the equilibrium potentials of charging and discharging. The experimental results of the SOC-OCV curves for LCO and LFP battery cells are shown in Figure 5.

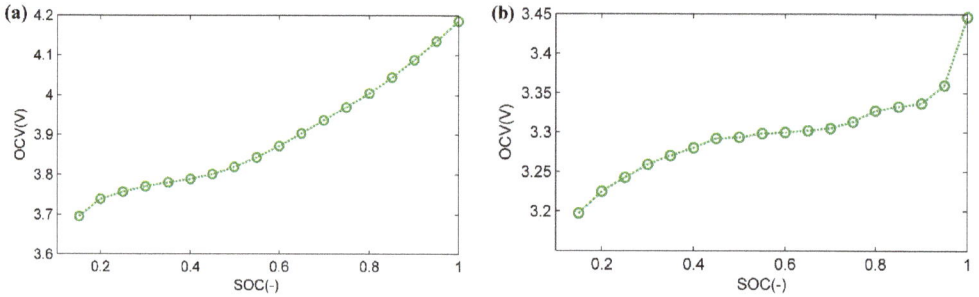

Figure 5. SOC-OCV curves: (**a**) LCO battery cell; (**b**) LFP battery cell.

Table 3. Model parameters.

LCO Battery Cell				
Model parameters	$R_s(\Omega)$	$R_p(\Omega)$	$C_p(F)$	$C_n(Ah)$
Values	0.187	0.046	1969	2.62
LFP battery cell				
Model parameters	$R_s(\Omega)$	$R_p(\Omega)$	$C_p(F)$	$C_n(Ah)$
Values	0.0048	0.0029	1186	4.29

2.3. Model Validation

In order to validate the battery models with identified parameters for LCO and LFP battery cells, the experimental test using Urban Dynamometer Driving Schedule (UDDS) driving cycle is conducted. UDDS is usually used for light duty vehicle testing under the city driving condition [32]. Therefore, the UDDS driving cycle is adopted in this study to simulate the battery dynamics under a realistic EV scenario. The model validation results for LCO and LFP battery cells are shown in Figures 6 and 7. Figure 6a,b shows the comparison profiles of the estimated terminal

136

voltage and measured terminal voltage for LCO battery cell. Figure 6c shows the corresponding voltage error and it can be seen that the maximum model error is around 0.02 V. Figure 6d shows the root mean squared error (RMSE) of the terminal voltage, which is less than 0.007 V. Similar results for LFP battery cell are shown in Figure 7. The maximum model error is within 0.01 V for LFP battery cell, while the RMSE is less than 0.004 V. According to the results, it can be concluded that the first order battery model is able to accurately capture the dynamic behavior of both Li-ion battery cells.

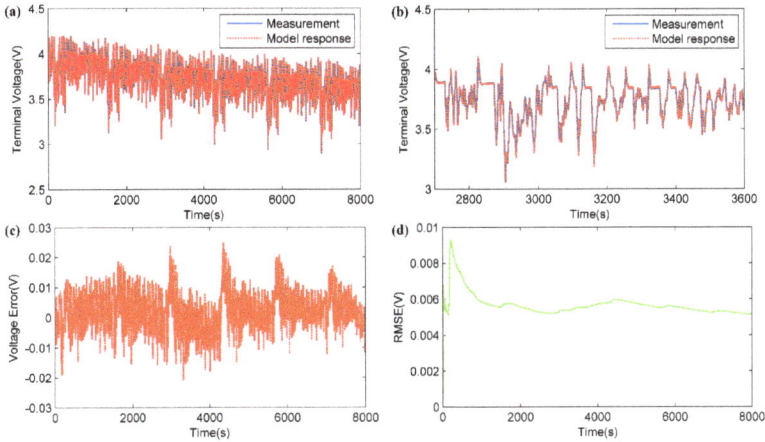

Figure 6. Model validation results of LCO cell: (**a**) Comparison of model output voltage and measured voltage; (**b**) Zoom plot; (**c**) Voltage error; (**d**) RMSE of voltage error.

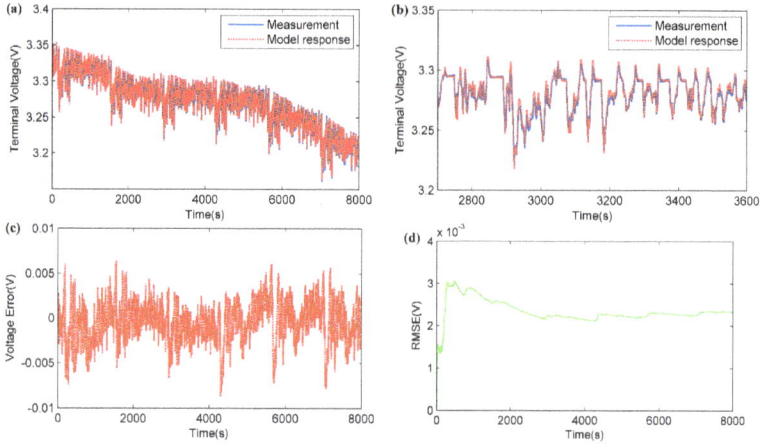

Figure 7. Model validation results of LFP cell: (**a**) Comparison of model output voltage and measured voltage; (**b**) Zoom plot; (**c**) Voltage error; (**d**) RMSE of voltage error.

3. Nonlinear Predictive Filter for SOC Estimation

In this section, the description of nonlinear predictive filter (NPF) is firstly introduced. Then, the NPF based SOC estimation using the first-order battery model is presented.

3.1. Nonlinear Predictive Filter

Nonlinear predictive filter (NPF) is a model based state estimation method implemented with nonlinear continuous-discrete time system. The state and measurement equations in NPF are given by Equations (7) and (8):

$$\dot{x}(t) = f[x(t), t] + g(t)d(t) \tag{7}$$

$$y(t_k) = h[x(t_k), t_k] + v(t_k) \tag{8}$$

where $x(t)$ is the state vector which needs to be estimated, f is system dynamic function, $d(t)$ is the model error which mainly represents the process noise and $g(t)$ is the model error distribution matrix. $y(t_k)$ is the system output sampled at time step t_k, h is the system measurement function, v is the measurement noise which is assumed to be an independent zero mean Gaussian white noise with:

$$E[v(t_k)] = 0, E[v(t_k)v(t_{k'})^T] = R\delta_{kk'} \tag{9}$$

where R is a positive-definite covariance matrix.

138

In order to estimate the system states, the model error $d(t)$ needs be obtained at every time step. A cost function consisting of the weighted sum square of the measurement-minus-estimate residuals plus the weighted sum square of the model correction term is established in order to get the mathematical expression of $d(t)$, as defined in Equaton (10):

$$J[d(t)] = 0.5[y(t+\Delta t) - \hat{y}(t+\Delta t)]^T R^{-1}[y(t+\Delta t) - \hat{y}(t+\Delta t)] + 0.5d(t)^T W d(t) \quad (10)$$

where $y(t) = y(t_k), y(t+\Delta t) = y(t_{k+1})$, Δt is the sampling interval, W is a positive semi-definite weighting matrix. A first-order Taylor expansion is used to approximate $\hat{y}(t+\Delta t)$, given in Equation (11):

$$\hat{y}(t+\Delta t) = \hat{y}(t) + \Delta t \dot{\hat{y}}(t) + 0.5\Delta t^2 \ddot{\hat{y}}(t) + \cdots \approx \hat{y}(t) + Z(\hat{x}(t), \Delta t) + \Lambda(\Delta t) S(\hat{x}(t)) d(t) \quad (11)$$

The solution of $d(t)$ is derived by minimizing the cost function $J[d(t)]$, given in Equation (12):

$$d(t) = -\left\{[\Lambda(\Delta t)S(\hat{x}(t))]^T R^{-1}\Lambda(\Delta t)S(\hat{x}(t)) + W\right\}^{-1} \\ \times [\Lambda(\Delta t)S(\hat{x}(t))]^T R^{-1}[Z(\hat{x}(t), \Delta t) + \hat{y}(t) - y(t+\Delta t)] \quad (12)$$

where $S(\hat{x}(t_k)), Z(\hat{x}(t_k), \Delta t), \Lambda(\Delta t)$ are intermediate matrices, given as follows:

$$S(\hat{x}(t)) = \begin{bmatrix} L_{g_1}L_f^{r_1-1}h_1(\hat{x}(t)) & \cdots & L_{g_l}L_f^{r_1-1}h_1(\hat{x}(t)) \\ \vdots & \ddots & \vdots \\ L_{g_1}L_f^{r_m-1}h_m(\hat{x}(t)) & \cdots & L_{g_l}L_f^{r_m-1}h_m(\hat{x}(t)) \end{bmatrix} \quad (13)$$

$$\Lambda(\Delta t) = \begin{bmatrix} \lambda_{11} & \cdots & 0 \\ \vdots & \lambda_{ii} & \vdots \\ 0 & \cdots & \lambda_{mm} \end{bmatrix}, \lambda_{ii} = \frac{\Delta t^{r_i}}{r_i!}, i = 1, 2, \cdots m \quad (14)$$

$$Z_i(\hat{x}(t), \Delta t) = \sum_{a=1}^{r_i} \frac{\Delta t^a}{a!} L_f^a h_i(\hat{x}(t)), i = 1, 2, \cdots, m \quad (15)$$

Here, $L_{g_j}L_f^{r_i-1}h_i(\hat{x}(t))$ is the Lie derivation, defined as:

$$L_f^0 h_i(\hat{x}(t)) = h_i(\hat{x}(t)) \quad (16)$$

$$L_f^n h_i(\hat{x}(t)) = \frac{\partial L_f^{n-1}h_i(\hat{x}(t))}{\partial \hat{x}} f[\hat{x}(t), t] \quad (17)$$

$$L_{g_j}L_f^n h_i(\hat{x}(t)) = \frac{\partial L_f^n h_i(\hat{x}(t))}{\partial \hat{x}} g_j(t) \quad (18)$$

139

The index r_i is the relative degree, which satisfies the following two constrained equations:

$$L_{g_j} L_f^{n_i} h_i(\hat{x}(t)) = 0 (i = 1, 2, \cdots m; j = 1, 2, \cdots, l, n_i < r_i - 1) \tag{19}$$

$$L_{g_j} L_f^{r_i-1} h_i(\hat{x}(t)) \neq 0 (i = 1, 2, \cdots m; j = 1, 2, \cdots, l) \tag{20}$$

Therefore, based on the measurement processed at time t_{k+1}, the new $d(t)$ in $[t_k : t_{k+1}]$ can be found. After that, the state estimates are propagated to time t_{k+1}.

The weighting matrix W in Equation (10) is derived as the inverse of the model error's covariance matrix D, given as follows:

$$W = D^{-1} = Cov[d(t)]^{-1} = E[(d(t) - E[d(t)])(d(t) - E[d(t)])^T]^{-1} \tag{21}$$

Assume that $d(t)$ is a stationary ergodic random process, then the covariance matrix D of the model error can be iteratively derived with a certain time interval; the steps of calculating W are listed as follows:

Step.1: Initialization: $W = E[(d(t_0) - E[d(t_0)])(d(t_0) - E[d(t_0)])^T]^{-1}$, for the total time length L, the time interval for updating W is defined as $l = L/r$, where r is the total iterations.

Step.2: For $t \in [t_{(k-1)l+1} : t_{kl}]$: estimate the model error $d(t)$ for $[t_{(k-1)l+1} : t_{kl}]$, and obtain the sequence of model error $[d(t_{(k-1)l+1}), \ldots, d(t_{kl})]$.

Step.3: For $t = t_{kl}$, update weighting matrix: $W = D_k^{-1}$, D_k is the covariance for $[d(t_{(k-1)l+1}), \ldots, d(t_{kl})]$.

Step.4: If $k < L/l$, return to step.2.

Based on the definitions and descriptions above, the workflow of NPF can be summarized as follows:

Step.1: Initialization: for $k = 0$:

(a) Set initial values: $\hat{x}(t_0) = E[x(t_0)]$, $W = E[(d(t_0) - E[d(t_0)])(d(t_0) - E[d(t_0)])^T]^{-1}$;

(b) Set weighting matrix update time interval: $l = L/r$.

Step.2: For every time step $k = 1, 2, \cdots$, doing the following:

(a) Estimate system output: $\hat{y}(t_k) = h[\hat{x}(t_k)]$;

(b) Calculate the intermediate parameter matrices $S(\hat{x}(t_k)), Z(\hat{x}(t_k), \Delta t), \Lambda(\Delta t)$;

(c) Estimate model error: $d(t_k) = - \left\{ [\Lambda(\Delta t)S(\hat{x}(t_k))]^T R^{-1} \Lambda(\Delta t)S(\hat{x}(t_k)) + W \right\}^{-1}$ $\times [\Lambda(\Delta t)S(\hat{x}(t_k))]^T R^{-1} [Z(\hat{x}(t_k), \Delta t) + \hat{y}(t_k) - y(t_{k+1})]$;

(d) Update state estimation from $\hat{x}(t_k)$ to $\hat{x}(t_{k+1})$ using discretized state equation: $\dot{\hat{x}}(t) = f[\hat{x}(t), t] + g(t)d(t)$.

Step.3: For $k = nl, 1 \leqslant n \leqslant r$

(a) Calculate covariance for $[d(t_{(k-1)l+1}), \ldots, d(t_{kl})]$: $D_k = \text{Cov}[d(t)]$, $t \in [t_{(k-1)l+1} : t_{kl}]$

(b) Update weighting matrix: $W = D_k^{-1}$

It can be seen that the process noise is determined online as a part of the solution without any prior assumption. As a result, the NPF method is robust to the process noise with any statistic distribution.

3.2. NPF Based SOC Estimation

In order to apply the NPF method to estimate the battery SOC, a continuous-discrete time model for the battery system is needed. According to the description of the equivalent circuit battery model in Section 2.1, the battery model in the continuous-discrete time form is formulated as:

$$\dot{x}(t) = \begin{pmatrix} \dot{SOC}(t) \\ \dot{U}_p(t) \end{pmatrix} = \begin{pmatrix} 0 \cdot SOC(t) + \frac{I_L(t)}{C_n} \\ -(R_p C_p)^{-1} U_p(t) + C_p^{-1} I_L(t) \end{pmatrix} + g \cdot \begin{pmatrix} d_1(t) \\ d_2(t) \end{pmatrix} \tag{22}$$

$$U_t(t_k) = U_{oc}[SOC(t_k)] + U_p(t_k) + I_L(t_k) R_s + v(t_k) \tag{23}$$

where the state vector $x(t)$ consists the components of $SOC(t)$ and $U_p(t)$, the model error vector $d(t)$ consists the components of $d_1(t)$ and $d_2(t)$ representing the errors of $SOC(t)$ and $U_p(t)$ respectively, and the battery's terminal voltage $U_t(t_k)$ is the system output sampled at t_k. An identity matrix $I_{2\times2}$ is assigned to the model error distribution matrix g. In addition, the input current $I_L(t)$ is treated as a known system parameter since we can sample the input current in every time step by the current sensor.

The expressions of the intermediate parameter matrices $S(\hat{x}(t_k))$, $Z(\hat{x}(t_k), \Delta t)$, $\Lambda(\Delta t)$ for the nonlinear battery system are given in Equations (24)–(26), which are derived according to Equations (13)–(15). The relative order r is 1 which can be obtained following the definitions in Equations (19)–(20).

$$S(\hat{x}(t_k)) = [L_{g_1} h(\hat{x}(t_k)), L_{g_2} h(\hat{x}(t_k))] = \left[\frac{\partial U_{oc}[SOC(t_k)]}{\partial SOC(t_k)}, 1\right] \tag{24}$$

$$Z(\hat{x}(t_k), \Delta t) = \Delta t \cdot L_f h(x(t_k)) = \Delta t \cdot \left(\frac{I_L(t_k)}{C_n} \cdot \frac{\partial U_{oc}[SOC(t_k)]}{\partial SOC(t_k)} - (R_p C_p)^{-1} U_p(t_k) + C_p^{-1} I_L(t_k)\right) \tag{25}$$

$$\Lambda(\Delta t) = \Delta t \tag{26}$$

The schematic diagram of the NPF based SOC estimation is shown in Figure 8. The $d_1(t)$ and $d_2(t)$ will be estimated at every iterative loop to predict the model errors of $SOC(t)$ and $U_p(t)$ in the next time step, then the battery state estimation of the next time step can be obtained based on the predicted model errors and battery dynamics.

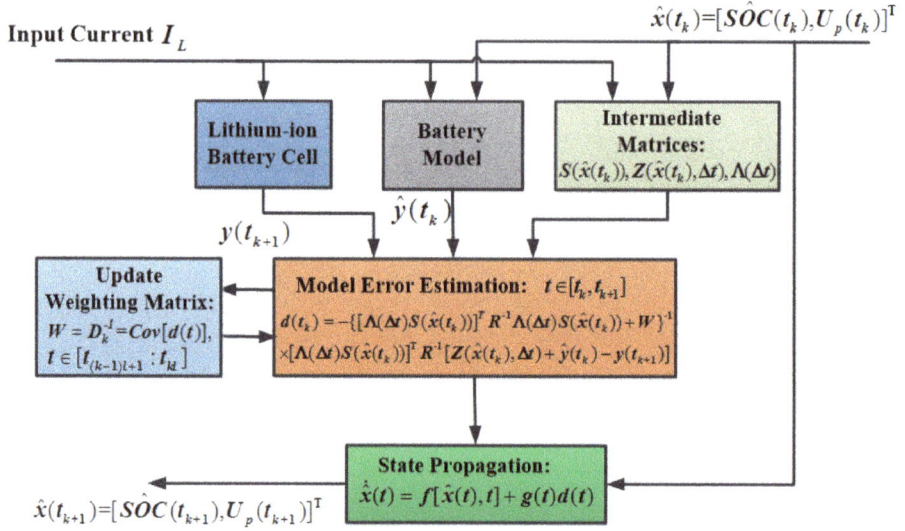

Figure 8. Schematic diagram of NPF based SOC estimation.

4. Results and Discussion

4.1. Part A: Evaluate the SOC Estimation of LCO Battery Cell

In this part, the experimental data is used to validate the performance of the NPF based estimation method for LCO battery cell. The model parameters in Table 3 are used in the estimation. The JC08 (Japanese Cycle 2008) driving cycle is applied as the loading profile to evaluate the proposed method. Figure 9 shows the SOC estimation results with the accurate initial SOC value, and Figure 10 shows the SOC estimation results with inaccurate initial SOC values.

Figure 9a shows the comparison profiles of the reference SOC and estimated SOC with the accurate initial SOC. Since the initial SOC is accurate and the integration time is not long, the SOC result obtained from coulomb counting is treated as an accurate estimation and thus taken as a reference. Figure 9b shows the estimation error between the reference SOC and estimated SOC and it can be seen that the maximum error is around 1%. Figure 9c,d shows the mean absolute error (MAE) and root mean squared error (RMSE) of SOC estimation, respectively. The results indicate that both of the MAE and RMSE are less than 0.4% at the end of estimation. Based on the discussion above, it can be concluded that the NPF method is able to accurately estimate the SOC of LCO battery cell.

142

Figure 10a shows the estimation results with inaccurate initial SOC values, while Figure 10b shows the corresponding errors. Two different initial SOC values (0.5 and 0.3) are used. It can be seen that the SOC estimates can quickly converge to the true solutions within several sampling steps for both initial SOC values. Therefore, the proposed NPF based SOC estimation method is robust to inaccurate initial values for LCO battery cell.

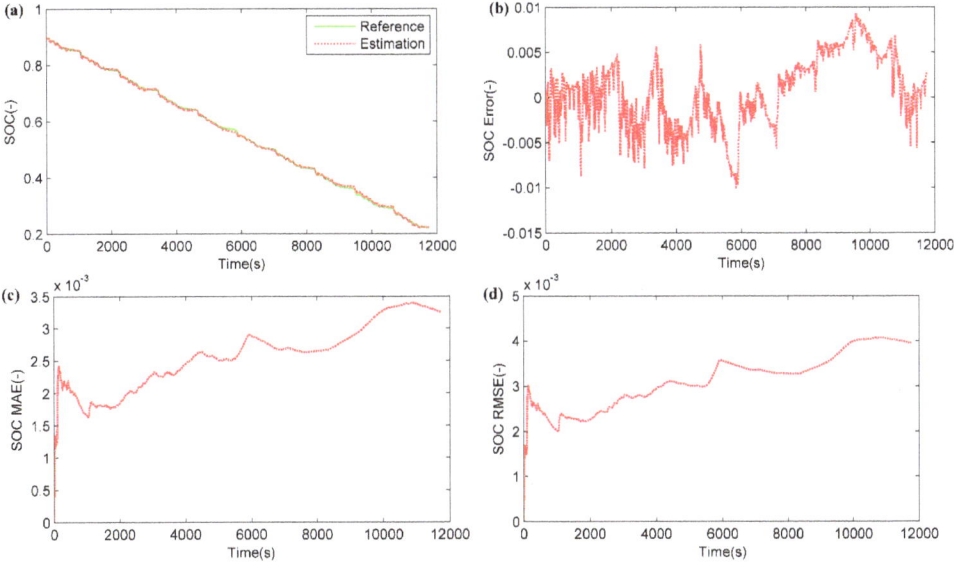

Figure 9. SOC estimation results with accurate initial SOC for LCO battery cell: (**a**) Comparative profiles of reference and estimated SOC; (**b**) Error of SOC estimation; (**c**) MAE of SOC estimation; (**d**) RMAE of SOC estimation.

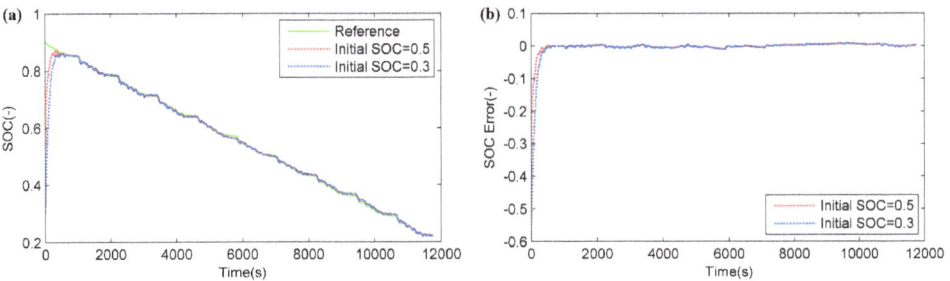

Figure 10. Estimation results with inaccurate initial SOCs for LCO cell: (**a**) Comparative profiles of reference and estimated SOC; (**b**) Error of SOC estimation.

4.2. Part B: Evaluate the SOC Estimation of LFP Battery Cell

In this section, the SOC estimation using NPF for LFP battery cell is conducted. Similarly, the JC08 driving cycle is applied as the loading profile. Figure 11 shows the SOC estimation results with the accurate initial SOC value. Figure 11a shows the comparison profiles of the estimated SOC and reference SOC, and Figure 11b shows the estimation error. It can be seen that the maximum estimation error is within 2%. Figure 11c,d shows the MAE and RMSE of SOC estimation, both of which are less than 0.9% at the end of estimation. Figure 12 shows the SOC estimation with inaccurate initial SOC values, where Figure 12a shows the comparison of the reference SOC and estimated SOC, and Figure 12b shows the estimation error. Similar to the LCO battery cell, two different initial SOC values are used. It can be seen from those results that the SOC estimation can also converge to the true solution within several sampling steps for both inaccurate initial values.

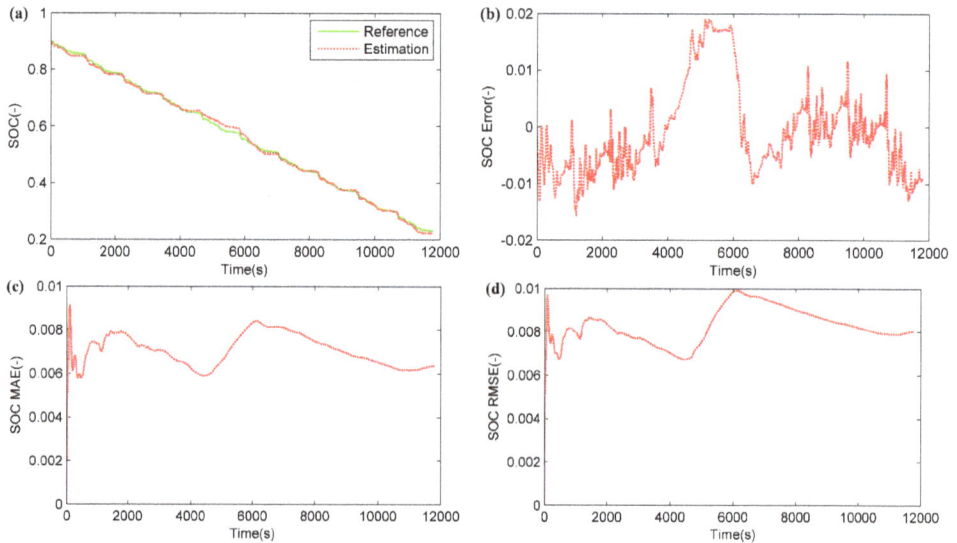

Figure 11. SOC estimation results with accurate initial SOC for LFP battery cell: (**a**) Comparative profiles of reference and estimated SOC; (**b**) Error of SOC estimation; (**c**) MAE of SOC estimation; (**d**) RMAE of SOC estimation.

Based on the estimation results of LCO and LFP battery cells, a comparison study of SOC estimation for LCO and LFP cells is conducted. The comparison results are listed in Table 4. Compared to LFP battery cell, the SOC estimation of the LCO battery cell is benefited by a 0.91% improvement in terms of the maximum error, and by 0.31% and 0.42% improvements in terms of MAE and RMSE. With respect to the convergence rate, the SOC estimation of the LCO battery cell converges much faster than that of the LFP battery cell with inaccurate initial SOC values. It seems that the estimation performance of the LCO battery cell is better than that of the LFP battery cell. The difference in the estimation performance can be mainly explained by the difference of the SOC-OCV relationships between these two cells, as shown in Figure 13. The comparison results of SOC-OCV curves are shown in Figure 13a,b. It is obvious that, compared to the LCO battery cell, the LFP battery cell has a much flatter SOC-OCV curve with smaller derivation. In other words, for the same OCV difference ΔOCV, the corresponding SOC difference ΔSOC_{LFP} of LFP battery cell will be much larger than the corresponding SOC difference ΔSOC_{LCO} of LCO battery cell. Therefore, the observability of the LFP battery cell is weaker than that of the LCO battery cell. As a result, for similar model error, the SOC error of the LFP battery cell will be larger than the SOC error of the LCO battery cell. Thus, the performance of NPF based SOC estimation method for LFP battery cells is worse than that for LCO battery cells due to weaker observability.

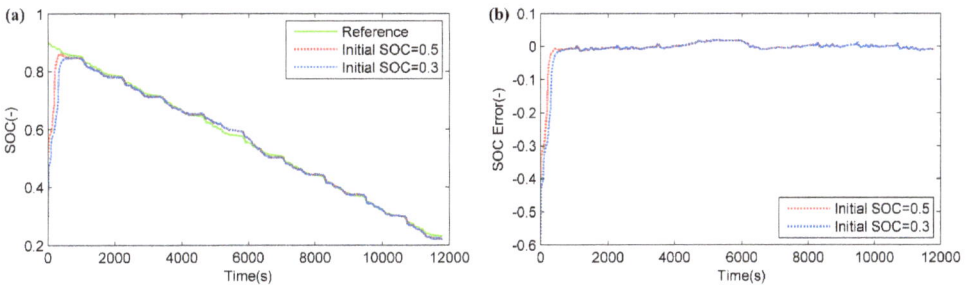

Figure 12. Estimation results with inaccurate initial SOC for LFP cell: (a) Comparative profiles of reference and estimated SOC; (b) Error of SOC estimation.

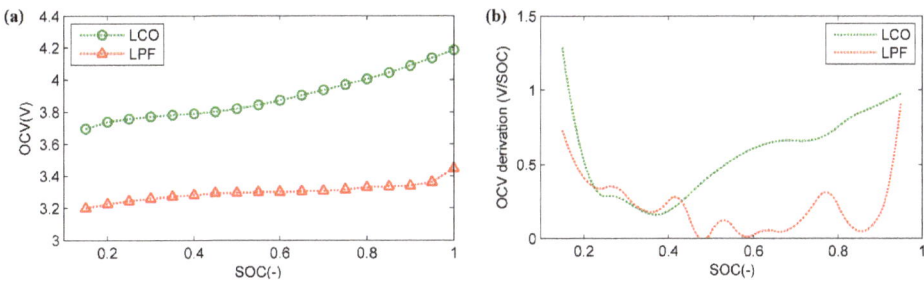

Figure 13. SOC-OCV curves for two cells: (a) Comparison of SOC-OCV; (b) Comparison of the derivation of SOC-OCV.

Table 4. Comparison of LCO and LFP battery cells.

Estimation accuracy	Maximum error	MAE	RMSE
LCO battery cell	1.01%	0.33%	0.39%
LFP battery cell	1.92%	0.64%	0.81%
Difference	0.91%	0.31%	0.42%

Convergence rate	Convergence time (Initial SOC = 0.5)	Convergence time (Initial SOC = 0.3)
LCO battery cell	302 s	427 s
LFP battery cell	392 s	556 s
Difference	29.8%	30.2%

4.3. Part C: Comparison with Extended Kalman Filter

In order to further evaluate the performance of proposed method, a comparison study is conducted between the NPF method and the well-established extended Kalman filter (EKF) method based on the same experimental conditions. The details of EKF for SOC estimation can be found in [15]. To facilitate the understanding of the SOC estimation using EKF, a summary of EKF is provided as follows:

$$\text{Nonlinear sate} - \text{space model} : \begin{cases} x_{k+1} = f(x_k, u_k) + w_k \\ y_k = h(x_k, u_k) + v_k \end{cases} \tag{27}$$

$$\text{Definitions} : \hat{A}_k = \left.\frac{\partial f(x_k, u_k)}{\partial x_k}\right|_{x_k=\hat{x}_k^+}, \hat{C}_k = \left.\frac{\partial h(x_k, u_k)}{\partial x_k}\right|_{x_k=\hat{x}_k^-} \tag{28}$$

Step.1: Initialization: For k = 0, set: $\hat{x}_0^+ = E[x_0]$, $p_0^+ = E[(x_0 - \hat{x}_0^+)(x_0 - \hat{x}_0^+)^T]$
Step.2: For k = 1,2, . . . n, do the following:
(a) State estimation time update: $\hat{x}_k^- = f(\hat{x}_{k-1}^+, u_{k-1})$
(b) Error covariance time update: $p_k^- = \hat{A}_{k-1}p_{k-1}^+\hat{A}_{k-1}^T + R_w$
(c) Calculate the Kalman gain: $L_k = p_k^-\hat{C}_k^T\left[\hat{C}_k p_k^-\hat{C}_k^T + Q_v\right]^{-1}$
(d) State estimation measurement update: $\hat{x}_k^+ = \hat{x}_k^- + L_k[y_k - h(\hat{x}_k^-, u_k)]$
(e) Error covariance measurement update: $p_k^+ = (I - L_k\hat{C}_k)p_k^-$
where w_k and v_k are independent, zero-mean, Gaussian noise processes with covariance matrices R_w and Q_v.

In comparison with the NPF based method which uses a continuous-discrete time model, the battery system in EKF is modeled in a discrete time form, given as follows:

$$\text{State equations} : \begin{cases} SOC_{k+1} = SOC_k + \frac{\eta\Delta t}{C_n}I_{L,k} + w_{1,k} \\ U_{p,k+1} = (1 - (R_pC_p)^{-1}\Delta t)U_{p,k} + \frac{\Delta t}{C_p}I_{L,k} + w_{2,k} \end{cases} \tag{29}$$

Measurement equations : $U_{t,k} = U_{oc}(SOC_k) + U_{p,k} + I_{L,k}R_s + v_k$ (30)

In this paper, two main aspects, estimation accuracy and convergence rate, are studied to show the advantages of the NPF based method.

To compare the estimation accuracy of these two methods, the SOC estimation with the accurate initial SOC is evaluated. Figure 14 shows the comparison of SOC estimation for LCO battery cell. Figure 14a shows the comparison of the reference SOC and estimated SOCs by NPF and EKF, while Figure 14b shows the corresponding errors. It can be seen that the NPF based method has better estimation accuracy than the EKF based method. The MAE and RMSE of SOC estimation also indicate that the NPF based method has better accuracy (shown in Figure 14c,d). Similarly, an improvement of estimation accuracy by the NPF based method can also be found for LFP battery cells, as shown in Figure 15. A summary of the estimation accuracy for these two methods is listed in Table 5.

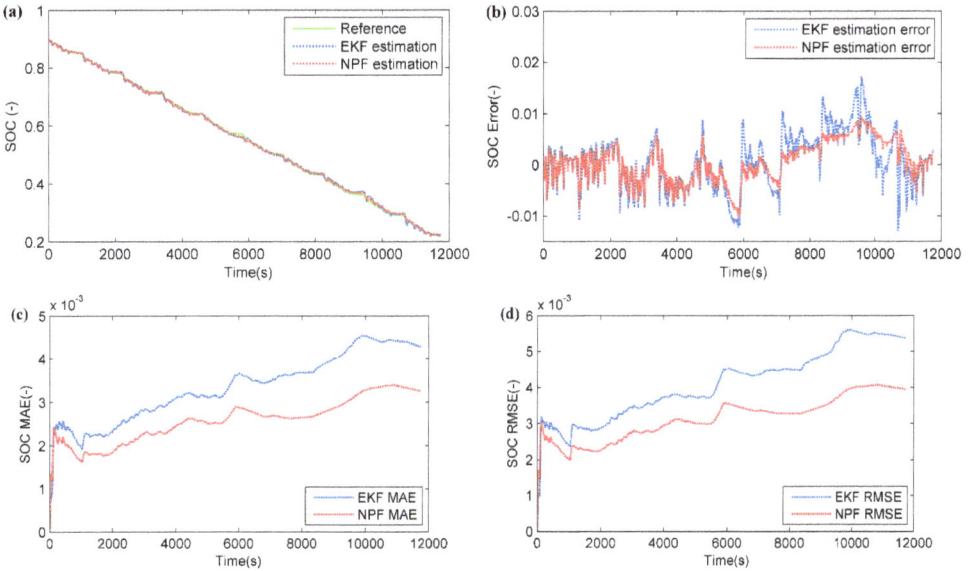

Figure 14. Comparison results of SOC estimation for LCO battery cell with accurate initial SOC: (**a**) SOC estimation; (**b**) Error of SOC estimation; (**c**) MAE of SOC estimation; (**d**) RMSE of SOC estimation.

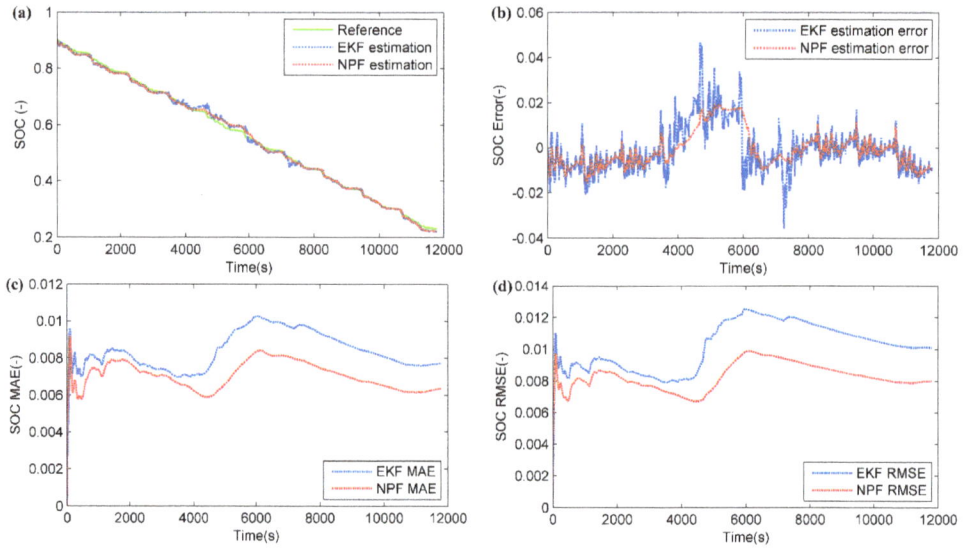

Figure 15. Comparison results of SOC estimation for LFP battery cell with accurate initial SOC: (**a**) SOC estimation; (**b**) Error of SOC estimation; (**c**) MAE of SOC estimation; (**d**) RMSE of SOC estimation.

Table 5. Comparison results of estimation accuracy.

Estimation Accuracy	Maximum Error	MAE	RMSE
LCO battery cell			
EKF	1.64%	0.43%	0.54%
NPF	1.01%	0.33%	0.39%
Improvement	38.4%	17.9%	27.8%
LFP battery cell			
EKF	4.66%	0.78%	1.01%
NPF	1.92%	0.64%	0.81%
Improvement	58.8%	17.9%	20.8%

To compare the convergence rate of these two methods, the SOC estimation with the inaccurate initial SOC is conducted. For convenience, the initial SOC value used in this study is set as 0.5 for both NPF and EKF. The results are shown in Figure 16 where Figure 16a,b shows the comparison for the LCO battery cell, and Figure 16c,d shows the comparison for the LFP battery cell. It can be seen that the NPF based method converges faster than the EKF based method for both cells. A summary of the convergence rate for these two methods is given in Table 6.

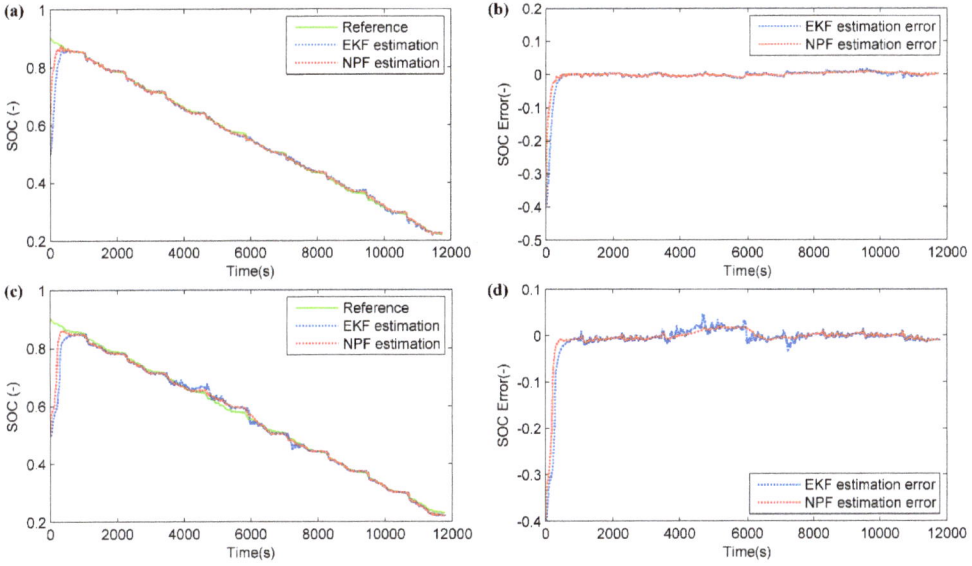

Figure 16. Comparison results of SOC estimation with inaccurate initial SOC: (a) SOC estimation of LCO battery cell; (b) SOC estimation error of LCO battery cell; (c) SOC estimation of LFP battery cell; (d) SOC estimation error of LFP battery cell.

Table 6. Comparison results of convergence rate.

Convergence Rate	Time
LCO battery cell	
EKF	439 s
NPF	302 s
Improvement	31.2%
LFP battery cell	
EKF	586 s
NPF	392 s
Improvement	33.1%

5. Conclusions

In this paper, the nonlinear predictive filter is developed to estimate the SOC of Li-ion batteries with two different chemistries. First, in consideration of the model accuracy and computational complexity, the first-order equivalent circuit battery model is adopted to characterize the dynamic performance of Li-ion battery cells. Second, a series of battery tests are conducted using two cells with different chemistries (LCO and LFP) to identify the model parameters. The model validation results based on UDDS driving cycle indicate that the battery models with identified

149

parameters are able to accurately represent the behavior of LCO and LFP battery cells. Third, the nonlinear predictive filter is introduced and developed to estimate the battery SOC using the first-order battery model. The JC08 driving cycle is applied for these two cells to evaluate the proposed method. According to the results, the NPF based method is able to estimate the battery SOC accurately. Besides, the proposed SOC estimation method has good robust performance to inaccurate initial values (quickly converging to the true solution within several sampling steps). Finally, the comparison of EKF and NPF based methods indicates that the proposed method has better estimation accuracy and faster convergence rate than the EKF based method for both cells.

Acknowledgments: The authors gratefully acknowledge the support from Nippon Chemi-Con Corporation.

Author Contributions: Yin Hua and Min Xu planned and drafted the main part of the paper. Mian Li revised the paper. Chengbin Ma and Chen Zhao participated in the experiment.

Conflicts of Interest: The authors declare no conflict of interest.

References

1. Xi, J.; Li, M.; Xu, M. Optimal energy management strategy for battery powered electric vehicles. *Appl. Energy* **2014**, *134*, 332–341.
2. Wang, B.; Xu, M.; Yang, L. Study on the economic and environmental benefits of different EV powertrain topologies. *Energy Conver. Manag.* **2014**, *86*, 916–926.
3. Lukic, S.M.; Cao, J.; Bansal, R.C.; Rodriguez, F.; Emadi, A. Energy storage systems for automotive applications. *IEEE Trans. Ind. Electron.* **2008**, *55*, 2258–2267.
4. Hammond, G.P.; Hazeldine, T. Indicative energy technology assessment of advanced rechargeable batteries. *Appl. Energy* **2014**, *138*, 559–571.
5. Garche, J.; Jossen, A.; Döring, H. The influence of different operating conditions, especially over-discharge, on the lifetime and performance of lead/acid batteries for photovoltaic systems. *J. Power Sour.* **1997**, *67*, 201–212.
6. Alzieu, J.; Smimite, H.; Glaize, C. Improvement of intelligent battery controller: State-of-Charge indicator and associated functions. *J. Power Sour.* **1997**, *67*, 157–161.
7. Xing, Y.; He, W.; Pecht, M.; Tsui, K.L. State of charge estimation of lithium-ion batteries using the open-circuit voltage at various ambient temperatures. *Appl. Energy* **2014**, *113*, 106–115.
8. Eddahech, A.; Briat, O.; Vinassa, J.M. Adaptive voltage estimation for EV Li-ion cell based on artificial neural networks state-of-charge meter. In Proceedings of the 2012 IEEE International Symposium on Industrial Electronics (ISIE), Hangzhou, China, 28–31 May 2012; pp. 1318–1324.
9. Salkind, A.J.; Fennie, C.; Singh, P.; Atwater, T.; Reisner, D.E. Determination of state-of-charge and state-of-health of batteries by fuzzy logic methodology. *J. Power Sour.* **1999**, *80*, 293–300.

10. Anton, J.C.A.; Nieto, P.J.G.; Viejo, C.B.; Vilan, J.A. Support vector machines used to estimate the battery state of charge. *IEEE Trans. Power Electron.* **2013**, *28*, 5919–5926.

11. Tsang, K.M.; Sun, L.; Chan, W.L. Identification and modelling of Lithium ion battery. *Energy Convers. Manag.* **2010**, *51*, 2857–2862.

12. He, H.; Xiong, R.; Guo, H.; Li, S. Comparison study on the battery models used for the energy management of batteries in electric vehicles. *Energy Convers. Manag.* **2012**, *64*, 113–121.

13. Plett, G.L. Extended Kalman filtering for battery management systems of LiPB-based HEV battery packs: Part 1. Background. *J. Power Sour.* **2004**, *134*, 252–261.

14. Plett, G.L. Extended Kalman filtering for battery management systems of LiPB-based HEV battery packs: Part 2. Modeling and identification. *J. Power Sour.* **2004**, *134*, 262–276.

15. Plett, G.L. Extended Kalman filtering for battery management systems of LiPB-based HEV battery packs: Part 3. State and parameter estimation. *J. Power Sour.* **2004**, *134*, 277–292.

16. Plett, G.L. Sigma-point Kalman filtering for battery management systems of LiPB-based HEV battery packs: Part 1: Introduction and state estimation. *J. Power Sour.* **2006**, *161*, 1356–1368.

17. Plett, G.L. Sigma-point Kalman filtering for battery management systems of LiPB-based HEV Battery packs: Part 2: Simultaneous state and parameter estimation. *J. Power Sour.* **2006**, *161*, 1369–1384.

18. He, Z.; Gao, M.; Wang, C.; Wang, L.; Liu, Y. Adaptive state of charge estimation for Li-ion batteries based on an unscented kalman filter with an enhanced battery model. *Energies* **2013**, *6*, 4134–4151.

19. Han, J.; Kim, D.; Sunwoo, M. State-of-charge estimation of lead-acid batteries using an adaptive extended Kalman filter. *J. Power Sour.* **2009**, *188*, 606–612.

20. Sun, F.; Hu, X.; Zou, Y.; Li, S. Adaptive unscented Kalman filtering for state of charge estimation of a lithium-ion battery for electric vehicles. *Energy* **2011**, *36*, 3531–3540.

21. Schwunk, S.; Armbruster, N.; Straub, S.; Kehl, J.; Vetter, M. Particle filter for state of charge and state of health estimation for lithium-iron phosphate batteries. *J. Power Sour.* **2013**, *239*, 705–710.

22. Charkhgard, M.; Farrokhi, M. State-of-charge estimation for lithium-ion batteries using neural networks and EKF. *IEEE Trans. Ind. Electron.* **2010**, *57*, 4178–4187.

23. Tian, Y.; Xia, B.; Wang, M.; Sun, W.; Xu, Z. Comparison Study on two model-based adaptive algorithms for SOC estimation of lithium-ion batteries in electric vehicles. *Energies* **2014**, *7*, 8446–8464.

24. He, H.; Qin, H.; Sun, X.; Shui, Y. Comparison study on the battery SoC estimation with EKF and UKF algorithms. *Energies* **2013**, *6*, 5088–5100.

25. Zou, Z.; Xu, J.; Mi, C.; Cao, B.; Chen, Z. Evaluation of model based state of charge estimation methods for lithium-ion batteries. *Energies* **2014**, *7*, 5065–5082.

26. Yuan, S.; Wu, H.; Yin, C. State of charge estimation using the extended kalman filter for battery management systems based on the arx battery model. *Energies* **2013**, *6*, 444–470.

27. Xia, B.; Chen, C.; Tian, Y.; Sun, W.; Xu, Z.; Zheng, W. A novel method for state of charge estimation of lithium-ion batteries using a nonlinear observer. *J. Power Sour.* **2014**, *270*, 359–366.
28. Crassidis, J.L.; Markley, F.L. Predictive filtering for attitude estimation without rate sensors. *J. Guid. Control Dyn.* **1997**, *20*, 522–527.
29. Hua, Y.; Cordoba-Arenas, A.; Warner, N.; Rizzoni, G. A multi time-scale state-of-charge and state-of-health estimation framework using nonlinear predictive filter for lithium-ion battery pack with passive balance control. *J. Power Sour.* **2015**, *280*, 293–312.
30. Hu, X.; Li, S.; Peng, H. A comparative study of equivalent circuit models for Li-ion batteries. *J. Power Sour.* **2012**, *198*, 359–367.
31. Abu-Sharkh, S.; Doerffel, D. Rapid test and non-linear model characterisation of solid-state lithium-ion batteries. *J. Power Sour.* **2004**, *130*, 266–274.
32. US Environmental Protection Agency. EPA urban dynamometer driving schedule (UDDS). Available online: http://www.epa.gov/oms/standards/light-duty/udds.htm (accessed on 3 April 2015).

Comparative Study of a Fault-Tolerant Multiphase Wound-Field Doubly Salient Machine for Electrical Actuators

Li-Wei Shi and Bo Zhou

Abstract: New multiphase Wound-Field Doubly Salient Machines (WFDSMs) for electrical actuators with symmetric phases are investigated and compared in this paper. With a comparative study of the pole number and pole arc coefficient, the salient pole topology of the three-phase, four-phase, five-phase, and six-phase WFDSMs with little cogging torque is presented. A new winding configuration that can provide symmetrical phases for the multiphase WFDSMs is proposed. Suitable fault-tolerant converters for the multiphase WFDSM are presented. With the simulated results in terms of the pole topology, flux linkage, back EMF and converters, it can be concluded that the pole numbers of the new five-phase WFDSM are very large. The high accuracy position sensors should be required to make the five-phase WFDSM commutate frequently and accurately at a high speed. The four-phase and the six-phase WFDSM can be divided into two isolated channels, and both of them have a good performance as a fault-tolerant machine. All of the investigations are verified by finite element analysis results.

Reprinted from *Energies*. Cite as: Shi, L.-W.; Zhou, B. Comparative Study of a Fault-Tolerant Multiphase Wound-Field Doubly Salient Machine for Electrical Actuators. *Energies* **2015**, *8*, 3640–3660.

1. Introduction

Today conventional aircraft are characterized by complex hydraulic nets. In order to reduce the weight of the pipelines, cylinders, pump, valves and switches of the hydraulic system, the aircraft is adopting more and more electrical systems in preference to others. Now, researchers and engineers have proved that electrical actuators can be used to reduce or to remove the traditional hydraulic, and mechanical systems in the next few years [1]. The more electric aircraft approach is widely discussed in the technical literature, which includes the following three main drives [2]:

- The starter-generator for the engine;
- The electrical actuators for the flight control;
- The electric machines for the fuel pump.

There are many different types of actuators in a conventional aircraft [3], such as the actuators in the wings and in the tail. In the hydraulic actuation system, the flight control is realized by a hydraulic pump and a hydraulic motor, several fluid pipelines and hydraulic actuators. Now, more and more electric machines are being used to replace or assist the hydraulic actuation system. For example, in the Boeing 787, the spoilers and the horizontal stabilizer flight controls are driven by electric machines in order to guarantee the operation in the case of a hydraulic failure.

A literature review reveals that several types of machine can be used as a drive motor for electrical actuators [4]. Among them, PM machines and Switched Reluctances Machines (SRMs) were abundantly studied in the past years, because of their very-high power density.

In [5], a five-phase PM brushless machine was developed for an aircraft flap actuator application, and the machine can endure the fault of one or two open phases or a phase short circuit. In [6], a PM fractional slot machine was designed, because the fractional slot windings have low mutual inductances between phases, which meet the magnetic isolation demands of phase windings for multiphase fault-tolerant PM synchronous machines [7]. Such fault-tolerant PM machines were also studied in [8,9].

It is necessary to remark that the actuators of an airplane have to work in very harsh ambient conditions, with temperature variations from $-60\,^\circ$C to $+70\,^\circ$C and the air pressure varies from almost 0 to 1 bar [2]. This harsh environment puts forward higher requirements for high performance PM materials. Furthermore, many of the electrical actuators care little about the torque ripple because the noise of the airplane is very high. Therefore, switched reluctances machines were investigated to drive the actuators in [10,11].

The Wound-Field Doubly Salient Machine (WFDSM) has the same rotor as the SRM that will not suffer from faults of the PM materials or brush faults of the wound-field synchronous motors. The WFDSM is derived from a doubly salient PM machine (DSPM) [12] by using field windings instead of permanent magnet excitation [13]. The WFDSM provides the excitation flux by the DC field windings instead of the PMs. Therefore, the output torque and speed can be adjusted by the field winding and phase windings. What's more, the phase windings of the WFDSM are isolated from each other, and it has low mutual inductances between phases, which reduces the negative influence of the faulty phase. It has broad application prospects in the fields that care little about the torque ripple, such as mining machinery, electrical actuators, and starter-generators.

For example, a WFDSM with two-section twisted-rotor was developed as a starter-generator for aerospace applications [14]. Recently, some new three-phase WFDSMs with new winding arrangements have been developed to take the place of traditional electric machines [15,16]. In the energy conversion area, a WFDSM worked as a DC generator was equipped in an EV range extender [17]. A prototype of a 24/32-pole WFDSM was developed as a low speed wind turbine generator [13].

Multiphase machines with more than three phases can be applied for high reliability applications because they can still run even with one or two open-circuited phases [18]. To improve the reliability of the WFDSM, a traditional four-phase WFDSM was studied in [17], which was similar to the 8/6-pole DSPM described in [19]. A five-phase WFDSM was developed as a generator in [20], which showed that it had good fault-tolerant characteristics. Therefore, the multiphase WFDSM is very suitable to be designed as a fault-tolerant machine.

However, the four-phase and the five-phase WFDSMs discussed above are traditional WFDSMs with their field windings wound around four and five stator poles, respectively. They have the disadvantage of phase asymmetry. The phase asymmetry of the three-phase WFDSM is not obvious, but we found that the phase asymmetry will increase with the number of phases. It was considered that the WFDSM cannot be designed with six phases, and there have been no reports of six-phase WFDSMs until now.

In this paper, new multiphase WFDSMs for electrical actuators with symmetric phases will be investigated and compared. With the comparative study of the poles number and pole arc coefficient, the salient pole topology of the WFDSMs that have little cogging torque will be presented. A new winding configuration to provide symmetrical phases will be proposed. Suitable fault-tolerant converters for the multiphase WFDSMs will be presented. With the comparison in terms of the pole topology, flux linkage, back EMF and converters, comparative conclusions will be proposed to select a multiphase WFDSM for electrical actuators.

2. Comparative Study of the Salient Pole Topology

2.1. Salient Pole Number

There is a wide range of possible combinations of the stator poles and the rotor poles. Nevertheless, only few combinations are suitable to be selected. To outline the pole combinations, the principle of the salient pole number should be studied first.

As we can see from the traditional three-phase WFDSM in Figure 1a, there are $6N$ stator poles and $4N$ rotor poles, where N is the number of element machines. Each phase coil is wound around one stator pole, and the field coils are wound around every three stator poles [17].

So the first law of the stator poles can be written as:

$$p_s = mi \tag{1}$$

where p_s should be an even number, and it stands for the number of stator poles. m is the phase number and i is a positive integer. If p_s is an odd number, the north field winding will not be equal to the south field winding, and the machine will generate an unbalanced magnetic force. Let p_r be the rotor poles number. The mechanical angle of one period β_r can be obtained as:

$$\beta_r = \frac{360°}{p_r} \qquad (2)$$

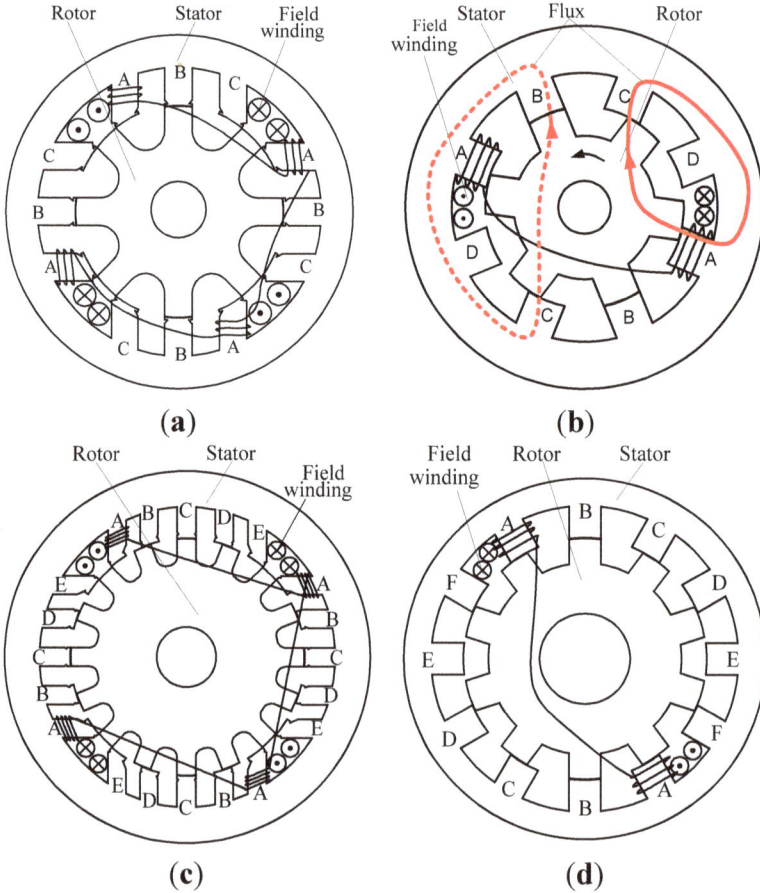

Figure 1. Structure of the three-phase and multiphase DSG. (**a**) 12/8-pole three-phase of WFDSM; (**b**) 8/6-pole four-phase; (**c**) 20/16-pole five-phase; (**d**) 12/10-pole six-phase.

The mechanical angle of each stator pole β_s is:

$$\beta_s = \frac{360°}{p_s} \qquad (3)$$

Because the adjacent poles have the adjacent phase windings, the mechanical angle between two phases can be expressed as:

$$\beta_\delta = \frac{360°}{m} \cdot \frac{1}{p_r} \tag{4}$$

and the difference between β_r and β_s is $\pm\beta_\delta$:

$$\beta_r - \beta_s = \pm\beta_\delta \tag{5}$$

From Equations (2)–(5), we can get the second law of the rotor and the stator poles:

$$\frac{p_s}{p_r} = \frac{m}{m \pm 1} \tag{6}$$

When $m = 3$, the elementary machine of three-phase WFDSM has six stator poles and four or eight rotor poles, which is called 6/4-pole machine or 6/8-pole machine [14]. In the same way, the elementary machine of a traditional four-phase WFDSM has an 8/6-pole or 8/10-pole structure. Table 1 gives the pole combinations of the WFDSM with different phases.

Table 1. The poles combinations of the WFDSM.

Phase number	Stator poles	Rotor poles	Example
Three-phase	6N	4N or 8N	12/8
Four-phase	4N	3N or 5N	8/6
Five-phase	10N	8N or 12N	20/16
Six-phase	6N	5N or 7N	12/10

2.2. Pole Arc Coefficient

The WFDSM stator is equipped with both field coils and phase coils. The self-inductance of the phase winding and the field winding change with the rotor positions and the pole arcs [21,22]. If the pole arc is not well-designed, the machine will generate torque ripples because the reluctance and the flux of the field winding will change with the rotor position [23]. In order to minimize the cogging torque caused by the mutative reluctance of the field winding, the self-inductance of the field winding should be constant when the rotor rotates. Overall, the increasing phase number should be equal to the decreasing phase number.

For the common inner-rotor motor, the rotor pole number is usually less than the stator pole number. Hence the rotor pole is generally wider than the stator pole. As the narrow pole of the stator and the rotor determines the increasing or decreasing mechanical angle of the phase inductance, the increasing mechanical angle of the phase inductance can be described as:

$$\beta_{working} = \frac{360°}{p_s} \cdot \alpha_s \qquad (7)$$

where α_s is the stator pole arc coefficient, which is the proportion of the stator pole arc length l_t and the pole pitch l_p.

$$\alpha_s = \alpha_s = \frac{l_t}{l_p} \qquad (8)$$

Let the electrical angle of one phase voltage waveform be θ:

$$\theta = \frac{360°}{p_s} \alpha_s p_r = x \frac{180°}{m} \qquad (9)$$

where x is the phase number that has a mutative self-inductance at any time, and $x \leqslant m$. If x is large, there will be more phases that can output torque or voltage, and the fault-tolerant ability of the machine will be strong. Because the phase number with increasing inductance should be equal to the phase number with decreasing inductance, it can be concluded that x should be an even number.

For the three-phase WFDSM, while $p_s/p_r = 3/2$, $\theta = 120°$. We can draw from Equation (7) and Equation (8) that:

$$\alpha_s = \frac{p_s}{3p_r} \qquad (10)$$

Therefore, to make the machine have no cogging torque, reluctance and flux of the field winding the pole arc coefficient of the three-phase WFDSM α_s is equal to 0.5.

While $p_s > p_r$, the rotor pole width is not generally thinner than the stator pole width. For the three-phase WFDSM, the mutual inductances of the field winding and the phase windings L_{pf} are shown in Figure 2. When $\alpha_r = 0.5$, the machine can be easily controlled by a BLDC controller, because one period can be divided into six equal parts like a BLDC machine. When $\alpha_r = 0.333$, the machine can output a large torque because the rotor poles and the stator poles are monospaced and the leakage flux is small. Overall, the stator and the rotor pole arc coefficient should comply with $\alpha_s = 0.5$ and $\alpha_r = 0.5$ or 0.333.

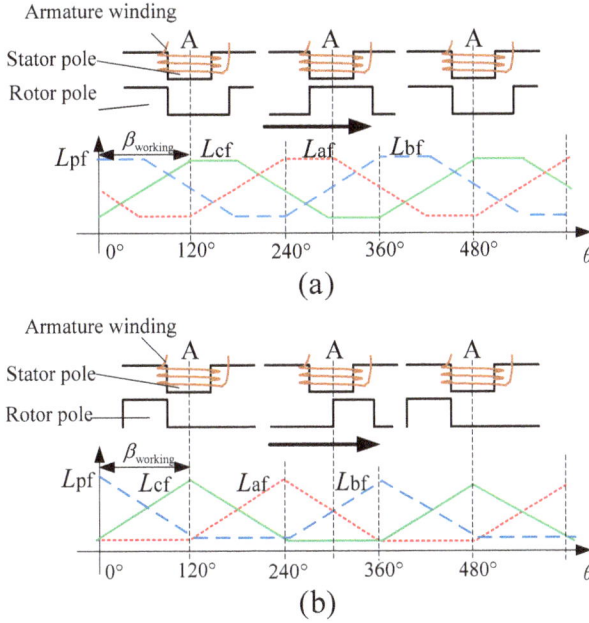

Figure 2. L_{pf} of the three-phase WFDSM with different α_r. (a) $\alpha_r = 0.333$; (b) $\alpha_r = 0.5$.

It can be concluded that the pole arc coefficients of three-phase WFDSM should comply with Equation (11):

$$
\left.
\begin{aligned}
\frac{p_s}{p_r} &= \tfrac{3}{2}: \alpha_s = 0.5;\ \alpha_r = 0.333\ \text{or}\ 0.5 \\
\frac{p_s}{p_r} &= \tfrac{3}{4}: \alpha_s = 0.25;\ \alpha_r = 0.333\ \text{or}\ 0.5
\end{aligned}
\right\}
\tag{11}
$$

However, the four-phase WFDSM should not be designed with $\alpha_r = 0.5$. As we can see from Figure 3, if $\alpha_r = 0.5$, there will be three changing inductances at any time, $x = 3$. With this structure, the reluctance of the field winding will change with the position of the rotor, which will generate a big cogging torque, as well as field winding back EMF.

In order to solve the problem of changeable field reluctance, a new four-phase WFDSM is proposed, whose pole arc coefficient complies with Equation (15):

$$
\left.
\begin{aligned}
\frac{p_r}{p_s} &= \tfrac{3}{4}: \alpha_s = 0.667;\ \alpha_r = 0.5 \\
\frac{p_r}{p_s} &= \tfrac{5}{4}: \alpha_s = 0.4;\ \alpha_r = 0.5
\end{aligned}
\right\}
\tag{12}
$$

159

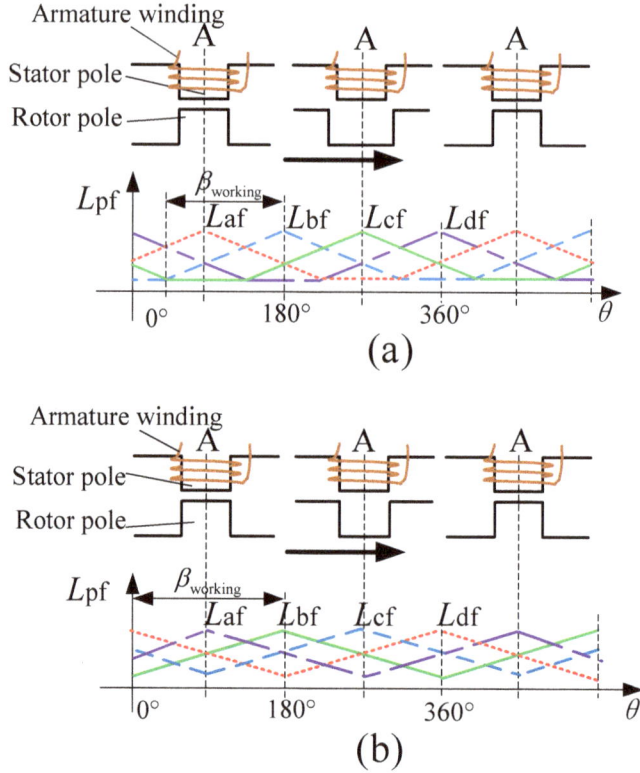

Figure 3. L_{pf} of the four-phase WFDSM with different α_r. (a) $\alpha_s = 0.5$, $\alpha_r = 0.375$; (b) $\alpha_s = 0.667$, $\alpha_r = 0.5$.

For this new machine, there are two increasing phase inductances and two decreasing phase inductances at any time, as shown in Figure 3b. The field inductance is steady, and it will not generate cogging torque. The phase voltage waveform electrical angle of this machine is 180°, and the phase number with mutative inductance at any time $x = 4$. Therefore, all of the four-phase windings can output torque at any time, which improves the fault tolerance of the machine.

Similarly, we can also deduce the pole arc coefficient of the other multiphase WFDSM, since the phase voltage waveform electrical angle of the five-phase WFDSM is 144°, and the angle of the six-phase WFDSM is 120°. In short, it can be newly concluded that the pole arc coefficient of the multi-phase WFDSM should comply with Equation (13).

In short, to reduce the torque ripple caused by the field winding of multi-phase WFDSM, the stator poles, rotor poles and pole arc should follow topology criteria as shown in Equations (1), (6) and (13).

The three pieces of topology criteria not only give a design basis for the WFDSMs with less than six phases, they can also be used to design PM doubly salient machines and hybrid excitation doubly salient machines. What is more, these topology criteria also provide a derivation example for the WFDSMs with more than seven phases.

$$
\left.
\begin{array}{l}
\text{three} - \text{phase,} \ \frac{p_s}{p_r} = \frac{3}{2}, \alpha_s = 0.5; \ \alpha_r = 0.333 \text{ or } 0.5 \\[4pt]
\text{three} - \text{phase,} \ \frac{p_s}{p_r} = \frac{3}{4}, \alpha_s = 0.25; \ \alpha_r = 0.333 \text{ or } 0.5 \\[4pt]
\text{four} - \text{phase,} \ \frac{p_s}{p_r} = \frac{4}{3}, \alpha_s = 0.667; \ \alpha_r = 0.5 \\[4pt]
\text{four} - \text{phase,} \ \frac{p_s}{p_r} = \frac{4}{5}, \alpha_s = 0.4; \ \alpha_r = 0.5 \\[4pt]
\text{five} - \text{phase,} \ \frac{p_s}{p_r} = \frac{5}{4}, \alpha_s = 0.5; \ \alpha_r = 0.4 \\[4pt]
\text{five} - \text{phase,} \ \frac{p_s}{p_r} = \frac{5}{6}, \alpha_s = 0.333; \alpha_r = 0.4 \\[4pt]
\text{six} - \text{phase,} \ \frac{p_s}{p_r} = \frac{6}{5}, \alpha_s = 0.4; \ \alpha_r = 0.333 \\[4pt]
\text{six} - \text{phase,} \ \frac{p_s}{p_r} = \frac{6}{7}, \alpha_s = 0.571; \alpha_r = 0.333
\end{array}
\right\}
\tag{13}
$$

Because the pole numbers and the switches of the converter increase with the phase number, the WFDSM with more than six phases is not suitable to be applied because of the weight and the cost of the converter is unacceptable, as well as the pole number. Therefore, their application prospects are not as broad as those of WFDSMs with less than six phases, because they are too complicated. This paper focuses on the WFDSMs with less than seven phases.

2.3. Simulation Results

Figure 4 shows L_{pf} and back EMF waveforms of the traditional multiphase WFDSMs, which are 12/8-pole three-phase, 8/6-pole four-phase 10/8-pole five-phase and 12/10-pole six-phase with the same stator pole arc coefficient $\alpha_s = 0.5$, and the rotor poles are as wide as the stator poles. The simulated result in Figure 4a shows that L_{pf} of the three-phase and four-phase WFDSM are consistent with the analysis result in Figures 2 and 3. The back EMF waveforms in Figure 4b show that the electrical angles of the phase voltage waveforms of the four machines are approximately 120°, 135°, 144° and 150°. This verifies the calculation results in Equation (8).

With the total torque formula given in Equation (14), we can see that there is a torque component $\frac{1}{2}i_f^2 \frac{dL_f}{d\theta}$ which has nothing to do with the phase current i_p. It can be called a cogging torque.

$$
T = \frac{1}{2}i_p^2 \frac{dL_p}{d\theta} + i_p i_f \frac{dL_{pf}}{d\theta} + \frac{1}{2}i_f^2 \frac{dL_f}{d\theta}
\tag{14}
$$

where i_p and i_f are the phase current and field current. In order to reduce the negative impact of the armature reaction, the number of turns of the field winding is much larger than the number of phase windings.

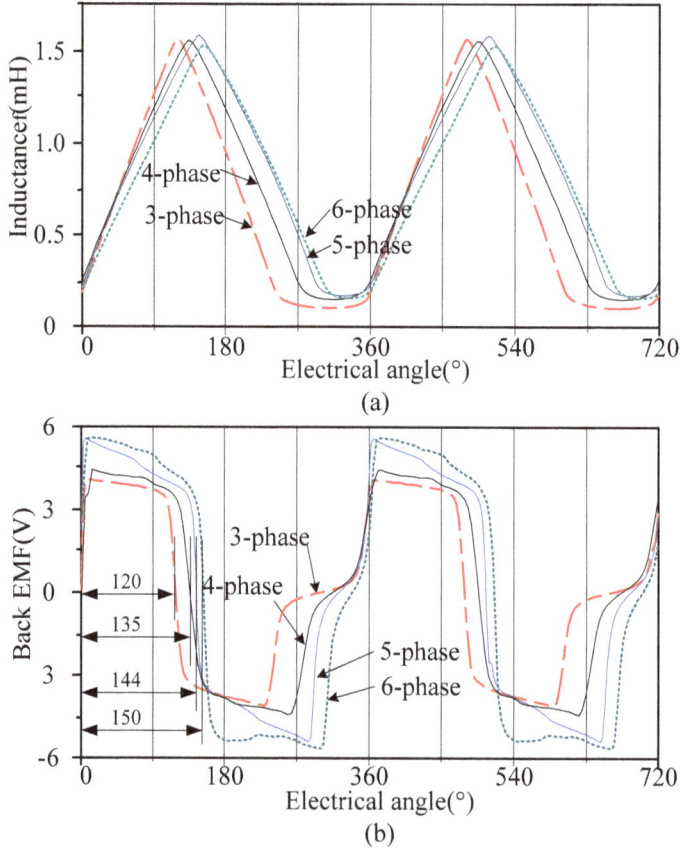

Figure 4. L_{pf} and u_p of multi-phase WFDSM. (**a**) L_{pf}; (**b**) The back EMF.

The self-inductance of the field winding L_f is much larger than the mutual inductance between the phase winding and field winding L_{pf}. Therefore, if L_f changes with the rotor position, and machine will generate a large cogging torque.

For the four-phase WFDSM with $\alpha_s = 0.5$, $\theta = 135°$, and $x = 3$, there are three inductances changing at any time. This situation does not conform to Equation (13). With the self-inductance of the field winding L_f waveform in Figure 5a, we can see that L_f will change with the rotor position, which will produce cogging torque ripples. If $\alpha_s = 0.667$ and $x = 4$, the self-inductance of the field winding L_f will be a constant, as shown in Figure 5b. Therefore, if x is an odd number, the machine will not generate cogging torque.

162

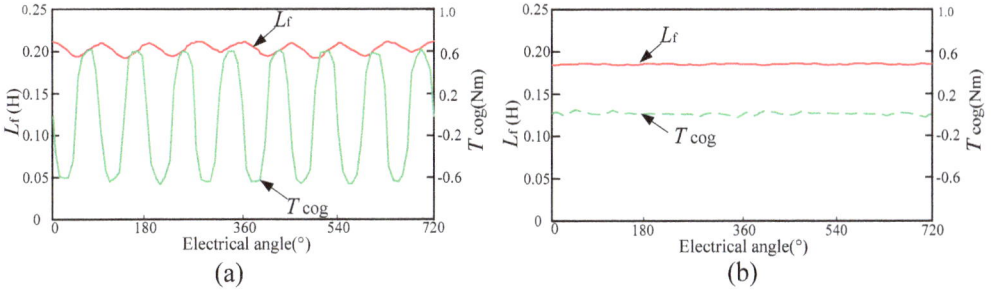

Figure 5. L_f of the four-phase WFDSM. (**a**) $\alpha_s = 0.5$; (**b**) $\alpha_s = 0.667$.

3. Comparative Study of the Symmetrical Phase Winding Configuration

3.1. The Symmetrical Phase Winding Configuration

As shown in Figure 1a, there are four field coils that are used to provide the magnetic field. Each field coil is wound around three stator poles. In Figure 1b, there are four stator poles in a field coil, which provides the magnetic field. Therefore, each excitation source of traditional m-phase doubly salient machine couples with m-phase coils. As the red lines show in Figure 1b, the magnetic circuit of phase A and D which are close to the excitation source is much shorter than phase B and C, and the inductance of phase A and D are larger than that of phase B and C. Overall the amplitudes of the inductance of the traditional four-phase WFDSM have the relationship given by Equation (15):

$$\max(L_{af}) = \max(L_{df}) > \max(L_{bf}) = \max(L_{cf}) \tag{15}$$

If the phase coils with short flux road are divided averagely, the total inductances added by the series coils will be equal [12]. Let j stand for the number of phase coils that coupled by a field coil. The stator poles can be calculated with:

$$p_s = m \cdot k = jP \tag{16}$$

where k and P are two natural numbers. Therefore, P_s is the least common multiple of m and j at least. Together with Equation (6), we can list the pole numbers of the four-phase WFDSM with different j in Table 2.

163

Table 2. The pole numbers of the four-phase WFDSM with different j.

j	Poles of an element machine		
	Four-phase	Five-phase	Six-phase
$j = 1$	8/6	10/8	12/10
$j = 2$	8/6	10/8	12/10
$j = 3$	12/9	30/24	12/10
$j = 4$	8/6	40/32	24/20
$j = 5$	–	10/8	30/25
$j = 6$	–	–	12/10

When $j = m$, the machine is a traditional WFDSM with asymmetric phases. If $j = 4$ in the five-phase WFDSM and six-phase WFDSM, the number of stator pole will be very large since it increases with the least common multiple of j and m. And if $j = 2$, the field winding coils will increase, which in turn increases the copper consumption of the field winding. A three-phase 6/4-pole variable flux reluctance doubly salient machine was reported in [16], which verified that the WFDSM can operate well when $j = 1$. With the same configuration in [16], every stator pole of the multiphase WFDSM is wound with a field winding.

In the traditional 8/6-pole WFDSM, as shown in Figure 6a, each field coil is wound around four stator poles. All the coils of phase A and phase D are nearby the field coil slots, and the coils of the other two phases are in the middle of the two field coil slots. Therefore, the total reluctance of phase B is larger than the reluctance of phase A.

The new 12/9-pole WFDSM has four field coils and twelve phase coils, which can be divided into four phases. As shown in Figure 6b, each field coil is wound around three stator poles, and the two neighboring field coils are in the opposite direction. Every phase has three coils, one is in the middle of the two field coil slots, and the other two phase coils are nearby the field coil slots. Therefore, the total reluctances of the four phases are equal. With the above analysis, the preferred configuration of the four-phase WFDSM with symmetry phases is with a 12/9-pole structure.

Similarly, we can draw the connected coils of the 30/24-pole five-phase WFDSM and the 12/10-pole six-phase WFDSM according to Table 2. Therefore, the elementary machine of the five-phase WFDSM with symmetry phases has a 30/24-pole structure.

In the six-phase WFDSM, if we wind the field coils around two or more stator poles, the six-phase WFDSM will still has the serious drawback of asymmetric phases, which will be verified by the simulation results in the next section Therefore, each stator pole of the 12/10-pole six-phase WFDSM should have a field coil if we want to get symmetrical phases.

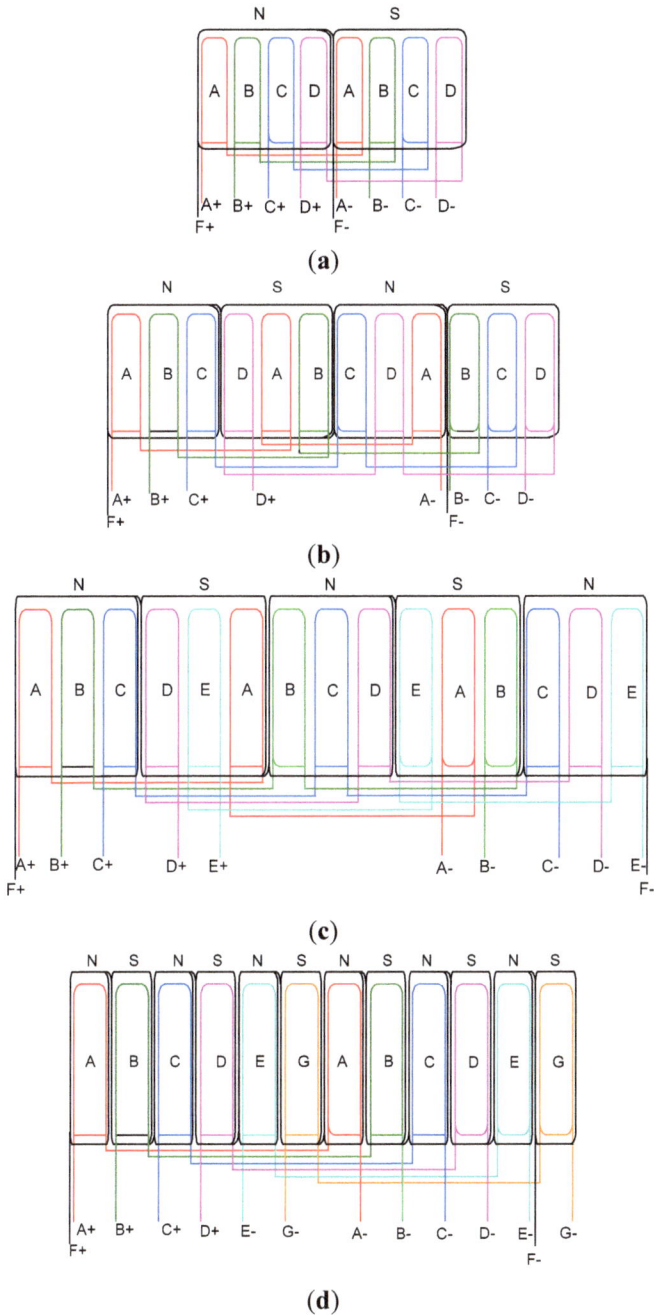

Figure 6. The connected coils of the multiphase WFDSM. (**a**) Traditional 8/6-pole four-phase; (**b**) new 12/9-pole four-phase; (**c**) 30/24-pole five-phase (half of the poles); (**d**) 12/10-pole six-phase.

165

3.2. Simulation Results

To compare the above multiphase WFDSMs, the simulation models of the traditional 8/6-pole and new 12/9-pole four-phase WFDSM were established. Figure 7a shows the flux of the 12/9-pole machine. Figure 7b shows the inductance between phase winding and field winding of the traditional 8/6-pole four-phase WFDSM. This verifies the formula of Equation (17). Figure 7c shows the same inductance of the new 12/9-pole four-phase WFDSG. It is shown that the amplitudes of the inductances have the relationship with:

$$\max(L_{af}) = \max(L_{df}) = \max(L_{bf}) = \max(L_{cf}) \tag{17}$$

The 2D-FEA results agree well with the theoretical analysis results. Figure 7d,e show the waveforms of the back EMF of the four-phase WFDSMs. The back EMFs of the traditional 8/6-pole WFDSM have slight difference in the amplitude and the shape of the waveform.

As the preferred five-phase WFDSM has a 30/24-pole structure, $j = 3$; its flux distribution is shown in Figure 8a. Figure 8b shows the inductances between the phase windings and the field winding of the traditional 10/8-pole five-phase WFDSM. The same inductance of the new 30/24-pole four-phase WFDSG is shown in Figure 8c. It shows that the traditional five-phase WFDSG has the disadvantage of asymmetric phases. The new WFDSG with its field coils wound around three poles can solve this problem. The theoretical analysis results are verified with these 2D-FEA results.

Figure 8d,e show the waveforms of the back EMF of the five-phase WFDSMs. It can be calculated that the back EMFs of different phases of the traditional 10/8-pole WFDSM have a difference of about 10.9%.

From the above analysis, we know that the traditional six-phase WFDSM with $j = 6$ has the phase asymmetry problem because the field coils is wound around six stator poles. We have set up the simulation models of the six-phase WFDSMs with different j. But even if we let $j = 5$, 4 or 3, there will be phase asymmetry problem.

If $j = 2$, the 12/10-pole six-phase WFDSM has a field coil wound around every two stator poles, and it also has the problem of asymmetric phases. As shown in Figure 9a, when the rotor pole is sliding to the pole of phase B, the back EMF of the phase B will be less than phase A because the pole of phase A has a lot of flux at this time. When the rotor pole is sliding to the pole of phase A, the back EMF of the phase A will be larger than phase B because the pole of phase B has no flux at this time. This is verified by the Figure 9d, which shows that the amplitudes of u_a and u_b are not equal. Therefore, the phases are still asymmetric even with $j = 2$.

(a)

(b)

(c)

(d)

(e)

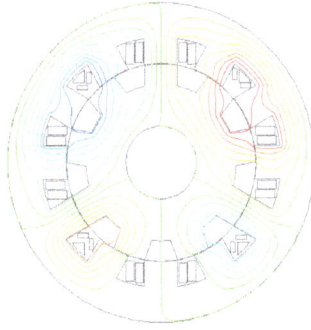

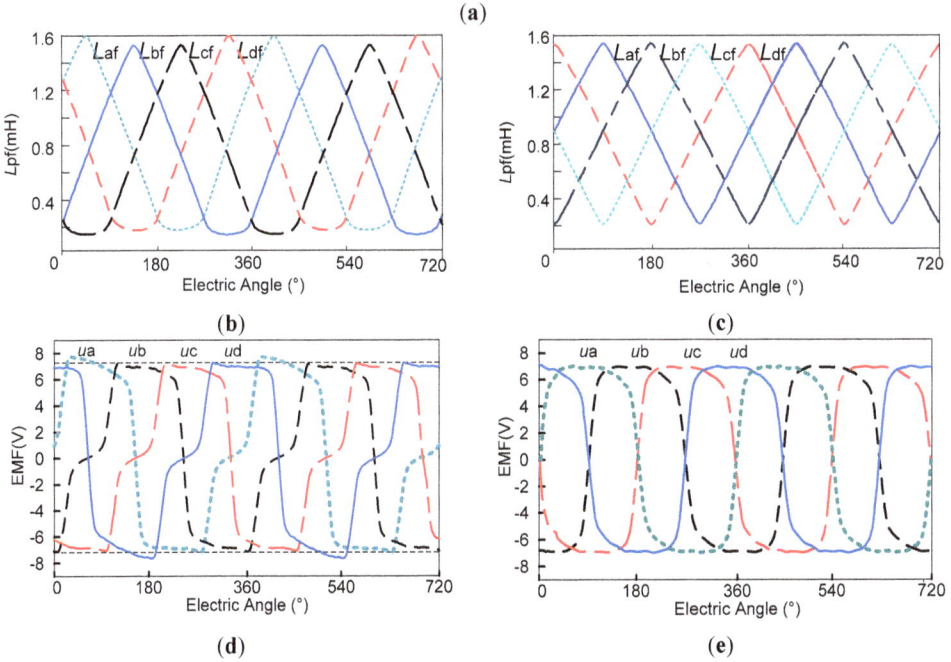

Figure 7. Comparison of the traditional and the new four-phase WFDSM. (**a**) The flux of the 12/9-pole WFDSM; (**b**) L_{pf} of the traditional 8/6-pole WFDSM; (**c**) L_{pf} of the new 12/9-pole WFDSM; (**d**) Back EMF of the 8/6-pole WFDSM; (**e**) Back EMF of the 12/9-pole WFDSM.

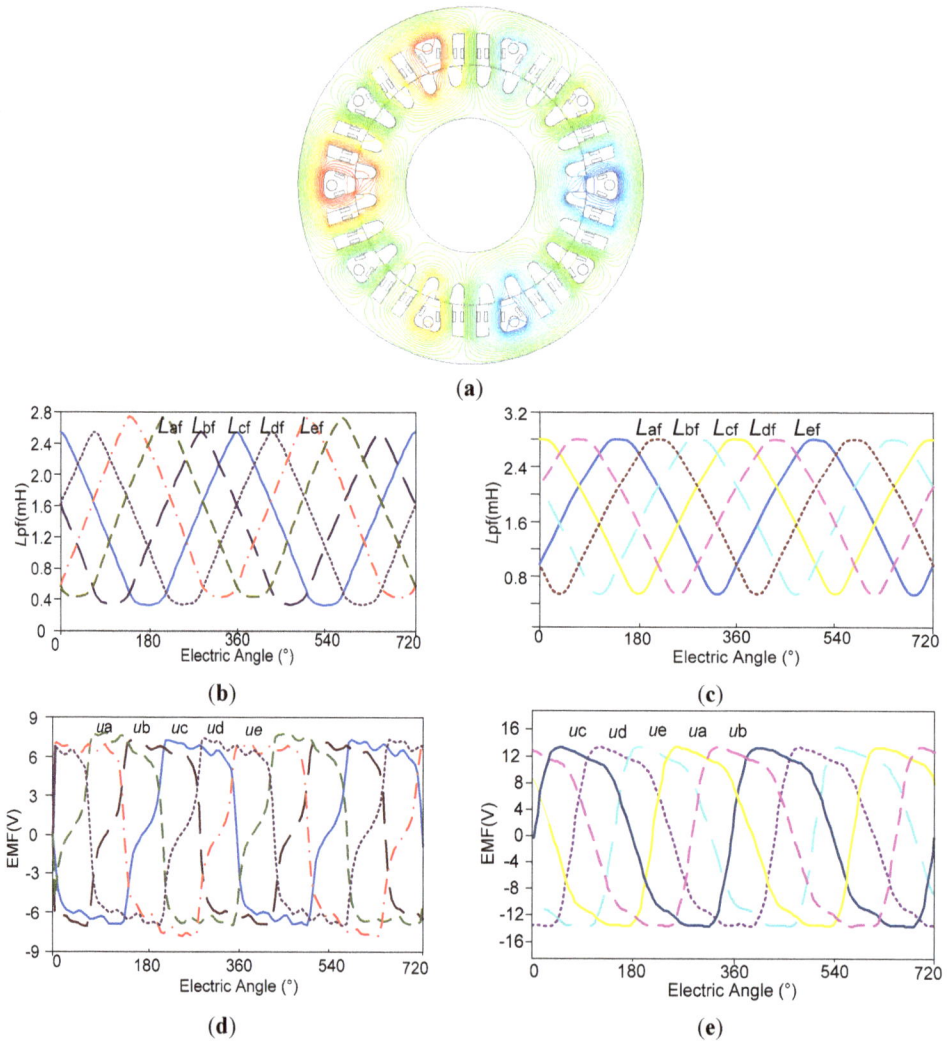

Figure 8. Comparison of the traditional and the new five-phase WFDSM. (a) The flux of the 30/24-pole WFDSM; (b) L_{pf} of the traditional 10/8-pole WFDSM; (c) L_{pf} of the new 30/24-pole WFDSM; (d) Back EMF of the 10/8-pole WFDSM; (e) Back EMF of the 30/24-pole WFDSM.

The flux of the six-phase WFDSM with $j = 1$ is shown in Figure 9b. This new WFDSG has its field coils wound around every stator pole, and solves the problem of phase asymmetry. The 2D-FEA results in Figure 9f verified this theoretical analysis. However, this new machine has a drawback of large copper loss, because there are field coils in every slot, and the resistance and the weight of the field winding will be increased.

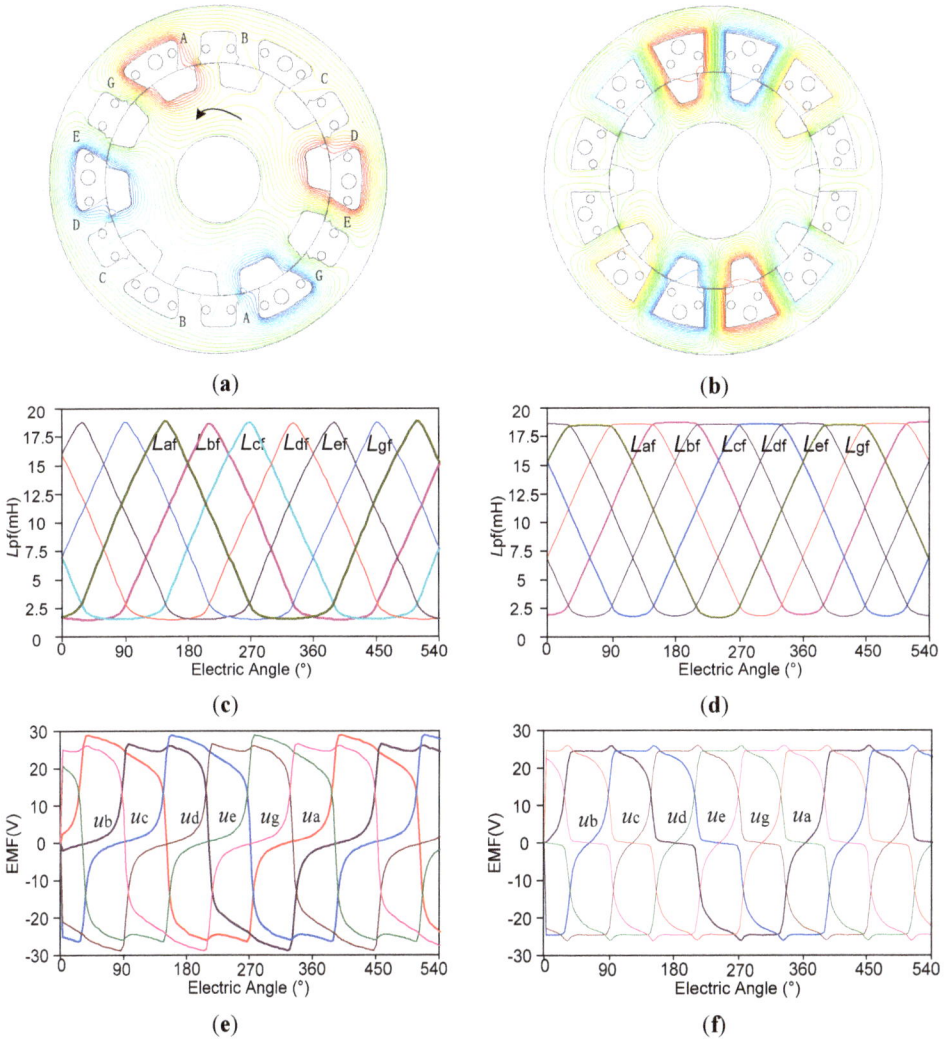

Figure 9. Comparison of the new 12/10-pole six-phase WFDSM. (**a**) The flux of the WFDSM with $j = 2$; (**b**) The flux of the WFDSM with $j = 2$. (**c**) L_{pf} of the WFDSM with $j = 2$; (**d**) L_{pf} of the WFDSM with $j = 1$; (**e**) Back EMF of the WFDSM with $j = 2$; (**f**) Back EMF of the WFDSM with $j = 1$.

4. Comparative Study of the Converter and Its Fault-Tolerant Performance

4.1. The Fault-Tolerant Converters

Multiphase machines can be divided into machines with prime number phases and machines with composite number phases. Because a composite number has at least one positive divisor other than 1 or the number itself, the machines with composite number phases can be divided into several channels which have little effect on each other [24]. For example, the four-phase WFDSM can be divided into two independent channels. However, the five-phase WFDSM has no positive divisor, and the machine may be susceptible to be suffer faults if the phase windings are connected together.

The fault-tolerant machine is usually equipped with a fault-tolerant converter. There are various types of fault-tolerant converters for different phase machines [25], and the four-phase converters will be discussed as an example in this paper.

The four-phase WFDSM is usually powered with a four-phase full bridge converter [26]. When there is a fault in one phase, the machine can keep on working because the fault is isolated from the other two phases, as shown in Figure 10a.

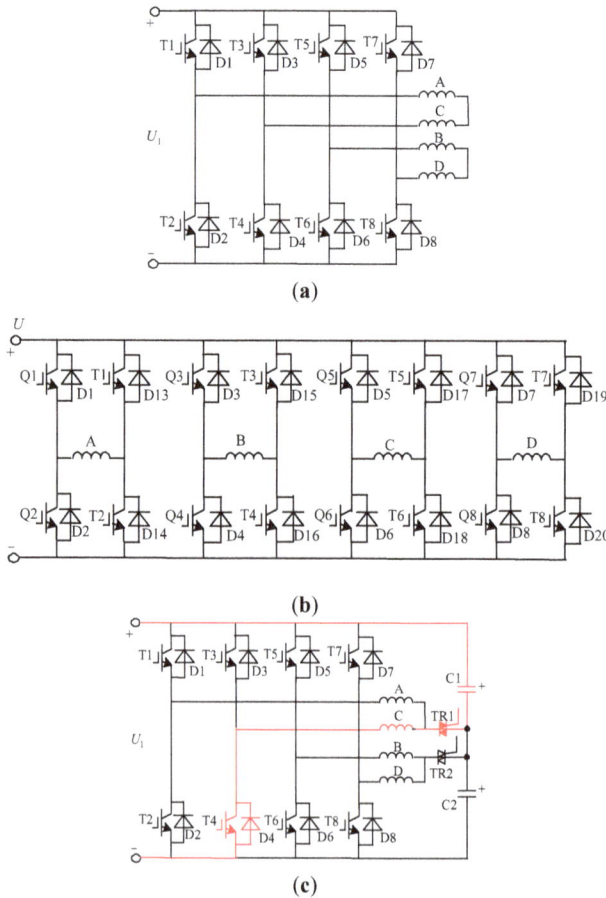

(a)

(b)

(c)

Figure 10. Comparison of the four-phase fault-tolerant converter. (a) The traditional four-phase converter; (b) The four-phase H bridges converter; (c) The half bridge four-phase converter.

170

The most excellent converter is the four-phase H bridges converter, as shown in Figure 10b, because all of the phases are isolated from each other. The machine can be designed as a modular machine, which offers potential fault-tolerant capability because the phase windings are isolated. When there is a fault in one phase, the other phases can keep on working without any infection from the fault phase. However, this converter design is not helpful to reduce the weight and the cost.

Figure 10c shows a half bridge four-phase converter. If phase A has an open circuit fault in this fault-tolerant converter, phase C of the machine can keep on operating with the help of split-phase capacitor C1. However, the five-phase WFDSM cannot be divided into two or three isolated channels, so it doesn't have such flexible fault-tolerant converters, and the faults of the machine may be susceptible to be infected if the phase windings are connected together.

As we can see from Figure 10, the switches increase with the phase number, and WFDSMs with more than six phases are not suitable because the weight and the cost are unacceptable. The WFDSMs with six phases or four phases can be divided into two isolated channels, which will improve the fault tolerance of the machine.

4.2. Fault-Tolerant Performance Comparison

The torque-angle characteristics of the multiphase WFDSMs are shown in Figure 11a. As we can see from the figure that the waveform of the 12/9-pole four-phase WFDSM is wider than the others, because the electrical angle of the phase voltage waveform of the four-phase WFDSM in Equation (8) is 180° and the phase voltage waveform electrical angle of the five-phase WFDSM is 144°, and the angle of the six-phase WFDSM is 120°.

There are four phases with mutative self-inductances at any time in all of the three WFDSMs, which can be described as $x = 4$ in Equation (8). If one phase of the four-phase WFDSM is isolated because of an open circuit or short circuit fault, the other three phases can keep on working. If the phase current maintains the same value because there is no fault, the fault-tolerant torque will be three-quarters of the normal torque. This is verified with Figure 11b.

To compare the fault-tolerant performance, the torque waveform with one phase open of the five-phase and the six-phase WFDSMs are shown in Figure 11b too. If the five steps in one period are named as five beats, then in the five beats of the five-phase WFDSM, there is a beat where the fault phase has no current. At this time, the output torque is equal to the normal torque. In the other beats, the fault-tolerant torque will be three-quarters of the normal torque too. The same result can be obtained with the six-phase WFDSM. This phenomenon is shown in Figure 11b. Because the WFDSMs are brushless DC machines, and their torque ripples are relatively large, which is mainly caused by the commutation torque ripple in the four-phase WFDSM. It should be noted that the commutation torque ripple of the WFDSM can

171

be reduced by the optimal control method, which is investigated in [27]. Besides the commutation torque ripples, the five-phase and the six-phase WFDSMs also have the torque ripples caused by the differences between fault-tolerant beats and normal beats.

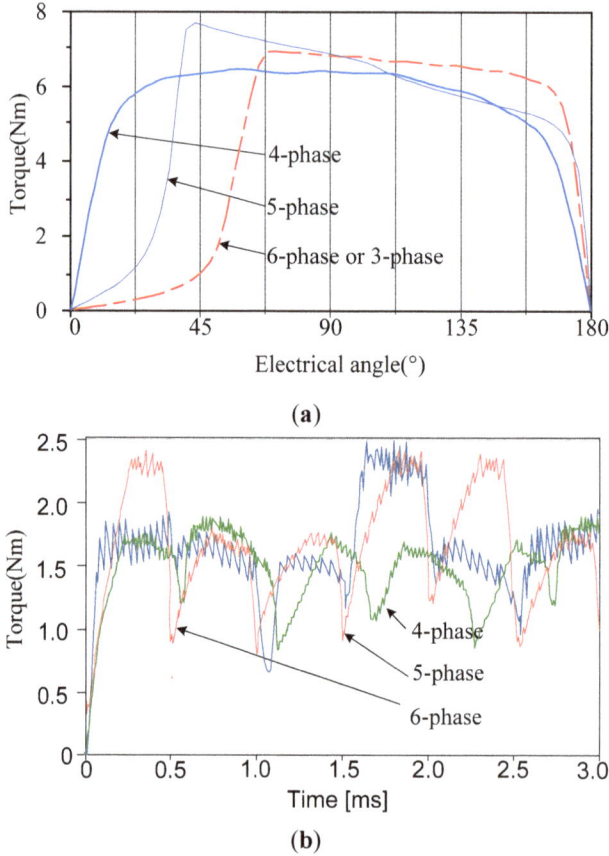

Figure 11. The torque of the multiphase WFDSMs. (a) The torque-angle characteristics; (b) The torque with one phase isolated in an H bridge converter.

What's more, as a brushless DC machine, the WFDSM is usually equipped with Hall sensors. Therefore, the five-phase WFDSM needs five Hall sensors, and the six-phase WFDSM needs only three Hall sensors, because phase A and phase D of the six-phase WFDSM have adverse EMF. The accuracy of the sensor should increase with the rotor pole number. Because the rotor pole number of the five-phase WFDSM with symmetrical phases is larger than the others, therefore, it is not an optimal solution for a fault-tolerant machine.

5. Conclusions

The four-phase WFDSM with $\alpha_s = 0.667$ has four changing inductance phases at any time, and its self-inductance of the field winding L_f will be a constant, which will not produce cogging torque. If it is used as a fault-tolerant machine, it has fewer switches in the converters than five-phase or six-phase machines.

The elementary machine of the five-phase WFDSM with symmetric phases has a 30/24-pole structure. It does not have cogging torque because the phase number with mutative inductance at any time is four. Different from the other five-phase machines with few poles, this five-phase WFDSM needs an expensive incremental encoder to provide sufficient position accuracy for the commutation control, but it can be a high performance fault-tolerant generator because it doesn't need position sensors.

If we wind the field coils around two or more stator poles, the six-phase WFDSM will have the serious drawback of asymmetric phases. The 12/10-pole six-phase WFDSM which has a field coil around each stator pole has symmetric phases, although it improves the copper loss of the field windings. Like the four-phase WFDSM, the six-phase WFDSM can be divided into two isolated channels, which will improve the fault tolerance of the machine. Because the switches of the converter and the pole numbers increase with the phase number, WFDSMs with more than six phases are not suitable because the weight and the cost of the converter are unacceptable, as well as the pole numbers.

Acknowledgments: This work was supported and funded by the National Natural Science Foundation of China (51477075), the Shandong Provincial Natural Science Foundation (ZR2014JL035), and the Fundamental Research Funds for the Central Universities (NP2015205).

Author Contributions: Bo Zhou conceived the idea of this paper, provide guidance and supervision; Li-Wei Shi implemented the research, performed the analysis and wrote the paper. All authors have contributed significantly to this work.

Conflicts of Interest: The authors declare no conflict of interest.

References

1. Villani, M.; Tursini, M.; Fabri, G.; Castellini, L. Electromechanical actuator for helicopter rotor damper application. *IEEE Trans. Ind. Appl.* **2014**, *2*, 1007–1014.
2. Boglietti, A.; Cavagnino, A.; Tenconi, A.; Vaschetto, S. The safety critical electric machines and drives in the more electric aircraft: A survey. In Proceedings of the 35th Annual Conference of IEEE Industrial Electronics, Porto, Portugal, 3–5 November 2009; pp. 2987–2995.
3. Bennett, J.W.; Atkinson, G.J.; Mecrow, B.C.; Atkinson, D.J. Fault-tolerant design considerations and control strategies for aerospace drives. *IEEE Trans. Ind. Electron.* **2012**, *5*, 2049–2058.
4. Garcia, A.; Cusido, J.; Rosero, J.A.; Ortega, J.A.; Romeral, L. Reliable electro-mechanical actuators in aircraft. *IEEE Aerosp. Electron. Syst. Mag.* **2008**, *8*, 19–25.

5. Villani, M.; Tursini, M.; Fabri, G.; Castellini, L. High reliability permanent magnet brushless motor drive for aircraft application. *IEEE Trans. Ind. Electron.* **2012**, *5*, 2703–2711.

6. Hao, L.; Du, H.Y.I.; Lin, H.; Namuduri, C. Design and analysis of PM fractional slot machine considering the fault operation. *IEEE Trans. Ind. Appl.* **2014**, *1*, 234–243.

7. Stewart, P.; Kadirkamanathan, V. Commutation of permanent-magnet synchronous AC motors for military and traction applications. *IEEE Trans. Ind. Electron.* **2003**, *3*, 629–630.

8. Du, Q.; Lu, E.; Shi, X. Design and analysis of five-phase tangent magnetic field permanent magnet generator for electric vehicle. *Int. J. Electric Hybrid Veh.* **2012**, *4*, 378–389.

9. Rottach, M.; Gerada, C.; Wheeler, P.W. Design optimisation of a fault-tolerant pm motor drive for an aerospace actuation application. In Proceedings of the 7th IET International Conference on Power Electronics, Machines and Drives, Manchester, UK, 8–10 April 2014; pp. 1–6.

10. Cossar, C.; Kelly, L.; Miller, T.J.E.; Whitley, C.; Maxwell, C.; Moorhouse, D. The design of a switched reluctance drive for aircraft flight control surface actuation. *IEE Colloq. Electr. Mach. Syst. More Electric Aircr.* **1999**, *11*, 1–8.

11. Hennen, M.D.; Hennen, M.D.; Heyers, C.; Brauer, H.J.; De Doncker, R.W. Development and control of an integrated and distributed inverter for a fault tolerant five-phase switched reluctance traction drive. *IEEE Trans. Ind. Electron.* **2012**, *2*, 547–554.

12. Gong, Y.; Chau, K.T.; Jiang, J.Z.; Yu, C.; Li, W. Design of doubly salient permanent magnet motors with minimum torque ripple. *IEEE Trans. Magn.* **2009**, *10*, 4704–4707.

13. Zhang, Z.; Yan, Y.; Tao, Y. A new topology of low speed doubly salient brushless DC generator for wind power generation. *IEEE Trans. Magn.* **2012**, *3*, 1227–1233.

14. Chen, Z.; Wang, H.; Yan, Y. A doubly salient starter-generator with two-section twisted-rotor structure for potential future aerospace application. *IEEE Trans. Ind. Electron.* **2012**, *9*, 3588–3595.

15. Liu, C.; Chau, K.T.; Zhong, J.; Li, J. Design and analysis of a HTS brushless doubly-fed doubly-salient machine. *IEEE Trans. Appl. Supercond.* **2011**, *3*, 1119–1122.

16. Liu, X.; Zhu, Z.Q. Electromagnetic performance of novel variable flux reluctance machines with DC-field coil in stator. *IEEE Trans. Magn.* **2013**, *6*, 3020–3028.

17. Yu, L.; Zhang, Z.; Chen, Z.H. Analysis and verification of the doubly salient brushless DC generator for automobile auxiliary power unit application. *IEEE Trans. Ind. Electron.* **2014**, *12*, 6655–6663.

18. Sui, Y.; Zheng, P.; Wu, F.; Yu, B.; Wang, P.F.; Zhang, J.W. Research on a 20-Slot/22-Pole five-phase fault-tolerant PMSM used for four-wheel-drive electric vehicles. *Energies* **2014**, *7*, 1265–1287.

19. Cheng, M.; Hua, W.; Zhang, J.; Zhao, W. Overview of stator-permanent magnet brushless machines. *IEEE Trans. Ind. Electron.* **2011**, *11*, 5087–5101.

20. Zhao, Y.; Wang, H.; Zhao, X.; Xiao, L. Characteristics analysis of five-phase fault-tolerant doubly salient electro-magnetic generators. In Proceedings of the IECON 2013—The 39th Annual Conference of the IEEE, Vienna, Austria, 10–13 November 2013; pp. 2668–2673.

21. Liu, X.; Zhu, Z.Q. Stator/Rotor pole combinations and winding configurations of variable flux reluctance machines. *IEEE Trans. Ind. Appl.* **2014**, *6*, 3675–3684.
22. Shi, J.T.; Liu, X.; Wu, D.; Zhu, Z.Q. Influence of stator and rotor pole arcs on electromagnetic torque of variable flux reluctance machines. *IEEE Trans. Magn.* **2014**, *11*.
23. Gaussens, B.; Hoang, E.; Barrière, O.D.; Saint-Michel, J.; Lecrivain, M.; Gabsi, M. Analytical approach for air-gap modeling of field-excited flux-switching machine: No-load operation. *IEEE Trans. Magn.* **2012**, *9*, 2505–2517.
24. Hill, C.I.; Zanchetta, P.; Bozhko, S.V. Accelerated electromechanical modeling of a distributed internal combustion engine generator unit. *Energies* **2012**, *5*, 2232–2247.
25. Bojoi, R.; Neacsu, M.G.; Tenconi, A. Analysis and survey of multi-phase power electronic converter topologies for the more electric aircraft applications. In Proceedings of the 2012 International Symposium on Power Electronics, Electrical Drives, Automation and Motion (SPEEDAM), Sorrento, Italy, 20–22 June 2012; pp. 440–445.
26. Chen, Z.; Chen, R.; Chen, Z. A fault-tolerant parallel structure of single-phase full-bridge rectifiers for a wound-field doubly salient generator. *IEEE Trans. Ind. Electron.* **2013**, *8*, 2988–2995.
27. Qin, H.; Wen, J.; Zhou, B.; Xue, H.H. Considerations of harmonic and torque ripple in a large power doubly salient electro-magnet motor drive. In Proceedings of the 2012 Asia-Pacific Symposium on Electromagnetic Compatibility (APEMC), Singapore, 22–24 May 2012; pp. 649–652.

Probabilistic Agent-Based Model of Electric Vehicle Charging Demand to Analyse the Impact on Distribution Networks

Pol Olivella-Rosell, Roberto Villafafila-Robles, Andreas Sumper and Joan Bergas-Jané

Abstract: Electric Vehicles (EVs) have seen significant growth in sales recently and it is not clear how power systems will support the charging of a great number of vehicles. This paper proposes a methodology which allows the aggregated EV charging demand to be determined. The methodology applied to obtain the model is based on an agent-based approach to calculate the EV charging demand in a certain area. This model simulates each EV driver to consider its EV model characteristics, mobility needs, and charging processes required to reach its destination. This methodology also permits to consider social and economic variables. Furthermore, the model is stochastic, in order to consider the random pattern of some variables. The model is applied to Barcelona's (Spain) mobility pattern and uses the 37-node IEEE test feeder adapted to common distribution grid characteristics from Barcelona. The corresponding grid impact is analyzed in terms of voltage drop and four charging strategies are compared. The case study indicates that the variability in scenarios without control is relevant, but not in scenarios with control. Moreover, the voltages do not reach the minimum voltage allowed, but the MV/LV substations could exceed their capacities. Finally, it is determined that all EVs can charge during the valley without any negative effect on the distribution grid. In conclusion, it is determined that the methodology presented allows the EV charging demand to be calculated, considering different variables, to obtain better accuracy in the results.

Reprinted from *Energies*. Cite as: Olivella-Rosell, P.; Villafafila-Robles, R.; Sumper, A.; Bergas-Jané, J. Probabilistic Agent-Based Model of Electric Vehicle Charging Demand to Analyse the Impact on Distribution Networks. *Energies* **2015**, *8*, 4160–4187.

1. Introduction

Electric vehicles (EVs) are presented as an alternative to current internal combustion vehicles powered by fossil fuels. Increasing oil prices, greenhouse gas emissions and environmental concerns of citizens boost interest in this technology. Energy supply from power networks is required and the impact on the distribution grids in a massive EV integration scenario has to be analyzed in detail [1]. Thus, studies about EV impact on power networks are needed to ensure the viability of the systems [2–4].

The EV charging demand model should allow the analysis of possible effects of this new demand supplied in present-day power networks.

In order to do so, an EV charging model should include specific characteristics for each case, such as mobility, and it should allow one to compare different cases. Moreover, it should consider probability distribution functions (PDF) to analyze the uncertainties of possible EV charges. In addition, this model should be designed to analyze the application of control strategies and enable their comparison.

Literature proposes models to calculate the demand with respect to vehicle, charging infrastructure, mobility, and social parameters. [5–9] use different parameters such as EV model, distance, and charging process among others to determine the EV charging demand.

1.1. EV Type

From the point of view of EV charging demand, EVs main characteristics are the vehicle type: Plug-in Hybrid Electric Vehicle (PHEV) or Battery Electric Vehicle (BEV), battery capacity, battery technology, EV range and energy consumption. Amjad *et al.* [10] expose an analysis about EV design considerations. Different authors only consider PHEV [2,3,5,11–15]. Others only BEV [16–20] or a combination of both [9,21–23]. Another option is to suppose average EV models, BEV and PHEV, with average characteristics like different authors do [5,6,18]. Pang *et al.* [24] simulate only two representative EV models: Chevy Volt (PHEV) and Nissan Leaf (BEV) and Valsera *et al.* [17] simulate Mitsubishi i-MiEV (BEV) only.

Soares *et al.* [9] proposed a stochastic model with mobility variables, but the vehicle characteristics are determined by a Gaussian distribution with standard values for the capacity, energy consumption and charging power of EVs.

The majority of papers simplify the EV model selection, but the capacity and the energy consumption are significant variables to be considered. The model presented proposes using real EV models and their technical data to define the battery capacity and energy consumption of each EV model. Moreover, the probability of each EV model is based on sales forecasting [25] to decide which EV model is more probable.

1.2. Battery and Charging Process

Regarding EV batteries, there are three variables linked: capacity (kWh), range (km) and energy consumption (kWh/km). [19,26] consider the battery characteristics of real models and [15,20,27,28] consider average battery characteristics. Moreover, it is important to take into account the relation between the power consumed and the State-of-Charge (SoC). Valsera *et al.* [17] determine a relation between EV model, battery characteristics (Li-ion, 50 Ah, 16 kWh and 330 V) and its charging process.

The charging process standards of IEC 61851 [29] from Europe and SAE J1772 [30] from the USA could also change the impact in the power system.

Maitra *et al.* [22] compare the impact of each SAE standard. The voltage level in Europe for slow charges is 230 V and a maximum current of 16 or 20 A. In Belgium, houses have a protection up to 20 A [31] and in Spain, the common protection is up to 16 A [17]. Valsera *et al.* use the power ratio of Mitsubishi i-MiEV when the initial SoC is 20% and the EV needs 4 h to reach 100%. Zhang *et al.* [32] use level 1 (120 V–15 or 20 A) in the studio located in the United States. To compare, Grenier *et al.* [33] use 230 V and 15 A and the study is located in New Zealand. The efficiency used in the studies is around 90%, as Collins *et al.* proposed [34] in 1983 and this assumption was recently confirmed by Shuang *et al.* [12] and Clement *et al.* [2,35].

Different authors, such as Clement *et al.* [2] and Guo *et al.* [36], use constant power profiles. On the other hand, Maitra *et al.* [22] consider variable power during the charging profiles. Qian *et al.* [37] propose a charging process model which links the power of the charger and SoC. Gao *et al.* [38] link the SoC and the charging time. Different authors use the specific EV charging profile of a real EV. For example, Qian *et al.* [37] and Lojowska *et al.* [39] use the charging profile of the Nissan Altra EV with a battery of 29 kWh, while Multin *et al.* [40] use a three-phase charging profile of Opel Meriva, which has a battery of 16 kWh.

1.3. Charging Infrastructure

Charging infrastructure parameters include the EV charging point's socket and availability to charge. The majority of works do not consider the EV infrastructure when calculating the EV charging demand. Inherent to this hypothesis is to neglect the effect of the queues at charging points by supposing there are enough charging stations, and the assumption of full compatibility between charging stations and EV connectors. Both could be reasonable in future scenarios with massive presence of EV, but could be a problem for fast chargers. García-Valle *et al.* [41] introduce the queue theory with exponential distribution function to simulate EV charging time and relate it to the maximum charging power of the EV.

1.4. Mobility

Mobility is the third key point of EV charging demand. There is a strong link between energy consumption of EV and urban mobility. For example, Keirstead *et al.* [42] reviewed the energy consumption in urban areas, including electric mobility.

Some authors employ the NHTS (National Household Travel Survey) to analyse the United States, such as [21,32,36,43–45]. In the United Kingdom, studies use NTS (National Travel Survey) and UKTUS (United Kingdom Time Use Survey), for instance [3,46,47]. In Germany, there is the MID (*Mobilität in Deutschland*) which Schroeder *et al.* [48] and Loise [20] apply. The MON (*Mobiliteitsonderzoek Nederland*) is utilised by Dutch studies, as Lojowska *et al.* [39]. The DTU *Transport, DTU. Transportvaneundersøgelsen* is used by Jull *et al.* [49] for a case study of Denmark.

In the case of Spain, there are different databases, for example *Dades Bàsiques de Mobilitat 2008* for Barcelona city [17] and *MOVILIA* for the whole Spain [50].

Metz [27] makes use of the Deutsches Mobilitätspanel to simulate 1000 mobility of household profiles and this includes day and time of departure and arrival, travel distance, vehicle used, and destination. Loise [20] makes projections of EV hourly charging profiles based on MID 2008.

The present work proposes that the reason of displacement be included to determine the destination and the instant of the day to displace. Due to that, it is possible to distinguish between professional and personal mobility.

1.5. Social

There are social variables related to the EV driver profile that could influence EV charging demand as GDP. Kelly *et al.* [43] analyze the EV charging demand considering the income, age and gender of drivers as well as the location (urban or rural). Sikai *et al.* [46] use the number of members of each household and the corresponding number of vehicles based on the UKTUS database. Valsera *et al.* [8] define the number of displacements, the number of houses, and the number of vehicles per house. The proposal of the present work is to combine these three approaches of the previous work to consider social aspects to calculate the EV charging demand.

1.6. Simulation Techniques

To define the characteristics of simulations, there are different details set out by each author. The first one is the data processing, after that the emulation of parameters and lastly, the driver behavior emulation.

Considering data processing, there are different types of simulation models to emulate the EV charging demand and the most used is agent-based. This type of model considers each EV driver autonomously defining the internal (e.g., energy consumption) and external (e.g., power demand to supply EV battery) variables. The bottom-up approach simulates the system coupling all the agents of the system. Different examples of agent-based and bottom-up approach studies are [44,51,52]. On the other hand, the bottom-down approach simulates the EV driver behavior with the average parameters [17,22].

As concerns the emulation of parameters, some models use deterministic variables and others stochastic ones. The deterministic approach considers just average values of parameters and stochastic models use probability distribution functions. The Monte Carlo technique is used to simulate stochastic variables in many applications and it is also used in modeling load, EV charging demand and distributed generation to determine their variability. The majority of studies set out a deterministic approach, but some of them include stochastic variables such as [3,17,20,36,39,46,47]. Some of them use Monte Carlo techniques to simulate the total demand.

EV driver behavior also influences the EV charging demand. This parameter is linked to time of day and location for EV charging, such as public stations between trips, at charging points at work or just home charging.

Venkatesan *et al.* [53] define user profiles related to estimated behavior in the function of mobility, current electricity price and price forecasting. Waraich *et al.* [51] use microsimulation techniques to emulate the driver behavior. Galus and Waraich [54,55] use MATSim (Multi-Agent Transport Simulation) and this tool allows the creation of more than a million connections between agents in transport issues. Balmer [56] uses evolutionary algorithms; Hedegaard *et al.* [57] propose using the Balmorel program to include distribution network, district heating, optimization, taxes and geographical data.

[58] proposes including the game theory to simulate the interaction between agents and including sale of electricity with V2G service. Smith *et al.* [59] use GPS data and EV metering to calculate the energy consumption and later to optimize the battery sizing of future PHEV.

The present work proposes combining some characteristics presented in literature. The methodology presented is a bottom-up approach to process the data with stochastic variables following the Monte Carlo formulation to emulate the parameters. And the driver behavior is defined in function of the range anxiety, the mobility needs and the energy price.

1.7. Power System Impact

Possible effects on power networks caused by EVs are related to power quality or grid saturation. The majority of studies analyze the voltage drop or transformer load, like Valsera *et al.* [8,17]. Clement *et al.* [2] include Joule losses and Maitra *et al.* [22] include overloading and unbalances. Kleiwegt *et al.* [60] propose a methodology to detect overloads in the course of a year. Moreover, vehicle-to-grid possibility is analyzed in many studies such as [26,61,62]. Another possible impact on the power system is economic and this is reviewed by Dallinger *et al.* [63]. The present work analyses the distribution network in terms of the HV/MV and MV/LV transformer capacities and the voltage of each node.

1.8. Contribution

The state-of-the-art analysis defined seven subjects to be determined in the EV charging demand problem formulation:

- EV type and model: the majority of current models simplifies this aspect with one model or an averaged model to represent a group of models.

- Battery and the corresponding charging process: according to the literature review, the main difference found in literature is the charging process. The most common simplification is to consider a constant power but the appropriate way is to consider the relation between the SoC and the power consumed.
- Power infrastructure: the majority of articles consider the AC slow charging and the current limit depends in function on the country analyzed.
- Mobility: the papers which consider it try to use the public data according to the country analysed.
- Social: the majority of the papers do not consider any economic or social variables.
- Simulation technique: the majority of papers take a bottom-down deterministic approach.analyzed
- How to analyze the impact on the power system: the majority of EV charging models avoid this issue and some of them try to optimize the EV charges to reduce some negative consequences.

The objective of this paper is to define a methodology based on agents to determine EV charging demand. The main contribution of this paper is to propose a methodology based on open data and combining social, technical and economic variables to calculate the EV charging demand and then determine the effects on the distribution networks. To do so, the parameters in literature were used separately; however, this paper proposes that all of them be combined in a single model in order to obtain more precise and realistic results. Figure 1 shows the relation among the variables that are implemented in the present model. For example, EV agents have a set of constant parameters as EV model (technical), place of residence (social), GDP (economic) and others, as well as variable parameters of mobility such as distance, day of the week and others.

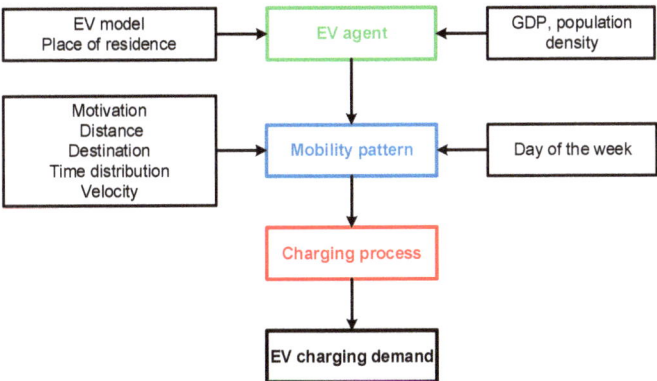

Figure 1. Basic scheme of EV charging demand parameters.

Finally, the result of this methodology leads to the charging process model for each EV agent, the total EV charging demand and consequently, it allows the impact on power networks to be analyzed. The methodology proposed uses all sources from public data and it is applied using statistics from the city of Barcelona.

The EV charging demand model is defined as the electric demand from EVs during a certain time period, such as a day or week, to supply their batteries. EV charging demand depends on EV user driving needs and it is linked to EV characteristics and mobility of users.

The methodology proposed in this paper is the Agent-Based Modeling and Simulation (ABMS). The main strengths and applications of ABMS are listed as follows:

- Heterogeneous individual components: EV model and mobility pattern of each EV owner.
- Flexible systems: to manage the charging demand of each EV.
- Influence of location: to consider the effects of the charging point location in the power network.
- Representation of social interactions: different types of EV owners could have different influences on the total system.

For these reasons, this methodology has been used for obtaining EV mobility patterns with an heuristic approach [64]. Furthermore, this methodology enables to simulate complex systems; for instance, load demand in power systems [7] or virtual power plants to include different types of agents [65]. Thus, agent-based modeling has been selected for this research.

In this work, the EVs are a set of agents that has been defined as autonomous entities with their attributes and their processes are dynamic and time-dependent [66,67]. It allows defining each EV driver as an agent considering the usage of each vehicle. Each agent is simulated individually including possible interactions through the relationships between agents. Section 2 describes the characteristics of the agent-based model to obtain the charging demand from EVs and their impact on the distribution network.

2. EV Charging Demand Model

According to the Figure 1, the parameters needed to model the EV charging demand can be clustered in three groups: the EV agent (Section 2.1), mobility pattern (Section 2.2) and the charging process (Section 2.3). All these parameters permit the determination of all charging processes needed to reach each destination.

2.1. EV Agent

In the model developed, every EV agent represents an EV driver and its vehicle. The EV agent attributes are the EV model, the mobility needs, and the charging preferences. The EV agent behaviors are the trips taken (mobility), their corresponding energy consumption from their battery, the energy consumed from the electricity network to charge the battery, and the charging decision. For instance, when EV agents reach their destination, their charging process begin depending on the EV agent preferences and the energy price. The EV agent states with their corresponding variables are: waiting, driving, and charging.

Moreover, there are two other agents that influence on EV agents behavior: the Electricity Retailer Agent, who determines the electricity price for each instant, and the EV Aggregator Agent, who control the EV charges to reduce the electricity price. In the scenarios A, B and C, explained in the Section 3, there is no EV aggregator and the price is determined by the Electricity Retailer Agent. In contrast, in the scenario D, also explained in Section 3, the price is determined by the EV Aggregator Agent and the Electricity Retailer Agent does not influence on EV agents.

The main rule is that each EV agent, after each trip, takes the decision of charging in function of the battery state-of-charge, the electricity price and, in scenario D, the signal from the EV aggregator. Moreover, before changing the state of an EV agent from waiting to driving state, it is necessary that the battery has enough energy to reach the destination. The EV agents structure, their relationships with other agents and their environment are shown in Figure 2. Note that there are two environments related to the EV agents: spatial distribution and electricity network. Furthermore, the electricity market is the environment of Electricity Retailer Agent and EV Aggregator Agent.

When the simulation begins, the system computes the EV agent mobility needs and the battery state-of-charge variation.

The first step to define the EV agents is the definition of EV agent groups (C_i) and their variables. For each group, it is necessary to define the number of agents (N), spatial distribution of influence and charging preferences. And the EV model of each agent is defined with variables EC_i, Aut_i, Cap_i, Ps_i and $Type_i$. The place of residence, defined in R_i, is considered for each agent, and this depends on the power network scenario and is modeled as a constant probability, based on public data such as [68]. R_i is linked with the charging point in home usage.

The PDF of each EV model is based on [25] data and it just considers passenger vehicles and $Type_i$. This data was filtered for the case study in relation to EV model characteristics and technical data available from automakers. It is shown in Figure 3.

In this model it is assumed that the PHEV drive is fully electric until the end of the energy stored in the battery, when they consume gasoline as hybrid electric vehicles. Other assumptions are exhibited in Section 2.4.

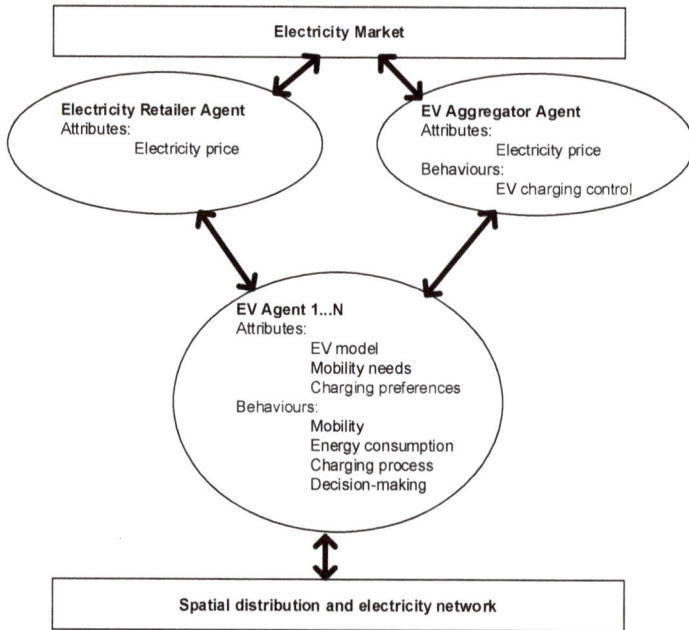

Figure 2. EV agent structure

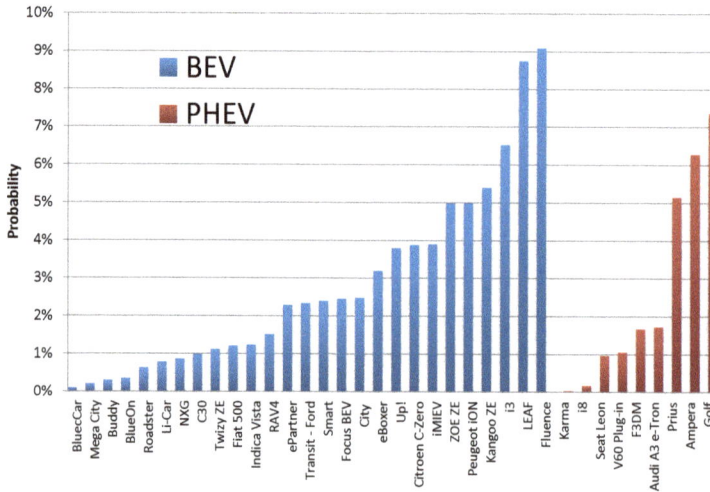

Figure 3. EV model probability distribution function EV_i. Based on [25] and adapted to Barcelona-Spain and automakers data.

2.2. Mobility Pattern

Mobility variables are assigned to each EV agent in order to model its mobility behaviour. Different mobility patterns are based on open data sources. The variables considered to define a mobility pattern are defined as follows:

- **Trips per day (S_i).** The total trips are determined using a probabilistic variable which is generated through a Poisson distribution function, which is defined as [69] proposes with Poisson parameter (λ) of Equation (1).

$$P(k, \lambda) = (e^{-\lambda}\lambda^k)/k! \tag{1}$$

This parameter is based on the average statistic value. It should ensure at least two trips per day and is defined by Equation (2).

$$S_i = 2 + \lambda \tag{2}$$

In the present study analyzed, $\bar{S}_i = 3.53$ trips/day are based on [70].

- **Distance (L_i) and Distance per trip (l_{ij}).** They are calculated using the exponential distribution function from public reports. Figure 4 shows cumulative exponential distribution functions of distance traveled per day from different countries and the relation between L_i and l_{ij} is shown the following equation:

$$L_i = \sum_{j=1}^{S_i} l_{ij} \tag{3}$$

In the case study analyzed, $\bar{L}_i = 83$ km/day is based on [71]. If $l_{ij} > 10$ km, the trip j is considered as metropolitan considering Barcelona characteristics.

- **Destination (D_{ij}).** The model considers the reason of displacement to determine the destination. The reasons considered for the case study are based on the destination of each trip: for personal issues and for commuting. It is strongly linked to grid node, where the EV is connected in relation to social data and mobility pattern. The destination is modeled with a constant PDF according to the power network topology.
- **Day of the week (d_i) and Time distribution (m_{ij}).** These parameters allow knowing when an EV consumes energy as a function of the EV user's motivation to travel on a specific day. It is implemented in a PDF, as shown in Figure 5 and Table 1 as an example applied in the case study.
- **Velocity (v_{ij}).** According to mobility data, velocity is modelled as a constant value, depending if the trip is urban or metropolitan. The average velocity from [70] and $v^{urban} = 22.2$ km/h and $v^{metrop} = 59.3$ km/h are applied.

185

- **Initial/Final time** (t_0, t_1). The relation between them is the average velocity (v_{ij}) and distance (l_{ij}). Each pair of time variables is grouped in the matrix Y_i, which stores the mobility data of an EV agent.

$$Y_i = \begin{bmatrix} t_0^1 & t_1^1 \\ \vdots & \vdots \\ t_0^{S_i} & t_1^{S_i} \end{bmatrix}$$

- **Social variables**. Regarding the case study, it is necessary to take into account different variables such as Gross Domestic Product (GDP) and population density to determine the total number of agents (N) that could charge the EV at the same connection point. C_i definition was described in Section 2.1 and applied in Section 3.1.

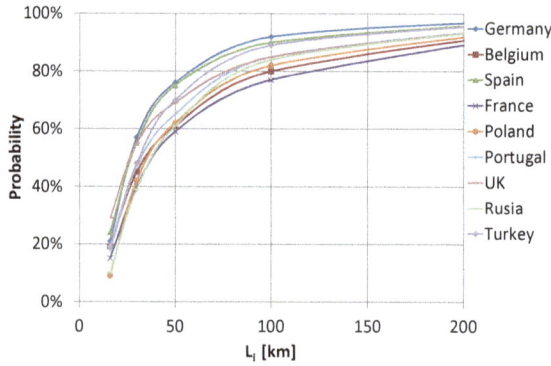

Figure 4. Probability distribution function of Distance L_i [71].

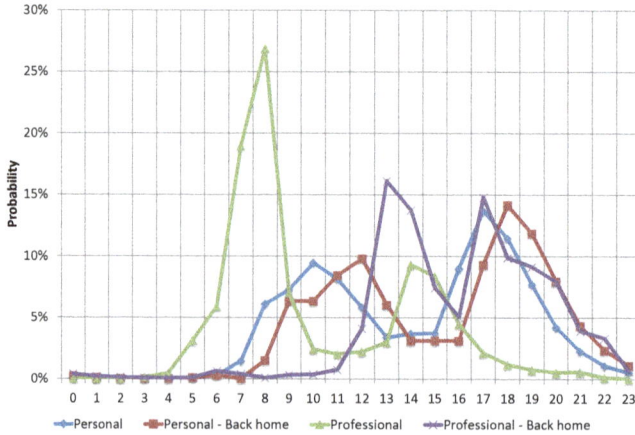

Figure 5. Probability distribution function of Time distribution m_{ij} [71].

Table 1. Time distribution considered in case study.

m_{ij}	Description
1	Personal
2	Personal—Back home
3	Professional
4	Professional—Back home

2.3. Charging Process

The charging process considered is slow charging—AC single-phase, depending on EV model, battery capacity, SoC, Energy required to arrive to next destination and time between displacements.

All the EV models are supposed to have Li-ion batteries and the slow charging process corresponds to a typical charging curve with two periods: constant period I and descendant period II [72]. The power rate Ps_i considered for charging is 3.7 kW (230 V, 16 A) because it is commonly available in residential and commercial areas in Europe [73] and it is also used by Marra *et al.* [72]. The charging process depends on initial SoC and energy required (E_{req}) in the process. Figure 6 shows the charging process of a battery with Cap_i and E_{req} of 16.5 kWh.

In this model, it is assumed that period I requires 50% of time for a full charge and period II finishes when the power output reaches 8% of Ps_i.

- Total energy (Battery capacity) is: $Cap = E_I + E_{II}$.
- μ and k are the exponential function parameters used in Equation (6).
- Total process efficiency considered is 90% [2].

The equations of EV charging process described before are:

- Period I is described by the following equations:

$$P_I(t) = Ps_i \tag{4}$$

$$E_I(t) = \int_0^a Ps_i dt \tag{5}$$

- Period II is described by the following equations:

$$P_{II}(t) = ke^{-\mu t} \tag{6}$$

$$E_{II}(t) = \int_a^b ke^{-\mu t} dt \tag{7}$$

where:

$$\mu = \frac{-ln(0.08)}{a} \tag{8}$$

$$k = \frac{Ps_i}{0.08} \tag{9}$$

$$c = 0.08Ps_i \tag{10}$$

The initial SoC depends on the EV agent consumption. In the first simulation, the battery starts fully charged.

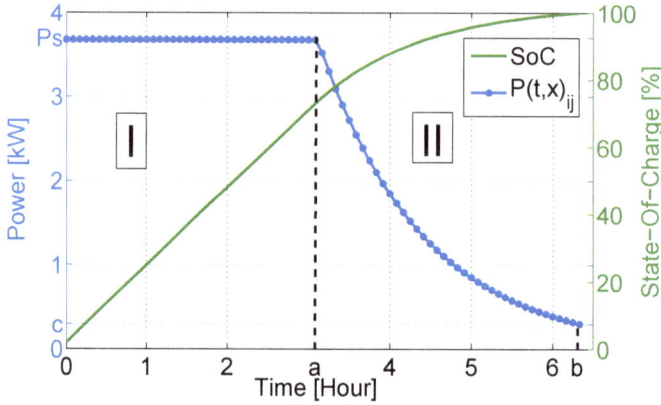

Figure 6. Slow charging profile—General scheme in relation to battery capacity. Based on [72].

2.4. Monte Carlo Simulation

Based on Figure 2, this paper proposes using the algorithm shown in Figure 7 to calculate the EV charging demand in a certain power network. This algorithm is based on Monte Carlo Methodology to include stochastic variables per agent and they are: R_i, S_i, L_i, l_{ij}, D_i, t_0, t_1 and EV_i. For this reason, it is necessary to define the number of iterations (T). Furthermore, to start the algorithm, it is necessary to define the number of agents (N) that charge the EV in the network analyzed. The time step used is 5 min.

The algorithm is used to define the EV agent group, the mobility variables and then the charging process for each EV agent.

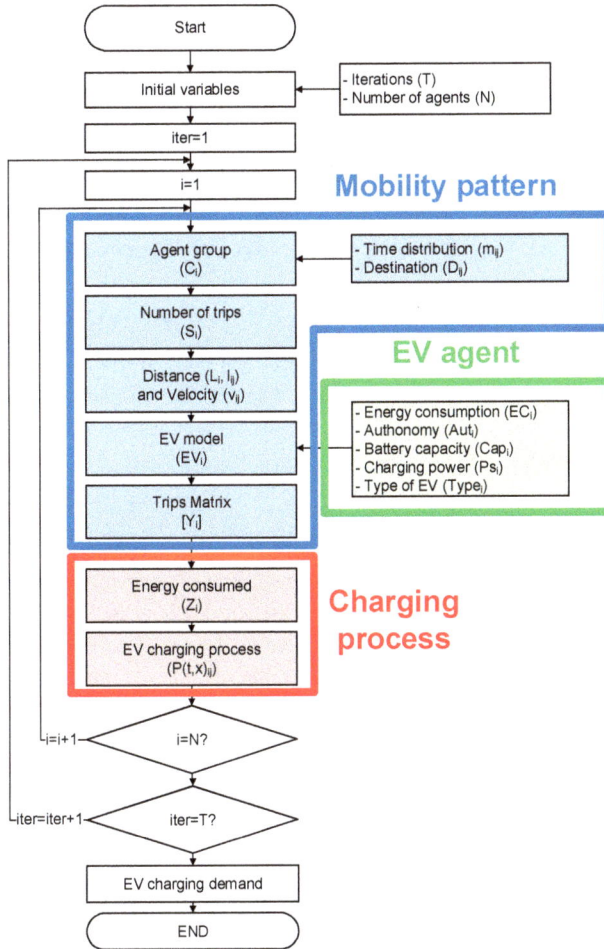

Figure 7. EV charging demand algorithm based on Monte Carlo.

3. Case Study

The proposed EV charging demand model is applied in a case study with a 37-node IEEE test feeder adapted to a typical distribution network and mobility data of Barcelona (Spain) [70]. The modeling of the case study was implemented in MATLAB and the power flow is solved by means of the Newton-Raphson method.

Four charging scenarios (A-D) were defined to model EV agent behavior, which are described in the following sections. The results are the energy (Z_i) and charging demand from EVs ($P(t, x)$) and the voltage profile in the distribution network.

3.1. Distribution Network

This case study is an adapted MV network 37-node IEEE test feeder, which is seen in Figure 8, and it applies Barcelona's mobility data. This network is adapted to a typical 25 kV MV network of Barcelona and the number of houses connected at the same MV/LV transformer [8]. In order to do that, it is necessary to consider social variables such as population density and technical regulation [74]. The maximum voltage drop permitted by the distribution system operator is 10% according to the EN 50160.

The total number of agents of group C_i is defined in relation to network topology and population density of different neighborhoods. According to social data from Barcelona and network branches, there are three zones: high, medium and low inhabitants per house and vehicles per inhabitant density. The farthest branch is linked with the high density zone. In this way, D_{ij} of group C_1 at the end of the day is the corresponding network node. In Barcelona, 38% of vehicles are driven each day and this percentage is used to determine active vehicles [70].

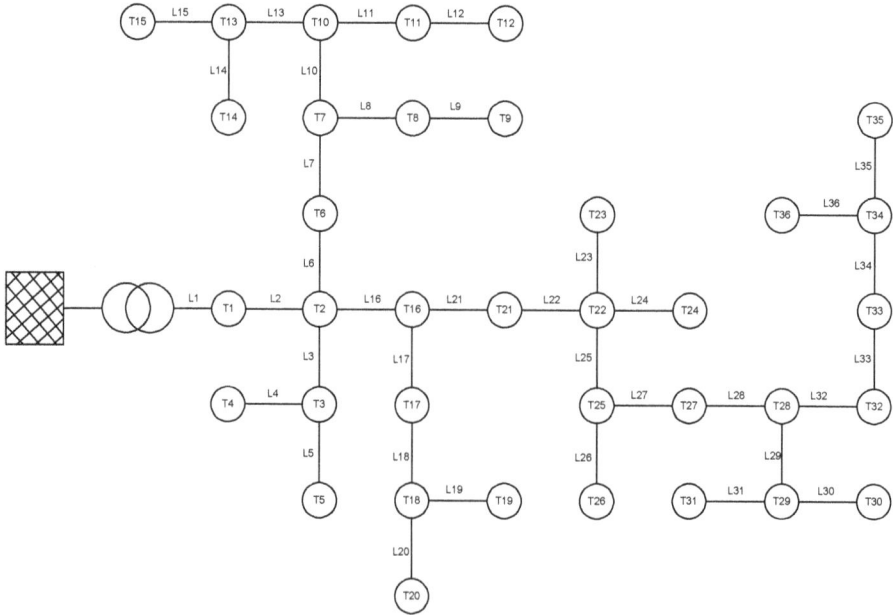

Figure 8. MV network - IEEE Test-feeder 37 node.

Table 2 shows calculations to get N of group C1.

Table 2. Number of agents C_1.

Zone	Nodes	Inhab./Hou.	Veh./Inhab.	Inhab.	Active Veh.
High	22–36	2.61	0.50	5016	950
Medium	3–5, 6–15	2.52	0.47	2541	448
Low	1, 2, 16–21	2.34	0.38	3288	471
				Total C1	**1870**

Load demand: Base load demand in this distribution network is based on system operator data [75] from national demand and it is adapted to network power capacity as 80% of HV/MV transformer power. Analyzing the consumption in Spain between 2007 and 2011, load demand used in the case study is from 17 December 2007, when the maximum energy demand reached 45,911 MWh between 18:00 and 19:00. This allows analyzing EV charging increase relative to this base load.

The load presented in Figure 9a is the base case, without EVs, of the distribution system analyzed. The peak demand is 10,640 kW and it occurs at 18:30. The load demand of the distribution system increases during the morning (8–10 o'clock), decreases during lunch time (13–16 o'clock) and increases during the evening (19–21 o'clock), when people come back home. The peak period is 79% higher than the valley period and the energy consumed during the course of a single day is 207.36 MWh. The voltage in the worst node is shown in Figure 9b; the minimum voltage is 0.9707 p.u. at 18:30 and the maximum is 0.9839 p.u. at 4:45. The voltage follows a similar behavior to the load demand. The lower limit of the voltage magnitude permitted by EN 50160 is 0.90 p.u.

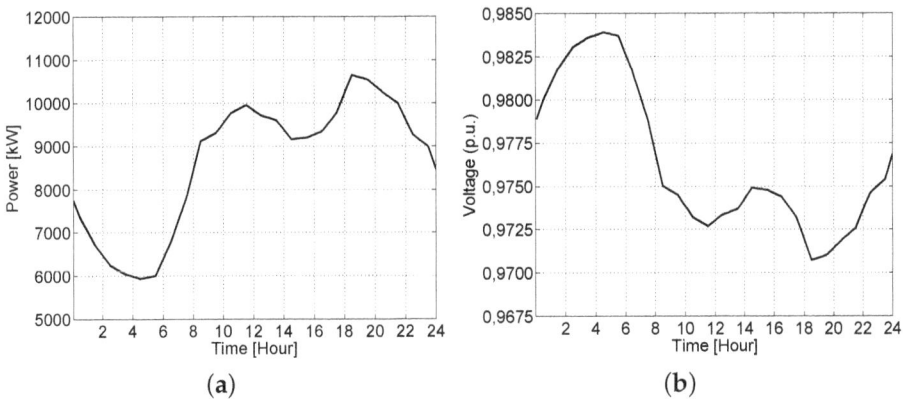

(a)

(b)

Figure 9. Residential and commercial demand without EVs. (a) Load demand; (b) Voltage drop.

191

3.2. Agent Profile

Six agent groups (C1–C6) were defined to consider mobility and residence. Mobility is divided between personal and professional reasons. According to the usual place where the EV is connected at the end of the day, three different areas of residence were defined: local, urban and metropolitan. Local area refers to the distribution network analyzed, urban refers to the city, and metropolitan is outside the city. Urban and metropolitan agents can plug in between displacements. On the other hand, local agents can charge at any time. Table 3 shows the main characteristics of each group. N is the number of EVs of each agent that charge their batteries in the case study network.

Each group has specific energy requirements for charging (E_{req}). Preferences are related to when to charge and they are described above relative to agent group definition. Regarding the E_{req} for each feasible charge between displacements, it is defined as the energy required to reach the next destination (D_{ij}) and distance (l_{ij}).

Mobility variables from Barcelona data [70] are implemented in the case study. S_i depends on agent group, d_i is the average weekday and L_i is according to [71].

Table 3. EV charging social characteristics in function of group.

C_i	m_{ij}	Active Veh.	N	Area	Preferences
C1	1 & 2	1870	561	Local	At-the-end
C2	1 & 2	449	135	Urban	Between disp.
C3	1 & 2	273	82	Metropolitan	Between disp.
C4	3 & 4	41	12	Local	At-the-end
C5	3 & 4	41	12	Urban	Between disp.
C6	3 & 4	10	3	Metropolitan	Between disp.
Total		2684	805		

3.3. Charging Scenarios

According to agent preferences, E_{req} and electricity market assumptions, four scenarios of EV charging demand are described, shown in Table 4.

Table 4. Table of charging scenarios.

Charging Scenario	Description	Range Anxiety
A—Intensive charge	As soon as possible	High
B—Plug-and-Play	Just at home	Medium
C—Tariff controlled	Off-peak tariff	Medium
D—Smart charging	With Aggregator	Low

Scenarios A and B consider constant electricity price for the whole day. In scenario A, EVs charge at the end of each trip due to the high range anxiety of EV agents. In scenario B, the EV agents have lower range anxiety and they charge the vehicle at home, when SoC is lower than 20% or lower than E_{req}. In scenario C it is considered that the EV agents have a Time-of-Use (TOU) tariff, special for EVs [76]. The cheapest period of this tariff begins at 1:00 am, based on the Spanish regulation [77], and then the EVs initiate the charge. The TOU tariff is an indirect control strategy to manage the EV charges. Scenario D considers one aggregator who manages all EV charges to consume the minimum power at the HV/MV transformer. This is based on an aggregator dedicated to reducing the impact in the transmission system, according to the Spanish regulation [77]. This scenario shows a direct control strategy to manage the EV charges and the aggregator offers lower electricity prices for EV agents.

3.4. Results

The following discussion presents the results of the four scenarios simulated. The analysis is focused on the EV demand, total demand and the voltage drop in the worst node. Due to the probabilistic design of the model, the results are variable and the plots show the variation between the maximum and minimum energy consumption. Furthermore, the plots also show the average consumption as the most probable value.

All scenarios are simulated considering that 30% of active vehicles are electric (N), based on maximum scenarios in [2,22,78]. EV_i PDF is based on [25]. What is also considered is that the EV agents with the value L_i greater than 100 km are only PHEV ($Type_i$).

The impact on power system is analyzed through voltage drop located in the farthest node, which is the 35. Figure 10 shows the minimum voltage per node during the whole day and the maximum voltage drop is located in node 35.

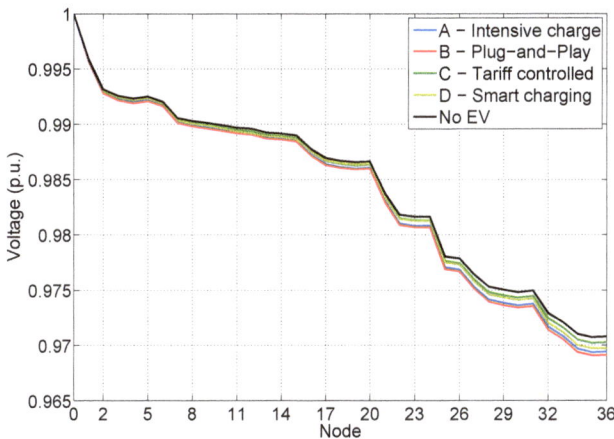

Figure 10. Voltage per node.

193

Iterations (*iter*). The standard deviations (std. dev.) of power demand are evaluated to determine the number of iterations (T) to obtain valid results. To do that, a simulation with 1200 iterations in scenario A for C1 group and with 30% of EVs was carried out.

Figure 11 shows the std. dev. around hour 21 and it varies during the first 100 iterations significantly; it is nearly stable from iteration 200 and is constant from iteration 600. The ideal should be to do 600 iterations for all the cases, but the computing time to do it is very high and the volume of results to be stored requires a huge amount of memory. For these reasons, it is not possible to simulate 600 iterations for all the scenarios and the number of iterations has to be lower. The std. dev. varies around 10 kW from iteration 100 and from iteration 200, the results are more stable than previously. According to this, the number of iterations applied in the case study is 200. Other instances and scenarios are also checked and they comply with the std. dev. analysis. The consumption variation is also checked and it behaves similarly to the std. dev.

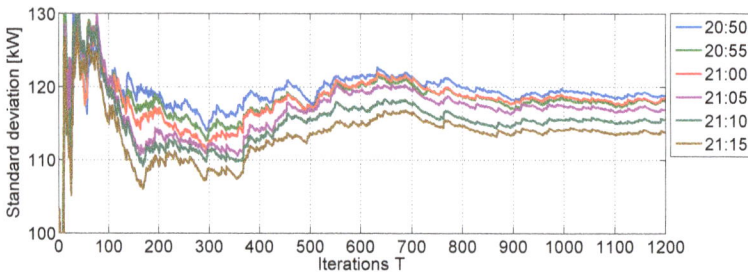

Figure 11. Standard deviation variation in function of iterations T.

3.4.1. A—Intensive Charge

EV charging demand: As is shown in Figure 12a, the EV charging demand presents two peaks with more consumption around 10:00 and 19:00. Both peaks are related to Barcelona's mobility pattern illustrated in Figure 5, which shows the same peaks: the peak during the morning is caused by professional mobility and the peak during the evening is caused by professional and personal back home reasons. The EV charging demand variability, the difference between the minimum and the maximum case, is significant in this scenario, and it can reach the 50% of the EV consumption as it occurs at 20:00.

The EV peak demand is near to 500 kW and the total peak demand is 11.04 MW, 3.75% higher than in the base case without EVs, as Figure 12b demonstrates. Furthermore, the peak during the morning is coupled with the residential and commercial demand. This is reflected in Figure 12b, where the active power increase is steeper from 6 to 12 hours due to the EV charging demand.

Impact on power system: Figure 12c shows that the minimum voltage in node 35 is 0.9694 p.u. and it is 0.13% lower than in the No EV case, which is higher than the lower limit of the standard of 0.9 p.u.

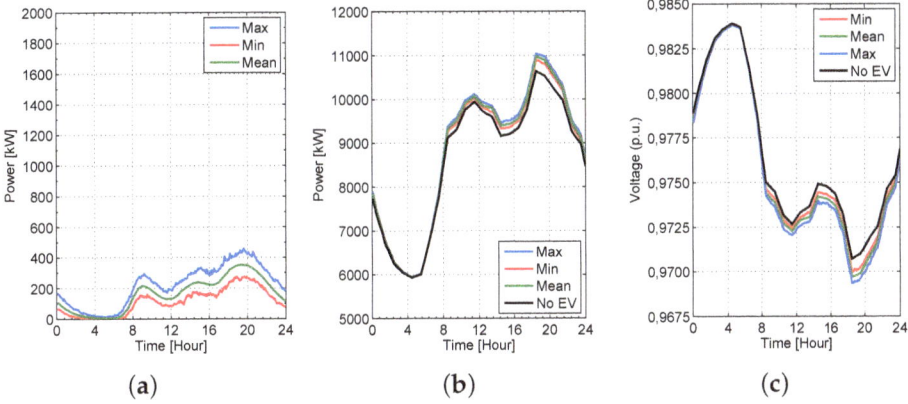

Figure 12. A—Intensive charge. (a) EV charging demand; (b) Total demand; (c) Voltage drop.

3.4.2. B—Plug-and-Play

EV charging demand: In this scenario, the EV agents prefer to charge at home, according to the back home time distributions (m_{ij}). As shown in Figure 13a, the first peak demand is lower than in scenario A because the agents do not charge at work. Moreover, the second peak demand is higher than before because the agents have not charged at work and the energy required by them is higher than in scenario A. In this scenario, the EV charging demand variability is also significant and it can reach the 33% of the EV consumption, as it occurs at 20:00.

As Figure 13b shows, this effect causes that the peak during the morning in the total demand is lower than the previous case. And the peak during the evening is higher due to the energy required and the maximum power consumed is 11,12 MW at 18:35 and the relative increase from the case without EV is 4.51%. Moreover, the power consumption during the night is higher than in case A, because the SoC of EV agents when they arrive at home is lower than previously.

Impact on power system: Figure 13c shows that the combination of the peak from the residential demand with the EV demand causes a higher voltage drop than scenario A, due to the different behaviors of the EV agents. The minimum voltage reached during the peak demand is 0.9691 p.u., 0.16% lower than the case without EV, and higher than the lower limit of 0.90 p.u.

195

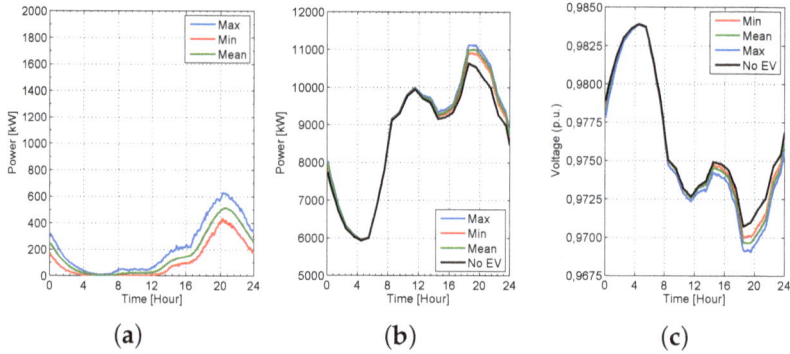

Figure 13. B—Plug-and-Play. (a) EV charging demand; (b) Total demand; (c) Voltage drop.

3.4.3. C—Tariff Controlled

EV charging demand: In this case, the TOU tariff causes that the EV agents begin to charge at 1:00, when the energy is cheaper. Therefore, the EV charging demand presents a peak of 1.86 MW at this moment due to the simultaneous EV charges, as seen in Figure 14a. What is more, the control reduces the EV charging demand variability.

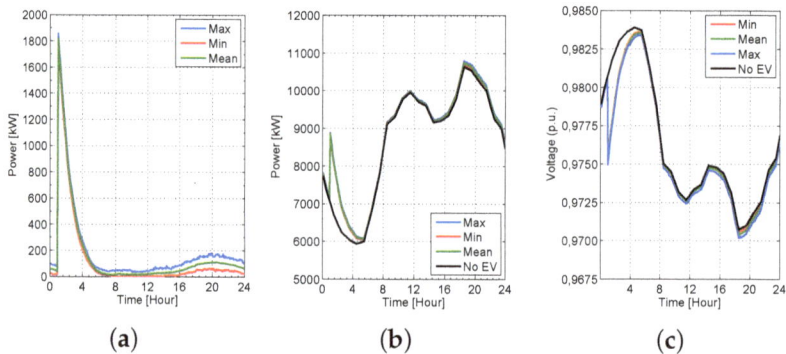

Figure 14. C—Tariff controlled. (a) EV charging demand; (b) Total demand; (c) Voltage drop.

The consumption during the rest of the day is related to the energy required (E_{req}) to reach the next destination (D_{ij}) and the low SoC of each EV agent. The maximum power consumed is 10.8 MW at 18:30, which means an increase of 1.5% from the original case.

Figure 14b shows that this EV peak happens during the off-peak period and the total demand increase is not significant. Despite this, the power generation gradient

196

could be a problem, which should be analysed from the point of view of the power generation and from the system stability point of view.

Impact on power system: The minimum voltage, shown in Figure 14c, is similar to the original case without EVs. The minimum voltage reached is 0.9702 p.u., 0.05% lower than without EVs, and higher than 0.90 p.u. The voltage variation at 1:00 could be a problem, which could be analyzed in a transient analysis.

3.4.4. D—Smart Charging

EV charging demand: Figure 15a shows the EV charging demand controlled by the aggregator which controls domestic EV charges. The EV charging demand is shifted to the valley period to reduce the consumption through the HV/MV transformer and to minimize the impact on the transmission system. According to this, the EV charges occur between 2 and 8 o'clock and the variability, the difference between the minimum and the maximum case, is very small.

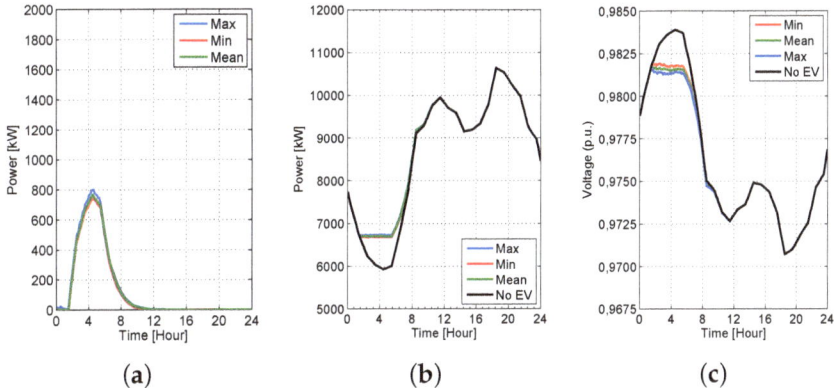

Figure 15. D—Smart charging. (**a**) EV charging demand; (**b**) Total demand; (**c**) Voltage drop.

Figure 15b shows that the total demand increases during the valley periods and the power consumption is constant at 6.6 MW. During the rest of the day, sporadic charges could occur, but the mean curve is near to the case without EVs.

Impact on power system: The minimum voltage is not increased by the EV charges, as is exhibited in Figure 15c. The voltage during the valley period is lower than in the original case according to the total demand, but this voltage is higher than during the peak hours, and the difference between the minimum voltages is 0.02%, and the minimum value of 0.90 p.u. is not reached.

The summary of all the scenarios is presented in Table 5. Voltage value is the minimum and it means the maximum voltage drop.

Table 5. Maximum results.

Scenario	EV Demand (Max) [kW]	Peak Time	Total Demand (Max) [kW]	Variation	Peak Time	Voltage (Min) [p.u.]	Variation
No EV		18:30	10640		18:30	0.9707	
A—Intensive charge	457	18:30	11040	3.76%	18:30	0.9694	−0.13%
B—Plug-and-Play	628	18:35	11120	4.51%	18:35	0.9691	−0.16%
C—Tariff controlled	1857	01:00	10800	1.50%	18.30	0.9702	−0.05%
D—Smart charging	799	04:30	10720	0.75%	18:35	0.9705	−0.02%

Box plots Figure 16a–d show total consumption in each node and this is compared to MV/LV transformer capacity. The results show that the nodes with less capacity could reach the nominal value in some cases, but the average value is under nominal power. In the case of scenario D, total demand never exceeds the nominal capacity of transformers, which means that there is enough capacity to supply the EVs.

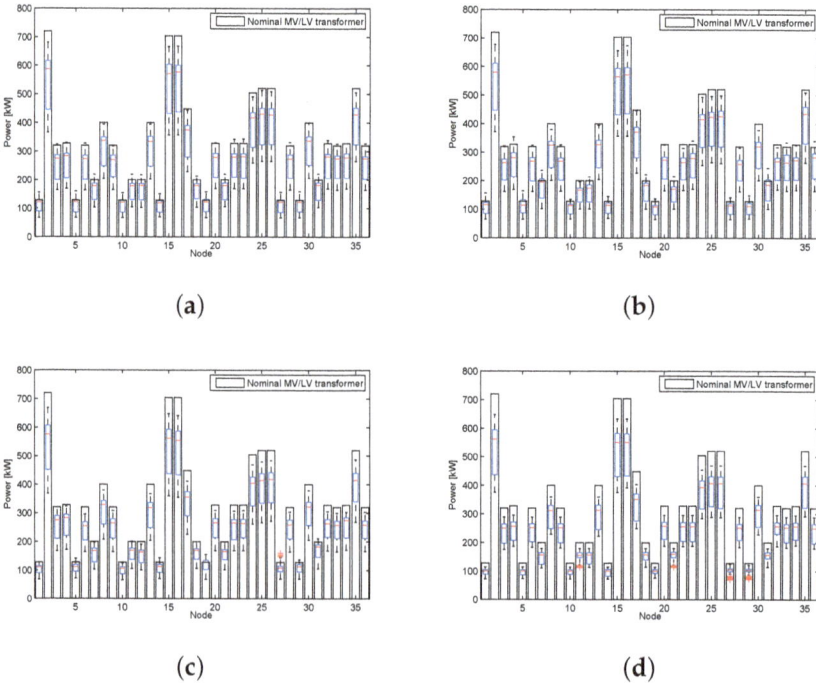

(a)

(b)

(c)

(d)

Figure 16. Total demand in each MV/LV transformer. (**a**) A—Intensive charge; (**b**) B—Plug-and-Play; (**c**) C—Tariff controlled; (**d**) D—Smart charging.

4. Conclusions

The probabilistic agent-based model (ABM) obtained in this paper allows the EV charging demand to be determined, taking into account different variables of EV characteristics such as battery capacity and energy consumption of each trip, economic and social attributes, mobility needs, and charging strategies of each agent. The model developed takes into account the interaction of these variables, allowing the obtainment of better accuracy in the results.

The probabilistic approach is useful to include the uncertainties related to the real behavior of EV users, like the time distribution and energy consumed on each trip. Therefore, the model permits the determination of the impact provoked on the grid by these uncertainties.

Moreover, the model proposed is a benchmark to compare case studies, such as different cities or areas in the same city. With this model, the weak regions of the grid or the areas with high EV density can be detected.

The case study presented shows that the uncertainties cause variability in the EV charging demand in scenarios without control on the EVs, as it is shown in scenarios A and B. In contrast, the consumption variability in scenarios with indirect and direct control on the EV charges, like scenarios C and D, respectively, is small.

The distribution feeder analyzed in the presented case study does not have a significant impact on the smart charging strategy (D) during the off-peak period and all EV agents can charge their EV. In contrast, some MV/LV transformers could exceed their nominal power in the scenarios without control. The voltage in all the scenarios is higher than the limit of 0.90 p.u. according to the EN 50160.

In further work, it could be very interesting to analyze the dynamic behavior of the system in case C during the connection of all EV at 1:00. Furthermore, the model permits analyzing the impact on distribution networks, but it can be applied for transmission and low voltage grids, too. Finally, this model could be applied and compared with a real distribution network with EVs to verify the accuracy of the model.

Acknowledgments: This research was co-financed by the Spanish Ministry of Economy and Competitiveness (MINECO) through the research project ENE2012-36337 "Micro redes con almacenamiento para la integración óptima de energías renovables y vehiculos electricos". This project has received funding from the European Union's Horizon 2020 research and innovation programme under grant agreement No 646476.

Author Contributions: Pol Olivella-Rosell, Roberto Villafafila-Robles and Andreas Sumper conceived the model; Pol Olivella-Rosell performed the simulations; Pol Olivella-Rosell, Roberto Villafafila-Robles, Andreas Sumper and Joan Bergas-Jané analysed the data; Pol Olivella-Rosell wrote the paper.

Conflicts of Interest: The authors declare no conflict of interest.

Nomenclature

Aut_i	Autonomy	C_i	Agent group
Cap_i	Battery capacity	d_i	Day of the week
D_{ij}	Destination	EC_i	Energy consumption
EV_i	EV model	i	Agent
$iter$	Iteration	j	Trip
L_i	Distance	l_{ij}	Distance per trip
m_{ij}	Time distribution	$P(t,x)_{ij}$	Charging process
Ps_i	Power supply	R_i	Place of residence
S_i	Number of trips	t_0, t_1	Initial/Final time
$Type_i$	Type of EV (BEV or PHEV)	v_{ij}	EV velocity
Y_i	Trips matrix	Z_i	Energy consumed

References

1. Bullis, K. Could Electric Cars Threaten the Grid? In *MIT Technology Review*; Massachusetts Institute of Technology (MIT): Cambridge, MA, USA, 2013.
2. Clement-Nyns, K.; Haesen, E.; Driesen, J. The Impact of Charging Plug-In Hybrid Electric Vehicles on a Residential Distribution Grid. *IEEE Trans. Power Syst.* **2010**, *25*, 371–380.
3. Huang, S.; Infield, D. The impact of domestic Plug-in Hybrid Electric Vehicles on power distribution system loads. In Proceedings of the 2010 International Conference on Power System Technology (POWERCON), Hangzhou, China, 24–28 October 2010.
4. Villafafila-Robles, R.; Girbau-Llistuella, F.; Olivella-Rosell, P.; Sudria-Andreu, A.; Bergas-Jane, J. Assessment of the impact of charging infrastructure for electric vehicles on distribution networks. In Proceedings of the 15th European Conference on Power Electronics and Applications, Lille, France, 2–6 September 2013.
5. Watts, R.A.; Letendre, S. *The Effects of Plug-in Hybrid Electric Vehicles on the Vermont Electric Transmission System*; In Proceedings of the Transportation Research Board 88th Annual Meeting, Washington, DC, USA, 11–15 January 2009.
6. Soãres, J.; Canizes, B.; Lobo, C.; Vale, Z.; Morais, H. Electric Vehicle Scenario Simulator Tool for Smart Grid Operators. *Energies* **2012**, *5*, 1881–1899.
7. Acha, S.; van Dam, K.H.; Shah, N. Modelling Spatial and Temporal Agent Travel Patterns for Optimal Charging of Electric Vehicles in Low Carbon Networks. In Proceedings of the IEEE Power and Energy Society General Meeting, San Diego, CA, USA, 22–26 July 2012.
8. Valsera-Naranjo, E.; Sumper, A.; Villafafila-Robles, R.; Martinez-Vicente, D. Probabilistic Method to Assess the Impact of Charging of Electric Vehicles on Distribution Grids. *Energies* **2012**, *5*, 1503–1531.
9. Soares, F.J.; Lopes, J.A.P.; Almeida, P.M.R.; Moreira, C.L.; Seca, L. A stochastic model to simulate electric vehicles motion and quantify the energy required from the grid. In Proceedings of the Power Systems Computation Conference (PSCC), Stockholm, Sweden, 22–26 August 2011.

10. Amjad, S.; Neelakrishnan, S.; Rudramoorthy, R. Review of design considerations and technological challenges for successful development and deployment of plug-in hybrid electric vehicles. *Renew. Sustain. Energy Rev.* **2010**, *14*, 1104–1110.

11. Peng, M.; Liu, L.; Jiang, C. A review on the economic dispatch and risk management of the large-scale plug-in electric vehicles PHEVs-penetrated power systems. *Renew. Sustain. Energy Rev.* **2012**, *16*, 1508–1515.

12. Gao, S.; Chau, K.T.; Wu, D.; Chan, C.C. Modeling and coordinated control for integrating electric vehicles into the power grid. In Proceedings of the 2011 International Conference on Electrical Machines and Systems (ICEMS), Beijing, China, 20–23 August 2011.

13. Sortomme, E.; Hindi, M.M.; MacPherson, S.D.J.; Venkata, S.S. Coordinated Charging of Plug-In Hybrid Electric Vehicles to Minimize Distribution System Losses. *IEEE Trans. Smart Grid* **2011**, *2*, 198–205.

14. Wang, J.; Liu, C.; Ton, D.; Zhou, Y.; Kim, J.; Vyas, A. Impact of plug-in hybrid electric vehicles on power systems with demand response and wind power. *Energy Policy* **2011**, *39*, 4016–4021.

15. Waraich, R.A.; Galus, M.D.; Dobler, C.; Balmer, M.; Andersson, G.; Axhausen, K.W. Plug-in hybrid electric vehicles and smart grids: Investigations based on a microsimulation. *Transp. Res. Part C Emerg. Technol.* **2013**, *28*, 74–86.

16. Rahman, S.; Shrestha, G.B. An investigation into the impact of electric vehicle load on the electric utility distribution system. *IEEE Trans. Power Deliv.* **1993**, *8*, 591–597.

17. Valsera-Naranjo, E.; Martinez-Vicente, D.; Sumper, A.; Villafafila-Robles, R.; Sudria-Andreu, A. Deterministic and probabilistic assessment of the impact of the electrical vehicles on the power grid. In Proceedings of the 2011 IEEE Power and Energy Society General Meeting, San Diego, CA, USA, 24–29 July 2011.

18. Mateo, A.R. Evaluación del impacto de los vehículos eléctricos en las redes de distribución. Master Thesis, Universidad Pontificia de Comillas—Escuela técnica superior de ingeniería ICAI, Madrid, Spain, 2010.

19. Druitt, J.; Früh, W.G. Simulation of demand management and grid balancing with electric vehicles. *J. Power Sources* **2012**, *216*, 104–116.

20. Loisel, R.; Pasaoglu, G.; Thiel, C. Large-scale deployment of electric vehicles in Germany by 2030: An analysis of grid-to-vehicle and vehicle-to-grid concepts. *Energy Policy* **2014**, *65*, 432–443.

21. Lyon, T.P.; Michelin, M.; Jongejan, A.; Leahy, T. Is smart charging policy for electric vehicles worthwhile? *Energy Policy* **2012**, *41*, 259–268.

22. Maitra, A.; Taylor, J.; Brooks, D.; Alexander, M.; Duvall, M. Integrating plug-in-electric vehicles with the distribution system. In Proceedings of the 20th International Conference and Exhibition on Electricity Distribution—Part 1, Prague, Czech Republic, 8–11 June 2009.

23. Lopes, J.A.P.; Soares, F.J.; Almeida, P.M.R. Identifying management procedures to deal with connection of Electric Vehicles in the grid. In Proceedings of the 2009 IEEE Bucharest PowerTech, Bucharest, Romania, 28 June–2 July 2009.

24. Pang, C.; Dutta, P.; Kezunovic, M. BEVs-PHEVs as Dispersed Energy Storage for V2B Uses in the Smart Grid. *IEEE Trans. Smart Grid* **2012**, *3*, 473–482.

25. Frost & Sullivan. Executive Analysis of Global Electric Vehicle Forecast; Technical Report; Frost & Sullivan: Mountain View, CA, USA, 2012.

26. Tomić, J.; Kempton, W. Using fleets of electric-drive vehicles for grid support. *J. Power Sources* **2007**, *168*, 459–468.

27. Metz, M.; Doetsch, C. Electric vehicles as flexible loads—A simulation approach using empirical mobility data. *Energy* **2012**, *48*, 369–374.

28. López, M.; Martín, S.; Aguado, J.; de la Torre, S. V2G strategies for congestion management in microgrids with high penetration of electric vehicles. *Electr. Power Syst. Res.* **2013**, *104*, 28–34.

29. *Electric Vehicle Conductive Charging System—Part 1: General Requirements*; IEC Standard 61851-1; International Electrotechnical Commission: Geneva, Switzerland, 2010.

30. *Vehicle Conductive Charge Coupler*; SAE Standard J1772; Society of Automotive Engineers: California Air Resources Board, CA, USA, 2009.

31. Geth, F.; Debreucker, S.; Clement, K.; Driesen, J. Charging Power Analysis for a Belgian Plug-in Hybrid Electric Vehicle Fleet. In Proceedings of the 5th IEEE Young Researchers Symposium, Leuven, Belgium, 29–30 March 2010.

32. Zhang, L.; Brown, T.; Samuelsen, G.S. Fuel reduction and electricity consumption impact of different charging scenarios for plug-in hybrid electric vehicles. *J. Power Sources* **2011**, *196*, 6559–6566.

33. Grenier, A.; Page, S. The impact of electrified transport on local grid infrastructure: A comparison between electric cars and light rail. *Energy Policy* **2012**, *49*, 355–364.

34. Collins, M.M.; Mader, G.H. The timing of EV recharging and its effect on utilities. *IEEE Trans. Veh. Technol.* **1983**, *32*, 90–97.

35. Clement-Nyns, K.; Haesen, E.; Driesen, J. The impact of vehicle-to-grid on the distribution grid. *Electr. Power Syst. Res.* **2011**, *81*, 185–192.

36. Guo, Q.; Wang, Y.; Sun, H.; Li, Z.; Xin, S.; Zhang, B. Factor Analysis of the Aggregated Electric Vehicle Load Based on Data Mining. *Energies* **2012**, *5*, 2053–2070.

37. Qian, K.; Zhou, C.; Allan, M.; Yuan, Y. Modeling of Load Demand Due to EV Battery Charging in Distribution Systems. *IEEE Trans. Power Syst.* **2011**, *26*, 802–810.

38. Gao, S.; Chau, K.; Chan, C.C.; Liu, C.; Wu, D. Optimal Control Framework and Scheme for Integrating Plug-in Hybrid Electric Vehicles into Grid. *J. Asia Electr. Veh.* **2011**, *9*, 1473–1481.

39. Lojowska, A.; Kurowicka, D.; Papaefthymiou, G.; van der Sluis, L. From transportation patterns to power demand: Stochastic modeling of uncontrolled domestic charging of electric vehicles. In Proceedings of the 2011 IEEE Power and Energy Society General Meeting, San Diego, CA, USA, 24–29 July 2011.

40. Multin, M.; Allerding, F.; Schmeck, H. Integration of electric vehicles in smart homes—An ICT-based solution for V2G scenarios. In Proceedings of the 2012 IEEE PES Innovative Smart Grid Technologies (ISGT), Washington, DC, USA, 16–20 January 2012.

41. Garcia-Valle, R.; Vlachogiannis, J.G. Letter to the Editor: Electric Vehicle Demand Model for Load Flow Studies. *Electr. Power Compon. Syst.* **2009**, *37*, 577–582.

42. Keirstead, J.; Jennings, M.; Sivakumar, A. A review of urban energy system models: Approaches, challenges and opportunities. *Renew. Sustain. Energy Rev.* **2012**, *16*, 3847–3866.

43. Kelly, J.C.; MacDonald, J.S.; Keoleian, G.A. Time-dependent plug-in hybrid electric vehicle charging based on national driving patterns and demographics. *Appl. Energy* **2012**, *94*, 395–405.

44. Stephens, T. An Agent-Based Model of Energy Demand and Emissions From Plug-In Hybrid Electric Vehicle Use; Technical Report; University of Michigan: Ann Arbor, MI, USA, 2010.

45. Weiller, C. Plug-in hybrid electric vehicle impacts on hourly electricity demand in the United States. *Energy Policy* **2011**, *39*, 3766–3778.

46. Huang, S.; Infield, D. The potential of domestic electric vehicles to contribute to Power System Operation through vehicle to grid technology. In Proceedings of the 44th International Universities Power Engineering Conference (UPEC), Glasgow, UK, 1–4 September 2009.

47. Huang, S.; Infield, D. Demand side management for domestic plug-in electric vehicles in power distribution system operator. In Proceedings of the 21st International Conference and Exhibitionon Electricity Distribution, Frankfurt, Geramny, 6–9 July 2011.

48. Schroeder, A.; Traber, T. The economics of fast charging infrastructure for electric vehicles. *Energy Policy* **2012**, *43*, 136–144.

49. Juul, N.; Meibom, P. Road transport and power system scenarios for Northern Europe in 2030. *Appl. Energy* **2012**, *92*, 573–582.

50. Mora, R.; Oyarzabal, J.; Cruz-Zambrano, M.; Gonzalez, A.; Corera, J. E-car and economic impact: Enhancing the smart grids. In Proceedings of the 2012 Workshop Integration of Renewables into the Distribution Grid, Lisbon, Portugal, 29–30 May 2012.

51. Waraich, R.; Galus, M.D.; Dobler, C.; Balmer, M.; Andersson, G.; Axhausena, K.W. Plug-in Hybrid Electric Vehicles and Smart Grid: Investigations Based on a Micro-Simulation. In Proceedings of the 12th International Conference on Travel Behaviour Research (IATBR), Jaipur, India, 13–18 December 2009.

52. Galus, M.D.; Andersson, G. Demand Management of Grid Connected Plug-In Hybrid Electric Vehicles (PHEV). In Proceedings of the Energy 2030 Conference, Atlanta, GA, USA, 17–18 November 2008.

53. Venkatesan, N.; Solanki, J.; Solanki, S.K. Residential Demand Response model and impact on voltage profile and losses of an electric distribution network. *Appl. Energy* **2012**, *96*, 84–91.

54. Galus, M.; Waraich, R.; Balmer, M.; Andersson, G.; Axhausen, K. A framework for investigating the impact of PHEVS. In Proceedings of the International Advanced Mobility Forum (IAMF), ZurichETH, Swiss Federal Institute of Technology, IVT Institute for Transport Planning and Systems, Geneva, Switzerland, 10–12 March 2009.

55. Galus, M.; Andersson, G. Integration of Plug-In Hybrid Electric Vehicles into energy networks. In Proceedings of the 2009 IEEE PowerTech Bucharest, Bucharest, Romania, 28 June–2 July 2009.

56. Balmer, M. Travel demand modeling for multi-agent traffic simulations: Algorithms and systems. Ph.D. Thesis, ETH Zurich, Zurich, Switzerland, 2007.

57. Hedegaard, K.; Ravn, H.; Juul, N.; Meibom, P. Effects of electric vehicles on power systems in Northern Europe. *Energy* **2012**, *48*, 356–368.

58. Mohsenian-Rad, A.; Wong, V.W.S.; Jatskevich, J.; Schober, R.; Leon-Garcia, A. Autonomous Demand-Side Management Based on Game-Theoretic Energy Consumption Scheduling for the Future Smart Grid. *IEEE Trans. Smart Grid* **2010**, *1*, 320–331.

59. Smith, R.; Shahidinejad, S.; Blair, D.; Bibeau, E.L. Characterization of urban commuter driving profiles to optimize battery size in light-duty plug-in electric vehicles. *Transp. Res. Part D Transp. Environ.* **2011**, *16*, 218–224.

60. Kleiwegt, E.; Lukszo, Z. Grid Impact Analysis of Electric Mobility on a Local Electricity Grid. In Proceedings of the 9th IEEE International Conference on Networking, Sensing and Control (ICNSC), Beijing, China, 11–14 April 2012.

61. Kempton, W.; Tomic, J. Vehicle-to-grid power fundamentals: Calculating capacity and net revenue. *J. Power Sources* **2005**, *144*, 268–279.

62. Zakariazadeh, A.; Jadid, S.; Siano, P. Multi-objective scheduling of electric vehicles in smart distribution system. *Energy Convers. Manag.* **2014**, *79*, 43–53.

63. Dallinger, D.; Wietschel, M. Grid integration of intermittent renewable energy sources using price-responsive plug-in electric vehicles. *Renew. Sustain. Energy Rev.* **2012**, *16*, 3370–3382.

64. ElBanhawy, E.; Dalton, R.; Thompson, E.; Kotter, R. A heuristic approach for investigating the integration of electric mobility charging infrastructure in metropolitan areas: An agent-based modeling simulation. In Proceedings of the 2012 2nd International Symposium on Environment Friendly Energies and Applications (EFEA), Newcastle upon Tyne, UK, 25–27 June 2012.

65. Vale, Z.; Pinto, T.; Morais, H.; Praça, I.; Faria, P. VPP's multi-level negotiation in smart grids and competitive electricity markets. In Proceedings of the IEEE Power and Energy Society General Meeting, San Diego, CA, USA, 24–29 July 2011.

66. Macal, C.; North, M. Tutorial on agent-based modelling and simulation. *J. Simul.* **2010**, *4*, 151–162.

67. Bonabeau, E. Agent-based modeling: Methods and techniques for simulating human systems. *Proc. Natl. Acad. Sci. USA* **2002**, *99*, 7280–7287.

68. Barcelona Council. Barcelona Statistical Guidelines and Districts; Technical Report; Department of Statistics: Barcelona, Spain, 2012.

69. Xu, Y. *Effective GPS-Based Panel Survey Sample Size for Urban Travel Behavior Studies*. Ph.D. Thesis, Georgia Institute of Technology, School of Civil and Environmental Engineering, Atlanta, GA, USA, 2010.

70. Institut d'Estudis Regionals i Metropolitans de Barcelona (IERMB) and Autoritat del Transport Metropolità (ATM). Enquesta de Mobilitat Quotidiana; Technical Report; IERMB & ATM: Barcelona, Spain, 2006.

71. Cetelem PARIBAS Grupo B. N. P. The electric vehicle and the europeans. (El coche eléctrico y los europeos); Technical Report; Cetelem PARIBAS Grupo B. N. P.: Madrid, Spain, 2012.

72. Marra, F.; Yang, G.Y.; Traholt, C.; Larsen, E.; Rasmussen, C.N.; You, S. Demand profile study of battery electric vehicle under different charging options. In Proceedings of the 2012 IEEE Power and Energy Society General Meeting, San Diego, CA, USA, 22–26 July 2012.

73. Valsera-Naranjo, E.; Sumper, A.; Lloret-Gallego, P.; Villafafila-Robles, R.; Sudria-Andreu, A. Electrical vehicles: State of art and issues for their connection to the network. In Proceedings of the 10th International Conference on Electrical Power Quality and Utilisation, Lodz, Poland, 15–17 September 2009.

74. *Load Forecasting for Low Voltage Supply*; ITC-BT-10 Stanadard; Ministerio de Industria, Energía y Turismo: Madrid, Spain, 2002.

75. Red Eléctrica de España—Spanish Transmission System Operator. Available online: http://www.ree.es (accessed on 7 July 2014).

76. Kostková, K.; Omelina, L.; Kyčina, P.; Jamrich, P. An introduction to load management. *Electr. Power Syst. Res.* **2013**, *95*, 184–191.

77. Royal Decree (Real Decreto) 647/2011, of May 9, which regulates aggregator activities (Gestor de carga) and Time-Of-Use tariff for EV. National Law; Ministerio de Industria, Energía y Turismo: Madrid, Spain, 2011.

78. Putrus, G.A.; Suwanapingkarl, P.; Johnston, D.; Bentley, E.C.; Narayana, M. Impact of electric vehicles on power distribution networks. In Proceedings of the IEEE Vehicle Power and Propulsion Conference, Dearborn, MI, USA, 7–10 September 2009.

Modeling and Simulation of DC Microgrids for Electric Vehicle Charging Stations

Fabrice Locment and Manuela Sechilariu

Abstract: This paper focuses on the evaluation of theoretical and numerical aspects related to an original DC microgrid power architecture for efficient charging of plug-in electric vehicles (PEVs). The proposed DC microgrid is based on photovoltaic array (PVA) generation, electrochemical storage, and grid connection; it is assumed that PEVs have a direct access to their DC charger input. As opposed to conventional power architecture designs, the PVA is coupled directly on the DC link without a static converter, which implies no DC voltage stabilization, increasing energy efficiency, and reducing control complexity. Based on a real-time rule-based algorithm, the proposed power management allows self-consumption according to PVA power production and storage constraints, and the public grid is seen only as back-up. The first phase of modeling aims to evaluate the main energy flows within the proposed DC microgrid architecture and to identify the control structure and the power management strategies. For this, an original model is obtained by applying the Energetic Macroscopic Representation formalism, which allows deducing the control design using Maximum Control Structure. The second phase of simulation is based on the numerical characterization of the DC microgrid components and the energy management strategies, which consider the power source requirements, charging times of different PEVs, electrochemical storage ageing, and grid power limitations for injection mode. The simulation results show the validity of the model and the feasibility of the proposed DC microgrid power architecture which presents good performance in terms of total efficiency and simplified control.

Reprinted from *Energies*. Cite as: Locment, F.; Sechilariu, M. Modeling and Simulation of DC Microgrids for Electric Vehicle Charging Stations. *Energies* **2015**, *8*, 4335–4356.

1. Introduction

Plug-in electric vehicles (PEVs), whose growth is predicted in future years, represent an important step to achieve low-carbon transport transition [1]. Nevertheless, PEV charging increases the real-time power consumption. Due to the high current needed and depending on when and where the vehicles are plugged-in, the charging stations causes issues and constraints in the public grid. The indirect greenhouse gases emitted by the charging stations depend on the mix of the spinning reserve generation capacity of the electric power system. The spinning reserve is composed mainly by fossils-fuel power plants that are on-line but unloaded and can respond to the demand [2]. In order to accommodate the PEVs' charging demand and requirements, the capacity of this spinning reserve should be expanded [3]. Furthermore, concerning the end-users, most of the end-users' preferences are to charge PEV when it is convenient rather than during periods of lesser demand. Thus, during the daytime, in many local areas the public grid could be strongly stressed by a PEVs power demand threshold.

On the other hand, the energy transition leads to the incitation of policies that support the expansion of renewable energy sources, such as distributed energy generation, and their integration. However, the increasing distributed energy generation reveals an increasing complexity for grid managers by requiring better quality and reliability to regulate electricity flows and less mismatch between electricity generation and demand. To overcome this grid issue, the distributed renewable generation tends to be in favor of self-consumption and therefore less stress is applied on the public grid. Therefore, the microgrid concept, aggregating renewable and traditional energy sources, storage, and public grid connection, aims to actively operate for power balancing of the electricity system. Microgrids are reliable and efficient options to increase the small scale renewable energy penetration while minimizing the energy cost [4–6], especially in urban areas. Due to the negative impact of the total and permanent energy injection, incentive pricing policies dedicated to self-consumption have recently been proposed in many countries.

Thus, facing the emergence of AC or DC microgrids [7,8], on the one hand, and the increasing number of PEV charging stations, on the other hand, one of the solutions is the local microgrid [9,10]. In this context, dedicated charging station infrastructures, equipped with photovoltaic (PV) sources that are the most commonly used renewable sources in urban areas, can be built. This is a matter of PV-based microgrids, *i.e.*, producer-consumer PEVs charging stations, aiming at achieving self-consumption. The PV sources' intermittency can be relieved by a storage system well integrated in the microgrid [11,12]. These local microgrids become an alternative to the total and permanent energy injection and reduce the dependence on the power grid [13–15].

Nowadays many studies focus on the design, operation, optimization, and scheduling of PEV charging stations. Currently the most commonly used PEV batteries are based on Li-ion technology. A fast charger for this technology is presented in [16]; the proposed bidirectional AC/DC converter allows a maximum of 25 kWh charging in one hour. PEV charging stations based on PV sources with optimal energy management are presented in several studies [17–21], however, the proposed systems do not take into account either storage ageing or grid power limitations for injection mode. Related to optimal charging and considering the public grid tariffs to buy or sell energy, a day-ahead scheduling method is studied in [22], in which, according to vehicle-to-grid strategy, a long term cost minimization model is proposed. An interesting strategy based on random vehicle charging and taking into account the excess power is presented in [17]. A car parking PV-integrated and grid-connected system is proposed as a charging station in [18], where the power is controlled by a real-time energy management algorithm. For smart buildings equipped with PV generators, a charging strategy based on the multi-agent approach is proposed in [19].

In [23] the charging of the PEVs is controlled in order to eliminate or reduce the stress on the grid but the impact of energy efficiency and the grid power limitation for injection mode are not discussed. In this case the PV system is connected to the DC link by employing a DC/DC boost converter.

An original PEV charging station topology based on PV system, superconducting magnetic energy storage, and the public grid is presented in [24]. All these components are respectively connected to a common DC bus with the corresponding converters. The proposed energy management strategy focuses on the voltage stability of DC bus and the energy transfer among these units. The building-integrated DC microgrid proposed in [25] aggregates wind and PV sources, hybrid storage, and public grid connection on a DC common bus, which offers fast or smart PEV charging. Compared with the existing droop controls, the authors propose droop curves as a function of the storage state-of-charge (SOC). However, control and architecture of the power electronic interface either needed or not needed to implement the PV system are not discussed in these references.

Through the use of a grid-tied neutral-point-clamped converter, a novel power architecture for a PEV DC charging station is proposed in [26]. It concerns a bipolar DC structure which reduces the step-down effort on the DC/DC fast chargers. Regarding the control, this paper proposes a balancing mechanism that allows handling any difference on the DC loads while keeping the midpoint voltage accurately regulated; however, the grid power limitation for injection mode is not taken into account.

Considering these different PEV charging station concepts, which aim to support direct PEV charging (*i.e.*, not for swapping batteries), it is noted that different requirements and constraints are taken into account to develop energy management strategies. Nevertheless, the studies mentioned above do not consider the overall topology efficiency. This work aims to improve the existing PEVs charging stations with a threefold scheme: (i) a novel DC microgrid power architecture to improve overall efficiency; (ii) an original and easily comprehensible model with the aid of Energetic Macroscopic Representations (EMR) and Maximum Control Structure (MCS); (iii) a realistic approach for considering the effects of electrochemical storage ageing and the public grid power injection limitations.

The proposed DC microgrid is based on a photovoltaic array (PVA), electrochemical storage, and grid connection; and it allows self-consumption according to PVA production and storage constraints, while the public grid is seen only as a back-up. To increase the overall efficiency, improve the utilization efficiency of PVA energy, and decrease the cost and losses of converters, as opposed to conventional power architecture designs, the PVA is coupled directly with the DC link without a static converter, which implies no DC voltage stabilization and less control complexity. It is also assumed that PEVs have direct access to their DC charger input. However, the fluctuation of PVA output and the change of charging power vary randomly and in this case will influence the power transfer among components of the PEV charging station. To cope with this problem while dealing with storage ageing, a real-time rules-based algorithm is proposed.

Following the presented considerations, the DC microgrid special power architecture for PEVs charging station is described in Section 2. The global energetic modelling of the DC microgrid based on EMR is presented in Section 3. Different energetic configurations of the available power and energy sources are given. Then, the DC microgrid control modeling using MCS, which is deduced from the EMR, is introduced in Section 4. The simulation results performed with MATLAB Simulink are given and discussed in Section 5. Conclusions and further discussions on advantages and limitations of the proposed control are presented in Section 6.

2. DC Microgrid Special Power Architecture for an Electric Vehicle Charging Station

The PEV charging station is designed based on the DC microgrid technology. As illustrated in Figure 1a, it is composed of a PVA, public grid connection, PEVs' batteries, and electrochemical storage. These components are connected directly or indirectly to the DC link. The proposed local grid connected DC microgrid can be car parking integrated, with PV panels installed on sun-shading roofs as shown in Figure 1b [27,28].

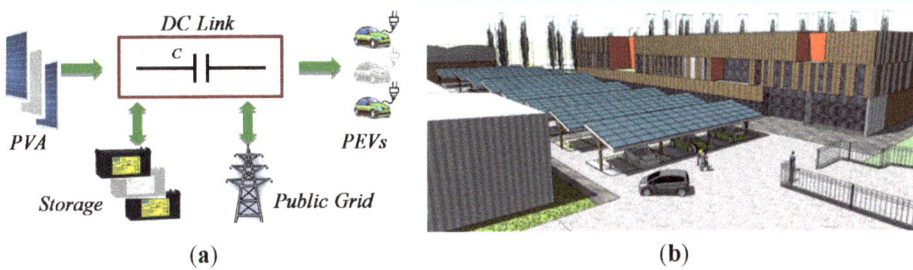

(a)   (b)

Figure 1. (**a**) DC microgrid for an electric vehicle charging station; (**b**) Computer graphic picture.

Expecting that the locally generated power is used where, when, and how it is generated, the DC microgrid energy management system operates with self-supply. The produced PV electricity is intended primarily for charging of PEVs. The storage represents an additional energy source to supply the PEVs or to absorb the excess of the produced PVA energy. The public grid connection is used as back-up and to sell the excess energy. For low PVA power, the needed complementary power to charge the PEVs is provided primarily by the storage, and then by the public grid. In contrast, if PVA power is higher than the power demanded by PEVs, the DC microgrid primarily feeds the storage and then injects excess power into the public grid. A bidirectional smart meter could be implemented for information about grid availability and dynamic pricing. Note that the operating mode called "vehicles to grid" is not developed in this work.

The electrical scheme of the DC microgrid power architecture is given in Figure 2. Since the energy efficiency may be improved by eliminating energy conversion, the PVA is directly connected to the DC link without a static converter. However, PEVs and storage are connected to the same DC bus link through their dedicated static converters whose switching functions are f_{PEVs} and f_{STO}, respectively. This power architecture is due to the fact that the DC link voltage is estimated to be about 1000 VDC; hence, in order to correctly feed the PEVs and to charge the storage, static converters are required. The public grid connection is carried out by a three-phase bidirectional AC/DC converter with the following switching functions: f_A, f_B and f_C.

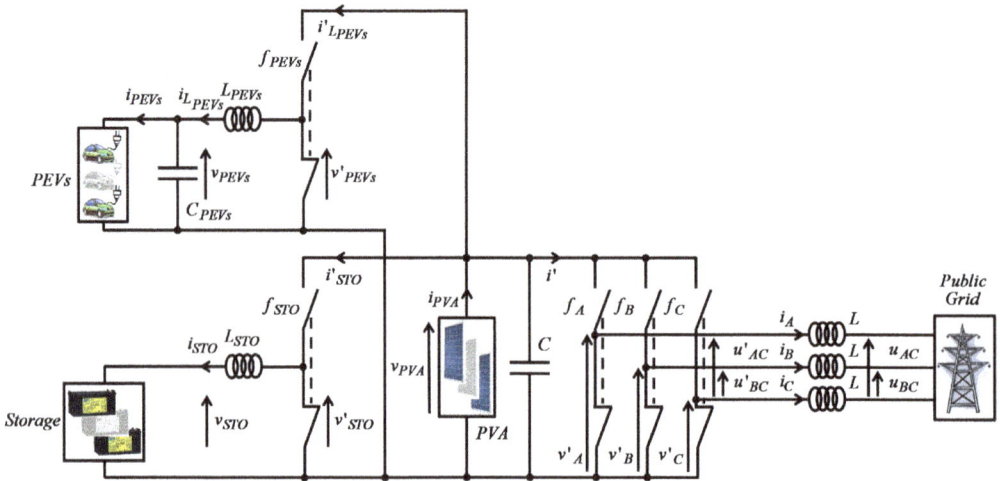

Figure 2. Electrical scheme of the DC microgrid power architecture.

All these switching functions are equal to 0 or 1. Inductors and capacitors are used in order to interface with the different elements. Assuming that there are no losses through the inductors, capacitors, and cables, and neglecting the total losses of power converter legs, the DC microgrid power balancing is given by Equation (1):

$$p_{PG} + p_{STO} = p_{PVA} - p_{PEVs} - v_{PVA} \cdot C\frac{dv_{PVA}}{dt} \qquad (1)$$

where p_{PG}, p_{STO} and p_{PEVs} are the power of the public grid, storage, and PEV charging system, respectively, v_{PVA} is the PVA voltage, and C the DC link capacitor. The power balancing shows that the adjustment variables are the public grid and storage powers, within their physical and functional limitations.

The required power reference p^* for power balancing is calculated by Equation (2):

$$p* = p_{PG}* + p_{STO}* \qquad (2)$$

where the public grid power reference $p_{PG}*$ and the storage power reference $p_{STO}*$ are calculated by Equation (3) according to the previously proposed strategy. For this, a distribution coefficient K is introduced and defined as given by Equation (3):

$$p_{STO}* = K \cdot p* \text{ with } K \in [0,1]$$
$$p_{PG}* = (1 - K) \cdot p* \qquad (3)$$

3. Energetic Macroscopic Representation Modeling of the Electric Vehicle Charging Station

The DC microgrid power architecture for the PEV charging station is modeled by EMR, which is an energy-based graphical tool for a global energetic view. As the PEV charging station is preeminently an energetic system, it is obvious that the considerations of energy should be emphasized in EMR. EMR modeling could be compared with Bond Graph formalism introduced successfully since 1950s, but bond graphs cannot help the control design directly, while EMR focuses on both system function and system structure. The EMR advantages lie in its clarity of physical concepts, as well as its functional modeling; EMR contributes significantly to the design of the control, and leads to energy management of systems.

EMR is a synthetic graphic tool that uses causal or functional representation. EMR allows a systematic approach to design all the interactions between the different subsystems of a complex system. The system representation is based on four basic elements as illustrated in Figure 3.

These pictograms represent: energy sources, accumulation elements, conversion elements, and coupling elements for energy distribution. They are interconnected following the action and reaction principle using exchange variable and respecting the integral causality. The instantaneous power exchanged between two elements is the result of the product of action and reaction variables represented by arrows (inputs and outputs).

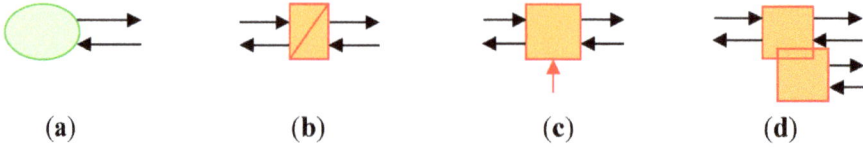

Figure 3. (**a**) Source of energy; (**b**) Element with energy accumulation; (**c**) Electrical converter (without energy accumulation); (**d**) Electrical coupling (without energy accumulation).

The integral causality considered in EMR defines an accumulation element by a time-dependent relationship between its variables, in which case the output is an integral function of the inputs. Other elements are described using relationships without time dependence. In order to respect the integral causality, specific associated rules are defined. Each element is internally described using transfer functions, mathematical relations, or other modeling tool. EMR allows the representation of multi-physic systems and the systematic deduction of control structures. The EMR formalism has been already used in many real applications [29–31]; the implementation is usually performed on MATLAB Simulink.

Following the electrical scheme of the DC microgrid given in Figure 2, all elements are electrically connected to DC link represented by the capacitor noted C. This electrical coupling is modeled by Equation (4):

$$\frac{dv_{PVA}}{dt} = \frac{i_{PVA} - i'_{STO} - i'_{L_{PEVs}} - i'}{C} \tag{4}$$

where i_{PVA}, i'_{STO}, $i'_{L_{PEVs}}$, and i' are the currents of PVA, storage system, PEV charging system, and the current of the inverter input side. The DC link is modeled by the EMR formalism as one accumulation element and one electrical coupling (without energy accumulation).

3.1. Photovoltaic Array Modeling

The proposed PEV charging station is based on 560 PV panels (60M250, SILLIA, Lannion, France), whose power is estimated to be 140 kW under standard test conditions. In consideration of the maximum 1000 V voltage, the 560 PV panels are arranged in parallel and series: 20 parallel branches of 28 serial panels by branch. By applying the EMR formalism, the PVA is modeled as a power source which provides the current i_{PVA} when the voltage v_{PVA} is imposed, *i.e.*, the power p_{PVA} ($p_{PVA} = v_{PVA} \cdot i_{P'VA}$). The voltage is imposed by a maximum power point tracking (MPPT) algorithm. However, for stringent cases, a limited PVA power control could also be applied. In this work, when a MPPT control is required, the well-known method "Perturb & Observe" (P&O) is used to extract the maximum power of PVA for any value of solar irradiance and PV cell temperature [32,33]. The evolution of p_{PVA} and i_{PVA} under the solar irradiance $g = 1000 \text{ W/m}^2$ and for different PV cell temperatures are shown as simulation results in Figure 4.

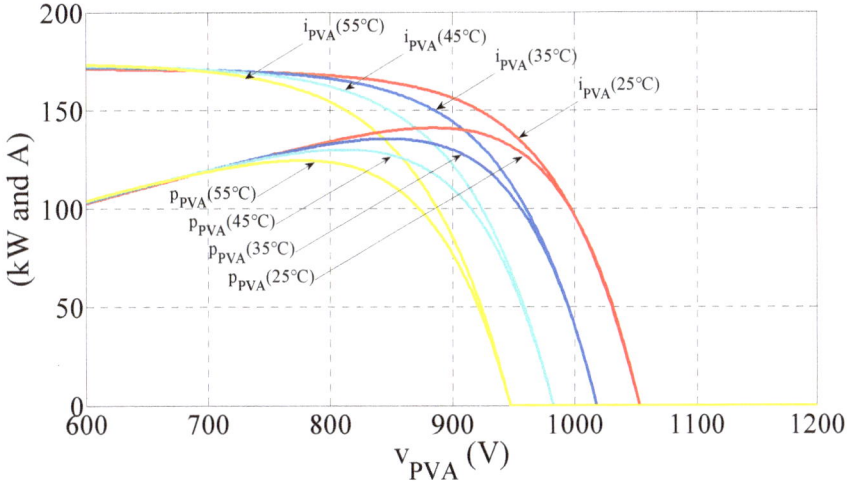

Figure 4. PVA power and current under 1000 W/m^2 irradiance and for different PV cell temperatures.

The PVA system is modeled by the EMR formalism as a power source, denoted *PVA*, to be connected to an electrical coupling element.

3.2. Modeling of PEVs Batteries Charging

In this work, the batteries of all PEVs are considered based on Li-ion technology. The Li-ion battery charging is performed by a constant current (CC) mode followed by a constant voltage (CV) mode, *i.e.*, CC/CV procedure [34]. During the CC mode, the charging current stays constant until the voltage rises to a cut-off voltage. During the CV mode, the voltage remains constant while the current drops. This CC/CV procedure is supposed to be controlled by a battery management system already integrated in PEV battery system.

To emulate the PEV charging, a CC/CV procedure has been applied to a 26650 LiFePO$_4$ cell (A123 Systems LLC, Waltham, MA, USA) and recorded. The PEV charging CC/CV profile is considered proportional to the profile recorded on one cell. According to some end-user's utilization models, the PEV charging profiles proposed in this study are:

- fast charging time for a period of about 30 min;
- average charging time for a period of about 1 h and 30 min;
- slow charging time for a period of about 4 h.

213

For one vehicle and for these three charging profiles, the PEV battery power, denoted as p_{PEV}, and its state of charge, noted SOC_{PEV}, are presented in Figure 5.

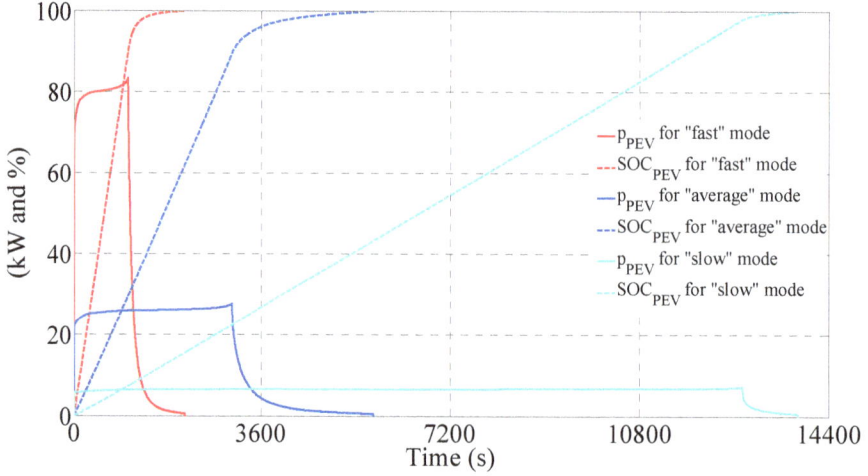

Figure 5. Power p_{PEV} and state of charge SOC_{PEV} for one PEV.

These characteristics are obtained by coupling 28 parallel branches of 120 serial cells; with the voltage of 3.6 V per cell. The total stored energy is about 24 kWh. Depending on the type of vehicles, state of charge of battery, charging voltage level, charging current level, and the demanded charging time, the total PEV charging power flow varies roughly. Regarding the electrical scheme presented in Figure 2, the PEV charging modelling is given by Equation (5):

$$v\prime_{PEVs} = f_{PEVs} \cdot v_{PVA} \Rightarrow v\prime_{PEVs} = m_{PEVs} \cdot v_{PVA}$$

$$i\prime_{L_{PEVs}} = f_{PEVs} \cdot i_{L_{PEVs}} \Rightarrow i\prime_{L_{PEVs}} = m_{PEVs} \cdot i_{L_{PEVs}}$$

$$m_{PEVs} = \tfrac{1}{T} \int_0^T f_{PEVs} \, dt \text{ with } m_{PEVs} \in [0 \, ; 1]$$

$$\tfrac{di_{L_{PEVs}}}{dt} = \tfrac{v\prime_{PEVs} - v_{PEVs}}{L_{PEVs}}$$

$$\tfrac{dv_{PEVs}}{dt} = \tfrac{i_{L_{PEVs}} - i_{PEVs}}{C_{PEVs}}$$

(5)

where $v\prime_{PEVs}$ and $i\prime_{L_{PEVs}}$ are the voltage and current of the PEV converter's output side, v_{PEVs} and $i_{L_{PEVs}}$ are the voltage and current of PEVs converter input side, m_{PEVs} the PEV control variable, i_{PEVs} the PEVs current, L_{PEVs} and C_{PEVs} the PEVs inductor and capacitor, respectively, T the period, and t the time variable.

In order to simplify the numerical simulation, the PEV charging system's assumptions considered in this study are:

- arbitrarily it is considered that 15 vehicles could be recharged simultaneously;
- for each PEV, charging operation starts from zero SOC_{PEV} limit;
- all PEVs remain in plug-in position to fulfill the battery (to achieve the complete CC/CV charging procedure).

Figure 6 shows PEVs flow evolution and corresponding absorbed power ($p_{PEVs} = v_{PEVs} \cdot i_{PEVs}$). The PEVs charging system is modeled by EMR formalism as a power source, noted *PEVs*, associated with two accumulation elements and one conversion element.

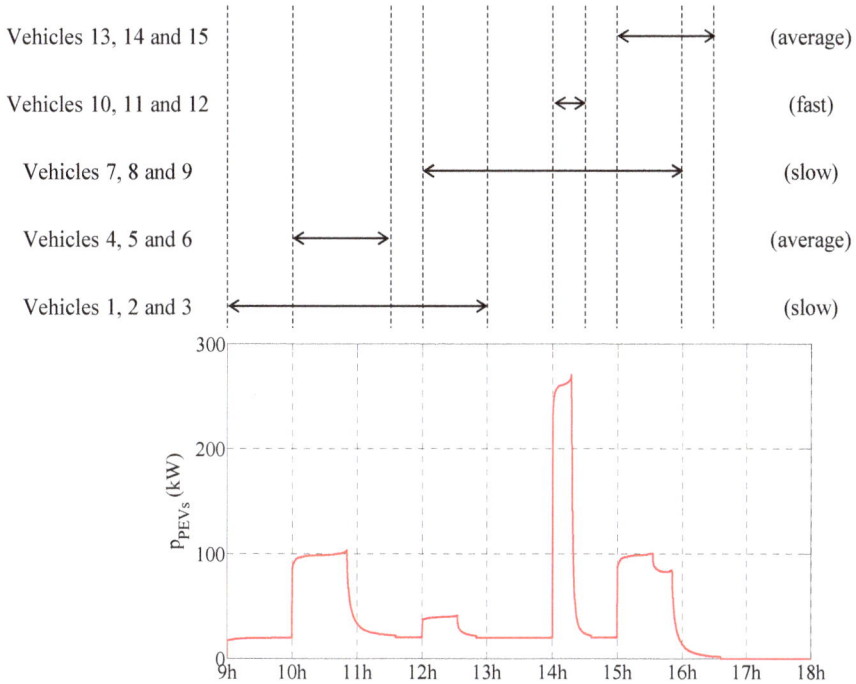

Figure 6. Plug-in electric vehicles' total power flow.

3.3. Storage Modeling

The electrochemical storage is a complementary energy source for the DC microgrid; it is a system composed by 35 serial batteries, characterized by 12V/230A each. Therefore, the whole storage system is characterized by 420 V/230 Ah. The storage system modeling equations are expressed by Equation (6):

$$v'_{STO} = f_{STO} \cdot v_{PVA} \Rightarrow v'_{STO} = m_{STO} \cdot v_{PVA}$$

$$i'_{STO} = f_{STO} \cdot i_{STO} \Rightarrow i'_{STO} = m_{STO} \cdot i_{STO}$$

$$m_{STO} = \frac{1}{T}\int_0^T f_{STO}\,dt \text{ with } m_{STO} \in [0\,;1] \tag{6}$$

$$\frac{di_{STO}}{dt} = \frac{v'_{STO} - v_{STO}}{L_{STO}}$$

where v'_{STO} and i'_{STO} represent the voltage and current of the storage DC link side (after converter), v_{STO} and i_{STO} are the voltage and current of the storage before the converter, m_{STO} the storage control variable, and L_{STO} the storage inductor.

The storage system is modeled by EMR formalism as a power source, denoted STO, associated with one accumulation element and one conversion element.

3.4. Public Grid Connection Modeling

The considered public grid is a low voltage network, characterized by 400 V phase-to-phase and 50 Hz. The public grid connection modeling is based on the inverter model. Thus, the simple voltages of the three phases, v'_A, v'_B and v'_C, and corresponding inverter switching functions, f_A, f_B, f_C are used as shown by Equation (7):

$$\begin{bmatrix} u'_{AC} \\ u'_{BC} \end{bmatrix} = \begin{bmatrix} v'_A - v'_C \\ v'_B - v'_C \end{bmatrix} \quad ; \quad \begin{bmatrix} v'_A \\ v'_B \\ v'_C \end{bmatrix} = \begin{bmatrix} f_A \\ f_B \\ f_C \end{bmatrix} v_{PVA}$$

$$\begin{bmatrix} v'_A \\ v'_B \\ v'_C \end{bmatrix} = \begin{bmatrix} f_A - f_C \\ f_B - f_C \end{bmatrix} v_{PVA} \Rightarrow \begin{bmatrix} v'_A \\ v'_B \\ v'_C \end{bmatrix} = \begin{bmatrix} m_A \\ m_B \end{bmatrix} v_{PVA} \tag{7}$$

$$\begin{bmatrix} m_A \\ m_B \end{bmatrix} = \frac{1}{T}\int_0^T \begin{bmatrix} f_A - f_C \\ f_B - f_C \end{bmatrix} dt \text{ with } \begin{bmatrix} m_A \\ m_B \end{bmatrix} \in [-1;1]$$

where u'_{AC} and u'_{BC} are the output's inverter voltages between the phases A and C, and B and C respectively, m_A and m_B the inverter control variables. The inverter current i' is expressed by using the currents of the three phases, i_A, i_B, and i_C, following Equation (8):

$$i\prime = f_A i_A + f_B i_B + f_C i_C \text{ with } i_A + i_B + i_C = 0$$
$$i\prime = (f_A - f_C) i_A + (f_B - f_C) i_B \tag{8}$$
$$i\prime = m_A i_A + m_B i_B$$

The connection between the inverter and the public grid is made by power lines, with no mutual inductor but a self-inductance L and an internal resistance R. The relationship between voltages and currents of power lines is given in Equation (9):

$$\begin{bmatrix} u'_{AC} \\ u'_{BC} \end{bmatrix} - \begin{bmatrix} u_{AC} \\ u_{BC} \end{bmatrix} = \begin{bmatrix} 2L & L \\ L & 2L \end{bmatrix} \frac{d}{dt} \begin{bmatrix} i_A \\ i_B \end{bmatrix} + \begin{bmatrix} 2R & R \\ R & 2R \end{bmatrix} \begin{bmatrix} i_A \\ i_B \end{bmatrix} \tag{9}$$

where u_{AC} and u_{BC} represent the public grid voltages between the phases A and C, and B and C, respectively. The public grid system is modeled by EMR formalism as a power source, denoted PG, associated with one accumulation element and one conversion element.

3.5. Energetic Macroscopic Representation of the DC Microgrid

Based on the model of each component described above, the global EMR of the DC microgrid designed for PEVs charging station is obtained and shown in Figure 7.

This system has six state variables (v_{PEVs}, $i_{L_{PEVs}}$, i_{STO}, v_{PVA}, i_A, and i_B) and four control variables (m_{PEVs}, m_{STO}, m_A and m_B). A control structure adapted to the operating system strategy is needed and described below.

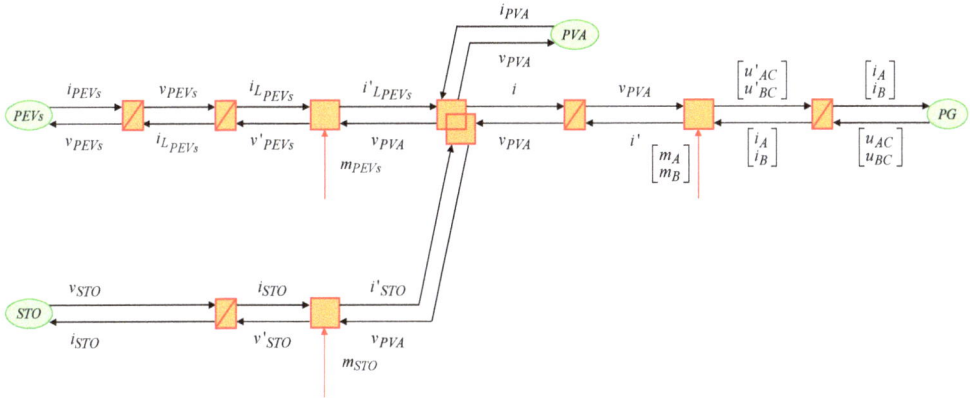

Figure 7. Energetic macroscopic representation of a DC microgrid for a PEV charging station.

4. DC Microgrid Control Modeling Using the Maximum Control Structure

The inversion-based control theory was introduced in 2000s. The control structure of a system is considered as an inversion model of the system. From EMR modeling, a MCS graphical representation can be obtained, which easily leads to a practical control structure. Generally, it is considered that the plant inversion is not robust to variations in the system parameters, but the MCS approach inherently assumes that the control structure can be obtained under the assumption that all variables are measurable. The merit of EMR/MCS modeling is highlighted by its systemic approach which involves the principle of holism that leads to studying the impact of subsystems association: Interconnected basic control blocks, dynamical performances, interest of the closed loop and so on.

To design control structures based on the EMR formalism, a corresponding Maximum Control Structure (MCS) can be deduced through specific inversion rules [28,29]. Indeed, the system's control computes the system's inputs for which the desired output is obtained. Therefore, system's control structure can be considered as an inversion model of the system. Thus, on one hand, the direct inversion, *i.e.*, without controller, is applied for items that are not time dependent, such as the conversion elements. But, on the other hand, as EMR formalism does not allow the derivative causality, a direct inversion of time dependent item is not possible. For this reason an indirect inversion, *i.e.*, the controller, is proposed for items that are time dependent. The accumulation elements are inverted using a close-loop control.

The MCS representation is based on three basic elements as shown in Figure 8: control bloc without controller, control bloc with controller, and bloc strategy.

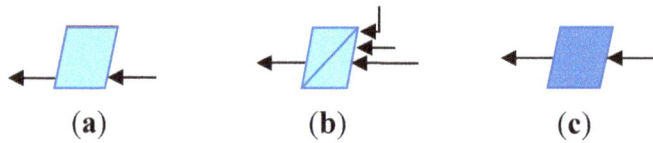

(a)   (b)   (c)

Figure 8. (a) Control bloc without controller; (b) Control bloc with controller; (c) Bloc strategy.

The DC microgrid control structure model is deduced from the EMR described in Section 3 and represented by MCS. The control variable m_{PEVs} imposes constant DC voltage (v_{PEVs}) across the terminals of the vehicles. Using the aforementioned specific inversion rules, it is obtained:

$$v\prime_{PEVs}* = CI_{PEVs}\left(i_{L_{PEVs}}* - i_{L_{PEVs}}\right) + v_{PEVs}$$

$$i_{L_{PEVs}}* = CV_{PEVs}\left(v_{PEVs}* - v_{PEVs}\right) + i_{PEVs} \qquad (10)$$

$$m_{PEVs}* = \frac{v\prime_{PEVs}*}{v_{PVA}}$$

where $v\prime_{PEVs}*$, $v_{PEVs}*$ are voltage references, $i_{L_{PEVs}}*$ is current reference, CI_{PEVs} is an integral proportional corrector with a bandwidth of 500 Hz, and CV_{PEVs} is a proportional corrector with a bandwidth of 50 Hz. The integral proportional corrector and the proportional corrector have been defined by the method of the pole placement.

The control variables m_A, m_B and m_{STO} impose a variable DC voltage v_{PVA} across the terminals of PVA. The value of this reference voltage ($v_{PVA}*$) is imposed by the MPPT algorithm. Inverting Equation (4) allows obtaining Equation (11):

$$i\prime * + i\prime_{STO}* = -CV\left(v_{PVA} * - v_{PVA}\right) + i_{PVA} - i\prime_{L_{PEVs}}$$ (11)

where $i^{\prime}*$, $i\prime_{STO}*$ and $v_{PVA}*$ are current reference and voltage reference respectively. The capacitor C is considered as a pure integrator and the disturbances related to i_{PVA} and $i\prime_{L_{PEVs}}$ are assumed to be compensated; therefore, CV is a proportional controller.

Based on the assumption that there are no losses in the system and considering the power balancing, the active power reference of the public grid, modeled on two-phase alpha-beta or stationary frame, is equal to:

$$p_{PG}* = v_{PVA} \cdot i\prime* = (1 - K) \cdot p* = v_\alpha i_\alpha * + v_\beta i_\beta*$$ (12)

with:

$$\begin{bmatrix} v_\alpha, i_\alpha \\ v_\beta, i_\beta \end{bmatrix} = [C_{23}] \begin{bmatrix} v_A, i_A \\ v_B, i_B \\ v_C, i_C \end{bmatrix} = \sqrt{\frac{2}{3}} \begin{bmatrix} 1 & -\frac{1}{2} & -\frac{1}{2} \\ 0 & \frac{\sqrt{3}}{2} & -\frac{\sqrt{3}}{2} \end{bmatrix} \begin{bmatrix} v_A, i_A \\ v_B, i_B \\ v_C, i_C \end{bmatrix}$$ (13)

where v_A, v_B and v_C are the single phase public grid voltages respectively (not shown in Figure 2).

The reactive power reference is imposed arbitrarily to zero and is defined by:

$$q_{PG}* = v_\alpha i_\beta * - v_\beta i_\alpha * = 0$$ (14)

Thus, the current references ($i_\alpha*$, $i_\beta*$) can be determined from Equations (12) and (14) as follows:

$$i_\alpha* = \frac{v_\alpha}{v_\alpha{}^2 + v_\beta{}^2} p_{PG}* = \frac{v_\alpha}{v_\alpha{}^2 + v_\beta{}^2}(1 - K) \cdot p*$$
$$i_\beta* = \frac{v_\beta}{v_\alpha{}^2 + v_\beta{}^2} p_{PG}* = \frac{v_\beta}{v_\alpha{}^2 + v_\beta{}^2}(1 - K) \cdot p*$$ (15)

To simplify the currents control, the regulation is performed in dq reference or rotary frame using the transition matrix given by Equation (16):

$$\begin{bmatrix} i_d* \\ i_q* \end{bmatrix} = [R(\theta)] \begin{bmatrix} i_\alpha* \\ i_\beta* \end{bmatrix} = \begin{bmatrix} \cos(\theta) & -\sin(\theta) \\ \sin(\theta) & \cos(\theta) \end{bmatrix} \begin{bmatrix} i_\alpha* \\ i_\beta* \end{bmatrix}$$ (16)

where θ is the phase shift between the voltages (v_A, v_B and v_C) and currents (i_A, i_B and i_C).

Based on current references, the expression of the reference control variables m_A* and m_B* is following Equation (17):

$$\begin{bmatrix} m_A* \\ m_B* \end{bmatrix} = \frac{[C_{23}]^{-1}\left([R(\theta)]^{-1}\left(CI\begin{bmatrix} i_d*-i_d \\ i_q*-i_q \end{bmatrix}\right)+\begin{bmatrix} v_\alpha \\ v_\beta \end{bmatrix}\right)}{v_{PVA}} \tag{17}$$

Knowing that a power balancing is always performed, the power reference of storage is equal to:

$$p_{STO}* = v_{PVA}\cdot i\prime_{STO}* = K\cdot p* = v_{STO}\cdot i_{STO}* \tag{18}$$

with:

$$i_{STO}* = \frac{p_{STO}*}{v_{STO}} = \frac{K\cdot p*}{v_{STO}} \tag{19}$$

Inverting Equation (6) allows obtaining Equation (20):

$$v\prime_{STO}* = CI_{STO}\left(i_{STO}*-i_{STO}\right)+v_{STO}$$
$$m_{STO}* = \frac{v\prime_{STO}*}{v_{PVA}} = \frac{CI_{STO}(i_{STO}*-i_{STO})+v_{STO}}{v_{PVA}} \tag{20}$$

where $v\prime_{STO}*$ and $i_{STO}*$ are voltage reference and current reference respectively, CI_{STO} is the same as CI_{PEVs} and has the same settings.

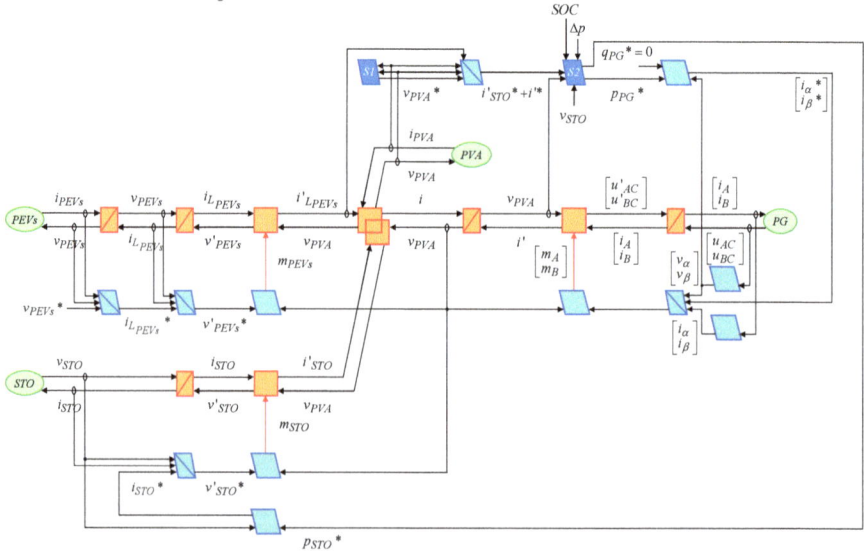

Figure 9. Maximum Control Structure for the DC microgrid for PEVs charging station.

The four control variables are known, thus the MCS of the DC microgrid could be presented in Figure 9. This graphical description gives particularly the PVA control strategy (P&O or limited power control), symbolized by S_1, and the secure operating system involving the storage and the public grid, symbolized by S_2.

5. Simulation Results

To validate the DC microgrid for PEVs charging station modeling approach, the system simulation is performed with MATLAB Simulink. The simulation results are obtained with the following values: PEV charging station voltage $v_{PEVs}{}^* = 200$ V, $C = C_{PEVs} = 10$ mF and $L = L_{PEVs} = L_{STO} = 1$ mH (with 1 mΩ internal resistance). The calculation step is 5 kHz. The storage SOC evolution is considered between 20% and 80%, i.e., the low limit and the high limit respectively. The system simulation is based on real experimental tests concerning: PV panel under real solar irradiation and its control, CC/CV charging procedure for one PEV battery cell, electrochemical storage charge/discharge and SOC calculation.

If the storage ageing has to be taken into account, a current limitation should be imposed provided the operation time occurs during the off-peak period for the public grid. In this way, the DC microgrid for the PEV charging station is secured thanks to the power supplied by the public grid. On the other hand, if the public grid power injection has to be limited for some off-peak period, the implemented control should be able to take into consideration this power grid timed limitation. Thus, the simulation results are presented following three cases: the storage current is not limited, the storage current is limited, and the public grid power injection and the storage current are both limited by the implemented control.

5.1. DC Microgrid Simulation Based on Storage without Current Limitation

Figure 10a shows the real solar irradiance and PV cell temperature during the day of 2 June 2014 at Compiegne, France. It was a sunny day but with a lot of passing clouds. The power evolutions of PVA, of PEVs charging, and the difference between these two powers, $\Delta p = p_{PVA} - p_{PEVs}$, are given in Figure 10b. Figure 10a,b prove that the P&O algorithm work correctly since p_{PVA} is proportional to solar irradiance evolution. Figure 10c presents the storage SOC and the distribution coefficient K evolution.

Figure 10d illustrates the variations of the power of public grid and storage. Figure 10e gives the storage current evolution facing to the current limitation of 230 A for the first charging curve of the considered storage.

For the period taken into account, it is observed that the strategy mentioned before is well respected at all the time. The PVA system is under MPPT control and the produced PVA energy is used primarily for self-feeding. Following the Figure 10b,d for $\Delta p > 0$ the storage and public grid receive energy, in contrast they provide when $\Delta p < 0$. In case of insufficient energy to supply the PEVs charging, the system security is provided by the storage system priorly. The storage charge and discharge operations have priority over the public grid. Once the storage has reached its upper or lower limit, the public grid takes over and provides or receives energy. The distribution coefficient K evolution shown in Figure 10c presents clearly the storage operation period for all durations when $K = 1$.

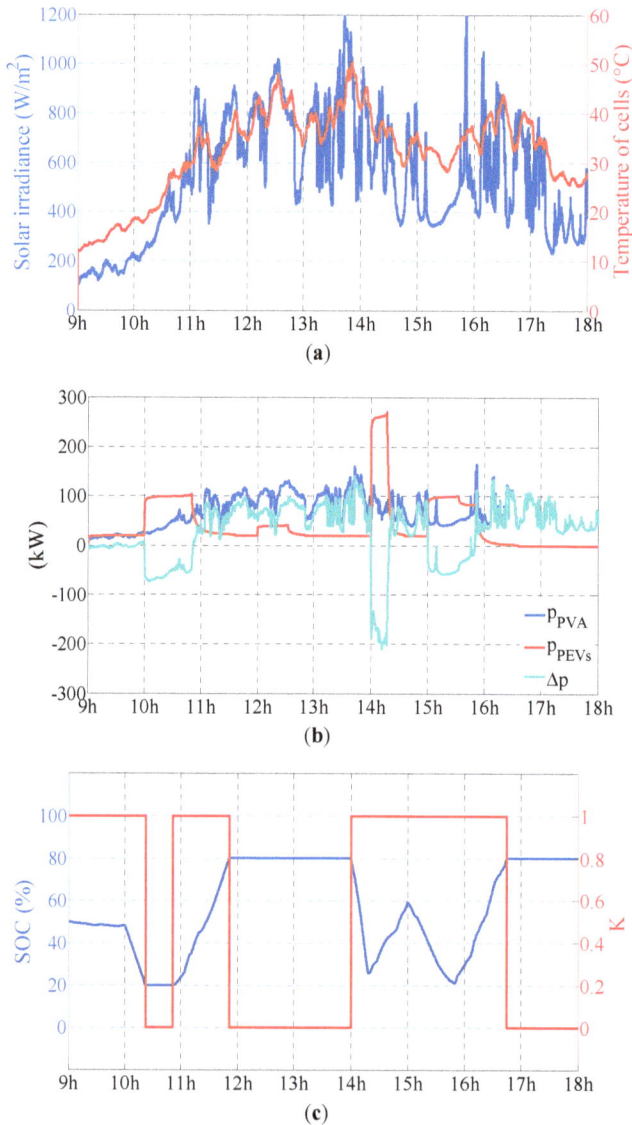

Figure 10. Cont.

222

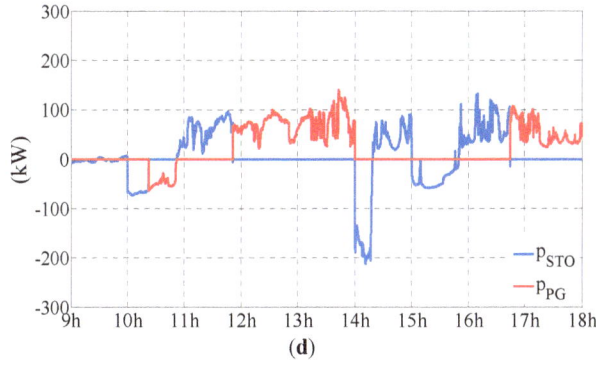

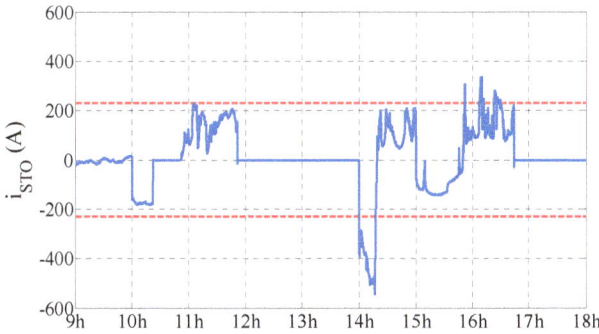

Figure 10. DC microgrid without storage current limitation case: (**a**) Solar irradiance and PV cell temperature; (**b**) Power evolutions of PVA, PEVs, and the difference between these two powers; (**c**) Storage SOC and distribution coefficient K; (**d**) Power evolutions of storage and public grid; (**e**) Storage current evolution.

These results show that no matter the sign and the amplitude of the difference of power between p_{PVA} and p_{PEVs}, the function of this system is secured.

5.2. DC Microgrid Simulation with Current Limitation for Storage

Based on the same meteorological conditions, this second simulation case presents the DC microgrid control with storage current limitation at 115 A while the PVA system is under MPPT control and the produced PVA energy is used primarily for self-feeding. Therefore, for this second case, the difference between PVA power and PEVs powers is the same as previously presented in Figure 10b. Figure 11 shows the storage current limitation case for a value of 115 A.

The distribution coefficient K given in Figure 11a does not take only binary values, but also fractional values between zero and one. Figure 11b shows clearly that during the current limitation periods the Δp is shared simultaneous between the storage and the grid, *i.e.*, for all durations when $0 < K < 1$. Figure 11c presents the storage current evolution facing to the 230 A limitation.

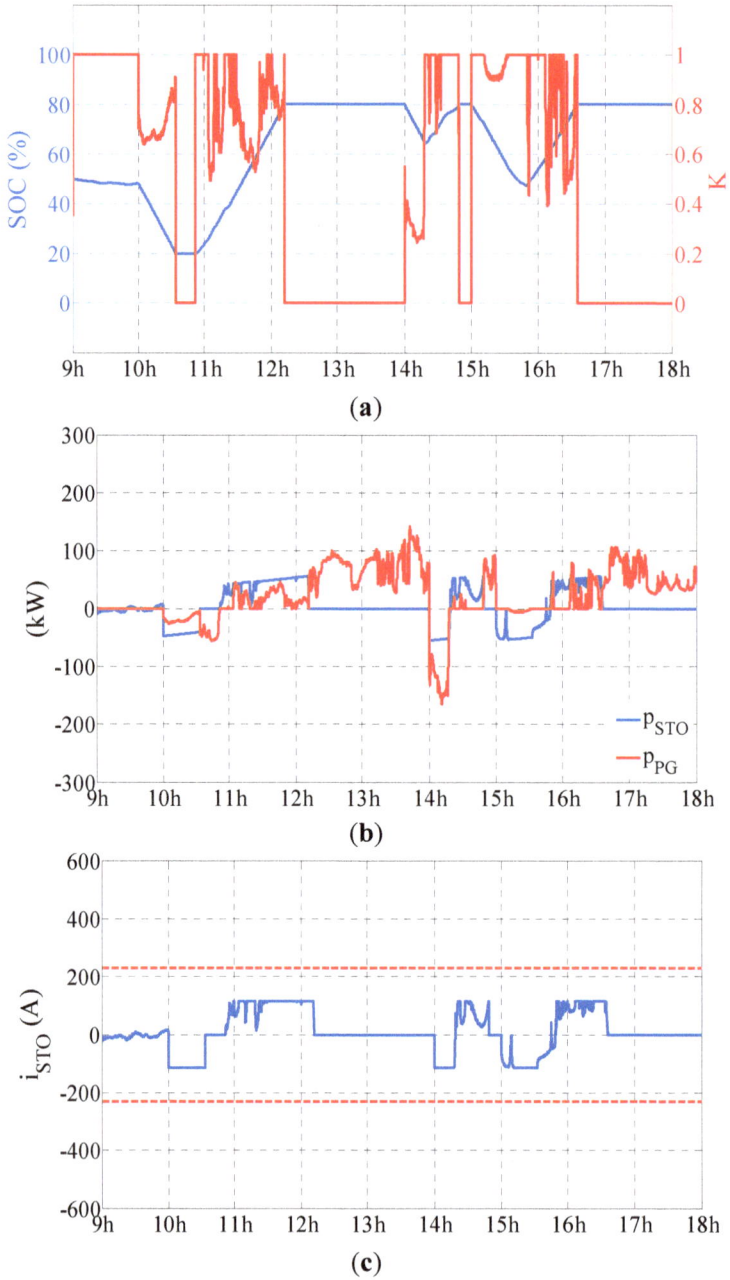

Figure 11. Storage current limitation case: (**a**) Storage *SOC* and distribution coefficient *K*; (**b**) Power evolutions of storage and public grid; (**c**) Storage current evolution.

5.3. DC Microgrid Simulation Based on Storage Current Limitation and Grid Power Injection Limitation

The same meteorological conditions are considered for this third case when the storage keeps the current limitation while the public grid power could be limited for injection. Figure 12 presents the storage current limitation for a value of 115 A and the public grid power limitation for the value of 80 kW between two time slots: 12:30–13:30 and 16:30–17:30.

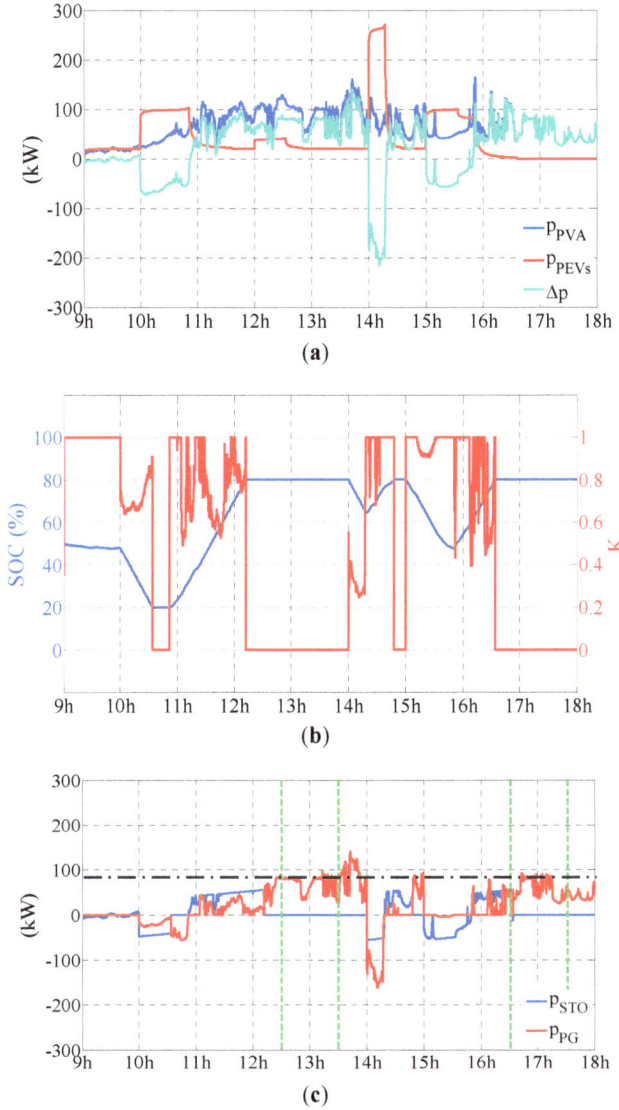

Figure 12. *Cont.*

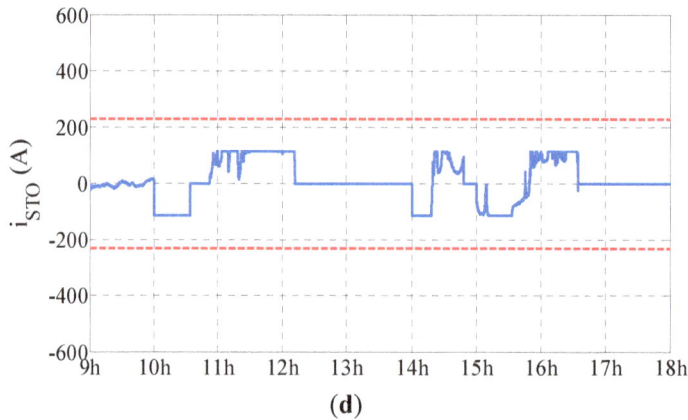

Figure 12. Storage current limitation and public grid power injection limitation case: (**a**) Power evolutions of PVA, PEVs, and the difference between these two powers; (**b**) Storage *SOC* and distribution coefficient *K*; (**c**) Power evolutions of storage and public grid; (**d**) Storage current evolution.

The power evolutions of PVA, of PEV charging, and the difference between these two powers, $\Delta p = p_{PVA} - p_{PEVs}$, are given in Figure 12a. Different from the cases presented above, due to the fact that the storage is full and the public grid power is limited at 80 kW, the PVA power also has to be shed between the same time slots: 12:30–13:30 and 16:30–17:30. Therefore, for this third case, the difference between PVA power and PEVs power, Δp, is not the same as that presented in Figure 10b.

During the mentioned time slots, a PVA power shedding algorithm is applied for the PVA control as developed in [35]. The PVA power limitation is operated according to PEVs power demand and the public grid power limitation. At the end of each limitation time slot, the classical P&O MPPT control of the PVA takes over.

The distribution coefficient *K* given in Figure 12b is almost the same as that given in Figure 11a, while the storage current evolution shown in Figure 12d is almost the same as that given in Figure 11c. These similarities can be explained by the fact that the time slots limitation of the public grid power occur during $K = 0$, *i.e.*, when the storage is not involved. Nevertheless, due to calculus implied by the simulation, here presented as new case, some very slight fluctuations make some very slight differences, almost imperceptible.

Figure 12c gives the evolution of the storage power and the public grid power. Compared to Figure 11b, this figure shows clearly the public grid power limitation that occurs during the considered time slots: 12:30–13:30 and 16:30–17:30.

Taking into account that the goal was to verify the feasibility of the suggested system control, finally, it can be stated that the overall DC microgrid system, as designed and modeled, responds satisfactorily to the outlined strategy.

However, in order to fully and correctly use the available storage, its control may be improved. In addition, in our further work, K values will be calculated according to techno-economic criteria such as weather, energy pricing, dynamic pricing, PEVs charging duration demand, *etc.*

6. Conclusions

Local renewable power production by microgrids becomes increasingly complex and selective for many types of loads. In this paper a DC microgrid special power architecture for a PEV charging station was presented. The proposed microgrid structure aims to maximize the efficiency of the PVA- produced energy by proposing self-feeding and direct coupling of the PVA to the DC link without a static converter. The study develops an original easy to comprehend EMR model that is very and helpful for the analysis of the DC microgrid operation and the energy management strategies.

The EMR formalism is one of the most efficient methodologies to describe a complex system based on scientific fields and allows obtaining a unified graphical representation and a physical modeling. Based on inversion rules applied to EMR model, the system's control structure is easily deduced using the MCS representation. So, by using the EMR formalism, the DC microgrid physical integral modeling based on the interaction principle becomes an attractive and very comprehensible graphical description. The inversion-based control structure represented by MCS allows also a quick graphical description.

The simulation results, given for one workday, show the feasibility of the DC microgrid control for different PEV charging scenarios. The results show that by using the PVA-produced energy within the microgrid, the power demand from the public grid can be lowered or avoided entirely. Furthermore, this study showed a realistic approach for considering the electrochemical storage ageing and the public grid power injection limitation. For larger amount of PEVs, the system will require improvements and optimization while taking into account constraints given by the PVA, the load and the public grid. Based on the simulation model described in this study, future research will focuses on the impact of more than fifteen PEVs and analyze the limits of this model of DC microgrid.

Author Contributions: Both authors have designed the system, performed the research project, and analyzed the data; both authors have contributed to writing and preparing revision of this manuscript.

Conflicts of Interest: The authors declare no conflict of interest.

References

1. Hajimiragha, A.; Cañizares, C.A.; Fowler, M.W.; Elkamel, A. Optimal transition to plug-in hybrid electric vehicles in Ontario, Canada, considering the electricity-grid limitations. *IEEE Trans. Ind. Electron.* **2010**, *57*, 690–701.

2. Jansen, K.H.; Brown, T.M.; Samuelsen, G.S. Emissions impacts of plug-in hybrid electric vehicle deployment on the US western grid. *J. Power Sour.* **2010**, *195*, 5409–5416.

3. Richardson, D.B. Electric vehicles and the electric grid: A review of modeling approaches, impacts, and renewable energy integration. *Renew. Sustain. Energy Rev.* **2013**, *19*, 247–254.

4. Patterson, B.T. DC, Come Home: DC microgrids and the birth of the "Enernet". *IEEE Power Energy Mag.* **2012**, *10*, 60–69.

5. Guerrero, J.M.; Chandorkar, M.; Lee, T.-L.; Loh, P.C. Advanced control architectures for intelligent microgrids—Part I: Decentralized and hierarchical control. *IEEE Trans. Ind. Electron.* **2013**, *60*, 1607–1618.

6. Schmitt, L.; Kumar, J.; Sun, D.; Kayal, S.; Venkata, S.S.M. Ecocity upon a Hill: Microgrids and the future of the European city. *IEEE Power Energy Mag.* **2013**, *11*, 59–70.

7. Lasseter, R.H. Smart distribution: Coupled microgrids. *Proc. IEEE* **2011**, *99*, 1074–1082.

8. Guerrero, J.M.; Vasquez, J.C.; Matas, J.; de Vicuna, L.G.; Castilla, M. Hierarchical control of droop-controlled AC and DC microgrids—A general approach toward standardization. *IEEE Trans. Ind. Electron.* **2011**, *58*, 158–172.

9. Sechilariu, M.; Wang, B.C.; Locment, F.; Jouglet, A. DC microgrid power flow optimization by multi-layer supervision control. Design and experimental validation. *Energy Convers. Manag.* **2014**, *82*, 1–10.

10. Sechilariu, M.; Wang, B.; Locment, F. Building integrated photovoltaic system with energy storage and smart grid communication. *IEEE Trans. Ind. Electron.* **2013**, *60*, 1607–1618.

11. Francesco, M.; Yang, G.Y.; Traeholt, C. EV Charging facilities and their application in LV feeders with photovoltaics. *IEEE Trans. Smart Grid* **2013**, *4*, 1533–1540.

12. Traube, J.; Lu, F.; Maksimovic, D. Mitigation of solar irradiance intermittency in photovoltaic power systems with integrated electric vehicle charging functionality. *IEEE Trans. Power Electron.* **2013**, *28*, 3058–3067.

13. Wi, Y.M.; Lee, J.U.; Joo, S.K. Electric vehicle charging method for mart homes/buildings with a photovoltaic system. *IEEE Trans. Consum. Electron.* **2013**, *59*, 323–328.

14. Sechilariu, M.; Wang, B.; Locment, F. Building-integrated microgrid: Advanced local energy management for forthcoming smart power grid communication. *Energy Build.* **2013**, *59*, 236–243.

15. Wang, B.C.; Sechilariu, M.; Locment, F. Intelligent DC microgrid with smart grid communications: Control strategy consideration and design. *IEEE Trans. Smart Grid* **2012**, *3*, 2148–2156.

16. Kuperman, A.; Levy, U.; Goren, J.; Zafransky, A.; Savernin, A. Battery charger for electric vehicle traction battery switch station. *IEEE Trans. Ind. Electron.* **2013**, *60*, 5391–5399.

17. Van Roy, J.; Leemput, N.; Geth, F.; Buscher, J. Electric vehicle charging in an office building microgrid with distributed energy resources. *IEEE Trans. Sustain. Energy* **2014**, *99*, 1–8.

18. Mohamed, A.; Salehi, V.; Tan, M.; Mohammed, O. Real-time energy management algorithm for plug-in hybrid electric vehicle charging parks involving sustainable energy. *IEEE Trans. Sustain. Energy* **2014**, *5*, 577–586.

19. Zhu, W.; Wang, L.F.; Dounis, A.I.; Yang, R. Integration of plug-in hybrid electric vehicles into energy and comfort management for smart building. *Energy Build.* **2012**, *47*, 260–266.

20. Byeon, G.; Yoon, T.; Oh, S.; Jang, G. Energy management strategy of the DC distribution system in buildings using the EV service model. *IEEE Trans. Power Electron.* **2013**, *28*, 1544–1554.

21. Jin, C.; Wang, P.; Xiao, J.; Tang, Y. Implementation of hierarchical control in DC microgrids. *IEEE Trans. Ind. Electron.* **2014**, *61*, 4032–4042.

22. Armstrong, M.; El Hajj Moussa, C.; Adnot, J.; Galli, A.; Rivière, P. Optimal recharging strategy for battery-switch stations for electric vehicles in France. *Energy Policy* **2013**, *60*, 569–582.

23. Goli, P.; Shireen, W. PV integrated smart charging of PHEVs based on DC link voltage sensing. *IEEE Trans. Smart Grid* **2014**, *5*, 1421–1428.

24. Liu, Y.; Tang, Y.; Shi, J.; Shi, X. Application of small-sized SMES in an EV charging station with DC bus and PV system. *IEEE Trans. Appl. Supercond.* **2015**, *25*, 5700406.

25. Strunz, K.; Abbasi, E.; Huu, D.N. DC microgrid for wind and solar power integration. *IEEE J. Emerg. Sel. Top. Power Electron.* **2014**, *2*, 115–126.

26. Rivera, S.; Wu, B.; Kouro, S.; Yaramasu, V.; Wang, J. Electric vehicle charging station using a neutral point clamped converter with bipolar DC bus. *IEEE Trans. Ind. Electron.* **2015**, *62*, 1999–2009.

27. Locment, F.; Sechilariu, M.; Forgez, C. Electric vehicle charging system with PV grid-connected configuration. In Proceedings of the IEEE Vehicle Power and Propulsion Conference, Lille, France, 1–3 September 2010.

28. Locment, F.; Sechilariu, M. DC microgrid for future electric vehicle charging station designed by energetic macroscopic representation and maximum control structure. In Proceedings of the IEEE International Energy Conference, Dubrovnik, Croatia, 13–16 May 2014; pp. 1454–1460.

29. Wei, W.; Bouscayrol, A.; Ming, C. Comparison of two different traction systems for subway application using Energetic Macroscopic Representation. In Proceedings of the IEEE Vehicle Power and Propulsion Conference, Seoul, Korea, 9–12 October 2012; pp. 984–989.

30. Lhomme, W.; Delarue, P.; Giraud, F.; Lemaire-Semail, B.; Bouscayrol, A. Simulation of a photovoltaic conversion system using Energetic Macroscopic Representation. In Proceedings of the European Power Electronics and Power Electronics and Motion Control Conference, Novi Sad, Serbia, 4–6 September 2012.

31. Yu, H.; Lu, R.; Wang, T.; Zhu, C. Energetic Macroscopic Representation based modeling and control for battery/ultra-capacitor hybrid energy storage system in HEV. In Proceedings of the IEEE Vehicle Power and Propulsion Conference, Dearborn, MI, USA, 7–11 September 2009; pp. 1390–1394.

32. Houssamo, I.; Locment, F.; Sechilariu, M. Experimental analysis of impact of MPPT methods on energy efficiency for photovoltaic power systems. *Int. J. Electr. Power Energy Syst.* **2013**, *46*, 98–107.

33. Houssamo, I.; Locment, F.; Sechilariu, M. Maximum power tracking for photovoltaic power system: Development and experimental comparison of two algorithms. *Renew. Energy* **2010**, *35*, 2381–2387.

34. Bergveld, H.J.; Notten, P.H.L.; Kruijt, W.S. *Battery Management Systems: Design by Modeling*; Philips Research Book Series; Kluwer Academic Publishers: Boston, MA, USA, 2002.

35. Wang, B.C.; Sechilariu, M.; Locment, F. A simple PV Constrained Production Control Strategy. In Proceedings of the IEEE International Symposium on Industrial Electronics (ISIE), Hangzhou, China, 28–31 May 2012; pp. 969–974.

Fast Charging Battery Buses for the Electrification of Urban Public Transport—A Feasibility Study Focusing on Charging Infrastructure and Energy Storage Requirements

Matthias Rogge, Sebastian Wollny and Dirk Uwe Sauer

Abstract: The electrification of public transport bus networks can be carried out utilizing different technological solutions, like trolley, battery or fuel cell buses. The purpose of this paper is to analyze how and to what extent existing bus networks can be electrified with fast charging battery buses. The so called opportunity chargers use mainly the regular dwell time at the stops to charge their batteries. This results in a strong linkage between the vehicle scheduling and the infrastructure planning. The analysis is based on real-world data of the bus network in Muenster, a mid-sized city in Germany. The outcomes underline the necessity to focus on entire vehicle schedules instead on individual trips. The tradeoff between required battery capacity and charging power is explained in detail. Furthermore, the impact on the electricity grid is discussed based on the load profiles of a selected charging station and a combined load profile of the entire network.

Reprinted from *Energies*. Cite as: Rogge, M.; Wollny, S.; Sauer, D.U. Fast Charging Battery Buses for the Electrification of Urban Public Transport—A Feasibility Study Focusing on Charging Infrastructure and Energy Storage Requirements. *Energies* **2015**, *8*, 4587–4606.

1. Introduction

Electric buses support the transition process towards a more sustainable public transport. The different electric bus systems in the market use basically the same traction system to convert electricity into propulsion. The supply of electricity is the defining difference between them. A simple and proven concept is for example the trolley bus. It is continuously connected to overhead wires, which cover the energy demand at any time. However, the overhead wire system causes high invest costs and maintenance efforts [1]. The bus is furthermore bound to certain tracks so that the level of flexibility is very low. Serial diesel hybrid buses generate the electricity onboard with a combustion engine and a generator. An all-day operation without refueling is manageable, due to the high energy density of diesel. A high level of flexibility is furthermore guaranteed, because no infrastructure is needed on the track. Fuel cell buses use hydrogen as their energy source. The available driving range is lower compared to diesel hybrid buses [2], but they still offer a high level of flexibility. The three concepts can always be combined with an energy storage inside the vehicle. Hybrid and fuel cell buses have for example energy storages for the recovery of braking energy [3–5]. The energy storage in hybrid buses can also be used for a partial emission free operation [6,7]. Furthermore, the battery of some hybrid buses can be charged externally. These concepts are also named plug-in hybrid bus or battery bus with range extender, depending on the size of the energy storage and the charging possibilities. Trolley buses can use energy storages for a partial operation without overhead wires [8].

Battery electric buses neither have a continuous power supply nor generate electricity onboard. Their energy is stored in the battery. The energy density of batteries is rather low compared to diesel or hydrogen [9]. The driving range of battery buses is therefore limited and the charging process requires a certain time. There are mainly two concepts for the charging of the battery, standard and fast charging [10]. Standard charging is performed with a moderate charging power mainly in the bus depot overnight and during longer brakes. This causes a high battery capacity and a high weight of the system, when the bus shall be operated the entire day [11]. Fast charging on the track during operation can reduce the battery capacity and therefore the weight significantly. However, the bus schedule must provide sufficient charging times at certain locations. The existing research in this field focusses mainly on the adaption of the vehicle scheduling on fixed predetermined charging infrastructure and vehicle configurations [12–14] or on the dimensioning of the battery capacity and charging infrastructure for a single bus route or standard driving cycle, without considering the vehicle scheduling in detail [11,15]. This works expands the scope from the secondly mentioned work to the entire bus network taking especially the influence of the vehicle scheduling on the system design into account. It is analyzed how and to what extend entire bus networks can be electrified with fast charging battery bus systems, without changing the existing bus routes and trips. This ensures a straightforward transition process from the conventional to an electrified bus fleet, because the operator does not have to adjust the already optimized operational planning. The results are discussed with the focus on the general feasibility and the required minimal battery capacity. Furthermore, the limits of this approach are shown and discussed based on examples.

2. Material and Boundary Conditions

This section highlights the processed raw data and the relevant boundary conditions for the analysis. The first part describes the considered bus network consisting of the routes and a certain set of service trips (trips on a bus route in regular passenger service), which are currently operated with conventional diesel buses. A common set of service trip types is identified and subsequently used in the energy consumption calculation. Part 2 focuses on current technological solutions for the fast charging process in public transport applications. Their characteristic data is shown and the modeling for the simulation is described. The third part introduces solutions for the positioning of fast charging stations and describes the concept used in this analysis.

2.1. Bus Network

The analysis is conducted for the bus network of Muenster, a medium sized city in Germany. The local bus operator "Stadtwerke Muenster" publishes the bus schedules online [16], but detailed vehicle schedules are not available to the public. Therefore, the trips on each bus route in regular passenger service, the so called service trips, are identified manually. The dataset of a workday is chosen, because it places the highest demand on the bus system referring to the number of vehicles and the service frequency. The prepared dataset is shown in Table 1. It consists of 1588 service trips, which have an accumulated driving distance of about 27,000 km per day. The service trips are performed on 23 different bus routes. Buses on the same route leave in 20 min intervals. Superposition of different routes is used to achieve a higher frequency in critical areas. Separate trips for demand response transport, e.g., school transport, are excluded from the scope of this analysis. Regional bus routes to suburbs are also not taken into account, because they are not operated by "Stadtwerke Muenster".

Table 1. Identified service trips for a workday.

Route	No. of service trip types	No. of service trips per day	Daily driving distance [km]	Route	No. of service trip types	No. of service trips per day	Daily driving distance [km]
1	4	98	2,297	13	2	90	583
2	5	88	1,558	14	2	91	1,063
3	1	45	550	15	2	94	1,977
4	1	45	562	16	4	91	1,768
5	3	95	2,201	17	7	90	1,184
6	4	116	1,847	80	2	16	224
7	7	93	2,019	81	2	16	453
8	4	92	1,700	82	7	17	312
9	3	91	1,678	83	2	16	323
10	6	90	1,962	84	2	16	279
11	2	92	1,392	85	2	16	419
12	2	90	653	Σ	76	1,588	27,003

The service trips on each bus route are clustered into 76 service trip types, which have a common course and duration. The number of service trip types differs depending on the considered bus routes. For bus routes which have two terminal stops and no shortenings during the day exist only 2 service trip types, representing the back and return trip. A higher value reveals that there are shortenings during the day, which use different terminal stops. The value 1 for route 3 and 4 indicates that the return trip is missing. In this special case, the return trip of bus route 3 is the trip of bus route 4. In the following analysis route 3 and 4 are combined to route 34. The service trip types are used in the simulation of the energy consumption. They can be transformed to individual service trips by adding a certain starting time.

Route 1–17 are operated until 8 p.m. and called day routes in the following. Afterwards the night routes 80–85 start their service. The night routes are considered as individual routes in this analysis, because their course differs from the day routes. The daily driven distance of the night routes is lower compared to the day routes due to the shorter operating time. Information about the vehicles serving the different bus routes is not available. Therefore, it is assumed that each bus route is served by articulated buses with a length of 18 m. This bus type dominates the fleet of "Stadtwerke Muenster" and is furthermore the largest bus operated by them. More detailed vehicle parameters are given in Section 3.1.

2.2. Fast Charging Systems

The fast charging systems available in the market are based on different coupling technologies. The concepts can be divided in two groups, which use conductive or inductive energy transfer. Conductive coupling devices are offered for example by ABB, Oprid, Schunk or Proterra. A further system is developed within the German research project SEB by the RWTH Aachen University. The conductive coupling devices enable a very high charging power of up to 500 kW, which is demonstrated for example by Proterra in the US [17]. The charging power of the inductive systems is about 200 kW for the Bombardier Primove system [18] and 120 kW for the system of Conductix Wampfler [19]. Table 2 shows the charging power of some exemplary systems.

Table 2. Examples of current fast charging systems [17–20].

Supplier	System	Technology	Charging power
Proterra	FastFill	conductive	500 kW
Bombardier	Primove	inductive	200 kW
ABB	TOSA	conductive	200 kW; 400 kW (15 s)
Conductix Wampfler	IPT charger	inductive	60–180 kW

The detailed coupling procedure is excluded from the scope of this analysis. Charging systems are modeled by their maximum continuous charging power and the duration of the coupling und decoupling process. The analysis takes into account charging powers from 100 kW to 500 kW in steps of 100 kW for simplicity reasons.

This subdivision reflects the currently available systems in the market. However, this value is the maximum charging power capability. The required charging power is determined in each case by the consumed energy and the available charging time.

2.3. Positioning of Fast Charging Infrastructure

Studies focusing on the positioning of charging stations for passenger vehicles have to predict the customer behavior in order to get information on the demand [21–23]. However, in the field of public transport buses, the operating conditions of the energy consumers are well known. The buses have a fixed route and the dwell time can be estimated based on the bus schedule and an expected delay. The fast charging can take place at the bus stops on the track, at the terminal stops and in the bus depot. Especially inductive solutions offer also the possibility to charge during driving, but this has not yet been implemented in practice for a public transport bus. The longest dwell time is usually located at the terminal stops, so that delays can be compensated and the bus driver can have a break according to the regulations of driving time. Furthermore, the terminal stops are usually located outside the city center, where the construction of a charging station can be carried out easier. The terminal stops appear therefore as a highly suitable location for the charging stations. In practice, the concept of fast charging at the terminal stops is implemented for example in the battery bus project in Vienna [24].

Figure 1. Localization of fast charging stations at the terminal stops in the surroundings of the city center.

For this analysis, it is assumed that all fast charging stations are located at the terminal stops of the bus routes in Muenster. The current dataset contains 44 individual terminal stops resulting in 44 charging stations. The charging stations located in the surroundings of the city center as well as the corresponding bus routes can be seen in Figure 1. The charging stations in the simulation are modeled by a predefined charging power for every demanding bus. A simultaneous charging of multiple buses with the full charging power is possible.

3. Calculation Method

The simulation is divided in 3 steps. First, the energy consumption of each service trip type is simulated based on the defined bus type and the geographical characteristics of the bus route. Secondly, the service trip types are combined to individual vehicle schedules based on the determined set of service trips including the available charging time at the terminal stops. Furthermore, the resulting power profiles are derived for every charging station and the entire network, which reveals the impact of simultaneous charging processes. In the third step, the required battery capacity is calculated for each bus route based on the given charging power.

3.1. Energy Consumption

The simulation of the energy consumption is performed according to the methodology of Sinhuber [11]. The proposed simulation model consists of a track and a vehicle model. The track model uses data from Openstreetmap on the course of the bus route and the position of the stops. The Shuttle Radar Topography Mission (SRTM) data of the NASA is used to calculate the height profile. It is assumed that the bus waits at every bus stop for 20 s and at every traffic light for 15 s. The energy demand for the traction system is calculated based on the driving resistances, which consists of air drag, rolling and climbing resistance. However, the air drag resistance is of minor importance, due to the low vehicle speed.

In addition to the energy consumption of the traction system, the consumption of the auxiliaries has to be taken into account [25,26]. Main consumers are for example the steering support, the compressor and the air condition. The interior heating can be realized by an electric heater, like a heat pump or a PTC heater, or by a conventional heating system which uses fossil fuels. However, especially the use of a PTC heater would lead to extreme energy demands, which intensively affects the outcomes of this analysis. A standard 18 m bus has for example an energy consumption without interior heating of about 2 kWh/km. Taking an average speed of 15 km/h into account, the average traction power can be calculated to 30 kW. The required power for the PTC heating system can equal this value in extreme conditions and therefore double the total energy consumption [27]. Heat pump systems have a lower energy consumption and they enable the use of waste heat from the traction system [28]. The measured system in the analysis of Cho *et al.* supplied a heating power of 30 kW with an electric energy consumption of 10 kW for the compressor. The use of a conventional heating system with fossil fuels enables heat generation without electricity. The conventional systems are used in many current battery bus projects and therefore chosen for this analysis. Hence, the energy consumption of the air conditioning system is dominated by the cooling scenario in the summer. The moderate climate conditions in Germany allow an energy efficient cooling concept. The German Transport Association recommends a cooling to a defined temperature difference between the vehicle interior and outside temperature, instead of cooling to a defined vehicle interior temperature [29]. The total power of the air condition system in the summer can therefore be limited to 6.75 kW for an 18 m bus. The dynamic behavior of the auxiliaries is not taken into account, because the main focus of the simulation is the overall energy consumption. The auxiliaries are therefore modeled by a constant load in the simulation.

The basic vehicle model, consisting of the mechanical and the traction system part, is parameterized and verified with data of the APTS Phileas bus. The data was gained within the "H2-Bus NRW" project [30]. The energy supply system of the "H2-Bus" consists of a fuel cell, double layer capacitors and a NiMH battery. In this analysis, these components are replaced by a lithium-ion battery, which is represented by its efficiency. Table 3 highlights the relevant parameters of the bus, which are used in the simulation. One important key parameter is the weight. It affects directly the rolling and climbing resistance and has therefore a strong impact on the energy consumption. This analysis focusses on the technical feasibility, which should be proven under the most challenging conditions. Thus, the maximum gross vehicle weight of 28 t is used in the simulation, which represents a fully packed vehicle. This value includes also the weight of the battery. The worst case scenario for the auxiliaries is represented by the summer condition, in which the vehicle interior is continuously cooled (6.75 kW) and the maximum continuous auxiliary power (2.25 kW) is demanded.

Table 3. Vehicle parameter.

Parameter	Value	Relevant for	Parameter	Value	Relevant for
Width	2.55 m	Cross section area for air drag calculation	Efficiency of the traction system	90%	Loss calculation
Height	3.44 m	Cross section area for air drag calculation	Efficiency of the battery system	95%	Loss calculation for charging and discharging
Maximum gross vehicle weight	28 t	Rolling resistance, climbing resistance	Max auxiliary power	9 kW	Energy consumption of the auxiliaries

3.2. Vehicle Scheduling and Grid Load Profiles

The technical feasibility of the electrification is analyzed individually for every bus route. A transition between the day and the night routes is therefore not taken into account. A lean algorithm is used for the vehicle scheduling to enable separate analysis of each route. Every bus serves only one route without any deadheading trips. After a service trip is finished, the bus waits at the terminal stop until the next service trip on the same route starts from the current stop. If there is no ongoing trip during the day, the bus will drive back to the depot. During the dwell time at the terminal stop, the bus has the possibility to charge its battery. The available charging time is calculated based on the resulting dwell time reduced by the average delay of the bus system and the required time for the coupling process of the fast charging system. The average delay for the bus system in Muenster is about 3 min [31]. The coupling time differs according to the considered system. This analysis compares different charging systems represented by the charging power, without focusing on the detailed coupling process. The coupling and decoupling time is set to 30 s. The resulting charging time can be calculated to:

$$T_{charge,av} = T_{dwell} - T_{delay} - 2\, T_{coupling} \qquad (1)$$

It is assumed that the battery of each bus is fully charged at the beginning of the shift, which means that the SOC (State of Charge) is 100%. This value is the upper limit and cannot be exceeded during the charging process. Within the charging process every bus tries to charge to 100% SOC. The required charging power can be calculated based on the energy demand from the traction system and the auxiliaries, the available charging time and the efficiency of the battery system:

$$P_{charge,req} = \frac{\Delta E_{demand}}{T_{charge,av}} \times \frac{1}{\eta_{bat}} \qquad (2)$$

The calculated charging power represents the value at the DC-output of the charging station and includes the losses of the battery system during charging. Furthermore, the efficiency of the charging system has to be taken into account in the calculation of the overall power demand:

$$P_{charge,grid} = \frac{P_{charge,req}}{\eta_{charger}} \qquad (3)$$

It is assumed that the efficiency of the charging system is 90%. The overall power demand represents the AC-input power of the charger and has to be covered by the electrical grid. All charging powers mentioned in the following correspond to this value.

The use of the available charging time in Equation (2) ensures that the entire dwell time is utilized. Another possibility would be to charge the bus with the maximum available charging power followed by a waiting period until the end of the break. This procedure is not recommended, because the losses increase with the applied charging power. Furthermore, the additional losses could raise the battery temperature, which results in an accelerated aging [32]. If the required charging power cannot be supplied by the charging station, the bus will charge with the maximum available power. In this case, the SOC will not reach 100%. The remaining energy deficit will be added to the energy demand of the next charging process. This causes an increase of the required battery capacity. At the end of the shift, the bus has additional 20 min charging time in order to compensate an energy deficit accumulated during the day. The charging time is limited because the buses arrive in 20 min intervals at the terminal stops. A remaining energy deficit is recharged in the bus depot overnight.

The resulting power of the individual charging processes is afterwards combined to a power profile for each charging station. This enables a spatially resolved analysis, which is important to discuss the electrical grid stability. The power profiles can furthermore be merged to an overall load profile for the entire bus network, which enables a conclusion about the simultaneities and the overlapping.

3.3. Battery Capacity

Different energy storages can be used for opportunity charging buses depending on the localization of the fast charging stations respectively the desired operating range and on the demanded charging power. Supercapacitors offer a very high charging power, but the energy density is rather low [33]. The operating range of the vehicle is therefore limited, so that charging stations at several bus stops along the route are required. The operating range can be increased when Supercapacitors are combined with a battery [15]. However, with the improvements in the lithium-ion battery technology their performance becomes sufficient for the opportunity charging application even without the use of Supercapacitors. Current battery systems in electric public transport buses are therefore mainly based on the lithium-ion technology (except of the 24 V lead acid batteries).

The capacity of a lithium-ion battery is not a constant value during its lifetime. It fades because of aging processes, which are time and usage depended [34]. The end of life (EOL) of a lithium-ion battery is usually defined as a remaining capacity of 80% or as a doubled internal resistance, whichever occurs first. The EOL conditions are important for the dimensioning of the battery capacity. To enable an unrestricted operation even at the EOL, a 20% reserve has to be taken into account. The fast charging application causes an additional reduction of the usable capacity due to the voltage limitation. At high SOCs the charging current has to be reduced in order not to exceed the upper voltage limit of the battery. This effect increases with the aging of the battery due to the increasing internal resistance [35]. The reduction of the current leads to an increase of the charging time, which contradicts the fast charging purpose. The upper region of the SOC can therefore not be used fast charging applications. This effect depends on the applied charging current and the used cell chemistry [35].

Figure 2 highlights the usable battery capacity at EOL of a typical NMC lithium-ion cell and of an ideal battery.

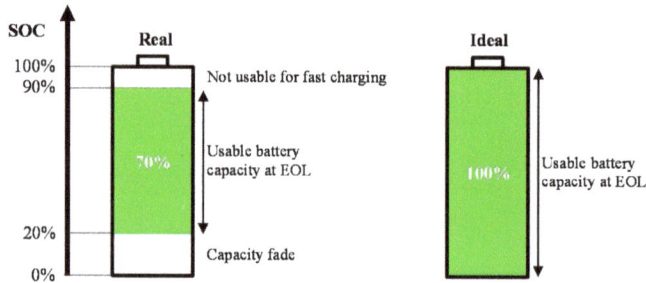

Figure 2. Usable battery capacity of an exemplary lithium-ion cell at EOL.

In this analysis, an ideal battery is used to achieve technological independence. The ideal battery can be charged to a SOC of 100% with the maximum charging current and the capacity does not fade during the lifetime of the battery. Nonetheless, the efficiency is taken into account in order to calculate realistic power profiles. The required capacity of the ideal battery is calculated based on the energy consumption of the service trips and the energy supply of the fast charging system. Therefore, the value differs according to the bus schedules. It is assumed that all buses, which are operated on the same bus route, have an identical battery system. The battery capacity must therefore cover the worst case scenario of the considered bus schedules on the route. The calculated capacity for the ideal battery can be transferred to real world conditions by multiplication with the oversizing factor. In the example in Figure 2 the factor is 1.43. It has been observed that the aging of a lithium-ion battery could depend on the applied cycle depth [36]. Therefore, it can be useful to use even higher oversizing factors to reduce the depth of discharge. However, this is always a tradeoff between the resulting weight of the battery system and the lifetime advantages.

4. Results and Discussion

The described bus network from Section 2 is analyzed with the calculation method described in Section 3 taking different charging power capabilities into account. The gained results are discussed with a broadening scope in this section, starting from the service trip over the route until the entire network scope. The first section focusses on the worst-case energy consumption of the service trip types on each bus route. The shown values reveal the minimum for the subsequent battery capacity calculation in the second part. The third part describes the proportion of routes, which can be electrified with a certain combination of charging power and battery capacity. The last part reveals the resulting power profiles of the charging stations and discusses the effects on the electrical grid.

4.1. Energy Consumption

The entire bus network of "Stadtwerke Muenster" can be fully described, in terms of energy consumption, by the 76 service trip types. The service trip types have a different course and therefore a different track profile consisting of elevation data, required stops, traffic signs *etc.* The energy consumption of the auxiliaries is determined by the travel time. Two simulations are conducted to reveal the influence of the auxiliaries on the overall energy consumption. In the first simulation the auxiliary power is set to zero, so that the outcome represents only the traction energy. The values range from 1.79 to 2.10 kWh/km with an average of 1.96 kWh/km. The distribution is highlighted in Figure 3 on the left. Sinhuber proposed a value of 0.072 kWh/km·t [11], which leads to an energy consumption of 2.016 kWh/km for a bus with a weight of 28 t. This matches the simulation results of this analysis.

The energy consumption increases significantly when the auxiliaries are taken into account. The distribution of the results from the second simulation with the maximum auxiliary power of 9 kW is shown in the right part of Figure 3. The minimum value is 2.26 kWh/km and the maximum 2.69 kWh/km. The average value is about 2.47 kWh/km, which is an increase by 26% compared to the average value of the traction without auxiliaries. This observation confirms the necessity of including the auxiliaries in the overall energy consumption simulation. The defined scenario for the auxiliaries represents the summer with an active interior cooling, which will be the worst case condition, if a conventional heating system with fossil fuels is used in the winter. An electric heating system would increase the energy consumption of the auxiliaries as described in Section 3.1.

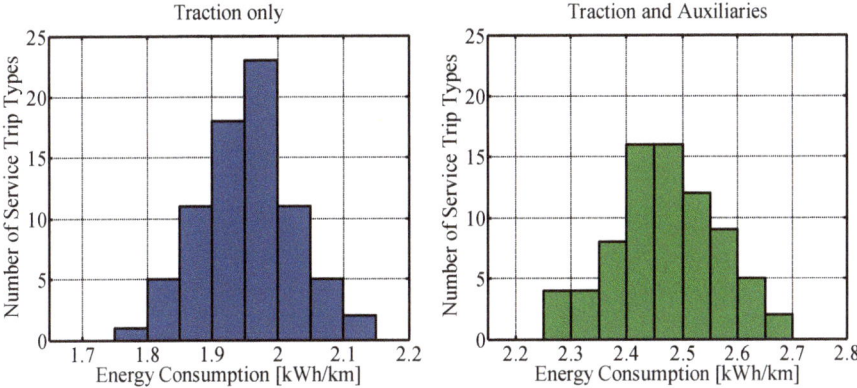

Figure 3. Histogram of the energy consumption of all service trip types.

Figure 4 gives an overview on the energy consumption of the individual bus routes. To enable a direct link to the required battery capacity for the bus route, only the service trip types with the highest energy consumption are shown. The values range from 18 kWh (route 13) to more than 70 kWh (route 7). These values represent the minimum requirements for the battery capacity of the bus routes and give a first indication on the electrification potential. However, it is not possible to draw a conclusion out of this data, unless the vehicle schedules are taken into account. As a consequence of the vehicle schedules, it is always possible that several short trips with minor energy consumption are combined without sufficient charging stops, so that the required battery capacity increases.

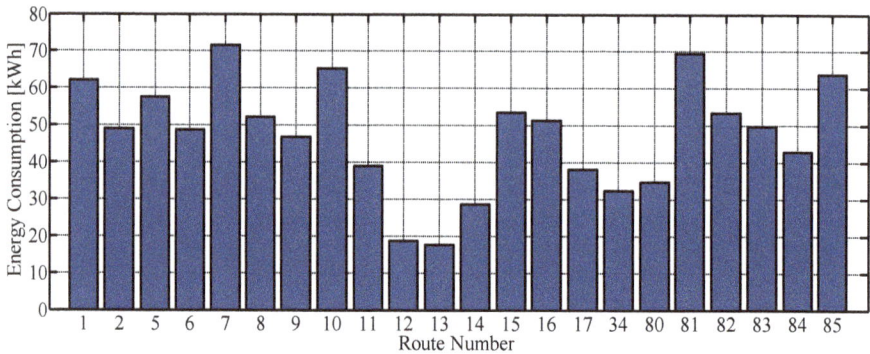

Figure 4. Highest energy consumption of the service trip types on the different routes.

4.2. Impact of the Charging Power on the Required Battery Capacity

The service trip types are combined to vehicle schedules according to the set of identified service trips and the methodology described in Section 3.2. In this section, the required battery capacity is analyzed separately for every route taking the worst case condition and different charging power capabilities into account.

Table 4 reveals the results of the calculation. The values shown for the charging power of 0 kW represent a full day operation without recharging. For the day routes they range from 433 kWh (route 12) to 715 kWh (route 7) and the values for the night routes range from 192 kWh (route 80) to 243 kWh (route 85). The differentiation between the day and the night routes is also clearly visible in the visualization shown in Figure 5. The different routes are shown on the x-axis, the charging power on the y-axis and the required battery capacity on the z-axis.

Table 4. Required usable battery capacity in kilowatt hours (kWh) for different charging power limits.

Route	Charging Power [kW]						Route	Charging Power [kW]					
	0	**100**	**200**	**300**	**400**	**500**		**0**	**100**	**200**	**300**	**400**	**500**
1	695	590	514	446	379	311	14	531	425	315	215	116	49
2	672	629	580	537	504	471	15	632	504	362	219	130	122
5	680	578	464	350	236	122	16	697	595	481	367	267	169
6	630	550	460	370	290	215	17	515	342	186	89	82	74
7	715	559	385	211	146	146	34	485	296	91	35	33	33
8	614	463	315	220	155	137	80	192	163	129	94	60	58
9	594	486	364	267	169	92	81	201	184	165	145	126	106
10	671	582	482	403	340	279	82	243	207	166	133	104	84
11	533	339	124	76	76	76	83	236	200	158	125	102	96
12	433	217	32	29	26	23	84	207	138	105	84	84	84
13	512	430	337	245	166	94	85	243	227	209	191	173	155

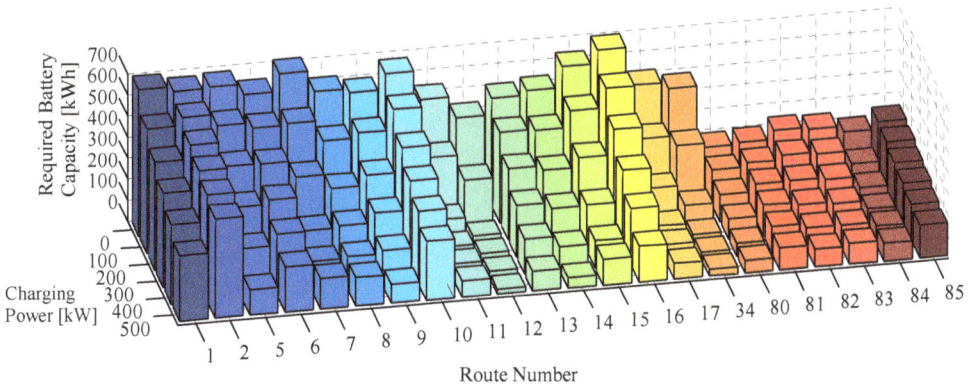

Figure 5. Required capacity of the ideal battery over charging power for every bus route.

The limiting factor for the installable battery capacity in an electric bus is the weight. Thus, it is always a tradeoff between passenger and battery capacity. The unloaded weight of an electric 18 m articulated bus is estimated to 14.5 t. It includes the weight of the electrical traction system, but not the weight of the battery system. Taking into account the defined gross vehicle weight in Section 3.1 of 28 t, the bus offers a theoretical passenger capacity of 180 people. The passenger capacity of conventional 18 m articulated buses range from 140 to 160 persons depending on the unloaded weight of the bus [37,38]. Current lithium-ion battery systems for fast charging applications achieve energy densities of about 100 Wh/kg on system level [39,40]. Even the battery system of the weight optimized BMW i3 is in this range [41]. The maximum installable battery capacity can be calculated by:

$$E_{max} = (m_{bus\ total} - n_{passengers} \cdot m_{passenger} - m_{bus\ unloaded}) \cdot \omega_{battery} \qquad (4)$$

With a total mass of 28 t, 140 passengers of 75 kg, an unloaded weight of 14.5 t and an energy density of 100 Wh/kg of the battery system the maximum installable battery capacity can be calculated to 300 kWh (ideal battery). Taking the oversizing factor of 1.43 into account reveals that the usable capacity at the EOL of the battery is limited to 210 kWh for a state-of-the-art lithium-ion system. Under this condition, only some night routes can be electrified without opportunity charging. A partly electrification of a small proportion of the day routes and a large part of the night routes requires at least a charging power of 200 kW.

The required battery capacity decreases with an increasing charging power. This effect is expected and can be observed for every route. The intensity of the decline differs among the routes. The required battery capacity for route 12 decreases very rapidly and reaches a value close to the minimum with a charging power of 200 kW. A further increase of the charging power has only a limited effect. This can also be observed for route 7, 11, 17, 34 and 84. Other routes like for example route 1 show a different behavior. The required battery capacity decreases even at the step from 400 kW to 500 kW. A further increase of the charging power is required, when the battery capacity should be minimized. The intensity of the decline is mainly caused by the available charging time. Therefore, it would be beneficial to express the slope based on the information on the dwell time of the vehicle schedules. However, this is not possible due to the limitation of the SOC of the battery. Additional charging time is not useful, when the battery is already charged to 100%. It is therefore mandatory to focus always on the resulting profiles instead of only on single values.

The results shown in Figure 5 indicate clearly that it is beneficial to focus on the entire vehicle schedules instead of focusing on single trips as in Figure 4. Route 2 has for example less than 50 kWh energy consumption per trip, which sounds manageable for an electrification. However, the route is operated heavily without sufficient charging time. The analysis of the vehicle schedules reveals that route 2 is the most problematic route to electrify. Another example is route 7 with an energy consumption per trip of more than 70 kWh. The analysis of the vehicle schedule reveals that the route can be operated with a charging power of 400 kW and a minimized battery capacity of 146 kWh, which points out that even routes with a high energy demand can be electrified if the dwell time at the terminal stop is sufficient.

The discussed values for the charging power describe the maximum power capability of the charging station. The power demand of the vehicles may be less than the offered charging power, depending on the energy consumption and the available charging time. Figure 6 highlights the distribution of the applied charging power for the 1588 service trips in the 500 kW scenario. The interval between 0 and 25 kW is dominated by the trips without any charging possibility. Due to the insufficient dwell time, no charging action can take place. A large part (46%) of the charging processes uses a charging power between 450 kW and 500 kW. This high proportion is caused by the definition of the charging process. Every vehicle tries to charge its batteries as fast as possible even if there is plenty of time at the next terminal stop. This issue can be solved with an intelligent control algorithm. The algorithm has to predict the future charging possibilities for the vehicle so that an optimization of the charging power can take place. The optimization criteria in this case would be the resulting life-cycle costs for the entire system taking the costs for the charging infrastructure as well as the costs for the battery system into account.

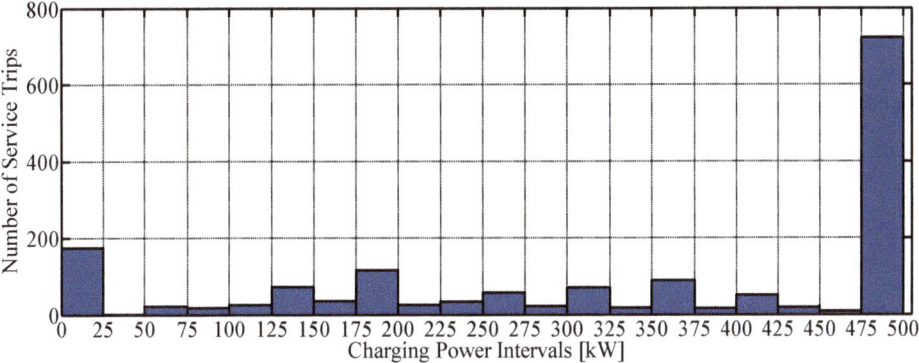

Figure 6. Histogram of the applied charging power for all 1588 service trips in the 500 kW scenario.

4.3. Electrification Level of the Bus Network

The electrification level reveals the percentage of a given bus network which can be electrified with a certain charging power and battery capacity. It is measured route wise and weighted based on the daily driven distance of each bus route. The weighting of the electrification level with the daily driven distance enables a direct connection with the CO_2 reduction potential, if CO_2 neutral electricity is used for the charging. An electrification level of 100% for a bus network means that all trips on all routes can be carried out with the defined charging power and battery capacity, even under worst-case conditions.

The described calculation procedure for the minimum battery capacity does not require that the whole energy consumed during operation is charged on the track. An energy deficit at the end of the shift is allowed. This leads to an additional charging overnight in the bus depot, where it can be done with lower power than the charging on the track due to the long dwell time.

The calculated electrification levels are shown in Figure 7. The night routes can easily be identified when the 0 kW curve is considered. Their required capacity of the ideal battery is less than 300 kWh, but the influence on the electrification level is rather low, due to their limited daily driving distance. The influence of the increasing charging power can be assessed based on the gap between the colored curves. The gap fades with an increasing charging power, because some routes have already reached the minimum value for the required battery capacity. An increase of the charging power from 400 kW to 500 kW has therefore only limited effects. In Section 4.2, route 2 was identified as the most problematic route to electrify. The influence of route 2 can also be observed in Figure 7. The electrification level above 95% is dominated by route 2. An increase of the charging power has only a minor impact due to the extremely limited charging time.

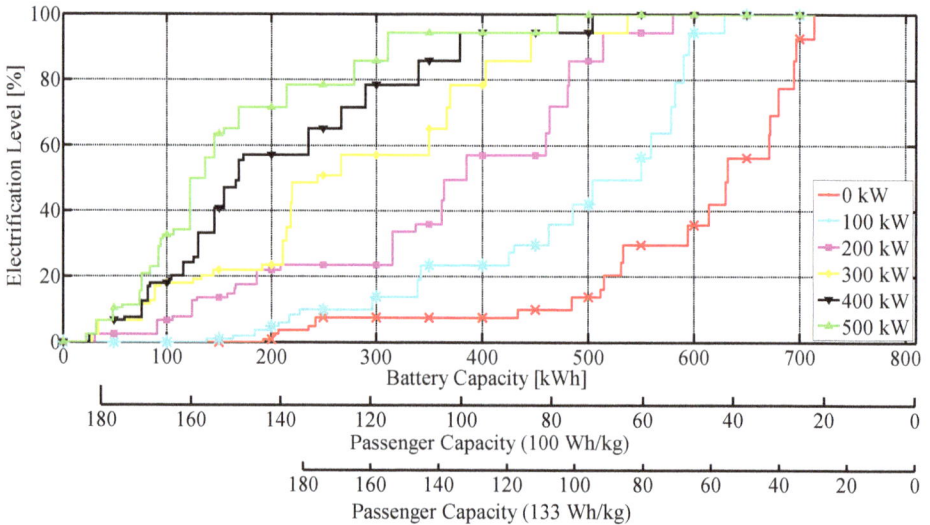

Figure 7. Electrification level of the bus network as a function of usable battery capacity and charging power.

The second and third x-axis in Figure 7 highlight the tradeoff between battery and passenger capacity. The resulting passenger capacity is calculated based on the parameters discussed in Section 4.2, taking the oversizing factor of 1.43 and a specific energy density of the battery system into account. The value of 180 passengers can be seen as the maximum, due to the available space inside the vehicle. The second x-axis bases on a specific energy density of 100 Wh/kg, which is a realistic value for current battery systems. It can be seen that an electrification level above

20% with a passenger capacity of 140 people can only be reached with a charging power of at least 200 kW. The electrification level can be increased to 50% with a charging power of 300 kW and a battery capacity of 220 kWh. A further increase of the charging power to 500 kW enables an electrification level of about 80%, with a passenger capacity of nearly 140 people. The electrification level exceeds 90% by decreasing the passenger capacity to 115 people and therefore increasing the installed battery capacity to 310 kWh. The third x-axis is also calculated based on the oversizing factor of 1.43, but the specific energy density for the battery system is set to 133 Wh/kg, which represents an estimation for the future technological development. Considering the increased energy density, an electrification level of 90% and a passenger capacity of over 150 people can be achieved with a charging power of 500 kW.

The analysis reveals that a significant electrification level and therefore a significant CO_2 reduction can be achieved with state of the art fast charging battery bus technology. This is possible even in the worst-case scenario without changing the current bus timetables. A defined electrification level can be reached by different combinations of charging power and battery capacity. However, this analysis proves the technical feasibility. The most economical solution has to be identified in life-cycle-cost calculations taking into account investment costs for infrastructure and vehicles, maintenance costs, replacement costs as well as energy costs. The result is always a tradeoff between battery capacity and charging power, which causes the lowest life-cycle costs for the entire system. It can for example be beneficial to increase the battery capacity in order to be able to reduce the charging power at critical locations or even to omit the construction of a certain charging station. Furthermore, an increase in the number of operated buses and an adaptation of the vehicle scheduling could extent the dwell times and therefore lower the required battery capacity and charging power. However, the additional costs for the further vehicles and drivers have to be compensated by the savings. This optimization problem is highly complex, because it addresses technical issues as the dimensioning of the components as well as the area of operations research in terms of vehicle scheduling and crew rostering.

4.4. Grid Load Profile

The electrification of bus networks with fast charging battery bus systems influences not only the transport sector. The high power demand of the buses cause also effects in the electrical grid. Figure 8 shows an exemplary power profile of the charging station at "Gallenkamp" with a maximum charging capability of 500 kW. The maximum charging power is not demanded in this case. The dwell time is sufficient for a complete charging of the battery even at a lower power of 475 kW. The shape of the power profile has a high dynamic. The sporadic peaks with a delta of

475 kW could cause problems in the electrical grid. The peaks can be equalized with peak shaving strategies. A stationary battery can for example buffer energy when no bus is charged and supply this energy afterwards in the charging process [42,43]. The moving average in Figure 8 gives a first indicative on the resulting power profile utilizing such a system. Another advantage of the peak shaving is that the grid connection costs for the bus operator could be reduced, because the operator has to pay a monthly fee for the installed power capability regardless of using time [44]. The optimal configuration for a peak shaving system can be determined by life-cycle-cost calculations taking the invest costs for the grid connection, the stationary storage and the monthly fee for the grid connection and consumed electricity into account.

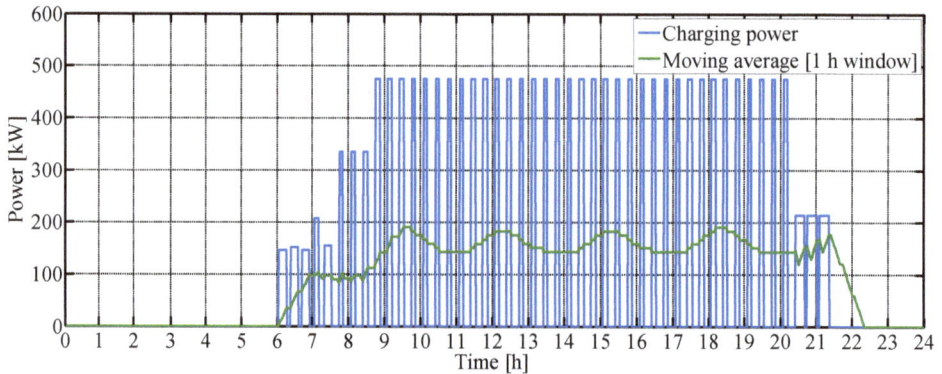

Figure 8. Power profile of the charging station at "Gallenkamp" with a charging capability of 500 kW.

The power profiles of all 44 charging stations can be merged to a power profile of the entire network, which is presented in Figure 9. The power profiles of the individual charging stations overlap, so that a continuous load is applied to the grid. However, the resulting power profile has still intensive fluctuations. The frequency is caused by the 20 min intervals of the bus schedules. The highest value of more than 9 MW is demanded in the evening, when the buses that serve the day routes finish their shift. The defined interval of 20 min for the charging at the end of the shift is longer than the usual dwell time. This leads to an increasing overlap between the individual charging processes. A decrease of this time frame and therefore a shift of the charging process to the bus depot can lower this peak.

The power profile of the entire network is not applied at only one single location. The individual charging stations, which are located throughout the city, are connected to the local electricity grid. The overall power profile has therefore in addition to the time also a geographical dependency, which has to be taken into account in the realization of the fast charging system in order to ensure the stability of the electricity grid.

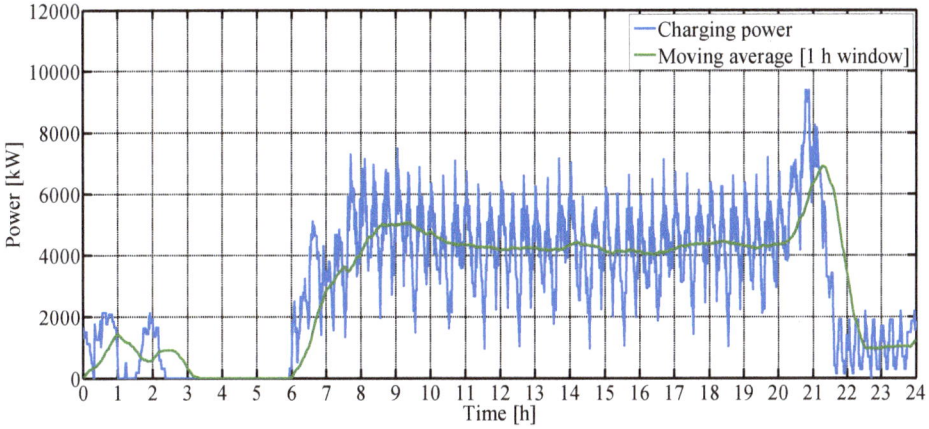

Figure 9. Power profile of the entire bus network with a charging capability of 500 kW per bus.

5. Conclusions

This work analyzed the existing bus network of the German city of Muenster regardless its electrification potential with fast charging battery buses. State of the art fast charging technology was presented and the locations of 44 fast charging stations were derived out of the bus network data. The energy consumption for individual service trip types was calculated for 18 m articulated buses and combined to vehicle schedules for every route. The analysis points out that it is necessary to focus on the entire vehicle schedules instead of individual trips, when the required battery size is calculated. It has been shown that 50% of the service trips can be electrified with a charging power capability of 300 kW and a usable battery capacity of 220 kWh. This is possible even under worst-case conditions using currently available battery systems and without any changes in the existing schedules. An increase of the charging power capability to 500 kW enables an electrification level of about 80%. The resulting power profiles for the charging stations have a high dynamic. Therefore, it can be beneficial to install additional hardware at the charging stations to equalize the load. The tradeoff between the required battery capacity and the passenger capacity was explained in detail. A reduction of the demanded passenger capacity enables an increase of the installable battery capacity, so that the required charging power can be reduced. This analysis proves the technical feasibility of the electrification with fast charging battery buses. Based on this, the cost-optimized electrification scenario can be derived in a life-cycle-cost calculation.

Acknowledgments: The work was conducted as part of the Climate-KIC research project LoCarUT funded by the European Institute of Innovation and Technology (EIT).

Conflicts of Interest: The authors declare no conflict of interest.

References

1. Brunton, L.J. Why not the trolleybus? In Proceedings of the IEEE Seminar on Electric, Hybrid and Fuel Cell Vehicles, Durham, UK, 11 April 2000; p. 5.
2. Wu, W.; Bucknall, R.W.G. Conceptual evaluation of a fuel-cell-hybrid powered bus. In Proceedings of the 48th Universities' Power Engineering Conference (UPEC), Dublin, Ireland, 2–5 September 2013; pp. 1–5.
3. Kellaway, M.J. Hybrid buses—What their batteries really need to do. *J. Power Sources* **2007**, *168*, 95–98.
4. Sasaki, M.; Araki, S.; Miyata, T.; Kawaji, T. Development of capacitor hybrid system for urban buses. *JSAE Rev.* **2002**, *23*, 451–457.
5. Xiong, W.; Zhang, Y.; Yin, C. Optimal energy management for a series–parallel hybrid electric bus. *Energ. Convers. Manag.* **2009**, *50*, 1730–1738.
6. Klingner, M.; Potthoff, U. The AutoTram: A system integration platform of multiple energy storage and transfer concepts. In Proceedings of the VDE Kongress 2010, Leipzig, Germany, 8–9 November 2010.
7. Wang, X.; He, H.; Sun, F.; Sun, X.; Tang, H. Comparative study on different energy management strategies for plug-in hybrid electric vehicles. *Energies* **2013**, *6*, 5656–5675.
8. Kühne, R. Electric buses—An energy efficient urban transportation means. *Energy* **2010**, *35*, 4510–4513.
9. Campanari, S.; Manzolini, G.; Garcia de la Iglesia, F. Energy analysis of electric vehicles using batteries or fuel cells through well-to-wheel driving cycle simulations. *J. Power Sources* **2009**, *186*, 464–477.
10. Sauer, D.U.; Rohlfs, W.; Sinhuber, P.; Rogge, M. Energy consumption, battery size, battery type and charging infrastructure—Optimal eÖPNV mobility through integral analysis. In Proceedings of the 4th VDV Conference Electric Buses—Market of the Future, Berlin, Germany, 18–19 February 2013.
11. Sinhuber, P.; Rohlfs, W.; Sauer, D.U. Study on power and energy demand for sizing the energy storage systems for electrified local public transport buses. In Proceedings of the IEEE Vehicle Power and Propulsion Conference (VPPC), Seoul, Korea, 9–12 October 2012; pp. 315–320.
12. Paul, T.; Yamada, H. Operation and charging scheduling of electric buses in a city bus route network. In Proceedings of the 17th IEEE International Conference on Intelligent Transportation Systems (ITSC), Qingdao, China, 8–11 October 2014; pp. 2780–2786.
13. Wang, H.; Shen, J. Heuristic approaches for solving transit vehicle scheduling problem with route and fueling time constraints. *Appl. Math. Comput.* **2007**, *190*, 1237–1249.
14. Zhu, C.; Chen, X. Optimizing battery electric bus transit vehicle scheduling with battery exchanging: Model and case study. *Procedia Soc. Behav. Sci.* **2013**, *96*, 2725–2736.

15. Mapelli, F.L.; Tarsitano, D.; Annese, D.; Sala, M.; Bosia, G. A study of urban electric bus with a fast charging energy storage system based on lithium battery and supercapacitors. In Proceedings of the 2013 Eighth International Conference and Exhibition on Ecological Vehicles and Renewable Energies (EVER 2013), Monte Carlo, Monaco, 27–30 March 2013; pp. 1–9.

16. Stadtwerke Muenster. Timetable of Muenster and the Region. Available online: https://www.stadtwerke-muenster.de/privatkunden/busverkehr/fahrplaninfos/ fahrplaene-netzplaene/fahrplantabellen.html (accessed on 10 February 2014).

17. Goldman, J. The proterra fast charge battery bus and charging station. In Proceedings of the International Workshop on Environment and Energy, San Diego, CA, USA, 2–4 November 2010.

18. Koebel, C. Update primove technology and projects. In Proceedings of the 5th VDV Conference Electric Buses—Market of the Future, Berlin, Germany, 17–18 February 2014.

19. Wechlin, M. Charging electric buses quickly and efficiently. Bus stops fitted with modular components make "Charge & Go" simple to implement. *Wireless Charging of Electric Vehicles*, 13 May 2013.

20. Prenaj, B. TOSA flash electric bus system: Experiences and perspectives after nine months of operation. In Proceedings of the 5th VDV Conference Electric Buses—Market of the Future, Berlin, Germany, 17–18 February 2014.

21. Cruz-Zambrano, M.; Corchero, C.; Igualada-Gonzalez, L.; Bernardo, V. Optimal location of fast charging stations in Barcelona: A flow-capturing approach. In Proceedings of the 10th International Conference on the European Energy Market (EEM 2013), Stockholm, Sweden, 27–31 May 2013; pp. 1–6.

22. Dong, J.; Liu, C.; Lin, Z. Charging infrastructure planning for promoting battery electric vehicles: An activity-based approach using multiday travel data. *Transport. Res. Part C Emerg. Technol.* **2014**, *38*, 44–55.

23. Wang, Y.-W. An optimal location choice model for recreation-oriented scooter recharge stations. *Transport. Res. Part D Transp. Environ.* **2007**, *12*, 231–237.

24. Wiesinger, P. Battery buses for Vienna. In Proceedings of the 4th VDV Conference Electric Buses—Market of the Future, Berlin, Germany, 18–19 February 2013.

25. Lajunen, A. Energy consumption and cost-benefit analysis of hybrid and electric city buses. *Transport. Res. Part C Emerg. Technol.* **2014**, *38*, 1–15.

26. López-López, Á.J.; Pecharromán, R.R.; Fernández-Cardador, A.; Cucala, A.P. Assessment of energy-saving techniques in direct-current-electrified mass transit systems. *Transport. Res. Part C: Emerg. Technol.* **2014**, *38*, 85–100.

27. Basile, R. Challenges for HVAC solutions in the electric bus—Status Quo. In Proceedings of the 4th VDV Conference Electric Buses—Market of the Future, Berlin, Germany, 18–19 February 2013.

28. Cho, C.-W.; Lee, H.-S.; Won, J.-P.; Lee, M.-Y. Measurement and evaluation of heating performance of heat pump systems using wasted heat from electric devices for an electric bus. *Energies* **2012**, *5*, 658–669.

29. Schmidt, M.; Classen, A. *Life-Cycle-Cost-optimierte Klimatisierung von Linienbussen. Teilklimatisierung Fahrgastraum - Vollklimatisierung Fahrerarbeitsplatz*; VDV Schrift 236-1; Verband Dt. Verkehrsunternehmen (VDV): Cologne, Germany, 2009.

30. Kaup, D.; Bouwman, R.; Schaedlich, G.; Sauer, D.U.; Lohner, A. H2 Bus NRW The Hybrid Electric Fuel-Cell Bus. In Proceedings of the 18th World Hydrogen Energy Conference - WHEC 2010, 16–21 May 2010.

31. König, D.; Oellers, F.-W.; Roes, J.; Schulte, R.; Wittenberg, H.-U. 2. Nahverkehrsplan Stadt Münster. Available online: http://www.muenster.de/stadt/stadtplanung/pdf/Nahverkehrsplan-Schlussfassung.pdf (accessed on 10 September 2014).

32. Schmalstieg, J.; Kabitz, S.; Ecker, M.; Sauer, D.U. From accelerated aging tests to a lifetime prediction model: Analyzing lithium-ion batteries. In Proceedings of the World Electric Vehicle Symposium and Exhibition (EVS27), Barcelona, Spain, 17–20 November 2013; pp. 1–12.

33. Zhu, C.; Lu, R.; Tian, L.; Wang, Q. The development of an electric bus with super-capacitors as unique energy storage. In Proceedings of the IEEE Vehicle Power and Propulsion Conference, Windsor, UK, 6–8 September 2006; pp. 1–5.

34. Vetter, J.; Novák, P.; Wagner, M.R.; Veit, C.; Möller, K.-C.; Besenhard, J.O.; Winter, M.; Wohlfahrt-Mehrens, M.; Vogler, C.; Hammouche, A. Ageing mechanisms in lithium-ion batteries. *J. Power Sources* **2005**, *147*, 269–281.

35. Rogge, M.; Rohlfs, W.; Sauer, D.U. Fast charging of battery buses and its impact on the aging of various lithium-ion cell types. In Proceedings of the 10th Symposium of Hybrid and Electric Vehicles, Braunschweig, Germany, 13–14 February 2013.

36. Ecker, M.; Nieto, N.; Käbitz, S.; Schmalstieg, J.; Blanke, H.; Warnecke, A.; Sauer, D.U. Calendar and cycle life study of Li(NiMnCo)O2-based 18650 lithium-ion batteries. *J. Power Sources* **2014**, *248*, 839–851.

37. MAN Truck & Bus AG. Lions's City datasheet. Available online: http://www.bus.man.eu/man/media/en/content_medien/doc/business_website_bus_master_1/Lions_City.pdf (accessed on 28 November 2014).

38. Mercedes-Benz. Citaro G technical information. Available online: http://www.mercedes-benz.de/content/media_library/hq/hq_mpc_reference_site/bus_ng/services_accessories/brochures/2013/citaro_g_tech/2081_0114Tech_Info_Citaro_G_EuroVI_EN_pdf.object-Single-MEDIA.tmp/2081_0114Tech_Info_Citaro_G_EuroVI_EN.pdf (accessed on 28 November 2014).

39. Akasol GmbH. Akasystem 45m Datasheet. Available online: http://www.akasol.com/fileadmin/Kundendaten/pdf/datenblatt/update_AKASYSTEM_09_2013_E/AKASOL_Datenblatt_AKASYSTEM_45M_E_09_2013.pdf (accessed on 27 November 2014).

40. Riegel, B. High-Voltage-Battery-Concepts for public transport application. In Proceedings of the 5th VDV Conference Electric Buses—Market of the Future, Berlin, Germany, 17–18 February 2014.

41. Schoewel, F. The High-Voltage Batteries of the BMW i3 and BMW i8. In Proceedings of the Advanced Automotive Battery Conference (AABC), Atlanta, GA, USA, 4–5 February 2014.

42. Jiang, G.; Zeng, X.; Xiao, S.; Chen, M.; Pu, S.; Zhou, N. Research on control strategy of energy storage buffer system for electric vehicle smart charging station. In Proceedings of the China International Conference on Electricity Distribution (CICED), Shanghai, China, 10–14 September 2012; pp. 1–5.

43. Mahmoodi, M.; McDonough, M.; Shamsi, P.; Fahimi, B. Peak shaving and minimum cost operation of an electric vehicle charging station based on Multi-port Power Electronic Interface. In Proceedings of the IEEE Transportation Electrification Conference and Expo (ITEC), Dearborn, MI, USA, 18–20 June 2012; pp. 1–5.

44. Pacher, A.; Stifter, M.; Übermasser, S.; Bletterie, B. Power supply analysis of a parking garage with a controlled charging of electric vehicles. In Proceedings of the 13th Symposium Energy Innovation, Graz, Austria, 12–14 February 2014.

Protection Principle for a DC Distribution System with a Resistive Superconductive Fault Current Limiter

Shimin Xue, Feng Gao, Wenpeng Sun and Botong Li

Abstract: A DC distribution system, which is suitable for access to distributed power generation and DC loads, is one of the development directions in power systems. Furthermore, it could greatly improve the energy efficiency and reduce the loss of power transportation. The huge short circuit current is always a great threat to the safety of the components, especially the capacitors and diodes. A resistive superconductive fault current limiter (SFCL), which could respond quickly once a fault happens and limit the fault current to a relatively low level, becomes a good solution to this problem. In this paper, the operational principle of the resistive SFCL is introduced first, and then, the DC short-circuit fault characteristic of the DC distribution system with the SFCL is analyzed and the effectiveness of the SFCL verified. In order to realize the selectivity of the protection in the DC distribution system with SFCL, a new transient current protection principle based on I_p (the peak value of the current) and t_p (the transient time that the current takes to reach its peak value) is proposed. Finally, a model of a 10-kV DC distribution system with an SFCL is established and simulated in PSCAD/METDC. Simulation results have demonstrated the validity of the analysis and protection principle.

Reprinted from *Energies*. Cite as: Xue, S.; Gao, F.; Sun, W.; Li, B. Protection Principle for a DC Distribution System with a Resistive Superconductive Fault Current Limiter. *Energies* **2015**, *8*, 4839–4852.

1. Introduction

Nowadays, more and more electric appliances, such as electric vehicles, LED lamps, mobile phones and computers, are becoming DC consumers with the development of power electronics technology. Scholars are now putting forward the DC distribution system based on the widely studied and applied distributed generation, which supplies DC power. Compared with the AC distribution system, a DC distribution system based on a voltage source converter (VSC) presents much more advantages, such as better power quality, larger power transportation capacity, higher reliability, being more economical, having lower energy waste, and so on [1–5]. Therefore, the research on DC the distribution system has drawn more and more attention.

However, in the DC distribution system based on a VSC, the Insulated Gate Bipolar Transistor (IGBTs) will be blocked from self-protection during the DC short-circuit fault. In this case, the VSC turns into an uncontrolled rectifier, as the freewheeling diodes will feed the fault [6]. In the meantime, the DC-link capacitance discharges, and the DC current rises rapidly to a relatively large value, which may be dozens of times the normal one and poses a considerable threat with respect to the safety issue. As a result, a current-limiting device is necessary, and the resistive superconductive fault current limiter (SFCL) is a good choice because of its fast response and low power loss characteristics [7].

Currently, SFCLs are mainly applied to the AC system. For example, a 220-kV saturated iron-core superconductive fault current limiter (SISFCL) has been installed in a high voltage transmission system in Tianjin, China. There are few studies about how resistive SFCL is to be used in the DC system, and some have evaluated its performance based on the effect of limiting the voltage and current only [8,9], while some studied the suitable location for the SFCL in the DC system considering the influence of the current only [7,9,10]. However, the influence of the SFCL on the fault characteristic and relay protection of the DC distribution system has had no research results till now. At present, most of the research about relay protection of the DC distribution system does not take the current limiters into consideration [11–18]. These protection schemes all apply a great short-circuit current and a low voltage. The impact of the SFCL on the protection of the DC distribution system needs to be further investigated, for introducing the SFCL will bring about problems with respect to protection.

In this paper, the fault characteristic of the DC distribution system with the SFCL and the influence of the SFCL on protection of the DC distribution system are studied. Then, a new transient current protection principle based on I_p (the peak value of the current) and t_p (the transient time that the current takes to reach its peak value) is introduced and investigated in detail. Finally, many fault simulations are completed to verify the accuracy of the protection principle.

2. Resistive Superconducting Fault Current Limiter

A resistive SFCL is composed of a superconducting cable and a shunt resistance. The equivalent model of the SFCL is a variable resistance R_{SFCL} and a shunt resistance in parallel, which is shown in Figure 1. The shunt resistance is necessary for reducing the overvoltage. Then, the resistive SFCL uses the transformation between the superconducting state and the normal resistive state to limit the increase of the current. When a fault happens, the resistive SFCL responds rapidly, so that the fault current is limited [8].

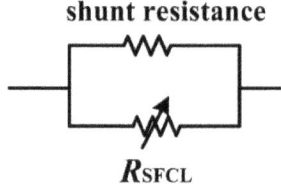

shunt resistance

R_{SFCL}

Figure 1. Structure of the resistive superconductive fault current limiter (SFCL).

In this paper, we choose Bi2212 as the material of the SFCL. The transformation of the resistive SFCL can be divided into three states as follows [8,19,20]:

(1) Superconducting state:

In this state, the current density going through the SFCL is below the critical value, and the resistance of the SFCL is very small. The electric field is:

$$E\,(J,T) = E_c \left(\frac{J}{J_c\,(T)}\right)^{\alpha} \tag{1}$$

where $E_c = 1$ μV/cm; $5 \leqslant \alpha \leqslant 15$; J is the current density and $J_c(T)$ is the critical current density, which is dependent on the temperature of the material. Additionally, the $J_c(T)$ is:

$$J_c\,(T) = J_c\,(77) \left(\frac{T_c - T}{T_c - T_0}\right) \tag{2}$$

where T_c is the critical temperature expressed in units of K; and T_0 is the initial temperature. Here, $T_c = 95$ K; $T_0 = 77$ K; and $J_c(77) = 1.5 \times 10^7$ A/m^2.

(2) Flux flow state:

When the current density exceeds the critical value, the SFCL enters this state, and the electric field starts increasing. Then, the resistance of the SFCL begins increasing, and as a result, the current starts to be limited and the temperature of the material begins to increase. Additionally, the increase of temperature makes the $J_c(T)$ decrease, so the electric field will increase continuously. In this state, the electric field is:

$$E\,(J,T) = E_0 \left(\frac{E_c}{E_0}\right)^{\frac{\beta}{\alpha}} \frac{J_c(77)}{J_c\,(T)} \left(\frac{J}{J_c\,(77)}\right)^{\beta} \tag{3}$$

where $E_0 = 0.1$ V/m is the electric field during the transition from the superconducting state to the flux flow state. $2 \leqslant \beta \leqslant 4$.

(3) Normal resistive state:

After the temperature gets higher than the critical temperature, the SFCL enters the normal resistive state. In this state, the resistance and the electric field are mainly dependent on the current density and the temperature. Additionally, the electric field is:

$$E(J, T) = \rho(T_c) J \frac{T}{T_c} \tag{4}$$

where $\rho(T_c)$ is the normal conducting resistivity, and here, $\rho(T_c) = 1 \times 10^{-6}$ $\Omega \cdot$m.

Overall, the resistance of the material is:

$$R_{sc} = \frac{E}{i_{sc}} l_{sc} \tag{5}$$

where the i_{sc} is the current going through the superconducting wire and l_{sc} is the length of the wire.

Furthermore, in all three states, the heat diffusion equation is:

$$\rho_v l_{sc} S C \frac{dT}{dt} = i_{sc}^2 R_{sc} - \frac{T - T_0}{\theta_{sc}} \tag{6}$$

where $\theta_{sc} = 1/[\kappa(l_{sc} \cdot 2\pi r + 2S)]$; $\kappa = 1.5 \times 10^3$ W/(K$\cdot$m^2); r is the radius of the superconducting wire and $C = 1.58T$ (in units of J/(kg$\cdot$K)). Here, the length of the superconducting wire is $l_{sc} = 200$ m, and the volumetric density is $\rho_v = 6$ g/cm^3. When the critical current is chosen, the cross-sectional area can be calculated by $S = I_c/J_c$.

The resistive SFCL model is established in PSCAD/EMTDC. In order to limit the surge current discharged by the capacitor, the resistance of the SFCL needs to become large enough quickly before the current gets to the peak value. In this paper, the length of the superconducting wire is assumed to be 200 m. In the DC distribution system, when a fault happens at 0.5 s, the variation of the resistance of the SFCL with respect to time is as shown in Figure 2.

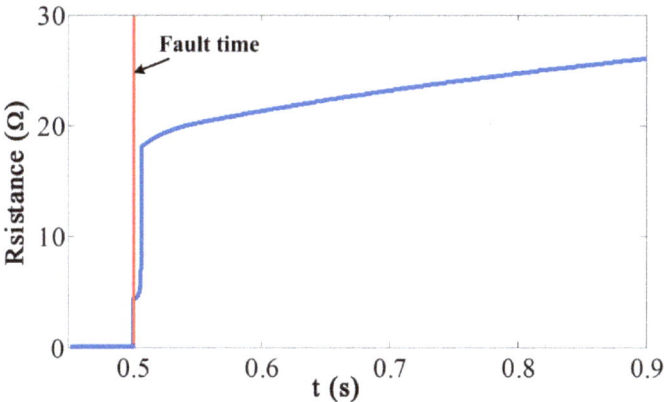

Figure 2. Variation of the resistance of the SFCL.

3. Analysis of Fault Characteristic in the DC Distribution System with the SFCL

When a DC short-circuit fault happens on the DC side of VSC, the IGBTs are blocked from self-protection. All of the diodes are blocked because of the reverse voltage. As a result, the AC system is separated from the DC side, and the AC current drops to zero [6]. If a SFCL is not used, the fault characteristic can be divided into four stages [16], which are shown in Figure 3. In the first stage, the capacitor discharges rapidly, and the DC current gets to a large value. This is a great threat to the safety of the capacitors. Additionally, in the third stage, the AC system is equivalent to a three-phase short circuit. These are unfavorable conditions for the system, and the large AC current is harmful for the diodes. Whether the third stage appears is decided by the damping property of the circuit. However, the installation of a resistive SFCL can limit the discharge of the capacitor and make the third stage disappear. Therefore, the resistive SFCL can limit the fault current on both the AC and DC sides, and the system will be safer. Then, the breaking capacity of the DC circuit breakers could be reduced. The fault characteristic under a particular situation of the simulation system is shown in Figures 4 and 5.

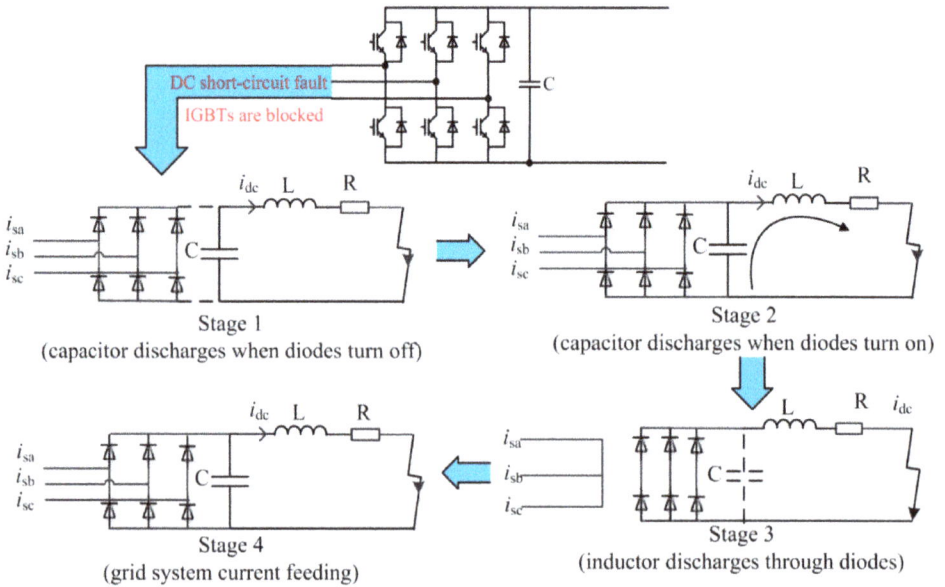

Figure 3. The four stages for the rectifier during a DC short-circuit fault without the SFCL.

As shown in Figures 4 and 5, the peak current is reduced from 14 down to 2.20 kA after the SFCL is installed, about an 84.3% decrease. The peak current is limited because of the fast operation of the SFCL. Furthermore, due to the SFCL, the time that the current takes from the beginning of the fault to the peak becomes shorter. In the steady state of the fault, the current is limited to a very low level with the increasing resistance of the SFCL. At this time, the current is close to 0.3 kA, even smaller than the normal value.

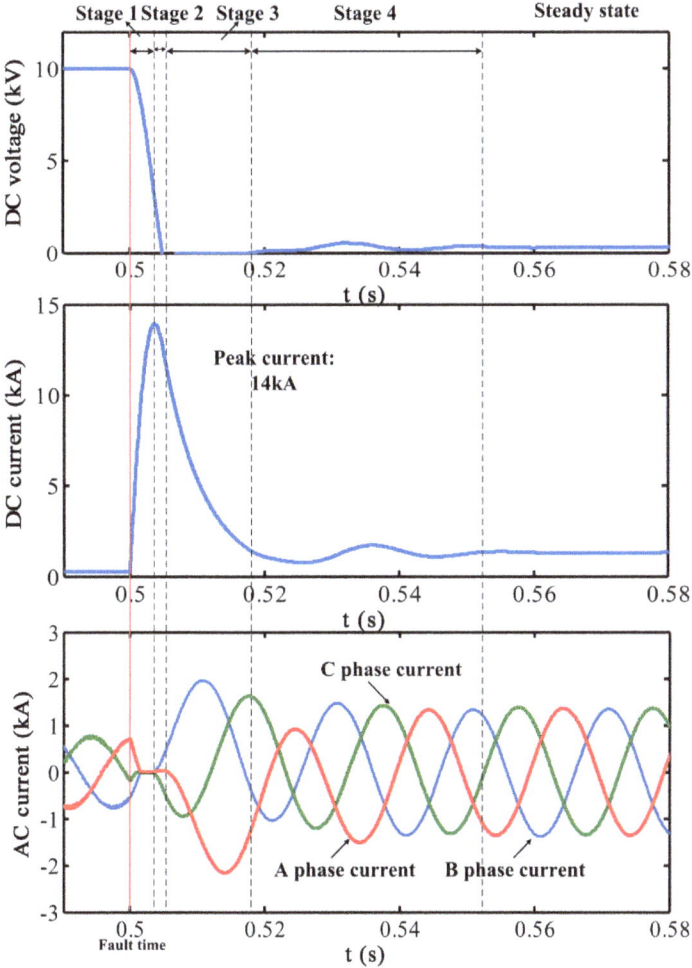

Figure 4. The fault characteristic of the voltage source converter (VSC) without the SFCL during a DC short-circuit fault.

In addition, the decrease of the DC voltage is also limited by the SFCL. The rate of descent is reduced greatly. Additionally, the steady value of the voltage rises from about 1 up to 4.4 kV, about a 77% increase. Furthermore, once the SFCL is installed, the voltage will remain at a relatively high value, and the third stage disappears. Therefore, the resistive SFCL overcomes the equivalent three-phase short

259

circuit on the AC side. For different distances of faults, the voltage of the steady state changes little.

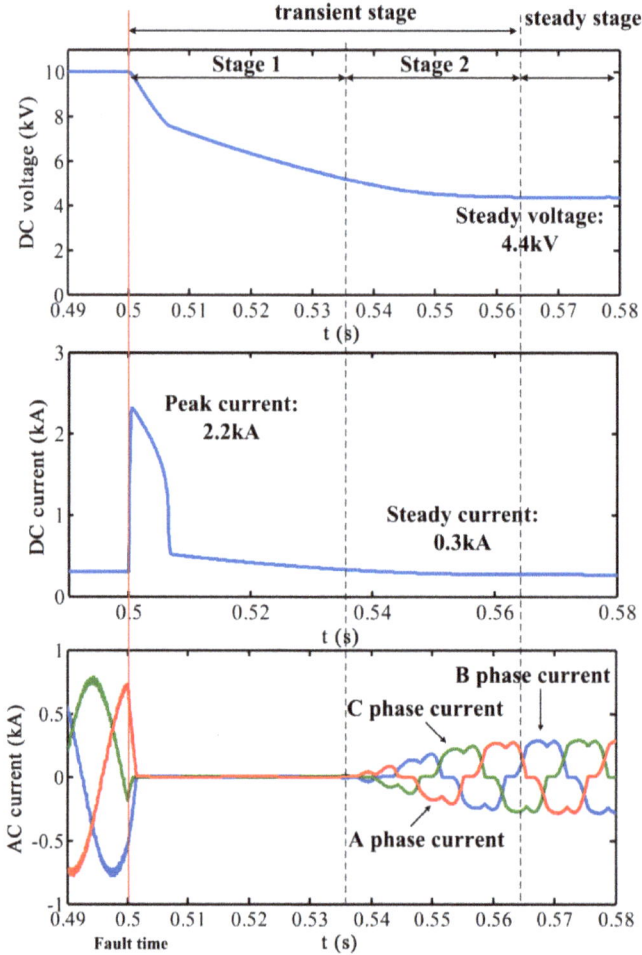

Figure 5. The fault characteristic of the VSC with the SFCL during a DC short-circuit fault.

The analysis of the fault characteristic is carried on in detail as follows.

3.1. The Transient State of the DC Short-Circuit Fault with the Resistive SFCL

Stage 1: In the initial period of the fault, when the resistive SFCL is included and the IGBTs are blocked, the capacity, the cable and the SFCL compose the discharge circuit on the DC side. During this period, the voltage begins to decrease, and the

DC current will rise to the maximum, then decrease. The equivalent circuit is shown in Figure 6.

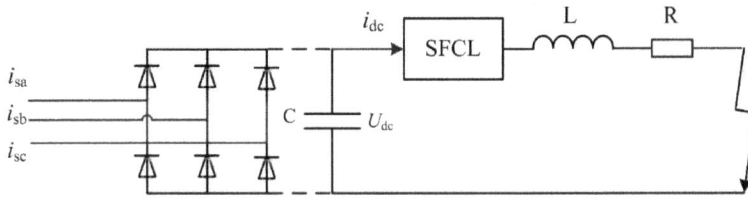

Figure 6. The equivalent circuit of the VSC in the initial phase of the fault.

According to the circuit, the voltage and current of the DC side can be calculated by:

$$\begin{cases} u_{dc} = (R + R_{SFCL})\,i_{dc} + L\frac{di_{dc}}{dt} \\ i_{dc} = -C\frac{du_{dc}}{dt} \end{cases} \tag{7}$$

where i_{dc} is the current of the circuit; u_{dc} is the voltage of the capacitor; C is the capacity; R and L are the equivalent parameters of the cable from the exit of the VSC to the fault position; R_{SFCL} is the resistance of the SFCL, which changes to nonlinear. Additionally, it is easy to know that the rate of descent of the voltage and peak current reduces greatly. The fault waveforms of Stage 1 are shown in Figure 5.

Stage 2: With the decrease of the DC voltage, the diodes start to turn on after the DC voltage is less than the AC voltage. The AC current begins to rise from zero. Then, the fault reaches the steady state gradually. The equivalent circuit at this stage is the same as the circuit in the steady state, which is shown in Figure 7. However, the discharge of the capacitor plays a major role at this stage. Additionally, the fault waveform of Stage 2 is shown in Figure 5.

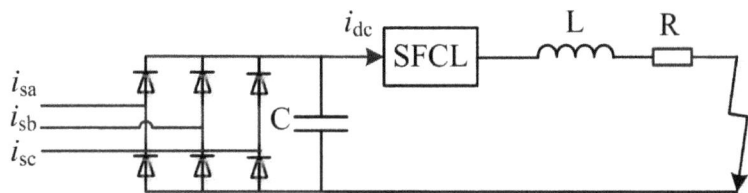

Figure 7. The uncontrolled rectifier.

3.2. The Steady State of the DC Short-Circuit Fault with the Resistive SFCL

In the steady state, the VSC changes into an uncontrolled rectifier, as shown in Figure 7. The output voltage is impacted by the flow angle of the diodes. Additionally, the voltage changes between 2.34 U and 2.45 U, where U is the RMS value of the phase voltage of the AC system.

Because the resistance of the SFCL is much greater than the impedance of the cable, the analysis can be done ignoring the inductance. Then, the flow angle (θ) of the diodes is influenced by ωRC, and the relationship is shown in Figure 8.

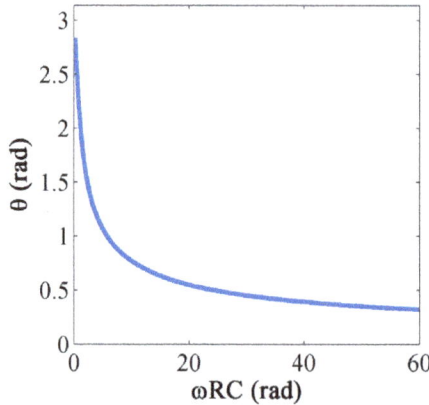

Figure 8. The flow angle (θ) *versus* ωRC.

From Figure 8, it is known that the flow angle (θ) will decrease, with the increase of ωRC. The critical condition is $\omega RC = \sqrt{3}$. When $\omega RC > \sqrt{3}$, the AC current of the uncontrolled bridge rectifier will be discontinuous, as shown in Figure 5. With the large increase of the resistance in the fault circuit, the current in the steady state may be close to or less than the value in the normal condition. Additionally, in the steady state of the fault, the voltage increases greatly compared with the condition that the SFCL is not installed. On the other hand, the current in the steady state of the fault cannot be used for protection; otherwise, the protection will be incorrectly tripped.

4. Protection Principle and Verification

4.1. Transient Current Protection and Its Coordination

Overcurrent protection using the peak current is widely used in the DC distribution system. As we know, before the integration of the SFCL, the overcurrent protection using the peak current could coordinate very well [16]. It can satisfy the requests for protection. However, after the integration of the SFCL, the difference among the peak currents of different fault positions becomes little. In this situation, if we still apply the overcurrent protection, the reliability coefficient will be too small to satisfy the request for sensitivity and reliability. Therefore, the overcurrent protection needs to be improved.

In this paper, it is assumed that the fault resistance of the DC short-circuit fault between the cables is very small. According to the analysis above, we can get the relationship between the fault position and peak current, as shown in Figure 9. If the SFCL is included, the peak current has an approximately linear relationship with the position of the fault. However, the peak current changes a little with respect to the increasing of the distance of the fault. However, the time that the current takes to reach the maximum varies greatly when a fault happens at different positions, as shown in Figure 10. Then, if the ratio of the peak current to the time is used for the protection, as shown in Figure 11, the sensitivity and reliability could be ensured.

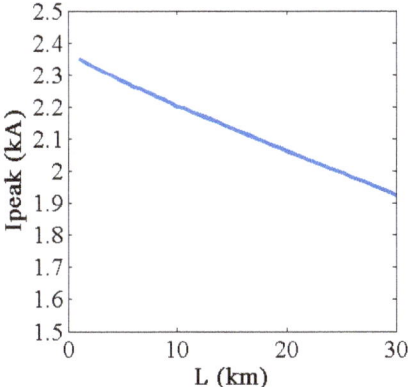

Figure 9. Peak currents of different fault positions.

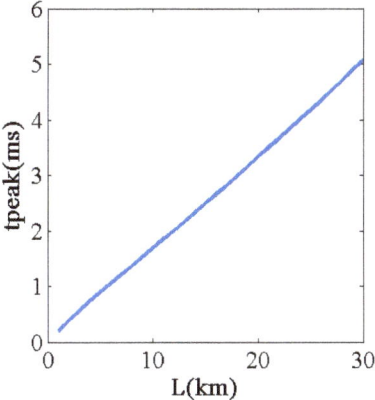

Figure 10. The time that the current takes to reach the peak *versus* the fault position.

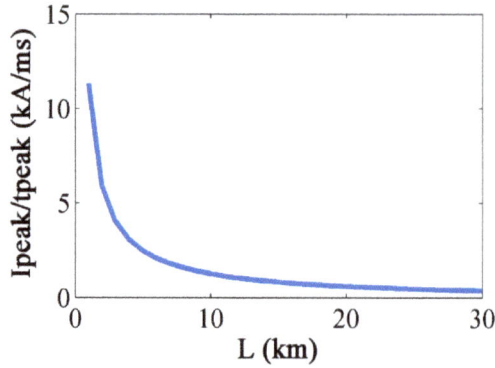

Figure 11. The ratio of the peak current to the time *versus* the fault position.

4.1.1. Transient Current Protection

Under the condition that the SFCL is installed, transient current protection is set according to the principle that the threshold is above the ratio of $I_{k\text{ peak end 1}}$ (the peak of the short circuit current at the end of the line) to $\Delta t_{k\text{ peak end 1}}$ (the time that the current takes to reach the maximum from the beginning of the fault). The operating criterion is:

$$\frac{I_{peak}}{\Delta t} > \left(\frac{I_{peak}}{\Delta t}\right)^{I}_{set\cdot 1} \tag{8}$$

The threshold is:

$$\left(\frac{I_{peak}}{\Delta t}\right)^{I}_{set\cdot 1} = K^{I}_{rel} \times \frac{I_{k\cdot peak\cdot end\cdot 1}}{\Delta t_{k\cdot peak\cdot end\cdot 1}} \tag{9}$$

where I_{peak} is the maximum of the short circuit current during the fault; Δt is the time that the current takes to reach the maximum from the start of the fault and K^{I}_{rel} is the reliability coefficient, which can be 1.2–1.3. Therefore, transient current protection can only protect about 80% or less of the line. However, the protection will operate without delay once the ratio exceeds the threshold.

4.1.2. Time-Limit under Voltage Protection Started by the Transient Current

Time-limit overcurrent protection coordinates with the downstream line by the time delay. The time delay can be 0.1 s or more in the DC system [16], considering the coordination with the downstream line and the operation time that the DC circuit breaker takes. However, when the SFCL is installed, the current has already been in the steady state at 0.1 s, and the value is close to the normal value after the fault. Therefore, traditional time-limited overcurrent protection cannot be applied directly to the DC system with the SFCL.

However, the DC voltage will decrease to a relatively low level at 0.1 s after the fault. When the resistive SFCL is installed, the voltage in the steady state changes little with the change of the fault position. Therefore, the under voltage protection alone cannot ensure the selectivity. Then, the transient current protection is combined with the under voltage protection, called the time-limit under the voltage protection started by the transient current. This kind of protection still protects the whole line and does not exceed the range of the transient current protection of the downstream line. It coordinates with the downstream line by a time delay. The starting value is the threshold of the transient current protection, and it can guarantee the selectivity of protection. After the time delay, whether the breaker operates is decided by the under voltage protection. The operating criterion is:

$$\left(\frac{I_{\text{peak}}}{\Delta t}\right) > \left(\frac{I_{\text{peak}}}{\Delta t}\right)^{\text{II}}_{\text{set·1}} \cap \left(t = t^{\text{II}}_{\text{set·1}}\right) \cap \left(U_{\text{dc}} < U^{\text{II}}_{\text{set·1}}\right) \tag{10}$$

The threshold is:

$$\left(\frac{I_{\text{peak}}}{\Delta t}\right)^{\text{II}}_{\text{set·1}} = K^{\text{II}}_{\text{rel}} \times \left(\frac{I_{\text{peak}}}{\Delta t}\right)^{\text{I}}_{\text{set·2}} = K^{\text{II}}_{\text{rel}} \times K^{\text{I}}_{\text{rel}} \times \left(\frac{I_{\text{k·peak·end·2}}}{\Delta t_{\text{k·peak·end·2}}}\right) \tag{11}$$

$$t^{\text{II}}_{\text{set·1}} = t^{\text{I}}_{2} + \Delta t \tag{12}$$

$$U^{\text{II}}_{\text{set·1}} = K^{\text{II}}_{\text{u·rel}} \times U_{N} \tag{13}$$

where $\left(I_{\text{peak}}/\Delta t\right)^{\text{II}}_{\text{set·1}}$ is the starting value of the under voltage protection; $\left(I_{\text{peak}}/\Delta t\right)^{\text{I}}_{\text{set·2}}$ is the threshold of the transient current protection of the downstream line; $K^{\text{II}}_{\text{rel}}$ is the reliability coefficient, which can be 1.1–1.2; $U^{\text{II}}_{\text{set·1}}$ is the threshold of the under voltage protection; U_{N} is the rated $K^{\text{II}}_{\text{u·rel}}$ is the reliability coefficient; $K^{\text{II}}_{\text{u·rel}} = 0.5 - 0.6$, t^{I}_{2} is the inherent opening time of the transient current protection and Δt is the time delay, always $\Delta t = 0.1$–0.3 s.

4.2. Case Study

A 10-kV DC distribution system with the resistive SFCL is simulated by PSCAD/EMTDC. The structure of the system is shown in Figure 12. It is assumed that DC short-circuit fault occurs at 0.5 s in the different positions of the DC side. The fault resistance is 0.001 Ω. Each of the lines in the system is 10 km long. The equivalent parameter of the line is $R = 0.078$ Ω/km and $L = 0.48$ mH/km. When the system operates in the normal condition, the currents that go through each line are $I_{\text{line1}} = 0.35$ kA, $I_{\text{line2}} = 0.252$ kA and $I_{\text{line3}} = 0.157$ kA.

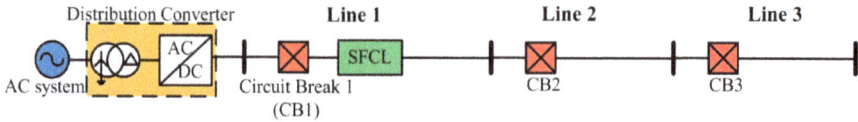

Figure 12. The structure of the DC distribution, including the SFCL.

According to the analysis above, the thresholds of protection are listed in Table 1. Because Line 3 is the last section of the transmission line, traditional overcurrent protection can be applied for Line 3. Therefore, the threshold can be several times the load current, and the instantaneous overcurrent protection protects the whole line. In this situation, the time-limited overcurrent protection is the backup protection of Line 3. Then, the principle of the protection mentioned above is verified in this system, and the result is listed in Table 2.

Table 1. The thresholds of protection.

Line		Parameters	Thresholds
Line1 (CB1)	I	transient current threshold	1.424 (kA/ms)
	II	transient current threshold under voltage threshold time delay threshold	0.812 (kA/ms) 5.0 (kV) 0.2 (s)
Line2 (CB2)	I	transient current threshold	0.738 (kA/ms)
	II	transient current threshold under voltage threshold time delay threshold	0.496 (kA/ms) 5.0 (kV) 0.2 (s)
Line3 (CB3)	I	instantaneous overcurrent threshold	1.5 (kA)
	II	time-limit overcurrent threshold time delay threshold	1.5 (kA) 0.2 (s)

From the result, it can be seen that the presented protection principle could clear the fault quickly and accurately. The transient current protection can protect about 80% of the length of the section. Additionally, it operates without a time delay. With the increasing of the distance of the fault, the rising of current slows down. Therefore, the operation time of the transient current protection becomes longer, but still operates very quickly. On the other hand, the time-limited under voltage protection started by the transient current can protect the whole length of the line with a time delay. It can operate at about 0.2 s or more after the fault. This kind of protection may operate slowly, but it can guarantee the selectivity between the upstream line and the downstream line.

Table 2. The action situation of each CB.

Line	Fault Location (Percentage of the Line)	The Action of Breakers (Time for Receiving Trip Instruction) (s)		
		CB1	CB2	CB3
Line1	30%	0.5010	#	#
	80%	0.5017	#	#
	100%	0.7021	#	#
Line2	30%	#	0.5027	#
	80%	#	0.5034	#
	100%	#	0.7038	#
Line3	30%	#	#	0.5026
	80%	#	#	0.5033
	100%	#	#	0.5039

Note: # is defined as that the circuit breaker (CB) of the Line will not operate.

Therefore, for the DC distribution system, including the SFCL, this principle of protection can operate within a short time for the feeder faults, and the coordination of the upstream and downstream protective relays is ensured.

5. Conclusions

A resistive SFCL has good performance for current limiting in the DC distribution system. It is suitable for a system with a high fault current and a high current rising speed. However, the integration of the SFCL has a serious influence on the coordination of the protection. In this paper, the influence of the SFCL on the protection and the fault characteristic is analyzed in detail. Then, the transient current protection is presented, and the time-limited under voltage protection started by the transient current is proposed. Finally, the principle of the protection is verified in PSCAD/EMTDC. The result demonstrates that this principle of protection can ensure the coordination of the upstream and downstream protective relays. Nevertheless, a DC short-circuit fault with fault resistance is not considered in this paper, which would be the main work in the future.

Acknowledgments: This work was supported in part by the National Natural Science Foundation of China under Grant 51207103, by the Tianjin Municipal Natural Science Foundation under Grant 14JCYBJC21000, by the Innovation Foundation of Tianjin University under Grant 2014XRG-0118, by the Specialized Research Fund for the Doctoral Program of Higher Education for New Teacher under Grant 20120032120085 and by National High Technology Research and Development (863 program) under Grant 2015AA050102.

Author Contributions: Shimin Xue researched the protection principle and drafted the article. Feng Gao provided the results of the simulation. Valuable comments on the first draft were received from Feng Gao, Wenpeng Sun and Botong Li. All four were involved in revising the paper.

Conflicts of Interest: The authors declare no conflict of interests.

References

1.	Starke, M.R.; Tolbert, L.M.; Ozpineci, B. AC *vs.* DC distribution: A loss comparison. In Proceedings of the IEEE/PES TDCE, Chicago, IL, USA, 21–24 April 2008; pp. 225–231.

2.	Kakigano, H.; Nomura, M.; Ise, T. Loss evaluation of DC distribution for residential houses compared with AC system. In Proceedings of the IPEC, Sapporo, Japan, 21–24 June 2010; pp. 480–486.

3.	Mahmoodi, M.; Gharehpetian, G.B.; Abedi, M. A suitable control strategy for source converters and a novel load-generation voltage control scheme for DC voltage determination in DC distribution systems. In Proceedings of the IPEC, Putra Jaya, Malaysia, 28–29 November 2006; pp. 363–367.

4.	Starke, M.R.; Li, F.; Tolbert, L.M.; Ozpineci, B. AC *vs.* DC distribution: maximum transfer capability. In Proceedings of the IEEE PES, Pittsburgh, PA, USA, 20–24 July 2008; pp. 922–927.

5.	Sannino, A.; Postiglione, G.; Bollen, M.H.J. Feasibility of a DC network for commercial facilities. *IEEE Trans. Ind. Appl.* **2003**, *39*, 1499–1507.

6.	Yang, J.; Fletcher, J.E.; O' Reilly, J. Short-circuit and ground fault analyses and location in VSC-based DC network cables. *IEEE Trans. Ind. Electron.* **2012**, *59*, 3827–3837.

7.	Khan, U.A.; Hwang, J.-S.; Seong, J.-K.; Lee, B.-W. Application and positioning analysis of a resistive type Superconducting Fault Current Limiter in AC and DC microgrids using Simulink and SimPowerSystem. In Proceedings of the ICEPE-ST, Xi'an, China, 23–27 October 2011; pp. 348–351.

8.	Manohar, P.; Ahmed, W. Superconducting Fault Current Limiter to mitigate the effect of DC Line fault in VSC-HVDC system. In Proceedings of the EPSCICON, Thrissur, Kerala, India, 3–6 January 2012; pp. 1–6.

9.	Lee, J.-G.; Khan, U.-A.; Hwang, J.-S.; Seong, J.-K.; Shin, W.-J.; Park, B.-B.; Lee, B.-W. Assessment on the influence of resistive superconducting fault current limiter in VSC-HVDC system. *Phys. C.* **2014**, *504*, 163–166.

10.	Hwang, J.-S.; Khan, U.A.; Shin, W.-J. Validity analysis on the positioning of superconducting fault current limiter in neighboring AC and DC microgrid. *IEEE Trans. Appl. Supercond.* **2013**, *23*.

11.	Ciezki, J.G.; Ashton, R.W. Selection and stability issues associated with a navy shipboard DC zonal electric distribution system. *IEEE Trans. Power Deliver.* **2000**, *25*, 665–669.

12.	Tang, L. Control and Protection of Multi-Terminal DC Transmission Systems Based on Voltage-Source Converters. Ph.D. Thesis, McGill University, Montreal, QC, Canada, 2003.

13.	Mesute, B.; Nikhil, R.M. Overcurrent protection on voltage-source-converter based multi-terminal DC distribution systems. *IEEE Trans. Power Deliver.* **2007**, *22*, 406–411.

14.	Tang, L.; Ooi, B.-T. Locating and isolating DC faults in multi-terminal DC systems. *IEEE Trans. Power Deliver.* **2007**, *22*, 1877–1884.

15.	Salonen, P.; Nuutinen, P.; Peltoniemi, P. LVDC distribution system protection: Solutions, implementation and measurements. In Proceedings of the EPE, Barcelona, Spain, 8–10 September 2009; pp. 1–10.

16. Xue, S.M.; Chen, C.C.; Jin, Y.; Li, Y.L.; Li, B.T.; Wang, Y. Protection for DC distribution system with distributed generator. *J. Appl. Math.* **2014**, *2014*.

17. Duan, P.; Xie, K.-G.; Zhang, L.; Rong, X. Open-switch fault diagnosis and system reconfiguration of doubly fed wind power converter used in a microgrid. *IEEE Trans. Power Electron.* **2011**, *26*, 816–821.

18. Choi, U.-M.; Jeong, H.-G.; Lee, K.-B.; Blaabjerg, F. Method for detecting an open-switch fault in a grid-connected NPC inverter system. *IEEE Trans. Power Electron.* **2012**, *27*, 2726–2739.

19. De Sousa, W.T.B.; Polasek, A.; Silva, F.A.; Dias, R.; Rio de Janeiro, B. Simulation and tests of MCP-BSCCO-2212 superconducting fault current limiters. *IEEE Trans. Appl. Supercond.* **2012**, *22*.

20. Langston, J.; Steurer, M.; Woodruff, S.; Baldwin, T.; Tang, J. A generic real time computer simulation model for superconducting fault current limiters and its applications in system protection studies. *IEEE Trans. Appl. Supercond.* **2005**, *15*, 2090–2093.

Metamodel for Efficient Estimation of Capacity-Fade Uncertainty in Li-Ion Batteries for Electric Vehicles

Jaewook Lee, Woosuk Sung and Joo-Ho Choi

Abstract: This paper presents an efficient method for estimating capacity-fade uncertainty in lithium-ion batteries (LIBs) in order to integrate them into the battery-management system (BMS) of electric vehicles, which requires simple and inexpensive computation for successful application. The study uses the pseudo-two-dimensional (P2D) electrochemical model, which simulates the battery state by solving a system of coupled nonlinear partial differential equations (PDEs). The model parameters that are responsible for electrode degradation are identified and estimated, based on battery data obtained from the charge cycles. The Bayesian approach, with parameters estimated by probability distributions, is employed to account for uncertainties arising in the model and battery data. The Markov Chain Monte Carlo (MCMC) technique is used to draw samples from the distributions. The complex computations that solve a PDE system for each sample are avoided by employing a polynomial-based metamodel. As a result, the computational cost is reduced from 5.5 h to a few seconds, enabling the integration of the method into the vehicle BMS. Using this approach, the conservative bound of capacity fade can be determined for the vehicle in service, which represents the safety margin reflecting the uncertainty.

Reprinted from *Energies*. Cite as: Lee, J.; Sung, W.; Choi, J.-H. Metamodel for Efficient Estimation of Capacity-Fade Uncertainty in Li-Ion Batteries for Electric Vehicles. *Energies* **2015**, *8*, 5538–5554.

1. Introduction

Electric vehicles (EVs) or hybrid electric vehicles (HEVs), which use lithium-ion batteries (LIBs) as their main energy source, are being widely adopted as a transportation innovation. During use, however, the batteries degrade, losing some of their capacity as they undergo charge and discharge cycles and eventually they will suddenly stop functioning. Therefore, top concerns for EVs are limited the battery life and potential battery failure on the road while in use, as observed in a study by the US-based Consumer Electronics Association (CEA) [1]. In order to prevent on-road failure and ensure safe and reliable operation, a battery management system (BMS) must provide functions to monitor the batteries' state of health (SOH) and predict the remaining life, thermal management, safety protection, charge control, cell balancing, and so on. While BMS technology in small-scale portable electronics such as cellular phones is relatively mature, it is not yet fully developed for EVs or HEVs due to the fact that the power and number of cells needed are hundreds of times greater due to the critical need to monitor the batteries' SOH [2].

The SOH represents the real-time physical condition of the battery and is usually defined by capacity fade, which typically occurs as the battery ages due to electrode degradation. The most influential factors for this degradation are temperature, charge/discharge rate, and depth of discharge. Considerable effort has been directed at developing a method for estimating the SOH [3–6]. The conventional approach has been to employ empirical models that make use of an equivalent circuit model (ECM) to mimic the battery dynamics. Electrochemical impedance spectroscopy (EIS) or direct current internal resistance (DCIR) tests have been widely used as a non-invasive method to estimate the changes in the internal parameters, which are the capacitance and resistance of the equivalent circuit. The estimated parameters are then correlated with the actual capacity and used as the indicator of capacity fade. The EIS measurement, however, is costly and not available aboard a vehicle. Moreover, the ECM does not provide insight into the physical and chemical phenomena driving the voltage dynamics of the cell, unless great care is taken to associate the parameters with specific electrochemical processes.

A more advanced SOH estimation approach that has recently gained attention is the use of a physics-based electrochemical model that solves coupled nonlinear partial differential equations (PDEs) in spatiotemporal coordinates that are related to the conservation of mass and charge in the solid and liquid phases [7–9]. Because the parameters used in this model have a physical interpretation, they are directly correlated with battery aging. The parameters can be estimated using battery-state data provided by the BMS, *i.e.*, the current, voltage, and temperature during the charge/discharge process. This approach has two advantages: (1) it does not require any extra means or interruption of the BMS, which enables online applications, and more importantly, (2) it provides a more accurate assessment of capacity fade than the ECM in view of diverse loading conditions from slow to rapid charging. Once capacity fade has been detected, it can be used to determine which cell, if any, of the battery pack needs to be replaced.

In parameter estimation, actual online measurements suffer from various uncertainties associated with the inaccuracy of battery-state data, inherent material variances, and harsh operating/environmental conditions. Due to the inability to account for these uncertainties, results obtained by deterministic optimization may give questionable results with regard to SOH estimation. In order to provide more reliable management, uncertainty should be incorporated into SOH monitoring by using probabilistic methods, which estimate parameters based on real-time battery-state data. The lower bound then can be estimated for the faded capacity under a given level of confidence. There have been numerous efforts in this direction, which addresses the uncertainty issue for SOH estimation in the recent years [3,4,10–12]. In the literature, most of the works were however based on the data driven approach and/or the empirical model which does not account for the physics associated with the degradation, hence, can be less insightful than the physics-based estimation. Only a few studies have been made in physics-based SOH estimation with uncertainty. Tong *et al.* [13] carried out a Markov Chain Monte Carlo (MCMC) simulation by generating samples that satisfied the probability distributions and then running a simulation for each sample to capture the probabilistic nature of parameter uncertainties. However, the parameter estimation was not conditional on battery-state data. In Ramadesigan *et al.* [14], the effective parameters and their uncertainties were estimated in the form of samples based on battery data, using the Bayesian approach and a mathematical reformulation of a porous electrode model. In their study, however, the computation cost to simulate the model with such large samples is not clearly stated; it is likely very high, which may limit its applicability to the on-board vehicle BMS.

Therefore, the purpose of this paper is to propose an efficient method to estimate battery capacity fade, including uncertainty factors. This method could then be easily integrated into the vehicle BMS. The pseudo-two dimensional (P2D) electrochemical model is employed, which simulates the battery state under the profile of the input current by solving a system of coupled nonlinear partial differential equations (PDEs). The reliability of this electrochemical model has been validated by experiments in which the simulated and measured voltage curves were compared under various charge and discharge conditions [15]. Rather than using data at discharge cycles, which usually undergo arbitrary conditions, this study uses the data at charge cycles, which tend to occur under a constant C-rate. By comparing the simulated and experimental voltage curves, the model parameters representing degradation are estimated by way of large samples to reflect their probabilistic nature.

In order to incorporate the uncertainties in the capacity fade estimation, probabilistic approach is needed, from which the confidence bounds can be determined. In that case, the results are given by the probability distributions instead of a deterministic value. In the practical implementation, a large number of samples (at least 5000) are necessary to represent the distributions, which means that the P2D model should be solved that number of times. Even though a single computation to solve a P2D model only takes a few seconds, it can take several hours to implement the whole number of P2D solutions, which is intractable from a practical viewpoint. Thus, a polynomial-based metamodel is developed to replace P2D electrochemical model. The output variable of the metamodel is the voltage at discrete time steps, and the input variables are the physical parameters such as diffusion coefficients and reaction constants. For the additional alleviation of the computational burden, only one dominant variable is selected as the input variable of the metamodel, and the

272

operating current condition is fixed as the charging cycle with constant C-rate. Note that the computational environment of vehicle BMS is extremely limited. Once the dominant parameter is estimated using the polynomial metamodel, a conservative capacity fade boundary can be determined in the vehicle's BMS at the 95% confidence level, which represents the safety margin reflecting the uncertainty.

The outline of this paper is as follows: in Section 2, the physical parameters of the LIB electrochemical model are estimated using a Bayesian-based probabilistic approach. The estimation is performed for five transport and kinetic factors that are the principal parameters affecting capacity fade. The estimation results reveal that one parameter dominates in determining capacity fade, which means that this parameter alone can be used to monitor SOH. The metamodel for the selected parameter is constructed in Section 3. The validity of the metamodel is proven through the comparison of voltage curves, and thus the metamodel with its acceptable error can replace the original electrochemical model. Section 4 shows the parameter estimation results obtained using the generated metamodel. The estimated results are again similar to those produced by the original electrochemical model. However, with the metamodel, the computational cost is reduced from 5.5 h to only a few seconds. This huge reduction in computational cost enables us to integrate SOH monitoring into a vehicle BMS, based on physical parameter estimation. Finally, Section 5 summarizes the paper's conclusions.

2. Parameter Estimation

This section presents the parameter estimation of the LIB electrochemical model. In more detail, the physical parameters of the P2D electrochemical model are estimated using a distribution based on Bayesian inference. The estimation is performed by comparing the voltage curves: (a) measured from the experiments and (b) obtained by solving the model equation. In this section, the LIB electrochemical model [16–18] is explained briefly. Then the MCMC method for parameter estimation is introduced. Finally, the estimation results are presented and discussed. From the parameter estimation results, the critical parameter of the LIB electrochemical model is determined.

2.1. Pseudo-Two Dimensional Electrochemical Model

The P2D electrochemical models that simulate the state of the LIB are well described in [16–18]. The model aims to obtain the output voltage curve with respect to time when applying the profile of the input current by calculating state variables such as electric potential φ, Li-ion concentration c, and molar flux j of Li at the surface of the spherical particles of active material. These state variables can be calculated by solving a system of coupled nonlinear PDEs [16–18], which represent the electrochemical phenomenon occurring in the LIB. The various numerical methodologies to solve a system of equations efficiently have been reported in [15,19,20]. The equations for the electrochemical model include the physical parameters θ, which represent the geometric parameters and transport and kinetic properties. Among the parameters, those that mainly affect capacity fade are selected as targets of the estimation for SOH monitoring. In this work, five transport and kinetic parameters are chosen per the study by [15]: (1) the liquid-phase diffusivity of Li-ion D_e, the solid-phase diffusivities of Li in the (2) positive D_{sp} and (3) negative D_{sn} electrodes, and the electrochemical reaction rate constants in the (4) positive k_p and (5) negative k_n electrodes. The effective solid-phase diffusivity was believed to be able to best describe the Li transport through the porous electrode as it is a function of almost all relevant parameters such as Li's molecular diffusivity, porosity and tortuosity. Similarly, the electrochemical reaction rate constant was considered to be able to represent the charge transfer across the SEI, which is also effective as it is inclusive of the true electrochemical reaction rate constant and the surface area available for electrochemical reactions [12]. The selected five transport and kinetic parameters are estimated by using MCMC method based on the measured voltage data.

2.2. The MCMC Approach for Parameter Estimation

The approach for parameter estimation is based on Bayesian inference, which updates a hypothesis as more observational data are acquired [21,22]. Specifically, the posterior probability density function (PDF) of the model parameter $\theta = \{D_e, D_{sp}, D_{sn}, k_p, k_n\}$, which is conditional on the measured data y, i.e., $P(\theta \,|\, y)$, can be calculated by Bayes' rule:

$$P(\theta \,|\, y) \propto L(y \,|\, \theta)\, p(\theta) \tag{1}$$

where $L(y \,|\, \theta)$ is the likelihood function of the data y conditional on θ, and $p(\theta)$ is the prior distribution of θ. The rule states that the degree of likelihood based on unknown parameters is given by the updated or posterior PDF conditional on the data y, which consists of the prior likelihood $p(\theta)$ and the likelihood of the data y. In this work, the data y become the vector of measured voltage V_k at discrete time intervals t_k ($k = 1, 2, \ldots, n$). The likelihood function for kth data y_k (i.e., $L(y_k \,|\, \theta)$) can be defined based on the assumption that the error of the data against the model follows the normal distribution:

$$L(y_k \,|\, \theta, \sigma) = \frac{1}{\sigma\sqrt{2\pi}} \exp\left\{ -\frac{1}{2} \frac{(y_k - y^k(\theta))^2}{\sigma^2} \right\} \tag{2}$$

where σ is the standard deviation and $y^k(\theta)$ is the model value corresponding to observed data y_k, which, in this work, is the electrochemical model voltage at interval

k. The equation represents the PDF value of the observation y_k with the mean being $y^k(\theta)$ and the standard deviation σ based on the assumption of normal distribution. The symbol | represents that the PDF of y_k is given conditional on the parameters $y^k(\theta)$ and σ. The joint likelihood function $L(y|\theta)$ of the voltage data y becomes the multiplier of the data at whole-number intervals:

$$L(y|\,\theta,\sigma) = \prod_{k=1}^{n} L(y_k|\,\theta,\sigma) = L(y_1|\,\theta,\sigma) \times L(y_2|\,\theta,\sigma) \times ... \times L(y_n|\,\theta,\sigma) \qquad (3)$$

The prior distribution for θ, and σ in this work is set as the uniform distribution function U between the upper bound Ub and the lower bound Lb:

$$p(\theta,\sigma) = U(Lb, Ub) \qquad (4)$$

Note that the standard deviation σ is also unknown and estimated from the process. As a result, the unknown parameters consists of the five model parameters $\theta=\{D_e, D_{sp}, D_{sn}, k_p, k_n\}$ and the standard deviation ion σ is also unknown and estimated from the process.y. The final joint posterior PDF of the unknown parameters is then revised from Equation (1) to:

$$P(\theta,\sigma|y) \propto L(y|\theta,\sigma)\, p(\theta,\sigma) \qquad (5)$$

From this point on, the unknown parameters will be denoted by a single symbol θ by including σ in it. The posterior PDF $P(\theta\,|\,y)$ can be effectively evaluated using the MCMC approach, which is widely used as a sampling method in modern computational statistics [21,22]. The MCMC approach is based on the fact that the PDFs built by the Markov Chain process converge on the actual distribution as the sample size increases. Among various sampling method for the MCMC approach, the Metropolis-Hastings (M-H) algorithm is the most representative method applied in this work [23]. The flowchart and illustration of the M-H algorithm are given in Figures 1 and 2, respectively. As shown in Figure 1, the M-H algorithm finds N large samples of the model parameter θ, using a process that includes variate generation and comparison. First, we start with an arbitrary initial sample θ. Then, a new parameter sample θ^* is obtained by calculating a random variate from the parameter sample in the previous step θ^{i-1} using the weighting vector w and randomly sampled value u from the uniform distribution between 0 and 1, i.e., $U(0,1)$. Next, the posterior PDF conditional on the measured data y is calculated at the variate θ^*. Then, the ratio of the posterior PDF of the new sample θ^* and the previous sample θ^{i-1} (i.e., $P(\theta^*|y)/P(\theta^{i-1}|y)$ is compared with the randomly sampled value u from the uniform distribution between 0 and 1, i.e., $U(0,1)$. If the ratio is larger than u, the sample at the current step θ^i becomes variate θ^*. Otherwise, the sample at the current step θ^i reverts to the value for the old sample at the previous step θ^{i-1}. By repeating the process, N samples of the model parameters are calculated, and the posterior PDF $P(\theta|y)$ is determined by counting the number of samples at each interval. The conceptual illustration of the M-H algorithm is presented in Figure 2.

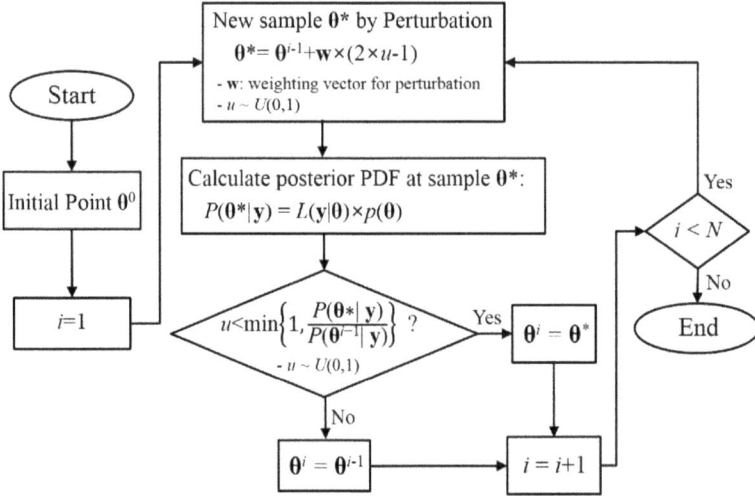

Figure 1. Flowchart of Metropolis-Hastings (M-H) algorithm for the parameter estimation using the Markov Chain Monte Carlo (MCMC) approach.

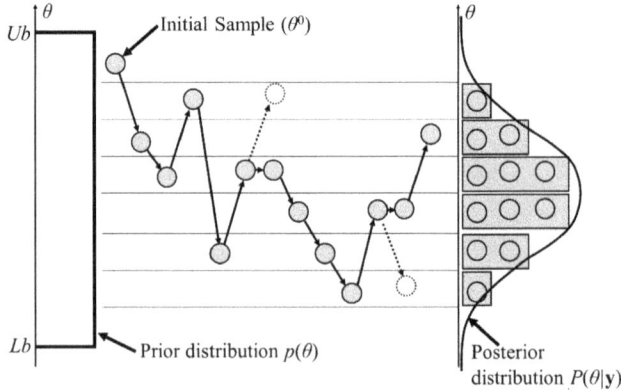

Figure 2. Illustration of Metropolis-Hastings (M-H) algorithm explained in Figure 1.

2.3. Parameter Estimation Result

The parameters for the electrochemical model are estimated using the MCMC approach with the M-H algorithm explained previously. The Li-ion cells used in this work are large-formatted with a nominal capacity of 42.5 Ah. Each cell consists of 21 positive electrodes and 22 negative electrodes; those are all two-sided. The active materials of the positive and negative electrodes are composite $LiNi_{1/3}Mn_{1/3}Co_{1/3}O_2$-$LiMn_2O_4$ (NMC-LMO) and natural carbon, respectively. The electrolyte consists of $LiPF_6$ salt in a tertiary solvent mixture of ethylene carbonate (EC), ethyl methyl carbonate (EMC) and diethyl carbonate (DEC). Each of the 21 positive electrodes is bagged by a separator and each of the 22 negative electrodes is sandwiched between the 21 positive electrode-containing separator bags. The entire assembly of positive and negative electrodes and separator is finally enclosed by a pouch. The cycling experiment is performed at 45 °C in 2C-rate condition with full charge and discharge cycles, which leads to capacity degradation. The reason for employing 2C-rate is because the rapid charge is usually made at 2C-rate. Full charging occurs until an end-of-charge voltage of 4.2 V is reached followed by constant voltage charging at 4.2 V until the current tapers down to 0.0235 C-rate. An hour later, the cycle continues with the constant current discharge at a 2C-rate down to a cut-off voltage of 2.5 V.

Note that although the discharge in this work occurs at a constant rate, the battery generally experiences arbitrary loading conditions. On the other hand, charging always takes place at a constant rate. In this sense, use of the data during charge is more desirable for estimation of parameters. The voltage data are collected every 10 s from beginning to end of the constant current charge process. Voltage profiles are taken at 200-cycle intervals between 200 and 2400 cycles. So the 12 data sets are obtained for the voltage profiles as shown in Figure 3. As the cycle progresses, the curve shifts to the left, which indicates that the time required to attain full charge is gradually reduced, resulting in capacity fade. Table 1 presents capacity fade percentage with respect to cycle number. Capacity fade occurs almost linearly as the cycling number increases.

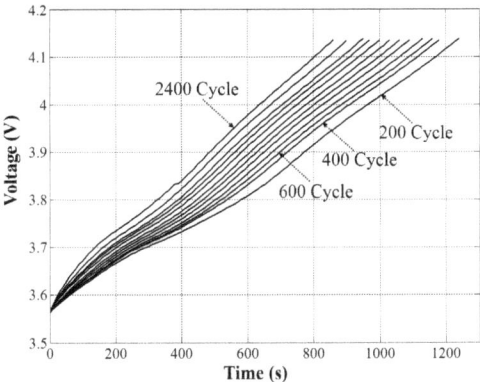

Figure 3. Voltage charge curve at various cycles in the battery experiment at 45 °C in 2C-rate condition with full charge and discharge.

Table 1. Capacity fade (%) with respect to the cycle number.

Cycle number	Fade %	Cycle number	Fade %
200	5.5	1400	19.7
400	9.0	1600	21.3
600	11.4	1800	23.3
800	13.6	2000	24.9
1000	15.8	2200	27.4
1200	17.7	2400	30.3

Based on the approach in Section 2.2, the unknown parameters at each given cycle are estimated based on the voltage data using the MCMC algorithm. The results are given in Figure 4 for the case at 1200 cycles. In the MCMC process, as was noted in the procedure at lines 22–32, the sampling starts with an arbitrary initial value. Although the MCMC technique is less affected by the initial values due to the unique algorithm of random walk which ensures the sampling convergence toward the target distribution, it is advised to choose the initial values at higher likelihood such as the mean or median of the distribution. In this study, to this end, the initial values of P2D model parameters are given by finding out the solution that minimizes the sum square error (SSE) between the data and the model:

$$\min_{\theta} SS_E = \sum_{k=1}^{n} (y_k - y^k(\theta))^2 \tag{6}$$

Then the initial value of the standard deviation is chosen from the square root of SSE, which is 0.004. The upper bound Ub and the lower bound Lb in Equation (4) are set as 0.4 and 5 times of the initial values of each parameter, respectively. The total number of samples in M-H algorithm is set as 10,000. The upper histograms in Figure 4 represent the PDF thus obtained. The lower plots represent plot the samples until the end of the MCMC iteration. Note that the first 1000 samples that are affected by the initial distribution are considered a burn-in period and are discarded [21]. The results as shown in this figure give more valuable information for the estimated parameters because they incorporate the uncertainties in both model and measurements, which is contrasted with the deterministic optimization that gives only point-estimated values.

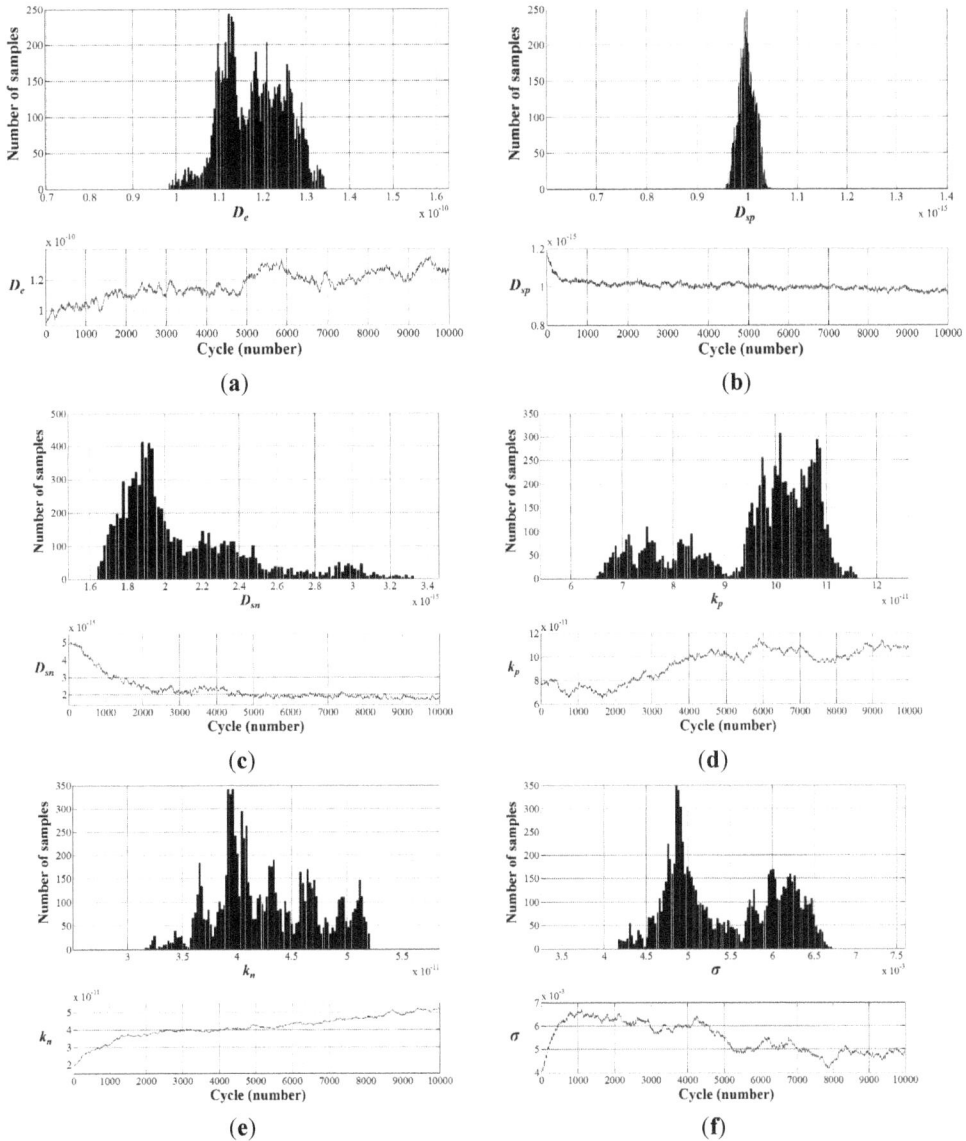

Figure 4. Parameter estimation as a result of voltage data at cycle 1200 (capacity fade 17.7%). The upper histograms show the number of samples of the M-H algorithm over the probable range, which represents the estimated PDF. The lower plots show the values of each parameter up to 10,000 iterations, as explained in Figure 2: (a) liquid-phase diffusivity D_e; (b) solid-phase diffusivity in the positive electrode D_{sp} (c) solid-phase diffusivity in the negative electrode D_{sn}; (d) electrochemical reaction-rate constant in the positive electrode k_p; (e) reaction-rate constant in the negative electrode k_n; (f) standard deviation σ of the likelihood function.

279

In comparing the estimated PDFs of each parameter, the PDF of D_{sp} is noted to be much narrower than that of the other parameters. This narrower PDF means that the uncertainty of the parameter D_{sp} is much lower than that of the other parameters for the given data set. The estimation using the MCMC approach is performed for all 12 data sets obtained from between 200 and 2400 cycles in 200-cycle increments. From the obtained histogram representing the PDF, the 95% confidence interval and mean value are calculated and plotted in Figure 5 as a function of the capacity fade %. The figure shows that the parameter D_{sp} decreases considerably with narrower confidence bounds than the others as capacity fade progresses. Thus there is a strong correlation between D_{sp} and capacity fade, indicating that D_{sp} is the dominant parameter affecting capacity fade for the given data set.

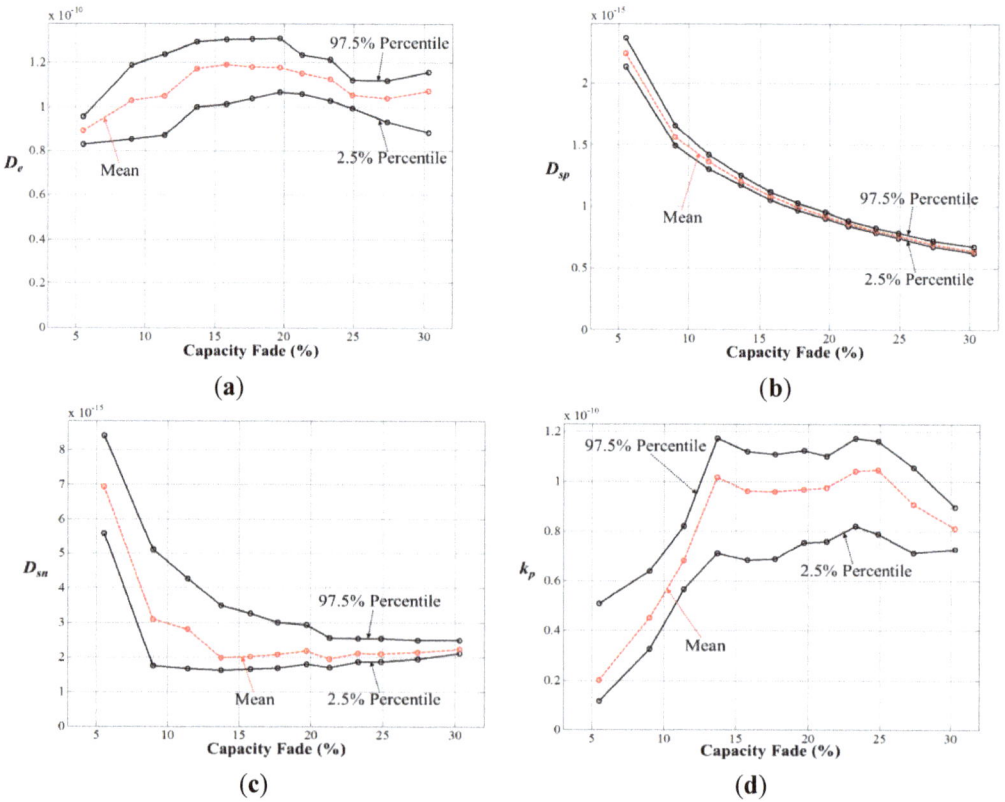

Figure 5. *Cont.*

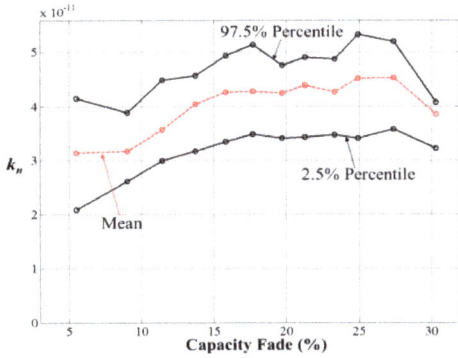

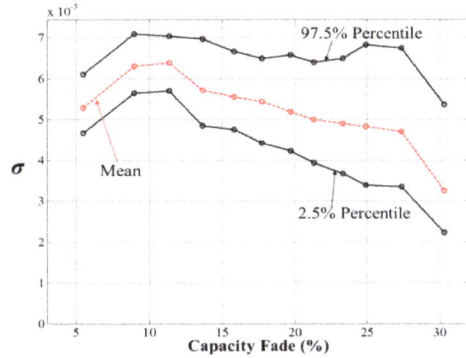

Figure 5. Interval estimation results of battery parameters as capacity fade progresses from 5.5% (cycle 200) to 30.3% (cycle 2400). The 95% confidence intervals and mean values are presented for each model parameter: **(a)** D_e; **(b)** D_{sp}; **(c)** D_{sn}; **(d)** k_p; **(e)** k_n; **(f)** σ.

3. Metamodel Generation

The metamodel approximates the original complex model with simplified, explicit functions. In this study, the metamodel can replace the complex electrochemical model to reduce computational cost during parameter estimation by using the MCMC approach. In this section, the Response Surface Method (RSM) used to generate the metamodel is explained. Then the generated model is validated by comparing the voltage curves of both metamodel and the original electrochemical model.

3.1. Response Surface Method

The RSM is a representative way to generate the metamodel [24]. The steps for generating the metamodel are as follows. First, output function values at multiple input sample points are evaluated from the original complex models. Considering the trend of output function values with respect to the input variables, the form of the approximated mathematical function is determined. Finally, the coefficients of the approximated function are found, based on the least-squares method.

In this study, the solid-phase diffusivity in positive electrode D_{sp} is the only input variable of the metamodel. The terminal voltage is set as the output function because it is the single most critical parameter for the change of the voltage curve as capacity fade occurs. Four other model parameters are fixed at their mean values, which are obtained using cycle-200 experimental data. By employing only one parameter D_{sp} instead of all five, the complexity of the metamodel can be greatly reduced. In order to validate the complexity reduction, P2D simulations are carried out every 200 cycles, first using all five estimated parameters and then using only

281

one estimated parameter D_{sp} with the other four fixed at their mean at cycle 200. Because the parameters and responses are all given by distributions, only the means of the voltage outputs are plotted against the experimental data for the sake of easy comparison. Figure 6 shows the results (a) using all the five model parameters and (b) using only D_{sp}.

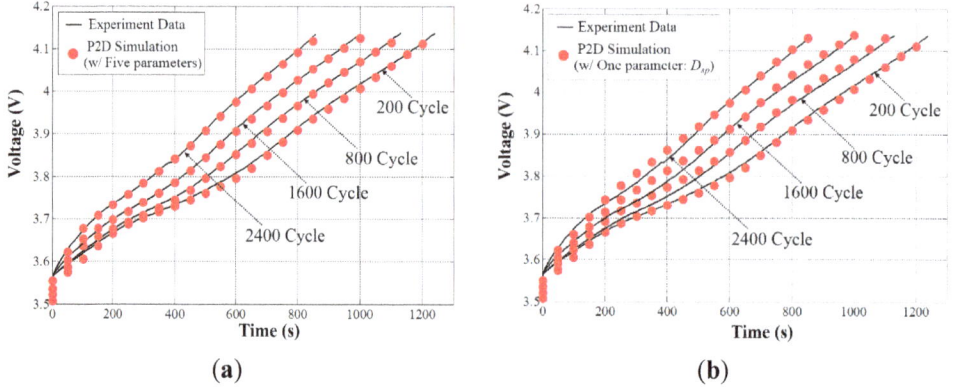

Figure 6. Comparision of the voltage obtained from the P2D model with experimental data. The dots represent the simulated result using the mean of estimated parameters: **(a)** all five model parameters estimated from experimental data; **(b)** only D_{sp} estimated while other parameters are fixed at mean value of cycle 200.

At each cycle i, the relative error e_i of the output voltage against the experimental data is defined as:

$$e_i = \frac{1}{n} \left[\sum_{k=1}^{n} \left| \frac{V_{ki}^* - V_{ki}}{V_{ki}} \right| \right] \times 100 \quad (\%) \tag{7}$$

where V_{ki}^* and V_{ki} are model output and experimental data at the time interval t_k ($k = 1 \sim n$) and ith cycle number, respectively. The error e_i is calculated at cycles 200, 800, 1600, and 2400 for each case, and the results are summarized in the first and second row in Table 2.

Table 2. Error e_i of the model with respect to experimental data.

	Cycle 200	Cycle 800	Cycle 1600	Cycle 2400
P2D w/5 parameters (Figure 6a)	0.195%	0.142%	0.110%	0.078%
P2D w/ D_{sp} (Figure 6b)	0.195%	0.332%	0.346%	0.262%
Metamodel (Figure 7)	0.210%	0.310%	0.351%	0.241%

The greatest error value, a maximum of 0.346%, is produced by only one parameter, as compared with the maximum value of 0.195% by the five parameters. However, the difference in magnitude is small enough to justify that the parameter D_{sp} can be used as the single input variable of the metamodel. The metamodel of the P2D electrochemical model is built using RSM. First, the response data are obtained using the original P2D electrochemical model, which is the terminal voltage V_{ki} at discrete time intervals t_k ($k = 1 \sim n$) and at m equally spaced discrete input variables $D_{sp;i}$ ($I = 1 \sim m$). In this study, the interval of the metamodel is 50 s, $n = 24$, and the end time is 1200 s. The number m is set as 10, with the lower and upper bounds $D_{sp;1}$ and $D_{sp;10}$ being $0.5 \, 10^{-15}$ and 2.5×10^{-15} respectively, considering the behavior of D_{sp} in Figure 5b. In Figure 8, the obtained response values V_{ki} are plotted as dots at the discrete values of the variable D_{sp} and time t. From the figure, it can be seen that the voltages at fixed intervals t_k decrease monotonically with respect to D_{sp}. So, a simple polynomial is introduced to represent voltage as a function of D_{sp} at each discrete interval.

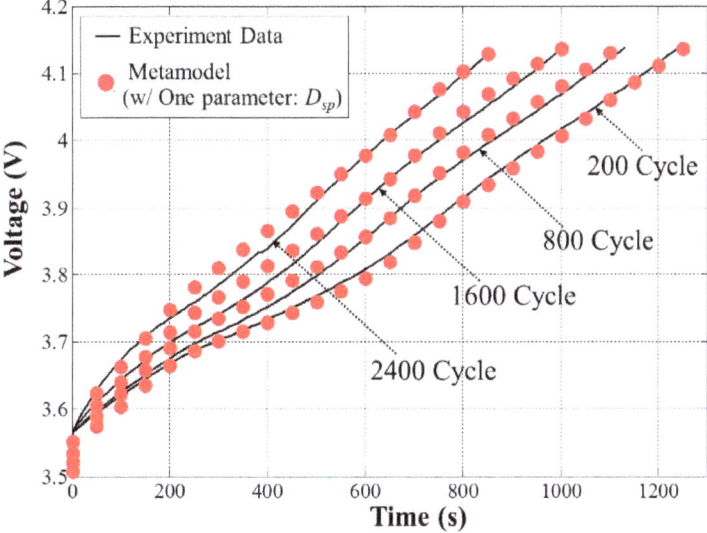

Figure 7. Comparision of the voltage obtained from the metamodel with experimental data.

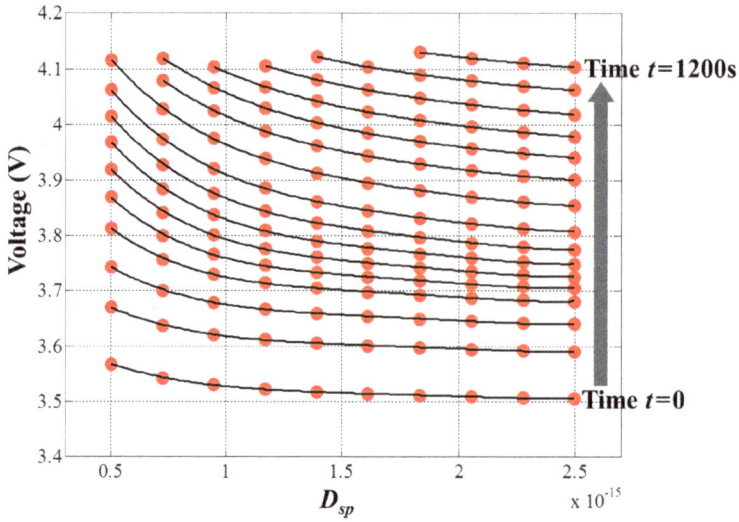

Figure 8. Output voltage with respect to input variable D_{sp} for various time intervals. The red dot represents the voltage V_{ki} at D_{sp} and interval $t_{k;i}$ as obtained from the P2D model; the solid black line represents the voltage curve $V_k(D_{sp})$ obtained using fourth-order polynomials in the metamodel.

After calculating the error between metamodel and original model, the order of the polynomials is determined to be fourth degree. Then the voltage curves at time t_k are represented as:

$$V_k(D_{sp}) = \sum_{p=1}^{4} a_{kp}D_{sp}^{p} = a_{k0} + a_{k1}D_{sp} + a_{k2}D_{sp}^2 + a_{k3}D_{sp}^3 + a_{k4}D_{sp}^4 \qquad (8)$$

In the above equation, the coefficients a_{kp} at time t_k, with p being the polynomial order, are calculated to match with the original response data, using the least-squares method. Because the polynomial function is linear with respect to the coefficients a_{kp}, a typical linear regression algorithm can be used, in which the m-by-5 matrix X is defined as:

$$X = \frac{\partial V_k(D_{e;i})}{\partial a_j} = \begin{bmatrix} 1 & D_{e;1} & \cdots & D_{e;1}^4 \\ 1 & D_{e;2} & & D_{e;1}^4 \\ \vdots & \vdots & & \vdots \\ 1 & D_{e;m} & \cdots & D_{e;m}^4 \end{bmatrix} \qquad (9)$$

Then, the 5-by-m matrix $\mathbf{a}$ can be obtained as:

$$\mathbf{a} = a_{kp} = \begin{bmatrix} a_{10} & a_{20} & \cdots & a_{n0} \\ a_{11} & a_{21} & & a_{n0} \\ \vdots & & & \vdots \\ a_{14} & a_{24} & \cdots & a_{n0} \end{bmatrix} = (\mathbf{X}^T\mathbf{X})^{-1}\mathbf{X}^T\mathbf{y} \tag{10}$$

where $\mathbf{y}$ denotes the m-by-n matrix composed of V_{ki} corresponding to ith input variable $D_{sp;i}$ at kth time interval t_k:

$$\mathbf{y} = V_{ki} = \begin{bmatrix} V_{11} & V_{21} & \cdots & V_{n1} \\ V_{12} & V_{22} & & V_{n2} \\ \vdots & & & \vdots \\ V_{1m} & V_{2m} & \cdots & V_{nm} \end{bmatrix} \tag{11}$$

In Figure 8, the solid black line represents the metamodel in which the voltage curve V_k is given by the fourth-order polynomial with respect to the input variable D_{sp}. This model is used instead of the original electrochemical model to gain computational efficiency during the MCMC sampling process.

3.2. Model Validation

The generated metamodel is validated by comparing the metamodel's output voltage with the corresponding experimental data for various cycles. The result of the comparison is presented in Figure 7, which shows that the metamodel describes the voltage curve of the experimental data well. To evaluate the accuracy of the metamodel against the original P2D model, the error e_i as defined by Equation (7) is calculated for the metamodel at the cycles 200, 800, 1600 and 2400, and the results are given in the third row of Table 2. The errors of the metamodel and the P2D model with one parameter show similar magnitude, with the maximum being about 0.35%.

4. Parameter Estimation Using the Metamodel

Based on the generated metamodel, the parameter estimation is again performed using the MCMC approach, and the estimation result is presented in Figure 9, in which the 95% confidence interval and mean value are plotted as a function of the capacity fade %. Please note that only one estimated parameter D_{sp} in Figure 9 includes all the sources of uncertainty. The various uncertainties are involved in one estimated parameter through the standard deviation σ. Here, the standard deviation σ is set as 0.05 V considering the uncertainty of the experimental data. Next, the estimated parameters are applied to the metamodel's voltage curve to obtain capacity fade in the form of distribution, from which the upper and lower bounds are calculated. The results in Figure 10 are given in terms of cycles at a 200-cycle increment. Using the metamodel based on fourth-order polynomials instead of the original P2D model that solves a system of nonlinear PDEs, the computational cost is tremendously reduced.

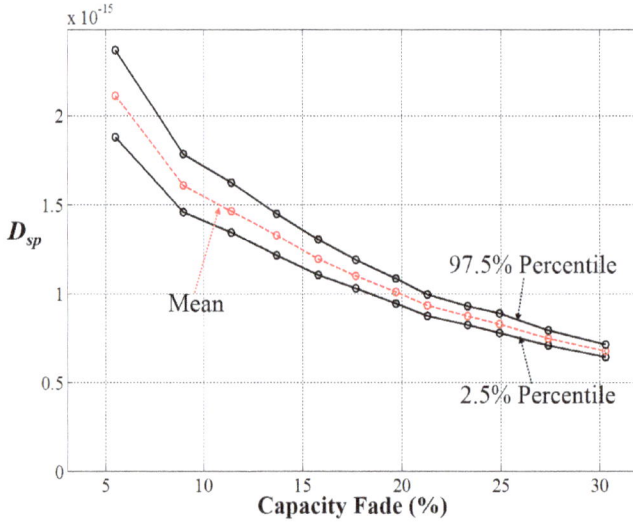

Figure 9. Parameter (D_{sp}) estimate using metamodel (see Figure 5b).

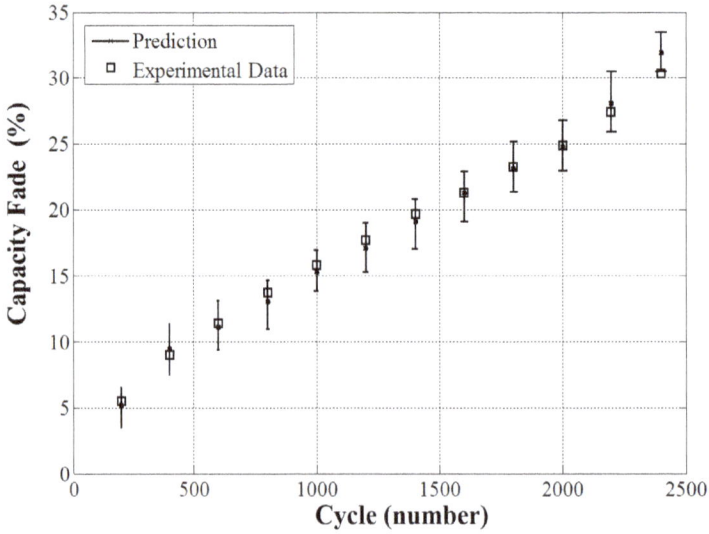

Figure 10. Capacity-fade estimation using the metamodel over the cycles.

Given 10,000 samples, the original model's computing time is 5.5 h (about 2 s per sample) on an octacore workstation with a 3.4 GHz processor and 16 GB of RAM. However, the same computation takes only a few seconds using the metamodel. Thanks to this reduction, the integration of the proposed approach into a vehicle BMS becomes a feasible option.

5. Conclusions and Future Work

This study proposes an efficient method of uncertainty estimation of capacity fade in LIBs, with the goal of integrating it into the BMS of electric vehicles. The physical parameters of the LIB electrochemical model are estimated using the Bayesian-based probabilistic approach. The estimation is performed for five transport and kinetic parameters that are known to affect capacity fade most significantly. Battery data from the full charge/discharge cycles with 2C-rate condition are utilized to implement the method. From the estimation, it is found that one parameter, the solid-phase diffusivity in the positive electrode, is much more responsible than the others for capacity fade. The metamodel is constructed in terms of this parameter in order to avoid the huge computations that occur in the P2D model for MCMC simulation. As a result, computation cost is reduced from 5.5 h to only a few seconds. The reduction of computation time allows the uncertainty estimation of the parameters in real time during battery use, which adds value to the BMS relative to safety and reliability.

This study considers only the case of 2C-rate full charge/discharge process to illustrate the method, in which the data at charge cycle are used for the capacity-fade estimation. The reason for 2C-rate is to accelerate the cycles. In practice, the rapid charge is usually made at 2C-rate. Therefore, the constructed metamodel works only under the same charge condition. In other words, as long as the parameters are estimated under the same charge condition during usage, the estimated results are reliable and represent the actual faded state at that cycle, regardless of the discharge condition the battery went through. If another C-rate (e.g., the normal 1C-rate) is used for charging, the metamodel can be constructed using the same procedure and applied to that condition. As a result, two models with different C-rate conditions can be installed in one BMS and can be used as appropriate to estimate capacity fade.

In this study, only the full-charge condition is addressed. In actual practice, however, the range of completely charged to completely discharged is never available, and a partial charge is more realistic. The proposed method is unable to estimate in this case, which is a challenge to be addressed in future work.

Author Contributions: All authors were involved in the development of ideas and the writing of this paper

Conflicts of Interest: The authors declare no conflict of interest.

References

1. Americans want to give electric vehicles a test drive. New CEA study reports. *CARS Magazine*, 23 August 2010.

2. Xing, Y.; Ma, E.; Tsui, K.L.; Pecht, M. Battery management system in electric and hybrid vehicles. *Energies* **2011**, *4*, 1840–1857.

3. He, W.; Williard, N.; Osterman, M.; Pecht, M. Prognostics of lithium-ion batteries based on Dempster-Shafer theory and the Bayesian Monte Carlo method. *J. Power Sources* **2011**, *196*, 10314–10321.

4. Williard, N.; He, W.; Osterman, M.; Pecht, M. Comparative analysis of features for determining state of health in lithium-ion batteries. *Int. J. Progn. Health Manag.* **2013**, *4*, 1–7.

5. Barre, A.; Deguilhem, B.; Grolleau, S.; Gérard, M.; Suard, F.; Riu, D. A review on lithium-ion battery ageing mechanisms and estimations for automotive applications. *J. Power Sources* **2013**, *241*, 680–689.

6. Rezvanizaniani, S.M.; Liu, Z.; Chen, Y.; Lee, J. Review and recent advances in battery health monitoring and prognostics technologies for electric vehicle (EV) safety and mobility. *J. Power Sources* **2014**, *256*, 110–124.

7. Chaturvedi, N.A.; Klein, R.; Christensen, J.; Ahmed, J.; Kojic, A. Algorithms for advanced battery-management systems. *IEEE Control Syst.* **2010**, *30*, 49–68.

8. Marcicki, J.; Canova, M.; Conlisk, A.T.; Rizzoni, G. Design and parametrization analysis of a reduced-order electrochemical model of graphite/LiFePO 4 cells for SOC/SOH estimation. *J. Power Sources* **2013**, *237*, 310–324.

9. Baek, K.W.; Hong, E.S.; Cha, S.W. Capacity fade modeling of a Lithium-ion battery for electric vehicles. *Int. J. Automot Technol.* **2015**, *16*, 309–315.

10. Hu, C.; Jain, G.; Zhang, P.; Schmidt, C.; Gomadam, P.; Gorka, T. Data-driven method based on particle swarm optimization and k-nearest neighbor regression for estimating capacity of lithium-ion battery. *Appl. Energ.* **2014**, *129*, 49–55.

11. Xiong, R.; Sun, F.; Chen, Z.; He, H. A data-driven multi-scale extended Kalman filtering based parameter and state estimation approach of lithium-ion polymer battery in electric vehicles. *Appl. Energ.* **2014**, *113*, 463–476.

12. Hu, C.; Jain, G.; Tamirisa, P.; Gorka, T. Method for estimating capacity and predicting remaining useful life of lithium-ion battery. *Appl. Energ.* **2014**, *126*, 182–189.

13. Tong, W.; Koh, W.Q.; Birgersson, E.; Mujumdar, A.S.; Yap, C. Correlating uncertainties of a lithium-ion battery—-A Monte Carlo simulation. *Int. J. Energ. Res.* **2015**, *39*, 778–788.

14. Ramadesigan, V.; Chen, K.; Burns, N.A.; Boovaragavan, V.; Braatz, R.D.; Subramanian, V.R. Parameter estimation and capacity fade analysis of lithium-ion batteries using reformulated models. *J. Electrochem. Soc.* **2011**, *158*, A1048–A1054.

15. Sung, W.; Shin, C.B. Electrochemical model of a lithium-ion battery implemented into an automotive battery management system. *Comput. Chem. Eng.* **2015**, *76*, 87–97.

16. Botte, G.G.; Subramanian, V.R.; White, R.E. Mathematical modeling of secondary lithium batteries. *Electrochim. Acta* **2000**, *45*, 2595–2609.

17. Gomadam, P.M.; Weidner, J.W.; Dougal, R.A.; White, R.E. Mathematical modeling of lithium-ion and nickel battery systems. *J. Power Sources* **2002**, *110*, 267–284.
18. Santhanagopalan, S.; Guo, Q.; Ramadass, P.; White, R.E. Review of models for prediction the cycling performance of lithium ion batteries. *J. Power Sources* **2006**, *156*, 620–628.
19. Subramanian, V.R.; Boovaragavan, V.; Ramadesigan, V.; Arabandi, M. Mathematical model reformulation for lithium-ion battery simulations: Galvanostatic boundary condtions. *J. Electrochem. Soc.* **2009**, *16*, A260–A271.
20. Ramadesigan, V.; Boovaragavan, V.; Pirkel, J.C.; Subramanian, V.R. Efficient reformulation of solid-phase diffusion in physics-based lithium-ion battery models. *J. Electrochem. Soc.* **2010**, *157*, 854–860.
21. Gelman, A.; Carlin, J.B.; Stern, H.S.; Rubin, D.B. *Bayesian Data Analysis*; Taylor & Francis: London, UK, 2014.
22. Andrieu, C.; Freitas, N.D.; Doucet, A.; Jordan, M.I. An introduction to MCMC for machine learning. *Mach. Learn.* **2003**, *50*, 5–43.
23. Dawn, A.; Choi, J.H. Improved MCMC method for parameter estimation based on marginal probability density function. *J. Mech. Sci. Technol.* **2013**, *27*, 1771–1779.
24. Arora, J. *Introduction to Optimum Design*; Academic Press: Salt Lake City, UT, USA, 2004.

Analysis and Design of a Permanent Magnet Bi-Stable Electro-Magnetic Clutch Unit for In-Wheel Electric Vehicle Drives

Wanli Cai, Chenglin Gu and Xiaodong Hu

Abstract: Clutches have been used in internal combustion vehicles and concentrated electric vehicles (EVs) to smoothen impulsion while starting and shifting. This paper proposes a permanent magnet bi-stable electromagnetic clutch unit (PMBECU) which is specially introduced into in-wheel EVs to make the rigid connection between hub and wheel more flexible. Firstly, the operation principle of the PMBECU is illustrated. Then, the basic magnetic circuit model is presented and analyzed, followed by optimal design of the main structural parameters by investigating the PM leakage flux coefficient. Further, according to the basic electromagnetic characteristics of the PMBECU, the current pulse supply is put forward, and the minimum pulse width which enables the operation of the PMBECU and its dynamic characteristics are analyzed by an improved finite element method. Finally, a prototype machine is manufactured and tested to validate all the analysis results.

Reprinted from *Energies*. Cite as: Cai, W.; Gu, C.; Hu, X. Analysis and Design of a Permanent Magnet Bi-Stable Electro-Magnetic Clutch Unit for In-Wheel Electric Vehicle Drives. *Energies* **2015**, *8*, 5598–5612.

1. Introduction

Electric vehicles (EV) have been intensively investigated recently as potential solutions for the growing problems of the energy crisis and environmental pollution [1–4], focusing on the drive form, electric motor, controller, battery, energy system, drive comfort, *etc.* Compared with centralized drive, the in-wheel EV drive is considered the more competent drive form for EVs in the near future [5–7], because of its merits of direct drive (no-gearbox), more flexible control strategy (torque at each wheel is independently controlled), high mechanical integrity (greatly different from conventional gasoline cars). However, the rigid connection between hub and motor, inevitably introduces mechanical shocks and electromagnetic impulsion during sudden start and stop processes, which can potentially harm the motor and controller and reduce drive comfort [8–10].

Referring to traditional gasoline cars, this electromechanical impulsion in in-wheel EV drives can be ameliorated by introducing a clutch between the hub and motor to make the rigid connection more flexible [11]. The simulation and experimental results of a conventional clutch between motor and load presented in [12,13] show that the starting current and jerk in clutch coupling starts under different idle speeds can be reduced to less than 1/2 compared to direct starting, and the impulsive back electromotive force to the controller can be eliminated by detached braking (the motor stops naturally after being disconnected from the braking load). Besides, in hybrid EVs, the conventional clutch has been used to cut off the engine or electrical machine while idling to avoid spin losses and extend the life cycle of the machine [14]. Moreover, in in-wheel driven EVs, clutches have been used to detach the motor from hub to reduce losses while coasting [15].

However, the conventional mechanical clutch system [16,17] is not suitable for the limited space available in a hub and suffers from a need for regular maintenance which makes it unsuitable for in-wheel EV drives. In addition, electromagnetic clutches [18,19], which can be easily manipulated by current control, are energy-consuming and also suffer from the problem of accommodating their shape in the hub. In other clutches [20] one encounters one or all of the aforementioned problems, thus are also not suitable options.

This paper proposes a permanent magnet bi-stable electromagnetic clutch unit (PMBECU), which is controlled by current and held by the PM in a steady state, and thus is energy-saving, and it also has a flat structure that makes its placement in a limited space viable. The clutch system is realized by assembling several PMBECUs around motors, combined with friction or jaw pairs.

As key parts of the clutch system, this paper focuses on the electromagnetic design and analysis of the PMBECU. The design and analysis of linear electromagnetic devices, such as electromagnetic valves [21], electric tools [22], oscillators [23,24], and switch gears [25–27], are mainly carried out by the finite element method (FEM). Likewise, aiming to satisfy the need to accommodate the clutch in the limited space available in the hub, the optimal design of the main structure parameters of the PMBECU are carried out by FEM which focuses especially on investigating the leakage flux coefficient of the PM. Moreover, in order to realize simple and reliable control of the operation of the PMBECU, the dynamic characteristics of the PMBECU are calculated by improved FEM, which shows that the low power capacitor pulse supply is very suitable. The influence of the temperature on the dynamic performance is also analyzed. The analysis method and results are finally validated by measurements taken on a prototype machine.

2. Operation Principle

The assembly of the PMBECUs to realize the flexible connection between hub and motor is shown in Figure 1a, and the structure of the PMBECU, in which two PMs with opposite polarities are mounted on each side of a rigid E-type ferromagnetic base, is shown in Figure 1b. The ferromagnetic mover is placed in two low-frictional slideways which are non-magnetic. Two coils are connected n series and wound around each slideway.

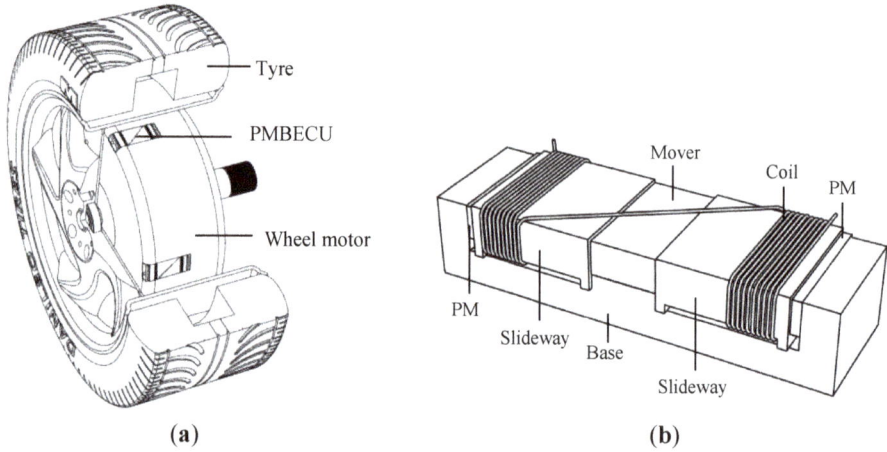

Figure 1. (a) Flexible connection of hub and motor; (b) Structure of the PMBECU.

The 2-dimensional (2D) analysis model of the PMBECU with its main structure parameters labeled is shown in Figure 2, where the right direction is prescribed as positive for force and movement variables.

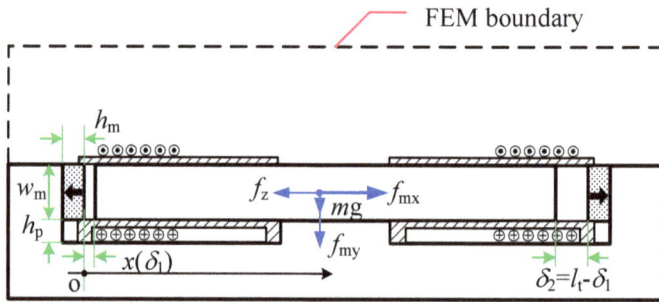

Figure 2. 2D analysis model of the PMBECU.

The flux line distribution of the PMBECU without current injected into the coils is shown in Figure 3a. Apparently, the mover is held by the left PM in a steady state without energy consumption. When current with a suitable orientation (*i.e.*, the current direction shown in Figure 2) and value accesses the coils, the mover is polarized, and the corresponding flux lines distribution is shown in Figure 3b. The mover will soon be propelled from the left steady state to the right by the resultant electromagnetic force. Meanwhile, the current is switched off automatically by the position sensor, and the mover is held by the right PM, again without any energy consumption, thus it is bi-stable.

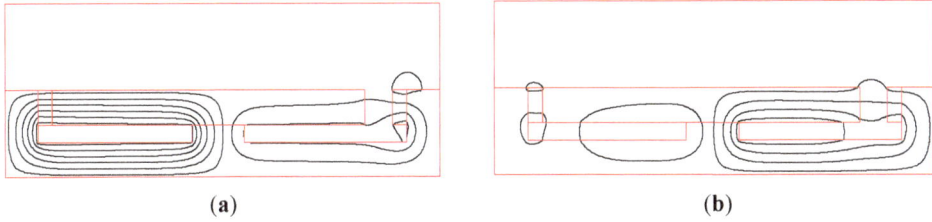

(a)   (b)

Figure 3. Magnetic flux lines distribution. (a) Steady state; (b) Action.

It is evident that the PMBECU has a flat structure thus is suitable for placement in a limited space, and the switchover between engagement and disengagement is electrically-controlled thus it can be conveniently manipulated, and only an instant current is required for switchover, but most time it is in a steady state which is held by a PM and thus is energy-saving.

3. Electromagnetic Design

3.1. Magnetic Circuit Model

According to the magnetic flux line distribution shown in Figure 3a, assuming the ferromagnetic material has infinite permeability and neglecting the contact air gaps, the magnetic circuit relations of the PMBECU under open circuit of the coils conditions can be expressed by a simplified magnetic network as shown in Figure 4.

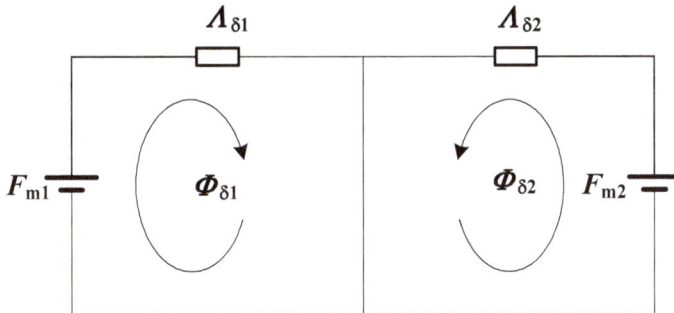

Figure 4. Simplified magnetic network.

The magnetic network comprises two independent branches, where F_{mj} ($j = 1, 2$), $\Lambda_{\delta j}$, and $\Phi_{\delta j}$ are the magneto-motive force furnished to the air gap by the PM, air

293

gap permeance, and magnetic flux pass through the pole face of the mover at each side, which are calculated by Equations (1)–(3), respectively:

$$F_{mj} = H_c h_m \frac{\delta_j}{\left(k_{\sigma j} \frac{h_m}{\mu_r} + \delta_j\right)} \tag{1}$$

$$\Lambda_{\delta j} = \mu_0 \frac{S_m}{\delta_j} \tag{2}$$

$$\Phi_{\delta j} = \frac{B_r S_m}{\left(k_{\sigma j} + \mu_r \frac{\delta_j}{h_m}\right)} \tag{3}$$

where B_r, H_c, and μ_r are remanence, coercivity, and relative permeability of the PM, h_m and S_m are the thickness and pole face area of the PM, δ_j is the air gap length as labeled in Figure 2, μ_0 is the permeability of air, and $k_{\sigma j}$ is the leakage flux coefficient which is defined as:

$$k_{\sigma j} = \frac{\Phi_{mj}}{\Phi_{\delta j}} \tag{4}$$

where Φ_{mj} is main magnetic flux through bottom face of PM.

The Maxwell stress tensors are given by the following equation [27]:

$$t_n = \left(B_n^2 - B_s^2\right) / (2\mu_0)$$
$$t_s = B_n B_s / \mu_0 \tag{5}$$

where B_n, B_s are the outer normal and tangential components of the flux density on the mover, respectively. Out of an infinite permeable surface, the flux density only has a normal component. Hence, combined with Equation (3), the holding force (horizontal) at steady state ($\delta_1 = 0$, $\delta_2 = l_t$, l_t is the travel length of the mover) can be approximately calculated by:

$$f_H = \frac{B_{\delta 1}^2 S_m}{2\mu_0} - \frac{B_{\delta 2}^2 S_m}{2\mu_0} = \frac{B_r^2 S_m}{2\mu_0} \left(1 - \frac{1}{\left(k_{\sigma 2} + \mu_r \frac{l_t}{h_m}\right)^2}\right) \tag{6}$$

With forces normalized to $f_b = 0.5 B_r^2 S_m / \mu_0$ (the same hereafter), the holding force is:

$$f_H = 1 - \frac{1}{\left(k_{\sigma 2} + \mu_r \frac{l_t}{h_m}\right)^2} \tag{7}$$

Apparently, the holding force of the PMBECU is determined by $k_{\sigma 2}$ (leakage coefficient at $\delta = l_t$), the ratio of travel length to thickness of the PM l_t / h_m, and the PM characteristics. Moreover, the leakage flux coefficient $k_{\sigma 2}$ is a function of the structure parameters, and can be calculated by Equation (4) after the magnetic flux derived from FEM analysis.

By increasing the current from 0, the electromagnetic force experienced by the mover can be obtained, and then the ideal threshold current i_T which critically enables the action of the mover can be obtained by FEM as well, corresponding to the horizontal electromagnetic force $f_{mx} = 0$. In this paper, current is all normalized to $i_b = H_c h_m / N$, where N is the number of turns for one coil.

3.2. Main Structure Parameters Design

The PMBECU works at steady state most of the time, which is reliably maintained by the holding force, thus the holding force is the most significant index. According to Equation (7), the leakage flux coefficient $k_{\sigma 2}$ at the detached side, which is a function of the structure parameters, has a great influence on the holding force. Moreover, the leakage flux coefficient determines the reasonable usage of the PM. Thus, the main structure parameters (as labelled in Figure 2), *i.e.*, the width of the PM w_m, the height from the PM to the base h_p, and the travel length of mover l_t, are optimized by studying $k_{\sigma 2}$, combined with accounting for the holding force and threshold current, where, other size ratios (proportioned to h_m) remain unchanged while one varies the parameters within the ranges $h_p / h_m = 1.2$, $w_m / h_m = 2.5, l_t / h_m = 2$.

The variation of $k_{\sigma 2}$ *versus* different structure parameters is shown in Figure 5. Figure 5a shows that the leakage flux coefficient increases quite slowly when h_p is 1.5 times bigger than h_m, hence h_p would better be within 1–1.5 times of h_m, which also indicates the PMBECU is capable of a flat structure. Likewise, w_m would better be around 2.5 times of h_m as seen in Figure 5b. Figure 5c shows the leakage flux coefficient $k_{\sigma 2}$ increases almost linearly with l_t, which shows no clear inflection point. But from Figure 5d, the holding force increases very slowly when l_t is 2 times larger than h_m, meanwhile, the threshold current keeps increasing, which makes the action of the mover harder. Hence, l_t within 1.5–2 times the thickness of the PM is more sensible.

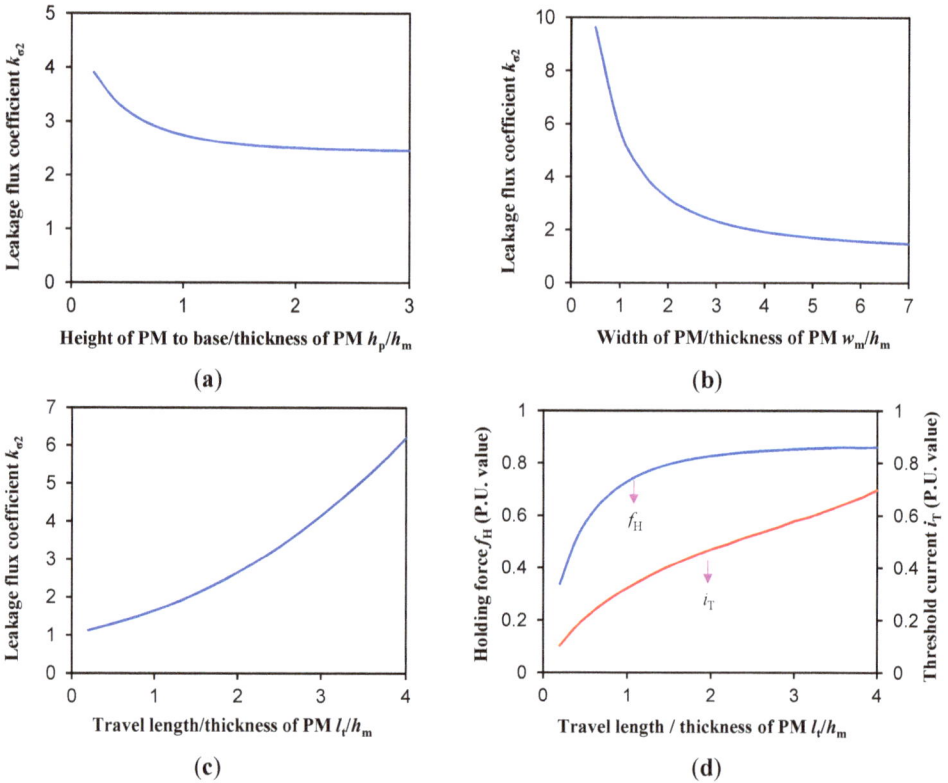

Figure 5. Opimization. (**a**) Height from PM to base; (**b**) Width of PM; (**c**) Travel length; (**d**) Travel length.

4. Dynamics Analysis

4.1. Electromagnetic Characteristics

Based on the aforementioned analyses, a PMBECU prototype designed with the main parameters listed in Table 1 is shown in Figure 6. Assuming the mover is fixed at different positions, changing the current (constant DC wave) in the coil, computing the magnetic field by FEM and the forces experienced by mover by (5), then the electromagnetic forces on the mover *versus* current i and displacement x are obtained as shown in Figure 7.

Parameter	Value	Parameter	Value
Thickness of PM h_m	2.5 mm	Width of base w_b	80 mm
Width of PM w_m	6 mm	Remanence of PM B_r	0.4 T
Length of PM l_m	20 mm	Coercivity of PM H_c	318 kA/m
Height of PM to base h_p	3 mm	Turns of coil N	60
Travel length l_t	4.8 mm	Mass of mover m	56 g

Table 1. Leading design parameters.

Figure 6. Prototype.

From Figure 7a, for open circuit conditions, the $i = 0$ horizontal force curve indicates that the PMBECU has two steady states held by the magnetic force from the PM, and an unstable equilibrium point (the half travel length location). When the mover exceeds this unstable point, the mover can be drawn to the other steady state automatically even if the current is switched off. Since the current increases to the ideal threshold current (enabling the action of the mover) $i = 0.49$, the mover starts moving. The maximum current in the coil is limited by the inflection point of the demagnetizing curve of the PM (critical point of irreversible demagnetization), which is $i = 0.77$ in this prototype.

In fact, because of the asymmetric structure in the vertical direction, the mover experiences a downward vertical electromagnetic force (as shown in Figure 7b) which introduces frictional resistance. Hence, accounting for friction, and other errors (material, model, measuring, *etc.*), the real threshold current i_T is bigger than the calculated value, which is $i_T = 0.52$ for the prototype. Moreover, to guarantee the performance of the PM, the maximum current should be limited to $i_M = 0.7$.

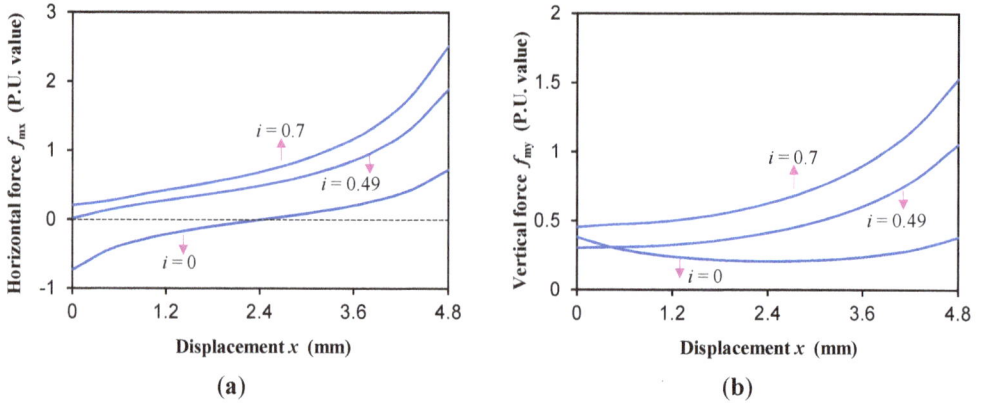

Figure 7. Electromagnetic forces. (**a**) Horizontal; (**b**) Vertical.

When the current is larger than the threshold current, the resultant positive horizontal force starts to drive the mover, and the force is a monotonously increasing function of the displacement. After moving through the middle point of the PMBECU, the mover can reach another steady state with the current switched off (*i.e.*, the pulse current sustains only half the travel length width). What's more, considering the inertial motion and variation of the kinetic friction coefficient, the pulse width of the current can be even smaller. Thus, a dynamics analysis of the PMBECU should be carried out.

4.2. Dynamics Equations and Analysis Method

Because of the motion symmetry of the PMBECU, only the movement of the mover from left to right is investigated. Supposing the static friction coefficient is equal to the kinetic friction coefficient, then the magnetic-kinematic coupled mathematic equations which determines the dynamics characteristics are described as:

$$f_{mx} - f_z = m\frac{dv}{dt} \tag{8}$$

$$v = \frac{dx}{dt} \tag{9}$$

$$f_{mx} = q(x, i), \ f_{my} = p(x, i) \tag{10}$$

$$f_z = \mu_s(f_{my} + mg) \tag{11}$$

where f_{mx} and f_{my} are the horizontal and vertical electromagnetic forces on the mover, f_z is the resisting force, v is the velocity of the mover, μ_s is the static friction coefficient which is 0.065 in this prototype (measured), and g is the acceleration constant of gravity.

The dynamics analysis of the PMBECU is to illustrate the coupling of the magnetic field and the movement. To cope with the varying friction resistance conditions of the PMBECU, and give consideration to the convenience of analysis of varied structure sizes, an improved FEM is proposed. As shown in Figure 8a, two l_t length rectangular areas (namely, the material variation area) in proximity to the PMs are established and uniformly meshed into n steps of quadrilateral shape, *i.e.*, the step length is $\Delta x = l_t/n$. The initial permeability of the left part and the right are set as iron (μ_{Fe}) and air (μ_0) respectively. As shown in Figure 8b, if the permeability of the first Δx meshes in the left material variation area is changed into μ_0 and the first Δx meshes at the right into μ_{Fe}, a Δx displacement of the mover is equivalently realized. Thus, a onetime mesh can cover the travel length displacement of the mover [23].

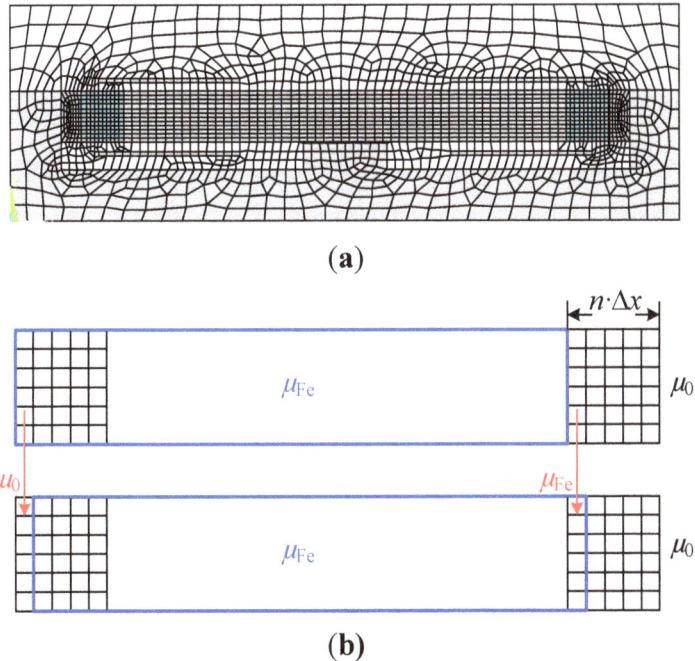

(a)

(b)

Figure 8. Onetime mesh technique. (a) Mesh; (b) Principle.

Further, by setting displacement as a known quality but time as an unknown variable, and calculating the time, velocity, and current before each time of material variation, the whole PMBECU movement process (*i.e.*, the dynamics characteristics of the PMBECU) can be solved by using only a onetime mesh. This improved FEM analysis flow chart is shown in Figure 9, where both the current change and resistance variation can be taken into account, which can be easily realized by commercial FEM software (e.g., ANSYS programmable design language). In this paper, at each side of the mover, the front part of the material variation area is finely meshed and the rear part is roughly meshed (because the front part of displacement takes much more time), so as to improve accuracy and reduce the amount of computation.

4.3. Minimum Driving Pulse Width

With the pulse threshold current accessed (in the calculations, the pulse width is given by the length of the displacement), the dynamics equations can be solved by the improved FEM, then the time pulse width, and finally the force and velocity curves *versus* displacement and different pulse widths, all can be obtained. The minimum pulse width t_w is the pulse width of the threshold current which critically enables the switchover of the PMBECU between engagement and disengagement, *i.e.*, the velocity of mover will be negative if the pulse width is less than t_w. The resultant force and velocity curve of the prototype, which vary with displacement and pulse width, are shown in Figure 10.

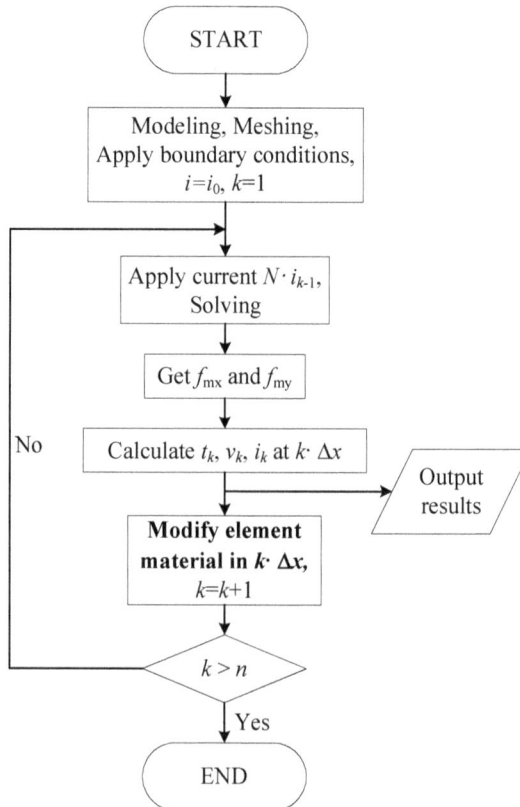

Figure 9. Flowchart for solving the dynamics by improved FEM.

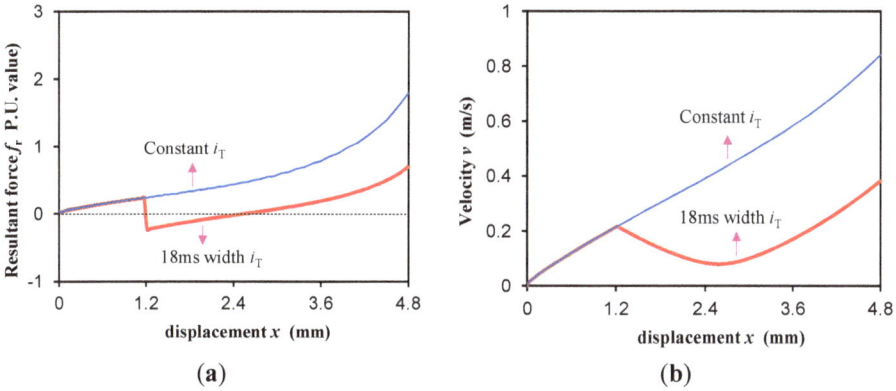

Figure 10. Dynamics characteristic under different pulse width. (**a**) Resultant horizontal force; (**b**) Velocity.

The smaller the width of the accessed pulse current, the more homogeneous the switchover process experienced by the mover will be. When the accelerating displacement is just longer than the decelerating displacement, it is the minimum current pulse width that accomplishes the switchover of the PMBECU between steady states, which is t_w = 18 ms in this prototype.

5. Experimental Validation

The experimental electric circuit as well as the experimental rig of the PMBECU are shown in Figure 11. First, the holding force at different positions was measured and compared to the FEM calculation results (there is an initial air gap δ_1 = 0.1 mm and a contact air gap in the mid part of δ_0 = 0.18 mm which had been accounted in the FEM model) as shown in Figure 12. The experimental results are a little smaller than the simulation results which is mainly attributed to the round corners of the PM, but it still shows an acceptable engineering accuracy.

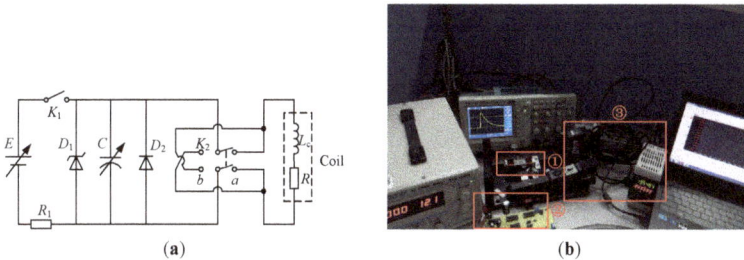

Figure 11. (**a**) Experimental circuit; (**b**) Experimental rig, ① Prototype, ② Capacitor supply, ③ Laser displacement transducer.

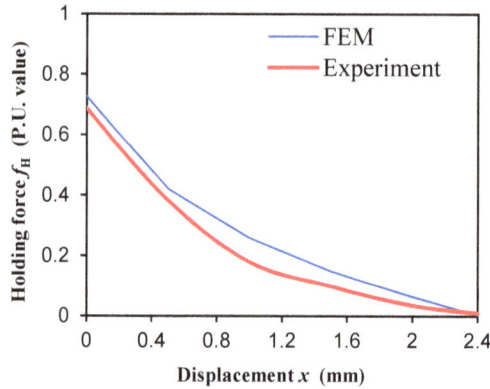

Figure 12. Comparison of the holding force at different positions.

The pulse current applied to coil is approximately generated by a capacitor discharge lower power pulse supply. By changing the capacitance (*i.e.*, changing pulse width), and tuning the charging voltage (keeping $i_M = 0.7$), the practicable minimum pulse width can be obtained, which is $t_w = 5.2$ ms in this prototype, and the corresponding discharge current curve is shown in Figure 13. Because the maximum discharge current is larger than the threshold current, and the continuous discharging current curve is superior to the rectangular pulse current, the minimum pulse width of the low power capacitor supply is much smaller.

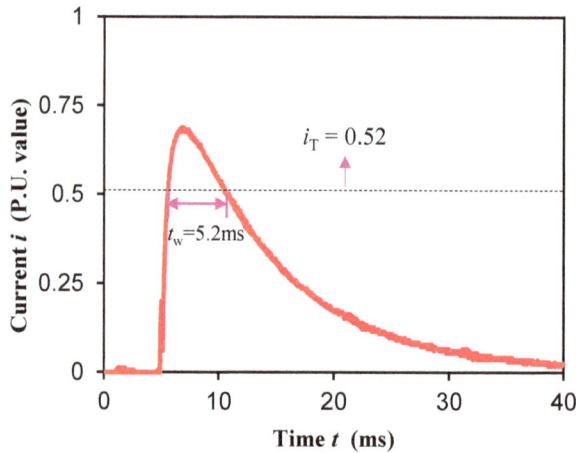

Figure 13. Discharge current curve of the low power supply which critically enables the switchover of the prototype.

In the dynamics experiments of the PMBECU (at an ambient temperature of 25 °C), the main electrical parameters are $C_b = 8.6$ mF, $R_b = 1.15$ Ω, and the

302

displacement of the mover is recorded by a laser displacement transducer. In fact, the dynamics characteristics and minimum pulse width of the low power capacitor supply also can be obtained by the improved FEM, with the current value at each step solved by the electric circuit equation.

The comparison between the improved FEM simulation and the experimental dynamics characteristics results are displayed in Figure 14, which shows a satisfactory agreement aside from the slight bounce of the mover, where the experimental velocity and dynamics force of the mover are derived from the differential and second order differential of the measured displacement curve. Thus, the improved FEM is an effective method for dynamics analysis of the PMBECU. What's more, compared to the dynamics characteristics under constant current, the force on the mover is evener, the velocity of the mover is steadier, and the control is much simpler (the current decays automatically without switching off by position detection), and thus represents the optimal power supply scheme for the PMBECU.

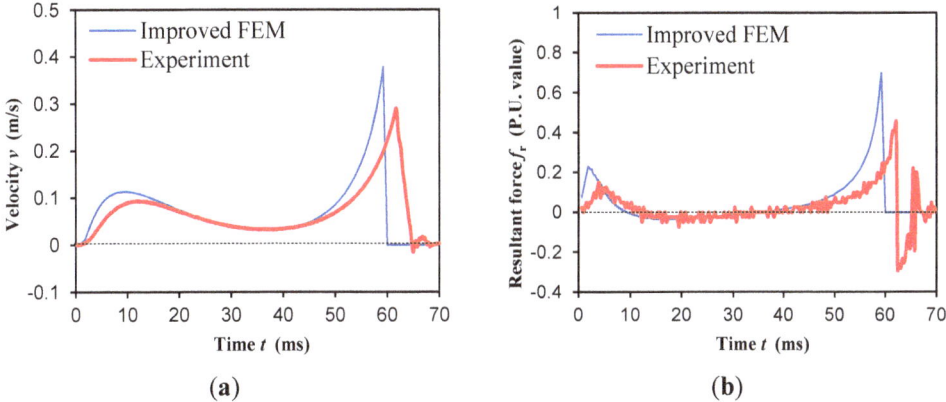

Figure 14. Comparison of dynamics characteristics under low power pulse supply. (a) Velocity; (b) Resultant horizontal force.

In in-wheel drive applications, the device suffers from harsh working conditions—vibration, temperature variation and EMI, etc. The vibration and EMI influence the mechanical and control reliability of the PMBECU, respectively. However, from the aspect of electromagnetic analysis, the temperature variation mainly changes the electromagnetic characteristics of the PMBECU. When the temperature rises, the resistance approximately increases 0.43% per degree Celsius (compared to resistance Rb at 25 °C). The corresponding variation of the maximum discharge current imax is shown in Figure 15a. As shown, when temperature is less than 0 °C, imax is 10% larger than iM which is about to irreversibly demagnetize the PM, thus indication that a NdFeB PM is a better choice than a ferrite PM. When the temperature is higher than 150 °C, imax is less than the threshold current iT even though the capacitor is infinite, which will disable the PMBECU, and thus should be avoided. Figure 15b shows the variation of the minimum capacitor Cmin (compared to Cb at 25 °C) which critically enables the work of the PMBECU at different temperatures. Four measurement points of experiments prove the validity of the simulation. From Figure 15b, when the temperature rises, the work of the PMBECU requires a bigger capacitor because of the reduction of imax. Hence, the capacitor size should be determined by the maximum working temperature.

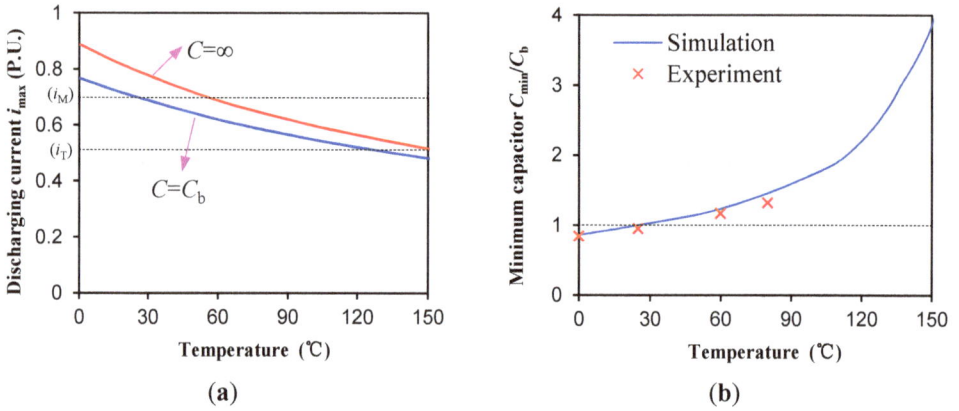

Figure 15. Electromagnetic characteristics variation with temperature. (**a**) Maximum discharge current; (**b**) Minimum capacitor enables the work of the PMBECU.

6. Conclusions

This paper proposes a permanent magnetic bi-stable electromagnetic clutch unit, which is introduced into in-wheel EV drives to turn the rigid connection between the hub and wheel into a more flexible form. The main structure parameters of the PMBECU are optimized by studying the leakage coefficient, holding force, and threshold current, which gives that the width of the PM, height from the PM to the base, and the travel length should better be around 2.5, 1.2, 2 times the thickness of the PM, respectively.

Based on the optimal structure parameters, a PMBECU prototype is fabricated. The basic electromagnetic characteristics indicate that the PMBECU is better controlled by pulse supply. Accordingly, an improved FEM is put forward to obtain the dynamics characteristics and minimum pulse width of the threshold current. The simulation results of both static force and dynamics characteristics are validated by experimental measurements on the prototype. Both the analysis and experimental results show that a low power capacitor supply is very suitable for the PMBECU, and the capacitor size should be determined under the maximum working conditions temperature. The analysis method and results lay a solid basis of the further design of whole clutch system.

Acknowledgments: This work was funded by National Natural Science Foundation of China (51377063).

Author Contributions: Wanli Cai conducted the simulations of FEM and experiments of the prototype and wrote the article. Xiaodong Hu did some experiments. Chenglin Gu proposed the basic idea and supervised the simulations and experiments.

Conflicts of Interest: The authors declare no conflict of interest.

References

1. Chan, C.C.; Bouscayrol, A.; Chen, K.Y. Electric, hybrid, and fuel-cell vehicles: Architectures and modeling. *IEEE Trans. Veh. Technol.* **2010**, *59*, 589–598.
2. Chau, K.T.; Chan, C.C.; Liu, C.H. Overview of permanent-magnet brushless drives for electric and hybrid electric vehicles. *IEEE Trans. Ind. Electron.* **2008**, *55*, 2246–2257.
3. Lee, C.; Liu, C.H.; Chau, K.T. A magnetless axial-flux machine for range-extended electric vehicles. *Energies* **2014**, *7*, 1483–1499.
4. Lee, H.D.; Sul, S.K.; Cho, H.S.; Lee, J.M. Advanced gear-shifting and clutching strategy for a parallel-hybrid vehicle. *IEEE Ind. Appl. Mag.* **2000**, *6*, 26–32.
5. Rahman, K.M.; Patel, N.R.; Ward, T.G.; Nagashima, J.M.; Caricchi, F.; Crescimbini, F. Application of direct-drive wheel motor for fuel cell electric and hybrid electric vehicle propulsion system. *IEEE Trans. Ind. Appl.* **2006**, *42*, 1185–1192.
6. Sakai, S.; Sado, H.; Hori, Y. Motion control in an electric vehicle with four independently driven in-wheel motors. *IEEE Trans. Mech.* **1999**, *41*, 9–16.
7. Ifedi, C.J.; Mecrow, B.C.; Brockway, S.T.M.; Boast, G.S.; Atkinson, G.J.; Perovic, D.K. Fault-Tolerant in-wheel motor topologies for high-performance electric vehicles. *IEEE Trans. Ind. Appl.* **2013**, *49*, 1269–1278.
8. Yang, Y.P.; Liu, J.J.; Wang, T.J.; Kuo, K.C.; Hsu, P.E. An electric gearshift with ultracapacitors for the power train of an electric vehicle with a directly driven wheel motor. *IEEE Trans. Veh. Technol.* **2007**, *56*, 2421–2431.
9. Usami, Y. Controller for Electric Vehicle. U.S. Patent 5896283, 20 April 1999.
10. Kasten, R.E.; Newendorp, B.C.; Lemmen, N.F. Variable Current Limit Control for Vehicle Electric Drive System. U.S. Patent 6492785, 10 December 2002.
11. Gu, C.L. Comprehensive review of electric vehicle PM wheel motors. *MICROMOTORS (China)* **2008**, *41*, 56–59.
12. Xiong, P.; Gu, C.L. Optimal idling speed control of direct-drive electric vehicle launch in consideration of drive comfort. In Proceedings of the 17th International Conference on Electrical Machines and Systems (ICEMS), Hangzhou, China, 21–24 October 2014; pp. 225–228.
13. Wu, D. Special Electromagnetic Clutch for a Novel Type of Transverse Flux Permanent Magnet Motor Direct Drive. Master's Thesis, Huazhong University of Science and Technology, Wuhan, China, May 2011.

14. Chen, L.; Xi, G.; Sun, J. Torque coordination control during mode transition for a series–parallel hybrid electric vehicle. *IEEE Trans. Veh. Technol.* **2012**, *61*, 2936–2949.
15. Camilleri, R.; Armstrong, P.; Ewin, N.; Richardson, R.; Howey, D.A.; McCulloch, M.D. The value of a clutch mechanism in electric vehicles. In Proceedings of the Conference of EVS27, Barcelona, Spain, 17–20 November 2013; pp. 1–11.
16. Balau, A.E.; Caruntu, C.F.; Lazar, C. Simulation and control of an electro-hydraulic actuated clutch. *Mech. Syst. Signal Process.* **2011**, *25*, 1911–1922.
17. Gauthier, J.P.; Micheau, P.; Rioux, R. Vehicle Clutch Control Method. U.S. Patent 8,744,709, 3 June 2014.
18. Ando, J.; Tsuda, T.; Ando, H.; Niikawa, Y.; Suzuki, K. Development of third-generation electronically controlled AWD coupling with new high-performance electromagnetic clutch. *SAE Int.* **2014**, *7*, 882–887.
19. Boules, N.M. Design analysis of electromagnetic particle clutch. In Proceedings of the Conference Record of the 1994 IEEE Industry Applications Society Annual Meeting, Denver, CO, USA, 2–6 October 1994; pp. 357–360.
20. Duan, G.H. *Clutch Structure Atlas*; National Defence Industry Press: Beijing, China, 1985.
21. Ohdachi, Y.; Kawase, Y.; Murakami, Y.; Inaguma, Y. Optimum design of dynamic response in automotive solenoid valve. *IEEE Trans. Magn.* **1991**, *27*, 5226–5228.
22. Kim, J.; Chang, J. A new electromagnetic linear actuator for quick latching. *IEEE Trans. Magn.* **2007**, *43*, 1849–1852.
23. Zhu, Z.Q.; Chen, X. Analysis of an E-core interior permanent magnet linear oscillating actuator. *IEEE Trans. Magn.* **2009**, *45*, 4384–4387.
24. Srairi, K.; Feliachi, M. Electromagnetic actuator behavior analysis using finite element and parametrization methods. *IEEE Trans. Magn.* **1995**, *31*, 3497–3499.
25. Kawase, Y.; Tatsuoka, S.; Yamaguchi, T.I. 3-D finite element analysis of operating characteristics of ac electromagnetic contactors. *IEEE Trans. Magn.* **1994**, *30*, 3244–3247.
26. Fang, S.H.; Lin, H.Y.; Ho, S.L. Transient co-Simulation of low voltage circuit breaker with permanent magnet actuator. *IEEE Trans. Magn.* **2009**, *45*, 1242–1245.
27. Woo, K.I.; Kwon, B.I. Characteristic analysis and modification of PM-type magnetic circuit breaker. *IEEE Trans. Magn.* **2004**, *40*, 691–694.

Effects of Scavenging System Configuration on In-Cylinder Air Flow Organization of an Opposed-Piston Two-Stroke Engine

Fukang Ma, Changlu Zhao, Fujun Zhang, Zhenfeng Zhao and Shuanlu Zhang

Abstract: In-cylinder air flow is very important from the point of view of mixture formation and combustion. In this direction, intake chamber structure and piston crown shape play a very crucial role for in-cylinder air pattern of opposed-piston two-stroke (OP2S) engines. This study is concerned with the three-dimensional (3D) computational fluid dynamics (CFD) analysis of in-cylinder air motion coupled with the comparison of predicted results with the zero-dimensional (0D) parametric model. Three configurations *viz.*, a flat piston uniform scavenging chamber, a flat piston non-uniform scavenging chamber and a pit piston non-uniform scavenging chamber have been studied. 0D model analysis of in-cylinder air flow is consistent with 3D CFD simulation. It is concluded that a pit piston non-uniform scavenging chamber is the best design from the point of view of tumble ratio, turbulent kinetic energy and turbulent intensity, which play very important roles in imparting proper air motion. Meanwhile a flat piston uniform scavenging chamber can organize a higher swirl ratio and lower tumble ratio which is important to improve the scavenging process.

Reprinted from *Energies*. Cite as: Ma, F.; Zhao, C.; Zhang, F.; Zhao, Z.; Zhang, S. Effects of Scavenging System Configuration on In-Cylinder Air Flow Organization of an Opposed-Piston Two-Stroke Engine. *Energies* **2015**, *8*, 5866–5884.

1. Introduction

The OP2S engine concept can be traced back to the late 1800s. Since then, many novel applications have been used in aircraft, ships and vehicles. In the first half of the 20th century, OP2S engines were developed in multiple countries for a wide variety of applications. However, modern emission regulations stopped widespread development of most two-stroke engines in the latter half of the 20th century [1]. In recent years, with the application of advanced design technology, modern analytical tools, materials and engineering methods, the emission problems no longer limit the successful design of a clean and efficient OP2S [2], so OP2S engines are once again attracting intensive attention to improve engine efficiency and emission performance [3–7]. Compared with conventional engines, OP2S engines have many fundamental advantages [8]. The opposed-piston structure characterized by two pistons reciprocating opposite to each other in a common cylinder, cancels the need for the cylinder head and valve mechanism, which leads to lower heat loss for a higher wall temperature with two piston crowns compared to a cylinder head, while the nearly symmetrical movement of opposed pistons leads to excellent engine balance, even for single cylinder configurations.

For conventional two-stroke gasoline engines, the serious fuel short circuit loss during the scavenging process results in poor fuel economy and high emission levels. OP2S gasoline direct injection (GDI) engines use uniflow scavenging and GDI technology to separate the injection and scavenging processes. For GDI engines, the air-fuel mixture is formed in the cylinder, so in-cylinder fluid dynamics play a crucial role in the mixture formation and combustion process. On one hand, in order to accelerate air-fuel mixing, high intensity turbulence is required from a micro perspective. On the other hand, in-cylinder air motion velocity is needed for forming a homogenous mixture from a macro perspective [9]. Swirl, tumble and squish are used to form the air-fuel mixture. For conventional four-stroke GDI engines, in-cylinder flow organization depends on intake duct structure, inlet valve shape, bore-stroke ratio and combustion-chamber shape [10,11]. The injector is installed on the cylinder head. Because injection happens in the intake process, mixing time is more than sufficient. For OPTS-GDI engines, mixture formation time is short since the fuel injection process mainly occurs during the compression process. Fluid motion inside the cylinder is inherently unsteady, turbulent and three dimensional. Gas motion is unstable during the scavenging and compression processes and breaks down into three dimensional turbulent motions. Therefore, a proper understanding of in-cylinder air motion organization and also the effect of intake chamber structure and piston configuration are required to improve mixture formation.

Nowadays, a number of cold and hot flow CFD simulation studies have been carried out to understand in-cylinder flow field, combustion and emission processes in IC engines [12–15]. These investigations have shown that complex flow structures like swirl, tumble and turbulence exist inside the engine cylinder, even after the closure of the intake valve [16]. Nordgren *et al.* [17] studied the in-cylinder air motion through experimental and theoretical methods *viz.*, PIV and CFD. Sweetland and Reitz [18] used the KIVA code to study the in-cylinder flow field during the intake and compression strokes using real intake port geometry and moving intake valves. They reported that turbulent kinetic energy estimated from the PIV photographs agreed well with the KIVA code. Gunasekaran and Ganesan [19] simulated the fuel-air interaction in a four stroke four valve direct injected spark ignition engine and reported that in order to achieve a combustible fuel-air mixture near the spark plug, proper air motion during induction and compression process is necessary. Rakapoulos *et al.* [20] investigated the three piston bowl geometries in diesel engines through CFD against a quasi-dimensional model by changing the ratio of piston bowl diameter to cylinder diameter. They concluded that both the models predicted similar cylinder pressure and temperature as well as axial and radial velocities.

Lin *et al.* [21], Shimoda *et al.* [22] and Dolak *et al.* [23] reported that even though the shape and design of the intake port play a predominant role in generating tumble, the combustion chamber geometry also plays a part in vortex formation, turbulence generation, combustion and emission processes in an internal combustion engine. In [24] the authors used numerical simulation to assess the influence of some intake duct geometrical parameters on the tumble motion generation during both the intake and the compression strokes to highlight the turbulence production process. In [25] the authors presented a theoretical model capable of describing the interaction between the squish velocity and the tumble velocity depending on the engine class. Ramajo *et al.* [26] reported results obtained running a mono-dimensional (1D) model developed for predicting in-cylinder tumble motion formation and breakdown till the appearance of high turbulence level close to TDC.

From the literature survey, it is clear that for conventional two-stroke gasoline engines the in-cylinder fluid flow is heavily dependent on the intake duct structure, inlet valve shape, bore-stroke ratio and combustion-chamber shape. However, limited research exists about the effect of intake chamber structure and piston bowl configuration on the in-cylinder flow characteristics. The aim of the paper is to introduce a new analysis approach to provide deep insight into the 0D parametric model development and 3D-CFD results performed to assess the intake chamber structure influence on the optimization of the in-cylinder flow organization during both the scavenging and compression processes. All the CFD simulations presented in the paper were performed by the AVL-Fire CFD code on an OPTS-GDI engine characterized by a unit displacement of 250 cm^3. The scavenging flow was changed during the analysis by changing the scavenging system configuration. The effects of in-cylinder air motion variations were evaluated in terms of the tumble ratio, swirl ratio, the turbulent kinetic energy and the squish characterization before inner dead center (IDC).

2. OP2S Engine Scavenging System

2.1. Engine Configuration

As shown in Figure 1, the OPTS-GDI engine is equipped with a GDI system and a "port-to-port" uniflow scavenging system, and its injector and spark plug are placed on the cylinder liner. The opposed crank-connecting rod mechanism is placed on both sides of cylinder body and a chain transmission mechanism is designed to realize the synchronized working of opposed pistons. On both sides of the cylinder liner there are gas ports, an intake ports on one side and an exhaust port on the other side. Intake ports are used to deliver fresh air into the cylinder, and exhaust port are used to remove exhaust gas from the cylinder. In the working process, the piston motion controls the opening and closing of ports. There are two pistons placed in the cylinder liner, and the combustion chamber is formed when the two pistons move to the closest position. The reciprocating movements of pistons are driven by the opposed crank-connecting rod mechanism, and the synchronous movements of opposed pistons are ensured by the synchronous chain. With the exception of accomplishing energy conversion, the reciprocating movement can accomplish air-exchange by combining the position with the air ports. The piston which controls the opening and closing of intake air ports is defined as the intake piston and the piston which controls the opening and closing of the exhaust air ports is defined as the exhaust piston. When the distance between two pistons is minimized, it is defined as the inner dead center (IDC); when the distance between two pistons is maximized, it is defined as the outer dead center (ODC). The structure parameters are shown in Table 1.

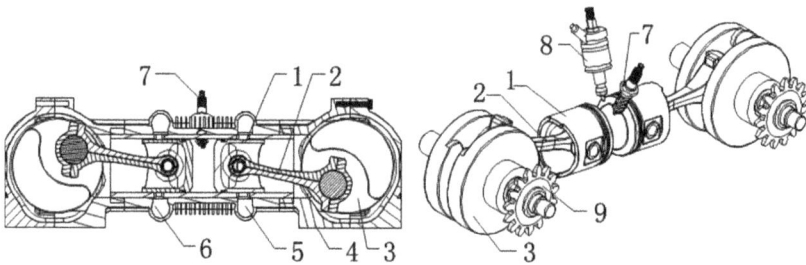

1-Piston, 2-Connecting rod, 3-Crankshaft, 4-Engine-block, 5-Intake port, 6-Exhaust port, 7-Spark plug, 8-Injector, 9-Chain wheel

Figure 1. OP2S-GDI engine.

Table 1. Engine specifications.

Parameters	Unit	Value
Bore	mm	56
Stroke	mm	49.5 (×2)
Connecting rod	mm	82.5
Effective compression Ratio	–	10.5
Engine speed	rpm	6000
Number of intake ports	–	10
Number of exhaust ports	–	10
Intake port height	mm	12
Exhaust port height	mm	14
Intake port circumference ratio	–	0.75
Exhaust port circumference ratio	–	0.6
Intake port radial angle	°	15
Exhaust port radial angle	°	0
Power	kW	15
Fuel consumption rate	g/kW·h	276

The exhaust port opens first before ODC and a blow down discharge process commences. The discharge period up to the time of the scavenging port opening is called the free exhaust period. The intake port opens also before ODC when the cylinder pressure slightly exceeds the scavenging pressure. When the cylinder pressure is less than the scavenging pressure fresh air enters the cylinder and the scavenging process starts. The intake port closes after the exhaust port closes, since the flow towards the intake port occurs continuously, and additional fresh air is obtained. The fuel injection starts after the exhaust port closes, which can avoid fuel short circuits. The spark plug fires before IDC, igniting the compressed air-fuel mixture which produces a powerful expansion of the vapor. On the expansion stoke, the opposed piston moves to ODC and the exhaust port is opened first.

2.2. In-Cylinder Air flow Organization

The intake chamber structure affects scavenging flow resistance, initial swirl, tumble level and air motion velocity. By the design of a non-uniform intake chamber structure, the velocity and flow rate of intake ports away from the inlet are decreased because of wall friction and structural mutation. With respect to intake ports close to the inlet, as flow momentum is small, the velocity and mass flow are bigger. On these ground, the simulation studied uniform and non-uniform inlets, respectively, with the same radial angle of 15° and flat pistons and pit pistons with the same compression ratio, as shown in Figure 2.

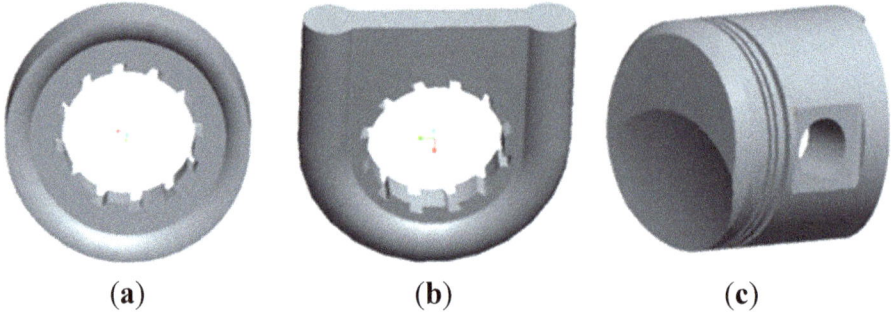

Figure 2. Intake chamber schemes and piston crown. (**a**) Uniform; (**b**) non-uniform; (**c**) pit piston.

As shown in Figure 3, for the uniflow scavenging system OP2S engine, in-cylinder swirl is formed by intake port radial angle and tumble is formed by the non-uniform intake chamber and pit piston. In addition, the pit piston scheme is used to produce squish around IDC.

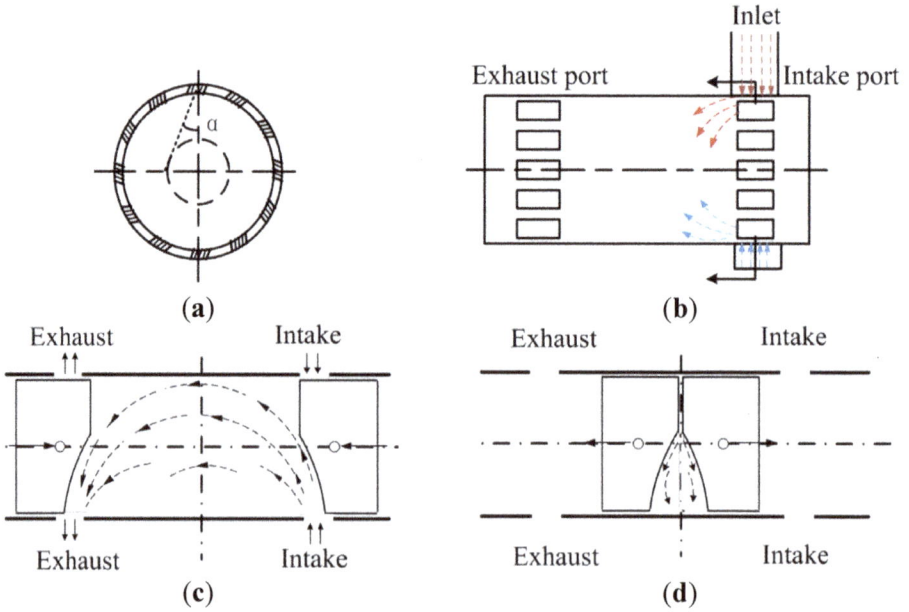

Figure 3. In-cylinder air flow organization. (**a**) Swirl organization by intake port radial angle; (**b**) tumble organization by non-uniform intake; (**c**) tumble organization by pit piston; (**d**) squish organization by pit piston.

3. Theoretic-Interpretative 0D Model

The current paper deals with the development of the 0D parametric model for qualitatively predicting how intake chamber schemes and piston crown affect the tumble, swirl and squish velocity value in the scavenging process and compression strokes, as shown in Figure 4.

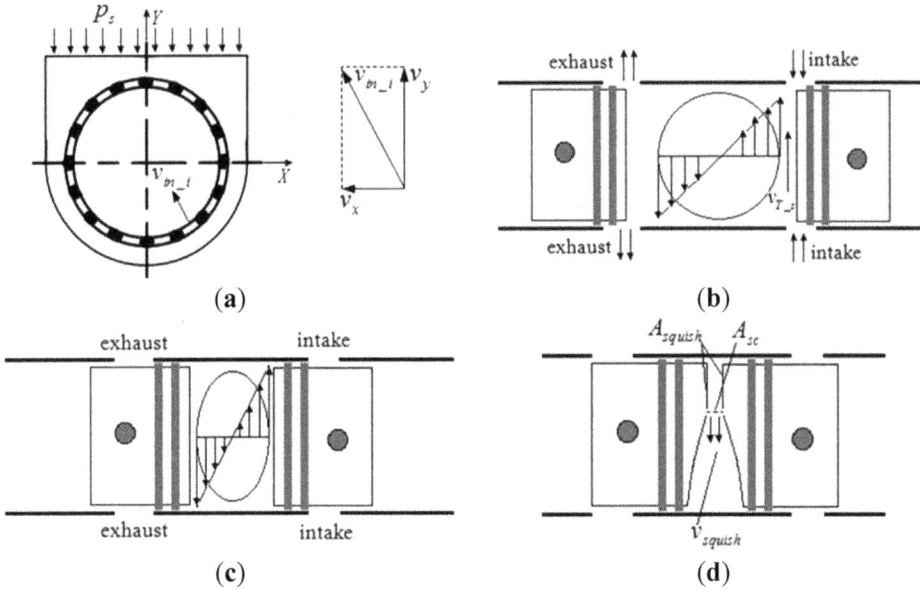

Figure 4. Sketch of in-cylinder air flow. (**a**) Intake chamber and intake velocity vector; (**b**) tumble velocity vector distribution in scavenging process; (**c**) tumble velocity vector distribute in compression process; (**d**) sketch of the squish motion.

3.1. Tumble Velocity of Non-Uniform Scavenging Process

Under the hypothesis of steady flow conditions through the intake ports in the scavenging process, the expression for air mass flow can be written as the following form:

$$\frac{dm_s}{d\phi} = \frac{\mu_s F_s}{6n}\sqrt{\frac{2gk}{k-1}} \cdot \frac{p_s}{\sqrt{RT}} \cdot \sqrt{\left(\frac{p_z}{p_s}\right)^{\frac{2}{k}} - \left(\frac{p_z}{p_s}\right)^{\frac{k+1}{k}}} \tag{1}$$

If the instantaneous flow velocity through the intake ports is v_{in} in the scavenging process, the expression volume mass flow can be written as the following form:

$$\frac{dV_s}{d\phi} = v_{in} \cdot F_s \tag{2}$$

313

By Equation (2), Equation (1) can be written as the following form:

$$v_{in} = \frac{\mu_s}{6n \cdot \rho} \sqrt{\frac{2gk}{k-1}} \cdot \frac{p_s}{\sqrt{RT}} \cdot \sqrt{\left(\frac{p_z}{p_s}\right)^{\frac{2}{k}} - \left(\frac{p_z}{p_s}\right)^{\frac{k+1}{k}}} \tag{3}$$

Because of the non-uniform scavenging process, every intake port has a different instantaneous flow velocity v_{in_i}, as shown in Figure 4a. The instantaneous flow velocity v_{in} can be seen as the mean velocity of every intake ports' instantaneous flow velocity.

If every intake ports' instantaneous flow velocity v_{in_i} directed along the cylinder radius and the number of ports is m, then v_{in_i} is the vectorial sum of the X axial component $v_{in\_i\_x}$ and the Y axial component $v_{in\_i\_y}$, as shown in Figure 4a. The sum of the X axial component v_x is zero but the sum of the Y axial component v_y is the tumble velocity v_{T_s} in the scavenging process, as shown in Figure 4b.

$$v_y = \sum_{i=1}^{m} v_{in\_i\_y} = v_{T\_s} \tag{4}$$

The more the degree of non-uniform scavenging process, the more the tumble velocity v_{T_s}. These considerations can be summed up by introducing the parameter K_1 proportional to the instantaneous flow velocity v_{in}. Equation (4) could be rearranged as reported below, considering the tumble velocity v_{T_s} in the scavenging process:

$$v_{T\_s} \propto K_1 \cdot v_{in} \tag{5}$$

The final expression of the tumble velocity v_{T_s} using Equations (3) and (5) becomes:

$$v_{T\_s} \propto K_1 \cdot \frac{\mu_s}{6n \cdot \rho} \sqrt{\frac{2gk}{k-1}} \cdot \frac{p_s}{\sqrt{RT}} \cdot \sqrt{\left(\frac{p_z}{p_s}\right)^{\frac{2}{k}} - \left(\frac{p_z}{p_s}\right)^{\frac{k+1}{k}}} \tag{6}$$

The tumble velocity v_{T_s} in the scavenging process is related to the intake port flow coefficient μ_s, so the in-cylinder tumble ratio increases first and then decreases.

3.2. The Tumble Velocity in Compression Strokes

During the compression stroke vortex deformation due to the reduction of the distance between the opposed pistons and the fluid vortex inertia I is reduced too so there is an acceleration of the vortex rotational speed ω_T—this is called "spin-up phase". This causes an increase of the Y axial component of the intake port instantaneous flow velocity and thus of the tumble velocity, as shown in Figure 4c. In

314

the compression process, the tumble velocity component is dissipated and turbulence is generated. The vortex angular speed equation is:

$$\omega_T \propto \frac{v_{T\_c}}{r_T} \tag{7}$$

At IPC the inertia angular momentum is:

$$J_{IPC} = I_{IPC} \cdot \frac{v_{T\_IPC}}{h_{IPC}} \propto I_c \cdot \frac{v_{T\_c}}{h_c} \tag{8}$$

The vortex inertia is defined as the fluid mass of the equivalent rotating solid body multiplied by the h squared, so it is possible to deduce the expression for the tumble velocity v_{T_c} in the compression process:

$$v_{T\_c} \propto v_{T\_IPC} \cdot \frac{I_{IPC}}{I_c} \cdot \frac{h_c}{h_{IPC}} \propto v_{T\_IPC} \cdot \frac{h_{IPC}^2}{h_c^2} \cdot \frac{h_c}{h_{IPC}} \propto v_{T\_IPC} \cdot \frac{h_{IPC}}{h_c} \tag{9}$$

The ratio of h_{IPC} to h_c lets us assess the degree of deformation of the tumble vortex during the compression stroke: the larger the ratio of h_{IPC} to h_c is, the larger the ratio of h_{IPC} is and the larger the degree of distortion of the tumble vortex is. The tumble ratio increases in the initial stage of the compression stroke and decreases before IDC because of the distortion of the tumble vortex.

3.3. Squish Velocity around IDC

Looking at Figure 4d, when the opposed pistons are approaching each other, fresh mixture is pushing forward the combustion chamber middle. It is possible to apply the mass conservation law through the squish outlet area A_{sc}:

$$V_{squish} \cdot A_{sc} \propto \frac{dh_s}{dt} \cdot A_{squish} \tag{10}$$

The ratio of the squish compression area to the cylinder cross section area is called K_2 and the relative velocity of the opposed pistons dh_s/dt before IDC is replaced by V_P. So the expression of the squish velocity is as follows:

$$V_{squish} \propto K_2 \cdot V_P \tag{11}$$

The squish velocity before IDC can be assessed by the ratio of squish compression area to cylinder cross section area K_2 and the relative velocity of the opposed pistons V_P.

4. Engine Modeling

4.1. CFD Model and Setup

The AVL-Fire software was used to build the CFD model for the working process simulation. Fame Engine is used to generate the moving meshes of the cylinder by defining moving selection, buffer selection, interpolation selection and the relative motion rules of the opposed pistons. Intake and exhaust chamber are generated by the non-moving meshes which are refined near the intake and exhaust ports, in order to accurately capture the significant flow gradients, as shown in Figure 5. The dynamic piston motion mesh of the intake and exhaust strokes has been treated according to realistic opposed piston motion rules. The scavenging calculation is from exhaust port opening (EPO) to intake ports closing (IPC), while the in-cylinder working process is from IPC to EPO. Mesh movement includes three parts: intake and exhaust piston and cylinder to simulate the gas motion during the entire working process. The $K - \varepsilon$ model was used in the calculation of turbulence.

The base engine for CFD analysis is the same for all three scavenging system configurations in Table 2. The mesh size is between 225,000 cells and 245,000 cells, varying slightly between the scavenging system configurations, including flat piston uniform scavenging chamber Scheme 1, flat piston non-uniform scavenging chamber Scheme 2 and pit piston non-uniform scavenging chamber Scheme 3.

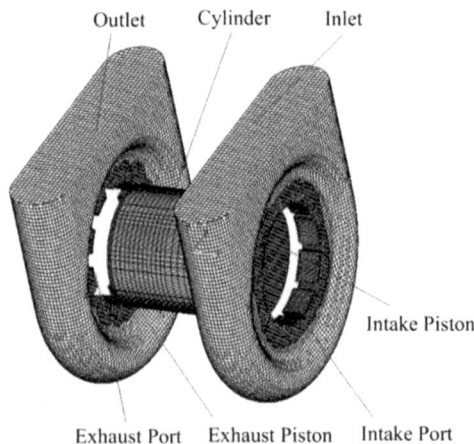

Figure 5. Computational grid.

Table 2. The various scavenging system configurations.

Scheme	Arrangement	Comments
1		Flat piston and uniform scavenging chamber
2		Flat piston and non-uniform scavenging chamber
3		Pit piston and non-uniform scavenging chamber

4.2. Boundary and Initial Conditions

The boundary conditions were chosen to reflect the physical conditions in the validation model and the prototype engine. A constant pressure boundary condition is used for both intake and exhaust ports. Mean scavenging pressure is taken as 1.2 bar and mean exhaust receiver pressure is taken as 1 bar. Frictional effects at the walls are not taken into account, *i.e.*, the smooth wall option is used for turbulent flow boundary conditions. The initial conditions in the cylinder for every scheme are extracted from the GT-Power software simulation. The flow field is initialized by specifying the temperature, pressure and turbulence intensity. By performance prediction, the initial pressure and temperature in the cylinder are computed in a scheme of 15 kW at an engine speed of 6000 rpm, which are the initial conditions for CFD. Initial temperatures of cylinder, intake chamber and exhaust chamber are given a value of 788 K, 322 K, and 634 K, respectively.

4.3. Model Validation

In order to investigate the mesh independence, two additional meshes are tested for Scheme 2. One with approximately 150,000 cells denoted "coarse" and one with approximately 237,000 cells denoted "medium". The reference mesh of 304,000 cells is referred to as "fine" [27]. The effect of mesh resolution is presented by comparing the radial profiles of the tangential velocity as shown in Figure 6. The profiles are sampled at the cylinder center cross section when the opposed piston is at the ODC. The comparison shows that the velocity profiles show good agreement and the medium mesh can be considered as the practical mesh.

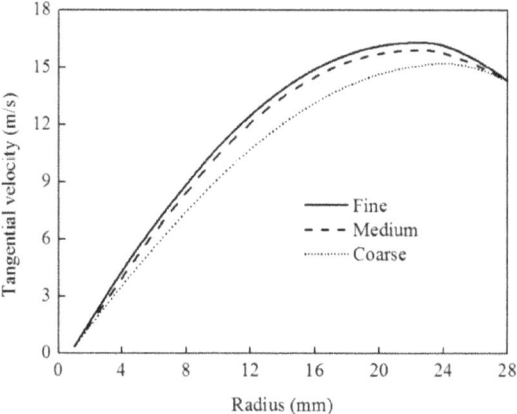

Figure 6. Comparison between different mesh resolutions.

A simulation running at 20% load of 1200 rpm is performed and a series of comparison validations of the one-dimensional in-cylinder working process, three-dimensional scavenging process and motored conditions were conducted, as shown in Figure 7. The simulation results agree with the experimental results in the scavenging process which indicates that the parameters are reasonably selected.

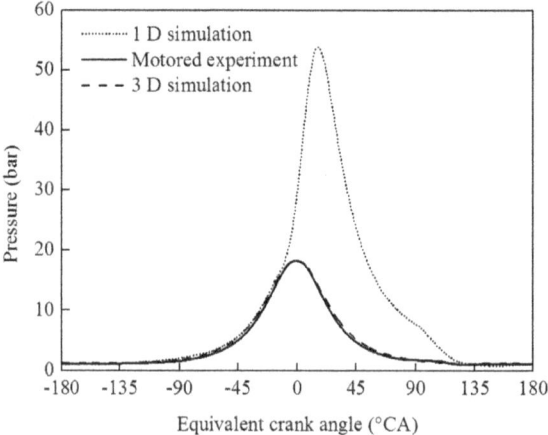

Figure 7. Comparison between in-cylinder pressures.

5. Simulation Results and Discussion

5.1. In-Cylinder Air Motion

In-cylinder flow is divided into a swirl component around the cylinder axis and a tumble component around the vertical cylinder axis, respectively. Figure 8 shows that tumble ratio changes using different scavenging system configurations at an engine speed of 6000 rpm. As shown in Figure 8a, for Schemes 1 and 2, the result shows that tumble is hardly possible in the uniform intake chamber scheme. In contrast, the in-cylinder tumble ratio corresponding to non-uniform intake chamber scheme shows a rising tendency throughout the initial phases of scavenging process and compression process, reaching its maximum value at 200 °CA and 310 °CA, respectively, while in the later phases of scavenging and compression process, the tumble ratio decreases due to tumble vortex crushing. As shown in Figure 8b for Schemes 2 and 3, two approaches can be adopted to organize tumble by the non-uniform intake chamber scheme and using different piston crown shapes. This simulation carried out a study on the flat piston and pit piston configurations with both using the same scavenging method. In-cylinder tumble changes remain similar in either flat and pit piston and the tumble ratio in Scheme 2 exceeds that in Scheme 1. Due to an earlier opening of exhaust ports corresponding to pit piston and the flow-guiding effect of the exhaust piston pit design, in-cylinder tumble ratio in Scheme 2 is bigger during the initial exhaust phase in which non-uniform exhaust plays the main role in tumble organization. As scavenging proceeds, the non-uniform intake chamber and flow-guiding effect generated by the intake piston pit intensify the tumble and tumble ratio increases accordingly. During the post-scavenging period after ODC, as in-cylinder air velocity reduces the in-cylinder tumble drops, reaching its minimum value at intake port closing. Due to the inertia effect caused by air motion and the directional organization of opposed pistons on in-cylinder flow, in-cylinder tumble ratio increases. As a result, tumble ratio in Scheme 2 exceeds that in Scheme 1. The result also shows that the tumble ratio decreases due to large-scale directional flow crushing in the compression process and tends to be uniform near the inner dead point.

In-cylinder swirl in uniflow scavenging is defined as an in-cylinder flow round the cylinder axis directionally mainly through intake port radial angle. Figure 9a shows that swirl the results of Schemes 1 and 2 show the influence of different intake chamber schemes on swirl ratio at the same intake radial angle: the swirl ratio organized by uniform intake chamber is obviously higher than that produced by a non-uniform intake chamber. The simulation study also proves that a non-uniform intake chamber with a coordinated intake port radial angle can produce inclined-axis tumble. The effect on in-cylinder swirl of different piston crown schemes is shown in Schemes 2 and 3 by the same scavenging chamber method. The result shows that in-cylinder swirl changes tend to be uniform and swirl ratio in Scheme 2 exceeds the value in Scheme 3. Due to an earlier opening of the intake ports for the pit piston, swirl in Scheme 3 forms earlier than in Scheme 2. Meanwhile, the circumferential flow is unevenly distributed, resulting in an increasing tumble momentum and a decreasing swirl momentum. As shown in Figure 9b, the swirl ratio in Scheme 3 is lower than that in Scheme 1 in the scavenging and compression process. After ODC in-cylinder swirl motion slows down at the end of scavenging.

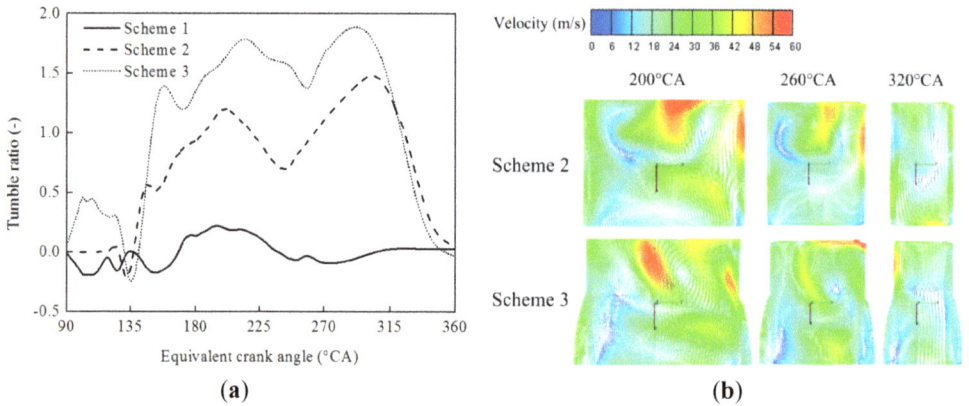

(a)

(b)

Figure 8. Variation of tumble ratio for different scavenging schemes; (**a**) Variation of tumble ratio with equivalent crank angle; (**b**) tumble velocity at different crank angle.

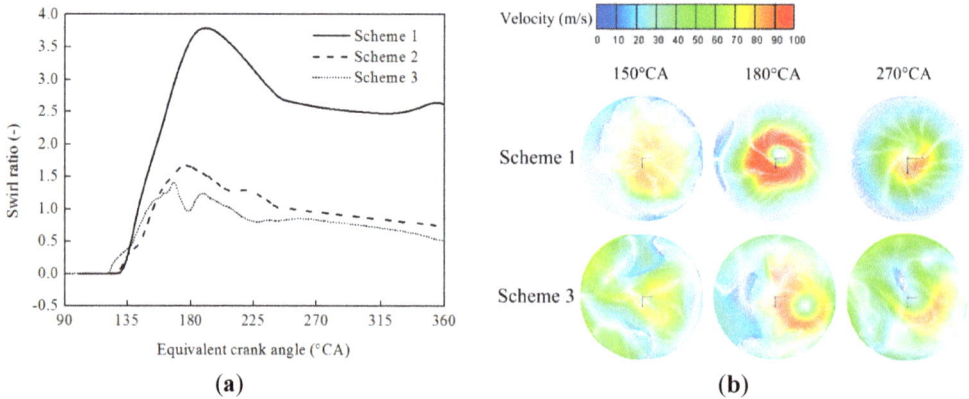

(a)

(b)

Figure 9. Variation of swirl ratio for different scavenging schemes; (**a**) Variation of swirl ratio with equivalent crank angle; (**b**) swirl velocity at different crank angle.

The comparison of squish around IDC for different opposed pistons characterized is shown in Figure 10. Figure 10a shows that the effect of flat piston on in-cylinder radial squish around IDC is small with a radial velocity approaching to zero, visible in Scheme 2. In contrast, Scheme 3 pit piston affects the squish around IDC significantly and the radial velocity is symmetric with respect to IDC. The SOC is at 340 °CA and Figure 10b shows that at SOC the piston crown with pit has a higher in-cylinder radial velocity.

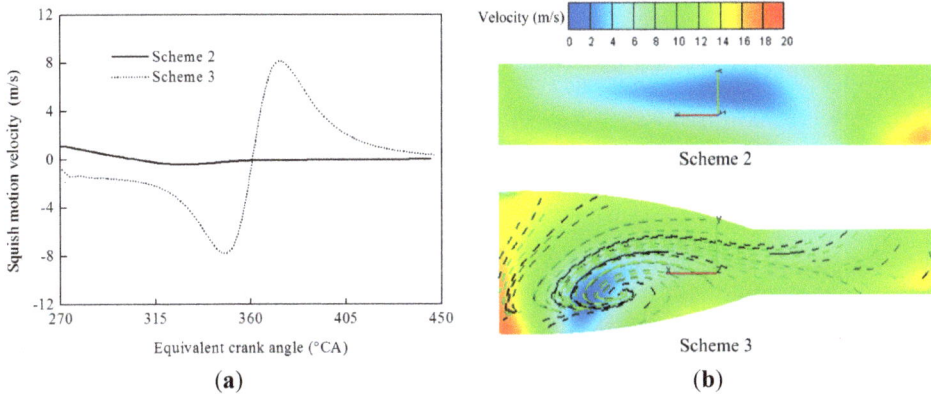

Figure 10. Variation of squish velocity for different piston crown shapes; (a) Variation of squish velocity with equivalent crank angle; (b) squish velocity 20 °CA before IDC.

5.2. Turbulence Kinetic Energy

Figure 11 shows that the variation of mean turbulence kinetic energy (TKE) for different scavenging system configurations at an engine speed of 6000 rpm. During the initial phase of the scavenging process, TKE is higher due to a high-frequency turbulence and is maximized around 190 °CA for the non-uniform intake chamber scheme. Though showing a downward trend in the compression process, the TKE decrease rate is small. A relatively small peak value is shown at 340 °CA, owing to high-frequency turbulence produced by tumble crushing, swirl decrease and sustained tumble. In contrast, due to a significant difference between the in-cylinder tumble and swirl components when a uniform intake chamber scheme is adopted, swirl continuously weakens by compression thereupon resulting in a decrease in the TKE level. TKE using non-uniform intake methods is higher than that obtained by uniform intake methods.

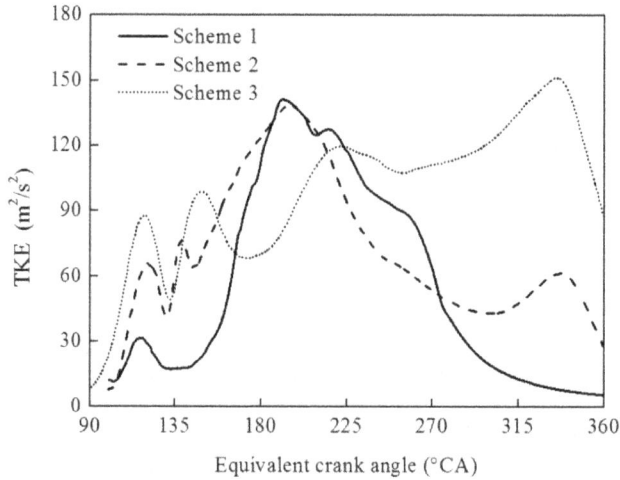

Figure 11. Variation of TKE for different scavenging schemes.

The simulation of Schemes 2 and 3 studied flat and pit pistons using the same intake chamber methods for both. The result shows that in-cylinder TKE intensifies in the scavenging process but TKE is lower in the compression process due to tumble shearing and dissipation. At SOC Scheme 3 has the highest TKE, which is 2.5 times higher than in Scheme 2. In the initial scavenging process, the in-cylinder TKE changes of the two schemes tend to be uniform and the TKE value in Scheme 3 exceeds that in Scheme 2, yet the changes in the two schemes diversify after ODC: variance in TKE is small since Scheme 3 is more advantageous than Scheme 2 due to a higher tumble ratio in Scheme 3. The simulation result indicates that during the initial and meta-phases of compression, the crushing and organization of in-cylinder tumble maintain a relatively high-level TKE. In contrast, Scheme 3 promotes the formation of internal tumble and swirl, produces a higher TKE in compression and keeps a rising tendency where squish is organized around IDC. Therefore, it can be concluded that continuous in-cylinder tumble and swirl form a relatively high TKE and an efficient squish motion further enhances TKE.

5.3. Scavenging Process

Figure 12 shows that the different scavenging system configurations affect scavenging efficiency and delivery ratio at 6000 rpm.

Compared with the uniform intake chamber scheme, the scavenging efficiency of the non-uniform intake chamber scheme is lower. During scavenging, the uniform intake chamber scheme can organize a higher swirl and produce a relatively high scavenging efficiency as well by virtue of an extremely low tumble level which avoids the blending of fresh charge and burnt gas. Because of the pit piston structure,

322

Scheme 3 allows an advanced and durable scavenging process compared with Scheme 2. The simulation result also shows that due to the characteristics of flat pistons which have a small influence on swirl, and the in-cylinder tumble is lower. As a result, the scavenging efficiency of Scheme 2 is higher than that of Scheme 3. Conversely, the pit piston design on the one hand aggravates non-uniform intake, but also hinders in-cylinder swirl organization resulting in a relatively lower swirl ratio. Meanwhile, Scheme 3 can promote in-cylinder tumble organization whereby a relatively higher tumble ratio is produced, thus decreasing scavenging efficiency. With respect to scavenging efficiency, the pit piston design is inferior to the flat piston one but similar in delivery ratio. During the meta-phase of the scavenging process, a larger tumble ratio hinders air flow and the delivery ratio is lower, while the delivery ratios of the two schemes tend to be uniform during the later scavenging process.

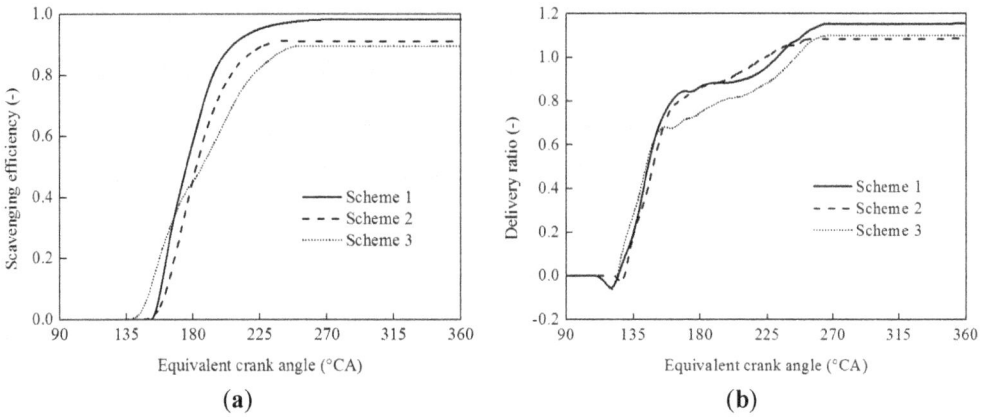

Figure 12. Variation of scavenging process for different scavenging schemes; (a) Scavenging efficiency; (b) delivery ratio.

Delivery ratio, trapping efficiency and scavenging efficiency were usually employed as evaluation index on the two-stroke scavenging system [28]:
    The delivery ratio:

$$l_0 = \frac{\text{mass of delivered air (or mixture) per cyclerefer}}{\text{reference mass}} \tag{12}$$

The reference mass is defined as displaced volume $\times$ ambient air (or mixture). Ambient air (or mixture) density is determined at atmospheric conditions or at intake conditions.
    The trapping efficiency:

$$\eta_{tr} = \frac{\text{mass of delivered air (or mixture) retained}}{\text{mass of delivered air (or mixture)}} \tag{13}$$

The trapping efficiency indicates what fraction of the air (or mixture) supplied to the cylinder is retained in the cylinder.

The scavenging efficiency:

$$\eta_{sc} = \frac{\text{mass of delivered air (or mixture) retained}}{\text{mass of trapped cylinder charge}} \tag{14}$$

The scavenging efficiency indicates to what extent the residual gases in the cylinder have been replaced with fresh air. When the reference mass in the definition of delivery ratio is the trapped cylinder mass (or closely approximated by it) then:

$$\eta_{sc} = l_0 \cdot \eta_{tr} \tag{15}$$

For the perfect scavenging model, trapping and scavenging efficiency vary with delivery ratio as follows:

$$\begin{array}{ll} \eta_{tr} = 1 & \eta_{sc} = l_0 \quad \text{for } l_0 \leq 1 \\ \eta_{tr} = 1/l_0 & \eta_{sc} = 1 \quad \text{for } l_0 > 1 \end{array} \tag{16}$$

For the perfect mixing model, trapping and scavenging efficiency vary with delivery ratio as follows:

$$\eta_{tr} = \frac{1}{l_0} \cdot \left(1 - e^{-l_0} \right)$$
$$\eta_{sc} = 1 - e^{-l_0} \tag{17}$$

As shown in Figure 13, by comparing the scavenging data of the three schemes with the results in reference [28], scavenging efficiency and delivery ratio of three schemes fall in between perfect scavenging and perfect mixing.

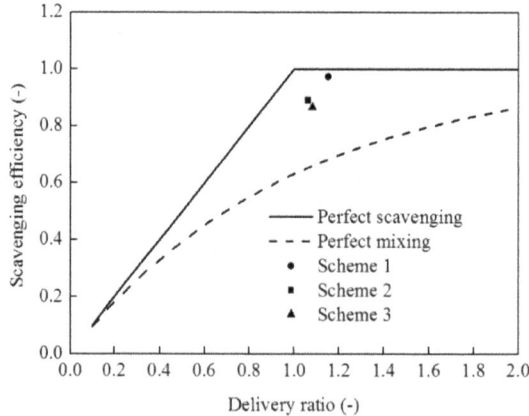

Figure 13. Comparison of the three scavenging schemes.

6. Conclusions

The important conclusions derived from the simulation of in-cylinder air motion in an OP2S engine with different scavenging system configurations can be summarized as follows:

(1) The coincidence of the 3D calculation results with the theoretical-interpretative results of the 0D model manifests the veracity of this approach.

(2) The swirl ratio organized by uniform intake chamber is obviously higher than that obtained by a non-uniform intake chamber, while the non-uniform intake chamber can organize inclined-axis tumble, which increases TKE around IDC.

(3) For the non-uniform intake chamber, the pit piston scheme is more beneficial to tumble vortex formation compared to all other scavenging schemes, with a maximum tumble ratio improvement of about 26% over the flat piston scheme.

(4) At 340 °CA, CFD results show that there is an increase in TKE of about 150% for a pit piston compared to that of a flat piston, which is due to squish effect of the pit piston. The uniform intake chamber scheme is not beneficial to the tumble vortex formation and the TKE is minimum before IDC.

(5) The scavenging efficiency and delivery ratio are higher for the uniform intake chamber scheme. It is concluded that a pit piston non-uniform scavenging chamber is a better choice to have good tumble ratio and TKE, but a flat piston uniform scavenging chamber is a better choice to have good swirl ratio, scavenging efficiency and delivery ratio.

Acknowledgments: The authors gratefully acknowledge the financial support by the foundation research funds of Ministry of Industry and Information Technology of the People's Republic of China.

Author Contributions: Fukang Ma and Changlu Zhao designed the simulation scheme; Fukang Ma and Zhenfeng Zhao performed the simulations; Zhenfeng Zhao and Shuanlu Zhang analyzed the data; and Fukang Ma and Fujun Zhang contributed to the editing and reviewing of the document.

Conflicts of Interest: The authors declare no conflict of interest.

Nomenclature

0D	zero-dimensional	m	number of intake port
1D	mono-dimensional	n	engine speed
3D	three-dimensional	ODC	outer dead center
A_{squish}	squish compression area	OP2S	opposed-piston two-stroke
CFD	computational fluid dynamics	PIV	particle image velocimetry
dh_s/dt	relative velocity of the opposed pistons	p_z	in-cylinder gas pressure

EPO	exhaust port opening	R	gas constant
F_s	area of intake port in different crank angle	r_T	radius of tumble vortex
GDI	gasoline direct injection	SOC	start of combustion
g	gravitational acceleration	T	gas temperature
h	distance between the opposed pistons	TDC	top dead center
h_{IPC}	half distance between the opposed pistons at IPC	V_p	relative velocity of the opposed pistons before IDC
h_c	half distance between the opposed pistons in compression process	V_{squish}	squish velocity
h_s	instantaneous squish height	v_{in}	instantaneous flow velocity of intake ports
I	fluid vortex inertia	v_{in_i}	instantaneous flow velocity of every intake ports
I_c	fluid vortex inertia in compression process	$v_{in\_i\_x}$	vectorial sum of the X axial component
I_{IPC}	the fluid vortex inertia at IPC	$v_{in\_i\_y}$	vectorial sum of the Y axial component
IDC	inner dead center	v_{T_c}	tumble velocity in compression process
IC	ignition combustion	v_{T_IPC}	the tumble velocity at IPC
IPC	intake ports closing	v_{T_s}	the tumble velocity
J	angular momentum	v_x	the X axial component
J_{IPC}	inertia angular momentum at IPC	v_y	the Y axial component
K_1	proportionality coefficient	ω_T	angular velocity of the equivalent rotating solid body
K_2	ratio of squish compression area to cylinder cross section area	μ_s	intake port flow coefficient
k	adiabatic exponent	ϕ	crank angle

References

1. Pirault, J.P.; Flint, M. *Opposed Piston Engines: Evolution, Use, and Future Applications*; SAE International: Warrendale, PA, USA, 2010.
2. Naik, S.; Johnson, D.; Koszewnik, J. *Practical Applications of Opposed-Piston Engine Technology to Reduce Fuel Consumption and Emissions*; SAE Technical Paper 2013-01-2754; SAE International: Warrendale, PA, USA, 2013.
3. Hofbauer, P. *Opposed Piston Opposed Cylinder (OPOC) Engine for Military Ground Vehicles*; SAE Technical Paper 2005-01-1548; SAE International: Warrendale, PA, USA, 2005.
4. Hirsch, N.R.; Schwarz, E.E.; McGough, M.G. *Advanced Opposed-Piston Two-Stroke Diesel Demonstrator*; SAE Technical Paper 2006-01-0926; SAE International: Warrendale, PA, USA, 2006.

5. Herold, R.E.; Wahl, M.H.; Regner, G. *Thermodynamic Benefits of Opposed-Piston Two-Stroke Engines*; SAE Technical Paper 2011-01-2216; SAE International: Warrendale, PA, USA, 2011.

6. Regner, G.; Herold, R.E.; Wahl, M.H. *The Achates Power Opposed-Piston Two-Stroke Engine: Performance and Emissions Results in a Medium-Duty Application*; SAE Technical Paper 2011-01-2221; SAE International: Warrendale, PA, USA, 2011.

7. Redon, F.; Kalebjian, C.; Kessler, J. *Meeting Stringent 2025 Emissions and Fuel Efficiency Regulations with an Opposed-Piston, Light-Duty Diesel Engine*; SAE Technical Paper 2014-01-1187; SAE International: Warrendale, PA, USA, 2014.

8. Regner, G.; Johnson, D.; Koszewnik, J. *Modernizing the Opposed Piston, Two-Stroke Engine for Clean, Efficient Transportation*; SAE Technical Paper 2013-26-0114; SAE International: Warrendale, PA, USA, 2013.

9. Zhao, F.; Lai, M.C.; Harrington, D.L. Automotive spark-ignited direct-injection gasoline engines. *Prog. Energy Combust. Sci.* **1999**, *25*, 437–562.

10. Arcoumanis, C.; Bae, C.S.; Hu, Z. *Flow and Combustion in a Fourvalve, Spark-Ignition Optical Engines*; SAE Paper 940475; SAE International: Warrendale, PA, USA, 1994.

11. Fan, L.; Reitz, R.D.; Trigui, N. *Intake Flow Simulation and Comparison with PTV Measurements*; SAE Technical Paper 1999-01-0176; SAE International: Warrendale, PA, USA, 1999.

12. Ismail, H.M.; Ng, H.K.; Gan, S. Evaluation of non-premixed combustion and fuel spray models for in-cylinder diesel engine simulation. *Appl. Energy* **2012**, *90*, 271–279.

13. Mikalsen, R.; Roskilly, A.P. A computational study of free-piston diesel engine combustion. *Appl. Energy* **2009**, *86*, 1136–1143.

14. Rakopoulos, C.D.; Kosmadakis, G.M.; Dimaratos, A.M. Investigating the effect of crevice flow on internal combustion engines using a new simple crevice model implemented in a CFD code. *Appl. Energy* **2011**, *88*, 111–126.

15. Jia, M.; Xie, M.Z.; Wang, T.Y. The effect of injection timing and intake valve close timing on performance and emissions of diesel PCCI engine with a full engine cycle simulation. *Appl. Energy* **2011**, *88*, 2967–2975.

16. Micklow, G.J.; Gong, W.D. Intake and in-cylinder flow field modeling of a four valve diesel engine. *Proc. Mech. Int. J. Automob. Eng.* **2007**, *221*, 1425–1440.

17. Nordgren, H.; Hildingsson, L.; Johansson, B. *Comparison between In-Cylinder Measurements, CFD Simulations and Steady Flow Impulse Torque Swirl Meter Measurements*; SAE Technical Paper 2003-01-3147; SAE International: Warrendale, PA, USA, 2003.

18. Sweetland, P.; Reitz, R.D. *Particle Image Velocimetry Measurements in the Piston Bowl of a DI Diesel Engine*; SAE Technical Paper 940283; SAE International: Warrendale, PA, USA, 1994.

19. Gunasekaran, E.J.; Ganesan, V. *Simulation of Fuel-Air Interaction in a Four Stroke Four Valve Direct Injected Spark Ignition Engine*; SAE Technical Paper 2007-01-0153; SAE International: Warrendale, PA, USA, 2007.

20. Rakopoulos, C.D.; Kosmadakis, G.M.; Pariotis, E.G. Investigation of piston bowl geometry and speed effects in a motored HSDI diesel engine using a CFD against a quasi-dimensional model. *Energy Convers Manag.* **2010**, *51*, 470–484.

21. Lin, L.; Shulin, D.; Jin, X. *Effects of Combustion Chamber Geometry on In-Cylinder Air Motion and Performance in DI Diesel Engine*; SAE technical paper 2000-01-0510; SAE International: Warrendale, PA, USA, 2000.

22. Shimoda, M.; Shigemori, M.; Tsuruoka, S. *Effect of Combustion Chamber Configuration on In-Cylinder Air Motion and Combustion Characteristics of DI Diesel Engine*; SAE Technical Paper 850070; SAE International: Warrendale, PA, USA, 1985.

23. Dolak, J.G.; Shi, Y.; Reitz, R.D. *A Computational Investigation of Stepped-Bowl Piston Geometry for a Light Duty Engine Operating at Low Load*; SAE Technical Paper 2010-01-1263; SAE International: Warrendale, PA, USA, 2010.

24. Falfari, S.; Brusiani, F.; Bianchi, G.M. *Assessment of the Influence of Intake Duct Geometrical Parameters on the Tumble Motion Generation in a Small Gasoline Engine*; SAE Technical Paper 2012-32-0095; SAE International: Warrendale, PA, USA, 2012.

25. Falfari, S.; Brusiani, F.; Bianchi, G.M. Numerical analysis of in-cylinder tumble flow structures-parametric 0D model development. *Energy Procedia* **2014**, *45*, 987–996.

26. Ramajo, D.; Zanotti, A.; Nigro, N. Assessment of a zero-dimensional model of tumble in four-valve high performance engine. *Int. J. Numer. Methods Heat Fluid Flow* **2007**, *17*, 770–787.

27. Sigurdsson, E.; Ingvorsen, K.M.; Jensen, M.V. Numerical analysis of the scavenge flow and convective heat transfer in large two-stroke marine diesel engines. *Appl. Energy* **2014**, *123*, 37–46.

28. Taylor, C.F. *The Internal-Combustion Engine in Theory and Practice: Thermodynamics, Fluid Flow, Performance*, 2nd ed.; MIT Press: Cambridge, UK, 1985; Volume 1, p. 584.

Credibility Theory-Based Available Transfer Capability Assessment

Yanan Zheng, Jin Yang, Zhaoguang Hu, Ming Zhou and Gengyin Li

Abstract: Since the development of large scale power grid interconnections and power markets, research on available transfer capability (ATC) has attracted great attention. The challenges for accurate assessment of ATC originate from the numerous uncertainties in electricity generation, transmission, distribution and utilization sectors. Power system uncertainties can be mainly described as two types: randomness and fuzziness. However, the traditional transmission reliability margin (TRM) approach only considers randomness. Based on credibility theory, this paper firstly built models of generators, transmission lines and loads according to their features of both randomness and fuzziness. Then a random fuzzy simulation is applied, along with a novel method proposed for ATC assessment, in which both randomness and fuzziness are considered. The bootstrap method and multi-core parallel computing technique are introduced to enhance the processing speed. By implementing simulation for the IEEE-30-bus system and a real-life system located in Northwest China, the viability of the models and the proposed method is verified.

Reprinted from *Energies*. Cite as: Zheng, Y.; Yang, J.; Hu, Z.; Zhou, M.; Li, G. Credibility Theory-Based Available Transfer Capability Assessment. *Energies* **2015**, *8*, 6059–6078.

1. Introduction

Available transfer capability (ATC) expresses a measure of transfer capability remaining in a physical transmission network for further commercial activity over and above already committed uses [1]. It is not only an important technical index for measuring the stability margin of the power grid, but also a useful tool for guiding transactions and market planning in the power market. Therefore it is very important to correctly assess ATC between different areas. The modern power system is a large scale dynamic system. As power systems have grown, their operation has become more complex with the introduction of more outside interferences. As a result, uncertainty factors are becoming prevalent in the assessment of ATC. Moreover, power industry reforms necessarily lead to an increasing number of market participants, which greatly changes the trade mode and operation control. These also produce more uncertainties in the ATC assessment. Hence how to correctly describe and fully consider uncertainties is the key issue of ATC assessment [2].

The uncertainty represents the unknown state of the future. The power system uncertainties mainly come from equipment outages, load changes, operation modes and so on. The concept of transmission reliability margin (TRM) was introduced by the North American Electric Reliability Council (NERC) in 1996 to consider the impacts of these uncertainties on ATC. It represents the amount of transmission transfer capability needed to ensure that the interconnected transmission network is secure under a reasonable range of uncertainties under the system conditions. Currently two approaches are usually used to deal with TRM [3,4]: (1) take a fixed percentage of the total transfer capability (TTC), such as 4% of TTC or reduce the limit of equipment parameters by a certain percent. This approach is easy to carry out but it is difficult to choose a reasonable percentage, and a rough result is achieved; (2) use stochastic methods [5,6] whereby according to the random distributions of the uncertainties, the ATC calculations are repeated using different uncertainty parameters, then the ATC distribution can be obtained, and TRM is the difference between the maximum value and the expected value of this ATC distribution. This type of approach only considers the power system randomness. However, power systems actually contain two types of uncertainties—randomness and fuzziness—and there a lot of uncertainties in power systems with both random and fuzzy features. For example, the failure of a generator is random, while its available output is fuzzy; similarly a transmission line failure is a random event, and its failure rate which is affected by many factors that embody fuzziness; the probability distribution of a load is random, but its distribution variance is fuzzy. Both randomness and fuzziness are important characteristics of the uncertainties in power systems, and each has great effects. Therefore, how to comprehensively describe these uncertainty features is an important potential difficulty in ATC assessment, but with little achievement reported in the state-of-the-art literature.

Although research on comprehensive evaluations that consider both randomness and fuzziness simultaneously has been constantly tried, a series of problems that cannot be easily overcome have made progress in this field very slow. The biggest obstacle is that there was no complete theory to support it in the field of basic mathematics until the credibility measure was proposed by Liu in 2002 and then the credibility theory was established [7–11]. The theory puts forward the concept of the opportunity space and mixed variables, allowing a comprehensive evaluation method for randomness and fuzziness to be set up.

Now the credibility theory has been applied to power systems [12–14]. According to the two-fold uncertainty combining randomness and fuzziness in power system operations, a novel operation risk assessment method based on credibility theory is presented; the expected value and the average chance measure of random fuzzy variables was used to build on index of operation risk in [12]. Considering both the randomness of force outage rates and the fuzziness of hydro energy, a comprehensive model of hydro-thermal generator maintenance scheduling based on credibility theory was established in [13]. In [14] a new methodological framework with chance-constrained random fuzzy programming, which evaluated the randomness of the forecasted load, the fuzziness of rivals' biddings strategies and price-demand elasticity, was developed for building optimal bidding strategies for generation companies with uncertainty theory-based risk management taken into account was presented and a hybrid intelligent algorithm with combined random fuzzy simulation, artificial neural network and genetic algorithm was proposed to solve the random fuzzy programming problem.

However, very limited work has considered both the randomness and fuzziness that actually coexist in the assessment of ATC. A chance-constrained random fuzzy programming method to consider TRM was introduced in [15], but it still treated the important factors which affected the reliable of transmission as random variables or ones that obeyed a normal distribution. Comprehensive research which considers both randomness and fuzziness at the same time has not been done. This paper builds for the first time comprehensive models of generators, transmission lines and loads according to their randomness and fuzziness features. No stochastic or possibilistic programming can single-handedly deal with these complex uncertainty models, therefore a random fuzzy simulation is also introduced for the first time to the assessment of ATC. A novel assessment method is proposed based on credibility theory. The bootstrap method and multi-core parallel computing technique are used to enhance the speed of processing. As the main contribution of this paper, the concept and approach have been tested on the IEEE-30-bus system and a real-life system.

The paper is organized as follows: the basic concepts of credibility theory with some application examples concerning power systems are introduced in Section 2, which includes four axioms, credibility measure, random fuzzy variables and their expected values and variances In Section 3. some subjects such as random fuzzy modeling of the main uncertainty factors, the ATC calculation model and the assessment indices, bootstrap method and multi-core parallel computing technique, and the ATC assessment of random fuzzy simulation are presented. A numerical simulation with the IEEE 30 bus system and an actual Chinese system aregiven to demonstrate the feasibility of the proposed models and methods in Section 4, followed by our conclusions in Section 5.

2. Credibility Theory

There are a large amount of uncertainties in the real world. They could be in results, categories, or their combinations. The result uncertainties are represented by probability methods. The corresponding theory and methods have already been well developed, whereas some other uncertainties affected by the weather, the environment and operational conditions are changing from time to time under real-time conditions. In other words, they may vary significantly under different conditions, so it is difficult to simply quantify the co-relationship among these uncertainties by randomness. Fortunately, they can be described using fuzzy words based on experienced operators' judgment (such as "most adverse", "fairly adverse" or "less adverse", *etc.*), which can be modeled using a fuzzy membership function [9,16–19]. Fuzzy theory and its methods have been popular yet challenging for many years. In 1975 Kaufmann proposed the concept of fuzzy variables for the first time [20]. Then in 1978 Zadeh put forward the possibility theory which depicted the possibility of occurrence of fuzzy events [21], and laid an important foundation for the development of fuzzy theory, although at that time incompatible counter-examples in fuzzy theory still existed, and fuzzy theory had not been verified axiomatically. The axiomatic system of fuzzy theory was finally verified by the Chinese mathematician Baoding Liu in the 21st century. Since then credibility theory was developed and now offers a strict theoretical foundation for modeling and solving problems with randomness and fuzziness.

2.1. Basic Concept

The following four axioms are the base of fuzzy theory. Let Θ be a non-empty set and ϕ an empty set. $P(\Theta)$ is the power set of Θ, $\wedge$ is the minimum operator and sup is the supremum operator. The possibility measure of the event A is defined as $P_{os}\{A\}$:

Axiom 1: $P_{os}\{\Theta\} = 1$.
Axiom 2: $P_{os}\{\phi\} = 0$.
Axiom 3: for any set $\{A_i\}$ in $P(\Theta)$.
$P_{os}\{\cup_i A_i\} = \sup_i P_{os}\{A_i\}$.
Axiom 4: if Θ_l is a non-empty set, $Posi\{\cdot\}$, $i=1,2,\ldots,n$ meets the first three axioms, and $\Theta=\Theta_1\times\Theta_2\times\ldots\times\Theta_n$, for any set $A\in P(\Theta)$, $P_{os}\{A\} = \sup_{(\theta_1,\theta_2,\cdots,\theta_n)\in A} P_{os1}\{\theta_1\} \wedge$
$P_{os2}\{\theta_2\} \wedge \cdots \wedge P_{osn}\{\theta_n\}$.

Definition 1: if P_{OS} meets the first three axioms, P_{OS} is defined as the possibility measure. $(\Theta, P(\Theta), P_{OS})$ is a possibility space. If A^c is the complement of A, the necessity measure N_{ec} is defined as $N_{ec}\{A\}=1 - P_{os}\{A^c\}$. Obviously, P_{OS} and N_{ec} are one pair of dual measures, so the credibility measure is defined as follows:

$$C_r\{A\} = \frac{1}{2}(P_{os}\{A\} + N_{ec}\{A\}) \tag{1}$$

When the possibility measure of a fuzzy event is 1, the event may not exist, but when the necessity measure of a fuzzy event is 0, this event may still exist. However, the fuzzy event must exist if its credibility is 1, otherwise it does not exist if its credibility is 0. The basis of credibility theory is the definition of a credibility measure, whose position equals one of the probability measures in probability theory.

Definition 2: A fuzzy variable ξ is a function from the possibility space $(\Theta, P(\Theta), P_{OS})$ to the real line R. The triangle fuzzy variable and the trapezoidal fuzzy variable are commonly used ones.

Definition 3: Let ξ be a fuzzy variable in a possibility space $(\Theta, P(\Theta), P_{OS})$, the membership function of ξ is:

$$\mu(x) = P_{os}\{\theta \in \Theta | \xi(\theta) = x\}, \quad x \in R \tag{2}$$

Through the above definition the following inversion formula can be proved: for any set B of real numbers, we have:

$$C_r\{\xi \in B\} = (\sup_{x \in B} \mu(x) + 1 - \sup_{x \in B^c} \mu(x))/2 \tag{3}$$

Example 1: The available output of a generator has an uncertain feature. Especially when green energy forms, such as hydropower, wind power, solar power and so on are vigorously developed, its unreliable forecasting characteristics and the weather dependency will increase the uncertainties. It is very difficult to get the exact output of a generator, but we can roughly use a triangle fuzzy variable ξ_G to represent its available output. Suppose the available output $\xi_G = (35, 50, 55)$ MW, we have the credibility measure of ξ_G:

$$C_r\{\xi_G \geq r\} = \begin{cases} 1, & r \leq 35 \\ \dfrac{65 - r}{30}, & 35 < r \leq 50 \\ \dfrac{55 - r}{10}, & 50 < r \leq 55 \\ 0, & 55 < r \end{cases}$$

Definition 4: Let ξ be a fuzzy variable. The expected value of ξ is defined as:

$$E_{fuz}[\xi] = \int_0^\infty C_r\{\xi \geq r\}\, dr - \int_{-\infty}^0 C_r\{\xi \leq r\}\, dr \tag{4}$$

To avoid possible occurrence of $\infty-\infty$, at least one of the two integrals is limited in the above Equation (4).

Example 2: The expected value of a triangle fuzzy variable (a, b, c) can be obtained:

$$E_{fuz}[\xi] = (a + 2b + c)/4$$

The expected value of a trapezoidal variable (a, b, c, d) is:

$$E_{fuz}[\xi] = (a + b + c + d)/4$$

2.2. Random Fuzzy Variable

Definition 5: A random fuzzy variable ε is a function from the possibility space $(\Theta, P(\Theta), P_{OS})$ to the set of random variables.

Example 3: There are two states of a generator: on state, and off fault state. They are random, while its available output is fuzzy, so the state of a generator can be represented using a two-point distribution random fuzzy variable ε. For instance, it can be represented as $P_{pro,G}(\varepsilon = \xi_G) = 0.99$, $P_{pro,G}(\varepsilon = 0) = 0.01$. ξ_G is a triangle fuzzy variable as defined in Example 1.

Definition 6: Let ε be a random fuzzy variable, the expected value of ε is defined as:

$$
\begin{aligned}
E_{pro-fuz}[\varepsilon] \quad &= \int_0^\infty C_r \left\{ \theta \in \Theta \,|\, E[\varepsilon(\theta)] \geq r \right\} dr \\
&- \int_{-\infty}^0 C_r \left\{ \theta \in \Theta \,|\, E[\varepsilon(\theta)] \leq r \right\} dr
\end{aligned}
\tag{5}
$$

Also to avoid the situation of $\infty - \infty$, at least one of the two integrals is limited in Equation (5).

Definition 7: Let ε be a random fuzzy variable and it has a limited expected value. The variance of ε is defined as:

$$
V[\varepsilon] = E[(\varepsilon - E[\varepsilon])^2]
\tag{6}
$$

Example 4: The random fuzzy variable $\varepsilon(\theta)$ defined in Example 3 is a two-point distribution random variable for each θ, where θ is a sample from a possibility space. The expected value of $\varepsilon(\theta)$ is:

$$
E_{pro}[\varepsilon(\theta)] = 0.99\xi_G(\theta)
$$

According to Definition 4, the expected value of a two-point distribution random fuzzy variable ε is:

$$
\begin{aligned}
E_{pro-fuz}(\varepsilon) \quad &= 0.99 E_{fuz}[\xi_G] \\
&= 0.99 \times ((35 + 2 \times 50 + 55)/4) = 47.025
\end{aligned}
$$

3. Credibility Theory-Based ATC Assessment Approach

3.1. Modeling Uncertainties in ATC Calculation

The assessment to ATC involves many uncertain factors, such as those in generation scheduling, in load forecasting and so on. Compared with previous methods, this paper comprehensively considers three main uncertain factors: random fuzzy models of generators, transmission lines and loads (to consider faults and fluctuations). It is worth noting that other uncertainty factors included in ATC calculation can also be treated in a similar way using credibility theory, and are potentially applicable to enrich and make the ATC calculation flexible by considering more factors. The corresponding random fuzzy models are built up as follows:

(1) For generators under the combined effects of the internal and external factors, there are two common states that are on-state and off-state. They are subject to a two-point random distribution according to the forced outage rate which comes from the statistics. When the generator is on-state, its output is greatly affected by many factors such as coal quality, reserve requirements, weather conditions and so on, so for a specific generator, its output is not an exact value and may vary significantly under different conditions. Although the equivalent forced outage rate [22] considers the conditions of outage and output reduction, it could not involve the conditions of overload operation, and since it depends on the statistics of a large number of historical data and the complex calculations, its value is hard to get for each generator in practical systems, especially for a newly installed generator, while through the simple fuzzy membership function of generator available output, which needs few parameters, the complex conditions can be described well by fuzzy variables, thus, the states of generators are suitable for being represented as discrete random fuzzy variables ε_G with two-point distribution:

$$
\begin{cases}
P_{pro,G}(\varepsilon_G) = \begin{cases} 1 - \lambda_G, & \varepsilon_G = \xi_G \\ \lambda_G, & \varepsilon_G = 0 \end{cases} \\[2ex]
F_{fuz,G}(\xi_G) = \begin{cases} \dfrac{\xi_G - a_{G,L}}{a_{G,M} - a_{G,L}}, & a_{G,L} \le \xi_G \le a_{G,M} \\[1.5ex] \dfrac{a_{G,H} - \xi_G}{a_{G,H} - a_{G,M}}, & a_{G,M} \le \xi_G \le a_{G,H} \\[1ex] 0, & Other \end{cases}
\end{cases}
\tag{7}
$$

where $P_{pro,G}$ is the state occurrence probability of the generator; $\varepsilon_G = \xi_G$ refers to the normal on-state, and $\varepsilon_G = 0$ is the off-state; the triangle fuzzy variable ξ_G is used to represents the fuzzy available output of a generator, and $F_{fuz,G}$ represents its membership function; $a_{G,L}, a_{G,M}, a_{G,H}$ are the minimum possible value, the most likely possible value and the maximum possible value of ζ_G, respectively. In this paper triangle fuzzy variables are used to represent the fuzzy states of the generator, transmission line and load, but other types of fuzzy variables such as trapezoidal fuzzy variables, can also be used according to specific conditions.

(2) The state of transmission line also obeys a two-point random distribution according to its failure rate. Conventionally, the longer the transmission line, the higher the failure rate in the same area and for the same voltage level, but practical experience indicates that the occurrence of faults in outdoor transmission lines is affected by many factors, including the weather conditions. Therefore the failure rate is a fuzzy value for a specific transmission line, which changes with the weather

conditions and the area considered [23]. The method presented in [24] where the failure rate can be modified through probability statistics is hard to apply in practical systems, so in order to comprehensively describe randomness and fuzziness on the whole, it is proper to representthe states of transmission lines as random fuzzy variables ε_B as follows:

$$
\begin{cases}
P_{pro,B}(\varepsilon_B) = \begin{cases} 1 - \xi_B, & \varepsilon_B = 1 \\ \xi_B, & \varepsilon_B = 0 \end{cases} \\
F_{fuz,B}(\xi_B) = \begin{cases} \frac{\xi_B - a_{B,L}}{a_{B,M} - a_{B,L}}, & a_{B,L} \le \xi_B \le a_{B,M} \\ \frac{a_{B,H} - \xi_B}{a_{B,H} - a_{B,M}}, & a_{B,M} \le \xi_B \le a_{B,H} \\ 0, & Otherwise \end{cases}
\end{cases}
\tag{8}
$$

where $P_{pro,B}$ is the state occurrence probability of the transmission line; $\varepsilon_B = 1$ expresses the normal on-state, and $\varepsilon_B = 0$ is the off-state; the triangle fuzzy variable ξ_B is used to represents the fuzzy failure rate of the transmission line, and $F_{fuz,B}$ represents its membership function; $a_{B,L}, a_{B,M}, a_{B,H}$ are the minimum possible value, the most likely possible value and the maximum possible value of ξ_B, respectively.

(3) As a main uncertainty factor in ATC assessment, the fluctuation of nodal load is considered to obey a normal distribution $N(\beta_L, \sigma_L)$ in traditional methods. Here the parameter β_L is the expected value of the distribution, which usually takes the predicted value of the nodal load. The parameter σ_L is the variance of the distribution, which shows the degree of deviation between the real value of the load and the forecasted one, and is usually determined according to the operator's experience. Actually, σ_L is not a fixed value, but a fuzzy one, so the fluctuation of nodal load has both randomness and fuzziness features, so the random fuzzy load ε_L is represented as follows:

$$
\begin{cases}
\varepsilon_L \sim N(\beta_L, \xi_L) \\
F_{fuz,L}(\xi_L) = \begin{cases} \frac{\xi_L - a_{L,L}}{a_{L,M} - a_{L,L}}, & a_{L,L} \le \xi_L \le a_{L,M} \\ \frac{a_{L,H} - \xi_L}{a_{L,H} - a_{L,M}}, & a_{L,M} \le \xi_L \le a_{L,H} \\ 0, & Otherwise \end{cases}
\end{cases}
\tag{9}
$$

where the triangle fuzzy variable ξ_L is used to represents the fuzzy variance of a nodal load, and $F_{fuz,L}$ represents its membership function; $a_{L,L}, a_{L,M}, a_{L,H}$ are the minimum possible value, the most likely possible value and the maximum possible value of ξ_L, respectively.

3.2. ATC Calculation Model

An improved repeated power flow method is adopted to calculate the ATC for each simulation state in this paper. The method starts from an initial operation point, then increases the load powers in the receiving area according to their proportion of the initial power, at the same time increases the generators' power outputs in the sending area according to their generation cost, until a certain constraint hits its limit, such as the node voltage exceeding its operating limit, or the transmission line being overloaded and so on. The net increase on the researched interface is the ATC of this simulation state. Such a conventional repeated power flow method only pays attention to system safety, but neglects the economics, so this paper introduces the optimal power flow [25,26] as shown in Equation (10) into every power flow computation of the repeated power flow method to comprehensively consider both safety and economy; it can get a more accurate results and better meet the real application needs in the power market:

$$
\begin{cases}
\min f \\
s.t. \quad P_g - P_d - V_x \Sigma V_y (G_{xy}\cos\delta_{xy} + B_{xy}\sin\delta_{xy}) = 0 \\
\quad\quad Q_g - Q_d + V_x \Sigma V_y (G_{xy}\cos\delta_{xy} - B_{xy}\sin\delta_{xy}) = 0 \\
\quad\quad P_g^{\min} \leq P_g \leq P_g^{\max} \\
\quad\quad Q_g^{\min} \leq Q_g \leq Q_g^{\max} \\
\quad\quad V_z^{\min} \leq V_z \leq V_z^{\max} \\
\quad\quad |S_l| \leq S_l^{\max}
\end{cases}
\tag{10}
$$

where the first two equality constraints are the active and reactive power balance equations and the remaining are the inequality constraints of generator output, node voltage and apparent power of the transmission lines.

3.3. ATC Assessment Indices

In order to accurately assess ATC and make a comparison with different methods, three assessment indices are proposed as follows:

(a)   The expected value of random fuzzy ATC—E_{ATC}—it comprehensively reflects the ATC of a power system.

$$
\begin{aligned}
E_{pro-fuzz,ATC} = \int_0^\infty C_r \{\theta \in \Theta | E[\varepsilon_{ATC}(\theta)] \geq r\} dr \\
- \int_{-\infty}^0 C_r \{\theta \in \Theta | E[\varepsilon_{ATC}(\theta)] \leq r\} dr
\end{aligned}
\tag{11}
$$

(b) The variance of random fuzzy ATC—V_{ATC}—it expresses the fluctuation of ATC and reflects the impacts of uncertainties on ATC:

$$V_{pro-fuz,ATC} = E[(\varepsilon_{ATC} - E_{ATC})^2] \qquad (12)$$

(c) Calculation time t: it reflects the efficiency of different ATC calculation approaches under the same initial conditions.

3.4. Parallel Algorithm with Bootstrap Method

The bootstrap method is a statistical analysis method adopted in numerical computation, which was for the first time put forward by Efron at Stanford University [27,28]. It makes full use of the information of samples and can provide an approximate distribution of the unknown parameters in an unknown system directly through (repeated) re-sampling in a small sample set. The bootstrap method is adopted in the random simulation in this paper, and as a result the sample size can be decreased remarkably and the processing speed is enhanced. ATC calculation for each simulation state is a complex nonlinear optimization problem. The primal-dual interior-point algorithm in Matpower4.0b4 [29] is applied to solve the problem, which makes full use of multi-core and multi-threading CPUs by parallel computing.

3.5. Random Fuzzy Simulation Based ATC Assessment

According to the above random fuzzy models of generators, transmission lines and loads, how to simulate these random and fuzzy uncertain factors is the key issue in ATC assessment. Obviously the conventional probability methods are invalid, and these models cannot be directly changed to deterministic equivalence problems, so the random fuzzy simulation is proposed to solve the ATC assessment with randomness and fuzziness. Firstly, according to the above models the fuzzy parameters are sampled, based on each fuzzy sample Monte Carlo random sampling is done for every random parameter, and as a result the possible states are simulated with the comprehensive consideration of both randomness and fuzziness.

Then the value of ATC for each simulation state is calculated by the improved repeated power flow method. The bootstrap method helps find the random distribution of ATC with fewer calculation results and enhances the processing speed. Lastly the comprehensive assessment indices can be obtained according to the random fuzzy computational framework. Figure 1 illustrates the procedure for assessing ATC. Detailed steps are described as follows:

(1) Read the initial parameters of generators, transmission lines and loads, build basic system information and set $e = 0$, $i = 1$.

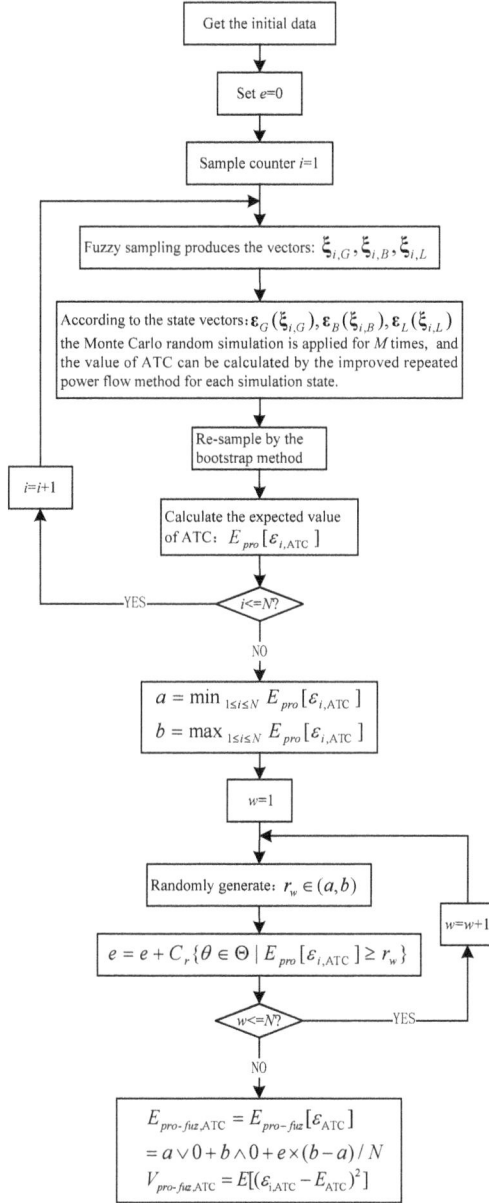

Figure 1. The flowchart of ATC assessment.

(2)   From the set Θ extract a θ_k which meets $P_{OS}\{\theta_k\} \geq \varepsilon$ (ε is a permissible small value making the sample space be bounded), get the variables of generators, transmission lines and loads, and produce a set of fuzzy sampling vectors: $\xi_{i,G}, \xi_{i,B}, \xi_{i,L}$.

339

(3) According to $\xi_{i,G}, \xi_{i,B}, \xi_{i,L}$ and the corresponding equipment random parameters, get the system state vectors: $''_G(\xi_{i,G}), ''_B(\xi_{i,B}), ''_L(\xi_{i,L})$, change the random fuzzy models of generators, transmission lines and loads to the random ones, then the fuzziness is eliminated. Then the Monte Carlo random simulation is applied M times, and the value of ATC can be calculated by the improved repeated power flow method for each simulation state.

(4) By the bootstrap method re-sample in the above obtained ATC values, and calculate their expected value of ATC. Figure 2 illustrates the bootstrap method procedure.

(5) Set sample counter $i = i + 1$, and repeat (2) to (4) for N times.

(6) Set $a = \min_{1 \le i \le N} E_{pro}[\varepsilon_{i,ATC}]$, $b = \max_{1 \le i \le N} E_{pro}[\varepsilon_{i,ATC}]$, and loop control variable $w = 1$.

(7) From the interval $[a, b]$ randomly generate r_w and calculate $e = e + C_r\{\theta \in \Theta | E_{pro}[\varepsilon_{i,ATC}] \ge r_w\}$.

(8) Set $w = w + 1$, and repeat (7) for N times.

(9) Lastly calculate the expected value and variance of ATC as follows:

$$E_{pro-fuz,ATC} = E_{pro-fuz}[\varepsilon_{ATC}] = a \vee 0 + b \wedge 0 + e \times (b-a)/N, \quad V_{pro-fuz,ATC} = E[(\varepsilon_{i,ATC} - E_{ATC})^2].$$

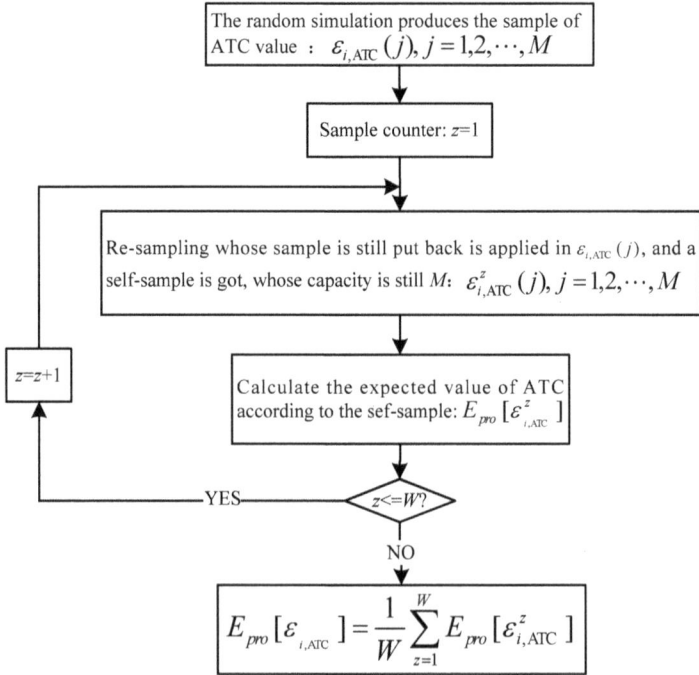

Figure 2. The flowchart of the bootstrap method.

4. Numerical Example

The simulation uses a Lenovo T400 (CPU: Core2 Duo 2.26 G; RAM: 3 G) laptop as the calculation platform. Matpower4.0b4 which provides the power flow calculation program is adopted as the simulation software under Matlab R2008a. The IEEE-30-bus system and an actual power system of two regions in Northwest China are used to demonstrate applications of the proposed models and algorithm.

4.1. IEEE-30-bus System

The single line diagram of the IEEE-30-bus system (the base capacity is 100 MVA) is shown in Figure 3. The system has 30 load nodes and 41 transmission lines, divided into three areas. The equipment parameters are given in [30]. This paper focuses on the ATC assessment from Area 1 to Area 2. The parameters of the random fuzzy simulation are set as follows: $N = 3000$, $M = 80$, $W = 10$. The proposed ATC assessment method based on credibility theory is verified in several scenarios. In Part 1, Part 2 and Part 3 only a single random fuzzy factor, such as random fuzzy generators or random fuzzy transmission lines or random fuzzy loads, is considered in each case to reflect its effect on ATC. In Part 4 all the above random fuzzy factors are considered simultaneously to compare the processing efficiency of each one.

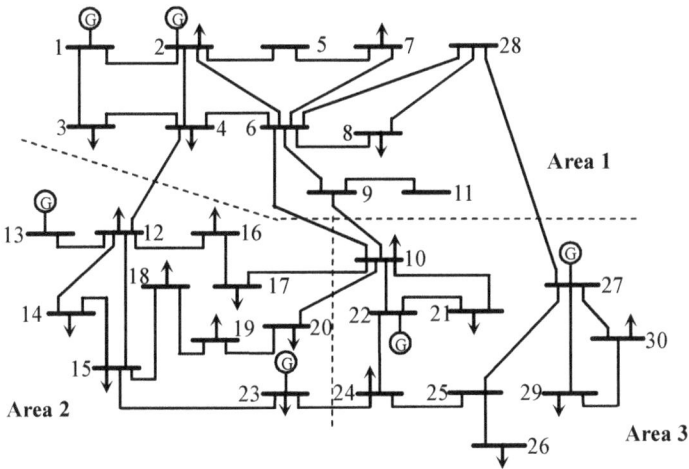

Figure 3. IEEE-30-bus system.

Part 1: Compatibility analysis between the proposed approach and the conventional Monte Carlo random simulation.

For Case A to Case F, the parameters of generators, transmission lines and loads are shown in Table 1, and the corresponding assessment results are listed in Tables 2–4. When the random fuzzy variables degenerate to random ones, the random fuzzy simulation becomes the Monte Carlo random simulation. In order to investigate the compatibility between the random fuzzy simulation and the Monte Carlo random simulation, the fuzziness of random fuzzy variables is set in a very small fluctuation range. As a result the random fuzzy variables can be approximately considered as random ones. As the results in Tables 2–4, the expected values and the variance of ATC given by the proposed method are nearly the same as the ones obtained by the Monte Carlo random simulation, and their maximum error is less than 12%, we can conclude that the proposed method is compatible with the Monte Carlo random simulation.

Table 1. The parameters of generators, transmission lines and loads for Cases A–F.

Method	Case	Generators		Transmission Lines	Loads
		λ_G	ξ_G	ξ_B	ξ_L
Monte Carlo random simulation (10,000 times)	A	0.01	1	None	None
	B	None	None	0.02	None
	C	None	None	None	0.02
Random fuzzy simulation	D	0.01	(0.9999, 1, 1.0001)	None	None
	E	None	None	(0.0199, 0.0200, 0.0201)	None
	F	None	None	None	(0.0199, 0.0200, 0.02001)

"None" means there is no fault or fluctuation.

Table 2. The results of Case A and Case D.

Case	$E_{pro\text{-}fuzz,\,\text{ATC}}$ **(MW)**	$V_{pro\text{-}fuzz,\,\text{ATC}}$ **(MW^2)**
A	8.5883	3.8085
D	8.4657	3.6602
Error (%)	−1.4275	−3.8939

Table 3. The results of Case B and Case E.

Case	$E_{pro\text{-}fuzz,\,\text{ATC}}$ **(MW)**	$V_{pro\text{-}fuzz,\,\text{ATC}}$ **(MW^2)**
B	9.7541	121.4598
E	10.8670	123.5610
Error (%)	11.4096	1.7300

Table 4. The results of Case C and Case F.

Case	$E_{pro\text{-}fuzz,\,\text{ATC}}$ **(MW)**	$V_{pro\text{-}fuzz,\,\text{ATC}}$ **(MW^2)**
C	11.3496	117.2293
F	11.7530	117.3173
Error (%)	3.5543	0.0751

Part 2: The comparison between the proposed assessment method and the traditional Monte Carlo simulation approach.

For Case J to Case I, the parameters of generators, transmission lines and loads are shown in Table 5. Tables 6–8 give the corresponding results. It can be seen from the results in Tables 6–8 that when both randomness and fuzziness factors of the uncertainty are considered, the variance index which reflects the fluctuation of ATC is changing. However, the fuzziness of generators has little effect on the variance (the error is only 0.5908%). In other words, it has little impact on ATC (the error is 1.9457%) and can be ignored, while the variance is greatly affected by the fuzziness of transmission lines and loads (the corresponding errors are 38.9011% and 123.2521%, respectively), so this fuzziness should be considered in the practical assessment in order to get a more accurate ATC.

Table 5. The parameters of generators, transmission lines and loads for Cases J–I.

Case	Generators		Transmission Lines	Loads
	λ_G	ξ_G	ξ_B	ξ_L
J	0.01	(0.700, 1.0000, 1.100)	None	None
H	None	None	(0.0100, 0.0200, 0.0600)	None
I	None	None	None	(0.0100, 0.0200, 0.0600)

"None" means there is no fault or fluctuation.

Table 6. The results of Case A and Case J.

Case	$E_{pro\text{-}fuzz, \text{ATC}}$ (MW)	$V_{pro\text{-}fuzz, \text{ATC}}$ (MW^2)
A	8.5883	3.8085
J	8.4212	3.8310
Error (%)	1.9457	0.5908

Table 7. The results of Case B and Case H.

Case	$E_{pro\text{-}fuzz, \text{ATC}}$ (MW)	$V_{pro\text{-}fuzz, \text{ATC}}$ (MW^2)
B	9.7541	121.4598
H	10.6504	168.7090
Error (%)	9.1890	38.9011

Table 8. The results of Case C and Case I.

Case	$E_{pro\text{-}fuzz, \text{ATC}}$ (MW)	$V_{pro\text{-}fuzz, \text{ATC}}$ (MW^2)
C	11.3496	117.2293
I	14.3061	261.7169
Error (%)	26.0494	123.2521

343

Part 3: The sensitivity analysis to the fuzzy influencing factors of ATC.

The above case studies show that the fuzziness of generators has little impact on ATC, so the following will mainly explore the impacts of the fuzziness of transmission lines and loads on ATC. In Case J all the parameters are the same as Case H except for the transmission line ones. The differences are that the states of some transmission lines (Lines 3, 6, 9, 10, 12, 13, 14, 15, 17, 18, 21, 25, 26, 31, 32, 35, 36, 37, 40, 41) are treated as random fuzzy variables ($\xi_B = (0.01,0.02,0.06)$) and the others are simulated only as random variables by the failure rate $\xi_B = 0.02$.

Case K differs from Case I in the load parameters. Case K sets some loads (4, 7, 8, 12, 18, 19, 20, 21, 23, 30) the random fuzzy variables ($\xi_L = (0.01, 0.02, 0.06)$) and the others the random variables with feature ($\xi_L = 0.02$), while on the basis of Case I, Case L reduces the fuzzy range of the load variance ($\xi_L = (0.01, 0.02, 0.04)$).

The corresponding assessment results are listed in Tables 9 and 10. From the results, it is indicated that: (a) by eliminating the fuzziness of some uncertainty factors, the variance of ATC is reduced, which means that the fluctuation of ATC is reduced, such as in Case J and Case K; (b) by reducing the fuzzy range of uncertainty factors, the variance and the fluctuation of ATC are changed, such as in Case L, so the fluctuation of ATC can be reduced by decreasing the fuzzy factors and the fuzzy range.

Table 9. The results of Case H and Case J.

Case	$E_{pro\text{-}fuzz,\,ATC}$ (MW)	$V_{pro\text{-}fuzz,\,ATC}$ (MW^2)
H	10.6504	168.709
J	10.5893	141.8906

Table 10. The results of Case I, Case K and Case L.

Case	$E_{pro\text{-}fuzz,\,ATC}$ (MW)	$V_{pro\text{-}fuzz,\,ATC}$ (MW^2)
I	14.3061	261.7169
K	14.3047	261.6058
L	13.0457	170.9183

Part 4: The comparison about the processing efficiency.

The processing efficiencies of different methods are compared here. In this simulation, all the uncertainty factors are considered, including the generators, the transmission lines and the loads. Table 11 gives the adopted methods for Case M to Case P, where the number of simulations is set as $N = 100$. The calculation time for each case is shown in Figure 4.

Table 11. The adopted methods of Case M, Case N, Case O and Case P.

Case	Bootstrap Method	Dual-core Parallel Computing Technique
M	√	√
N	×	√
O	√	×
P	×	×

"√" means the method is adopted and "×" means the method is not adopted. If the bootstrap method is used, set $M = 80$, $W = 10$; otherwise set $M = 800$, $W = 0$.

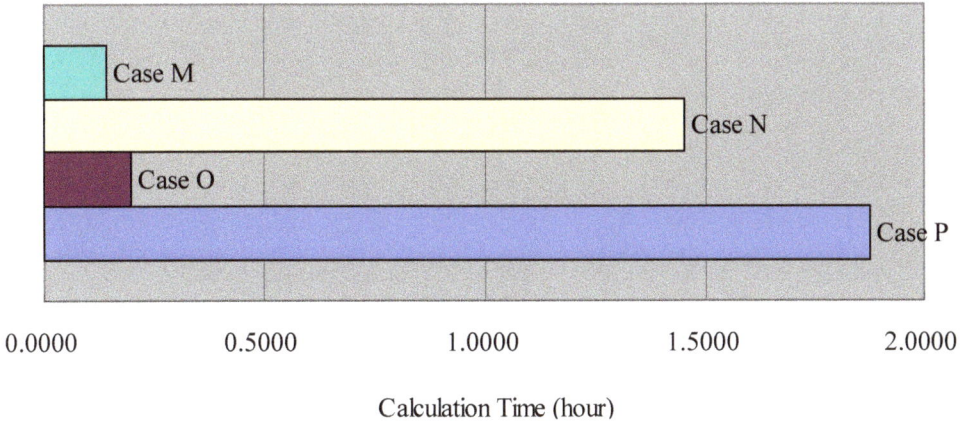

Figure 4. The comparison of calculation times.

From Figure 4, the calculation time of Case M is only about one tenth that of Case N, and about seven tenths that of Case O. Therefore, the bootstrap method can greatly improve the processing speed, while the multi-core parallel computing technique can reduce the calculation time to some extent. Despite the help of the bootstrap method and the multi-core parallel computing technique, due to the restrictions of the calculation platform, the calculation time of the proposed method is still considerable. This is due to the time-consuming optimal power flow. Therefore to improve the optimal power flow and further optimize the efficiency of the multi-core parallel computing will be an important future task.

4.2. An Actual Power System in Northwest China

The proposed ATC assessment approach is next applied in an actual power system of two regions in Northwest China. This is a 750 kV planning network for 2020 (the base capacity is 1000 MVA), whose single line diagram is shown in Figure 5. The total active power generation and load are 18,079 MW and 9857 MW, respectively. This paper focuses on the ATC assessment from Area 1 to Area 2. Case

345

Q and Case R are studied. In Case Q a Monte Carlo random simulation (10,000 times) is applied. The proposed approach is used in Case R, and all the uncertainty factors are considered, including the generators, the transmission lines and the loads. The parameters of Case Q and Case R are shown in Table 12, which are based on the historical data in the two regions and the experience of dispatchers. The parameters of random fuzzy simulation are set as follows: $N = 3000$, $M = 80$, $W = 10$.

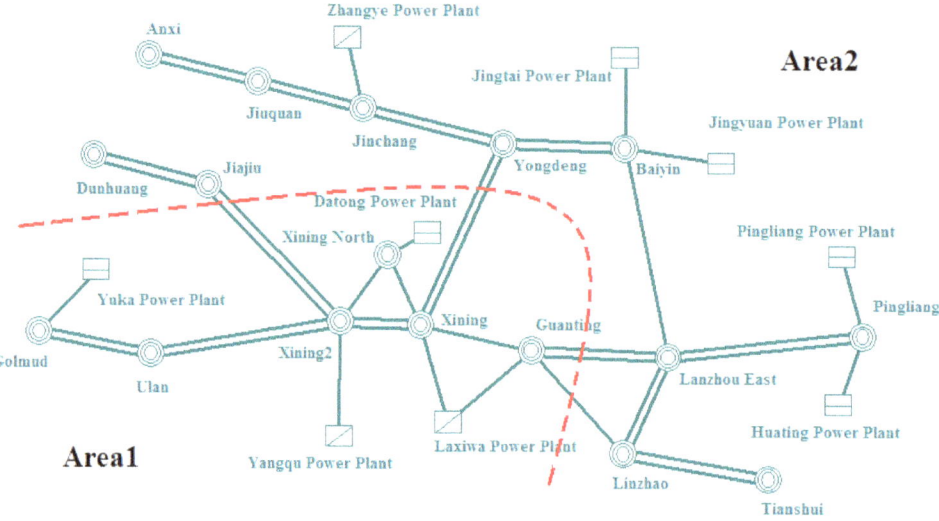

Figure 5. 750 kV planning network of two regions in Northwest China.

The corresponding assessment results are listed in Table 13. When both randomness and fuzziness of generators, transmission lines and loads are considered, the system uncertainty grows significantly. In Case Q and Case R, the variance index increases by about 11%, and the expected value of ATC is reduced by 272 MW, or about 6%. The results calculated using the proposed approach are consistent with current empirical judgments of Transmission Network Operators (TNOs), which for China TNOs, is that the maximum ATC does not exceed 4200 MW. The reduction value is large for the network, and if the fuzziness is not taken into consideration together with randomness, the system maybe becomes unstable in the peak period. Therefore the proposed ATC assessment approach can more comprehensively consider assessment risk, and supply accurate information.

Table 12. The parameters of generators, transmission lines and loads for Case Q and Case R.

Method	Case	Generators		Transmission Lines	Loads
		λ_G	ξ_G	ξ_B	ξ_L
Monte Carlo random simulation (10,000 times)	Q	0.01	1	0.02	0.02
Random fuzzy simulation	R	0.01	(0.9400, 1, 1.1400)	(0.0100, 0.0190, 0.0400)	(0.0100, 0.0190, 0.0400)

Table 13. The results of Case Q and Case R.

Case	$E_{pro\text{-}fuzz,\,ATC}$ (MW)	$V_{pro\text{-}fuzz,\,ATC}$ (MW^2)
Q	4417	4,702,842
R	4145	5,241,506
Error (%)	−6.1591	11.4540

5. Conclusions

To consider the two-fold uncertainties in the ATC assessment, this paper proposes a novel ATC assessment approach based on credibility theory, whereby according to the features of both randomness and fuzziness, the corresponding credibility models of generators, transmission lines and loads are comprehensively built up for the first time; then the random fuzzy simulation is applied in the ATC assessment; the bootstrap method and the multi-core parallel computing technique are adopted to enhance the computation speed. By testing on the IEEE-30 bus system and an actual system in China, the viability of the proposed models and algorithm is verified. Preliminary research is done on the sensitivity analysis to the fuzzy influencing factors of ATC and as a result, it gives a way to reduce the fluctuation of ATC.

Randomness and fuzziness are two general features in power systems. Compared with the traditional Monte Carlo random simulation, which only considers the randomness, the proposed method can coherently consider both uncertainties. Random variables that can perhaps adopt different values with certain probabilities can get the same results of the ATC assessment, but this modeling is not sensible. For example, for a generator it means to improve its forced outage rate. However, when the generator is in an on-state, its available output is greatly affected by many external factors, and not a fixed value, so it cannot be described by improving the forced outage rate, and a random variable is not appropriate, while our proposed random fuzzy variable can more suitably describe both the random forced outage rate and the fuzzy available output of a generator at the same time. The proposed method can effectively reduce the assessment risk, and supply accurate information for the mid- and long-term planning of power systems. This will have a better potential of development and application in power source and grid planning and operation. However, the proposed method can only be used with detailed parameters which refer to both randomness and fuzziness. For multiple area power systems, different TSOs share limited information. Therefore to deal with this case, we have carried out some exploratory research on building clustering models of generators, transmission lines and loads, using a multi-agent approach. This will be reported in future publications.

Author Contributions: The paper was a collaborative effort between the authors. Yanan Zheng performed the simulation, and designed and conceived the article. Jin Yang contributed to the English editing. Zhaoguang Hu, Ming Zhou and Gengyin Li gave some academic advice.

Conflicts of Interest: The authors declare no conflict of interest.

Nomenclature

Θ	Nonempty set.
ϕ	Empty set.
$P(\Theta)$	Power set of Θ.
$\wedge$	Minimum operator.
$\vee$	Maximum operator.
P_{os}	Possibility measure of fuzzy event.
N_{ec}	Necessity measure of fuzzy event.
C_r	Credibility measure of fuzzy event.
μ	Membership function of fuzzy variable.
B	Borel set.
sup	Supremum.
E_{fuz}	Expected value of fuzzy variable.
E_{pro}	Expected value of random variable.
$E_{pro\text{-}fuz}$	Expected value of random fuzzy variable.
R	Set of real numbers.
$(\Theta, P(\Theta), P_{OS})$	Possiblity space.
$P_{pro,G}$	State occurrence probability of generator.
$P_{pro,B}$	State occurrence probability of transmission line.
ε_G	Random fuzzy state of generator.
ε_B	Random fuzzy state of transmission line.
ε_L	Random fuzzy nodal load.
λ_G	Forced outage rate of generator.
ξ_G	Fuzzy available output of generator.
ξ_B	Fuzzy failure rate of transmission line.

ξ_L	Fuzzy variance of a nodal load.
$F_{fuz,G}$	Membership function of ξ_G.
$F_{fuz,B}$	Membership function of ξ_B.
$F_{fuz,L}$	Membership function of ξ_L.
$a_{*,L}$	Minimum possible value.
$a_{*,M}$	Most likely possible value.
$a_{*,H}$	Maximum possible value.
β_L	Load forecasting value.
f	Electricity purchase cost.
P_g	Active power output of the generator g.
P_g^{max}, P_g^{min}	Upper and lower limits of P_g.
Q_g	Reactive power output of the generator g.
Q_g^{max}, Q_g^{min}	Upper and lower limits of Q_g.
P_d	Active load of the node d.
Q_d	Reactive load of the node d.
V_z	Voltage of the node z.
V_z^{max}, V_z^{min}	Upper and lower limits of V_z.
S_l	Apparent power of the transmission line l.
S_l^{max}	Maximum value of S_l.
G_{xy}	Conductance of the branch from node x to y.
B_{xy}	Susceptance of the branch from node x to y.
δ_{xy}	Voltage phase angle difference of the branch from node x to y.
ε_{ATC}	Random fuzzy value of ATC.
$E_{pro\text{-}fuz,ATC}$	Expected value of random fuzzy ATC.
$V_{pro\text{-}fuz,ATC}$	Variance of random fuzzy ATC.
t	Calculation time.
N, M, W	Sampling times.

References

1. North American Electric Reliability Council (NERC). *Available Transfer Capability Definitions and Determination*; NERC: Swindon, UK, 1996.
2. Power Systems Engineering Research Center (PSERC). *Electric Power Transfer Capability: Concepts, Applications, Sensitivity and Uncertainty*; PSERC Publication: Tempe, AZ, USA, 2001.
3. Sauer, P.W. Alternatives for calculating transmission reliability margin (TRM) in available transfer capability. In Proceedings of the Thirty-First Hawaii International Conference, Kona, HI, USA, 6–9 January 1998; p. 89.

4. Ou, Y.; Singh, C. Assessment of Available Transfer Capability and Margins. *IEEE Trans. Power Syst.* **2002**, *17*, 463–468.

5. Shin, D.-J.; Kim, J.-O.; Kim, K.-H.; Singh, C. Probabilistic approach to available transfer capability calculation. *Electr. Power Syst. Res.* **2007**, *77*, 813–820.

6. Akbari, T.; Rahimikian, A.; Kazemi, A. A multi-stage stochastic transmission expansion planning method. *Energy Convers. Manag.* **2011**, *52*, 2844–2853.

7. Liu, B.; Liu, Y.-K. Expected value of fuzzy variable and fuzzy expected value models. *IEEE Trans. Fuzzy Syst.* **2002**, *10*, 445–450.

8. Liu, B. Toward fuzzy optimization without mathematical ambiguity. *Fuzzy Optim. Decis. Mak.* **2002**, *1*, 43–63.

9. Liu, B. *Uncertainty Theory: An Introduction to Its Axiomatic Foundations*; Springer-Verlag: Berlin/Heidelberg, Germany, 2004.

10. Liu, B. A survey of credibility theory. *Fuzzy Optim. Decis. Mak.* **2006**, *5*, 387–408.

11. Liu, B.; Liu, Y.-K. Expected value operatot of random fuzzy variable and random fuzzy expected value models. *Int. J. Uncertain. Fuzziness Knowl.-Based Syst.* **2003**, *11*, 195–215.

12. Feng, Y.; Wu, W.; Zhang, B.; Li, W. Power system operation risk assessment using credibility theory. *IEEE Trans. Power Syst.* **2008**, *23*, 1309–1318.

13. Feng, Y.; Wu, W.; Zhang, B.; Sun, H.; Wang, S. Hydro-thermal generator maintenance scheduling based on credibility theory. *Proc. Chin. Soc. Electr. Eng.* **2006**, *26*, 14–19.

14. Ma, X.; Liu, J.; Wen, F. Random-fuzzy programming model for developing optimal bidding strategies in the uncertain environment. *Proc. Chin. Soc. Electr. Eng.* **2009**, *29*, 77–83.

15. Wu, J.; Wu, Q.; Chen, G.; Liang, Y.; Zhou, J. Fuzzy random chance-constrained programming for quantifying the transmission reliability margin. *Autom. Electr. Power Syst.* **2007**, *31*, 23–28.

16. Kwakernaak, H. Fuzzy random variables I. *Inf. Sci.* **1978**, *15*, 1–29.

17. Kwakernaak, H. Fuzzy random variables II. *Inf. Sci.* **1979**, *17*, 253–278.

18. Zhou, J.; Liu, B. Analysis and algorithms of bifuzzy systems. *Int. J. Uncertain. Fuzziness Knowl.-Based Syst.* **2004**, *12*, 357–376.

19. Liang, R.-H.; Liao, J.-H. A fuzzy optimization approach for generation scheduling with wind and solar energy systems. *IEEE Trans. Power Syst.* **2007**, *22*, 1665–1674.

20. Kaufmann, A. *Introduction to the Theory of Fuzzy Subsets*; Academic Press: Waltham, MA, USA, 1975.

21. Zadeh, L.A. Fuzzy sets as a basis for a theory of possibility. *Fuzzy Sets Syst.* **1978**, *1*, 3–28.

22. IEEE Power & Energy Society. *IEEE Standard Definitions for Use in Reporting Electric Generating Unit Reliability, Availability, and Productivity*; ANSI/IEEE Std 762-1987; IEEE Power & Energy Society: Piscataway, MJ, USA, 1998.

23. Feng, Y.; Wu, W.; Zhang, B.; Sun, H.; He, Y. Short-term transmission line maintenance scheduling based on credibility theory. *Proc. Chin. Soc. Electr. Eng.* **2007**, *27*, 65–71.

24. Ding, M.; Dai, R.; Hong, M.; Xu, X. Simulation to the weather condition affecting the reliability of transmission network. *Autom. Electr. Power Syst.* **1997**, *21*, 18–20.

25. Duman, S.; Güvenç, U.; Sönmez, Y.; Yörükeren, N. Optimal power flow using gravitational search algorithm. *Energy Convers. Manag.* **2012**, *59*, 86–95.
26. Mahdada, B.; Bouktir, T.; Srairi, K.; EL Benbouzidc, M. Dynamic strategy based fast decomposed GA coordinated with FACTS devices to enhance the optimal power flow. *Energy Convers. Manag.* **2010**, *51*, 1370–1380.
27. Efron, B.; Tibshirani, R.J. *An Introduction to the Bootstrap*; Chapman and Hall/CRC Press: London, UK; Boca Raton, FL, USA, 1993.
28. Khosravi, A.; Nahavandi, S.; Creighton, D. A neural network-GARCH-based method for construction of prediction intervals. *Electr. Power Syst. Res.* **2012**, *96*, 185–193.
29. Zimmerman, R.D.; Murillo-Sánchez, C.E.; Thomas, R.J. MATPOWER's extensible optimal power flow architecture. In Proceedings of the Power and Energy Society General Meeting, Calgary, AB, Canada, 26–30 July 2009; pp. 1–7.
30. Shaaban, M.; Liu, H.; Li, W.; Yan, Z.; Ni, Y.; Wu, F. ATC calculation with static security constraints using benders decomposition. *Proc. Chin. Soc. Electr. Eng.* **2003**, *23*, 7–11.

Battery Design for Successful Electrification in Public Transport

Susanne Rothgang, Matthias Rogge, Jan Becker and Dirk Uwe Sauer

Abstract: Public transport is an especially promising sector for full electric vehicles due to the high amount of cycles and predictable workload. This leads to a high amount of different vehicle concepts ranging from large batteries, designed for a full day of operation without charging, to fast-charging systems with charging power up to a few hundred kilowatts. Hence, many different issues have to be addressed in the whole design and production process regarding high-voltage (HV) batteries for buses. In this work, the design process for electric public buses is analyzed in detail, based on two systems developed by the research projects Smart Wheels/econnect and SEB eÖPNV. The complete development process starting, with the demand analysis and the operating scenario, including the charging routine, is discussed. This paper also features details on cell selection and cost estimations as well as technical details on the system layout, such as the management system and passive components as well as thermal management.

Reprinted from *Energies*. Cite as: Rothgang, S.; Rogge, M.; Becker, J.; Sauer, D.U. Battery Design for Successful Electrification in Public Transport. *Energies* **2015**, *8*, 6715–6737.

1. Introduction

Currently, all large vehicle manufacturers are gradually electrifying their fleets either with full (BEV) or hybrid electric vehicles (HEV) to meet the new European CO_2 emission regulations for 2015 and 2020 [1]. This results in new requirements for batteries. At the same time, batteries will remain the most expensive components of the vehicles for the next 10 to 20 years, though battery cell costs have been rapidly decreasing recently [2]. Cycle lifetimes of up to 1000 cycles at a depth of discharge (DoD) of 80% are needed for passenger vehicles [3]. An average vehicle lifetime of about 12 years in Germany defines the needed calendric lifetime. Aging studies such as [4] show that the cycle lifetime of automotive cells is more than sufficient to fulfill these passenger vehicle requirements. Moreover, in the currently available vehicles, a relatively large driving range is guaranteed. The BMW i3 for example is designed for up to 150 km pure electric driving, and the Tesla Model S promises even a maximum driving range of up to 480 km to meet customers' range needs. At the same time, the battery capacity actually used is comparatively low; 95% of all single drives in Germany are, for example, shorter than 42 km [5] and the average daily driven mileage in the United Kingdom is only around 40 km [6]. Hence, a good part of the battery remains mostly unused, resulting in second use concepts for grid stabilization (see e.g., [7–9]) in order not to waste the expensive systems due to calendric aging. Calendric aging occurs no matter whether the battery is cycled or not [4,10].

As for the situation for electric buses in public transport, here, the daily mileage as well as the load is precisely known for each respective bus route. Moreover, the average annual usage is, in Germany for example, with 50,000 km [11] and a vehicle lifespan of 12 years, significantly more demanding than for conventional passenger vehicles. In an example system explained later on, the equivalent full cycles per year are 1200. Hence, the battery cells are effectively used regarding their possible cycle lifetime. This makes public transport the most promising sector for economic benefit from electrified vehicles in the near future.

This paper will therefore provide insight into the complete development process of an electric public transport bus. It will start with an introduction about the development process itself. In the first main step a typical usage profile and a suitable reference system will be introduced. Based on this, possible system layouts and charging regimes will be discussed and compared with each other, also highlighting existing systems on the market. This will be followed by important design choices such as the topology within the battery and its monitoring system. Moreover, the passive components and thermal management needed for a reliable battery system will be addressed, taking the available mounting space, volume and costs into account. In that discussion, experience gained from two developed prototype systems will be exploited for further details. In the research project "Smart Wheels," a small inner-city bus was electrified (Mercedes Benz Sprinter City 65). In "SEB eÖPNV" a standard 12 m low floor bus is electrified and equipped with a fast charging system of up to 500 kilowatts.

2. Steps in the Development of Electric Buses

With cities already paying extra money for their public transport systems, the success of electric buses requires aiming for minimal life cycle costs (LCC) in order to be competitive with diesel buses. This can only be achieved by choosing components, especially the battery system, through a structured design process. In Figure 1 the general steps in the development process are shown, and they will be explained in more detail in later sections.

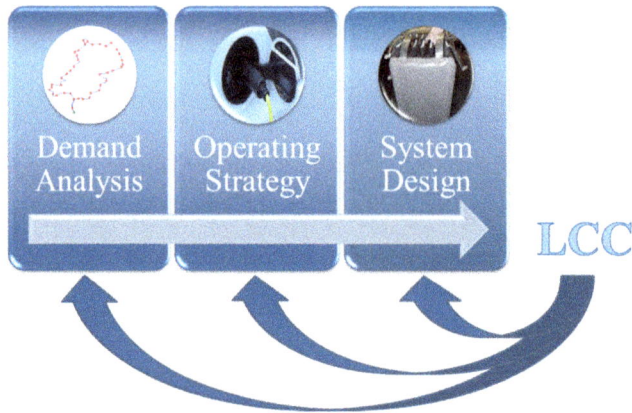

Figure 1. Steps in the development process of an electric bus.

Several assumptions have to be made when calculating the LCC of buses. The main parameters that should be considered for diesel and electric buses are given in Table 1. Moreover, ISEA provides an online tool to estimate the LCC costs based on parameters introduced below, so that it can be run for different configurations during the demand analysis, and on choosing an operating strategy [12]. As the operating scenario, a usage period of 12 years with an annual mileage of 50,000 km shall be considered according to the numbers given in section 1. An interest rate of 4% is assumed. Regarding the inflation rate, especially for electric buses, a higher market penetration in the future has to be assumed. Hence, different inflation rates for diesel and electric buses should be used. The overall maintenance costs of electric buses will be less for electric vehicles due to the low-maintenance electric machine, less braking effort and no oil changes. However, they will strongly depend on the installed infrastructure and charging technique. Using these parameters, a comparison of an electric to a diesel bus can be done assuming a battery size of 90 kilowatt-hours as was installed in the SEB bus. Parameterized by these data, an electric bus would be more expensive within the first 7 years due to the high investment costs but become cheaper than a diesel bus in the following years due to the high cost-saving potential compared to high diesel costs. The costs per kilometer for diesel buses rise from 1.27 Euros per kilometer in the first year up to 1.78 Euros per kilometer in the 12 year. Electric buses in this configuration, in contrast, have higher initial costs of 1.64 Euros per kilometer, which decline in the following years to 1.48 Euros per kilometer in year 12. Hence, it can be seen that electric buses are potentially competitive with diesel buses in terms of total cost of ownership. At the same time it is obvious that this strongly depends on the costs for the battery systems and their lifetime, as they are the main cost-pusher for electric buses. In Table 2 the influence of the battery size on the costs per kilometer is shown in order to emphasize this fact.

Therefore, within the demand analysis, detailed knowledge of the expected energy consumption as well as the needed peak driving power of the specific bus route has to be obtained in order to size the battery appropriately. As also shown by [13] it is not sufficient to use a standard drive cycle but rather the real operation scenarios and their occurrence probability [14]. Based on that, the operating strategy can be selected, taking the effort for the infrastructure as well as the boundary conditions of the chosen vehicle type into account. The electrical, mechanical and thermal development have to take place in the system design process. Life cycle costs have to be taken into account during all steps, allowing for feedback loops if a prior decision turns out not to result in a cost-optimized system.

Table 1. LCC parameters for electric (90 kWh) and diesel buses.

Parameter	Diesel		Electric	
Acquisition costs	240,000 € bus		300,000 € bus (without battery)	
			800 €·kW/h battery	
			225,000 € charging infrastructure	
Residual value & multiuse	12,000 €	12 years	15,000 € bus	12 years
			0 € battery	5 years
			4500 € infrastructure	20 years
			Charging infrastructure can be used for several buses/lines ▶ only calculate the infrastructure costs proportionally in e.g., 5 buses with a charging station at each end of the line	
Diesel/Electricity costs and price increase	1.2 €/L	6.1%/a (Germany)	0,15 €/kWh	3.2%/a (Germany)
Inflation rates	2%		−3% bus	
			−12% battery (see section 3.2)	
			0% infrastructure	
Maintenance and additional costs	15,000 €/a		Depending on installed infrastructure	
Consumption	40 L/100 km		152 kWh/100 km	
Efficiency	-		90% battery 93% charging infrastructure	

Table 2. Price per kilometer development for diesel and electric buses taking all above explained costs into account and showing the impact of battery costs on the overall costs per kilometer.

Bus Type	1	2	3	4	5	6	7	8	9	10	11	12
Diesel (€/km)	1.27	1.30	1.34	1.38	1.42	1.47	1.51	1.56	1.61	1.66	1.72	1.78
Electric 90 kWh (€/km)	1.64	1.64	1.65	1.66	1.66	1.52	1.52	1.53	1.54	1.55	1.48	1.48
Electric 180 kWh (€/km)	1.96	1.97	1.97	1.98	1.99	1.69	1.69	1.70	1.71	1.72	1.57	1.57

3. System Layout Options

Table 3 gives the important parameters of a 12 m full electric public transport bus, which is used in SEB eÖPNV. Moreover, the parameters of the Smart Wheels inner city bus are given. While the SEB bus is a standard public transport vehicle and therefore of interest for all routes, the Smart Wheels bus is designed for use cases such as operation in narrow streets in old city centers as well as shuttle operation. Figure 2 shows the two busses.

Based on these parameters a vehicle model can be parameterized to calculate the energy demand. The main auxiliaries have to be approximated depending on whether the full bus is electrically heated or only contains an air-conditioned driver's cabin. This results in a range of 4.5 to 20 kilowatts of auxiliary loads. Hence, the heating and cooling concept has an enormous impact on the continuous energy consumption and must be considered during system design.

Table 3. Reference system for a 12 m full electric bus used to calculate the energy consumption on different bus routes.

Parameter	Reference Bus (SEB)	Smart Wheels
width	2.55 m	1.99 m
height	3.2 m	2.71 m
length	12 m	7.02 m
average time at bus stop	0.5 min	
max. driving power	200 kW	120 kW
admissible total weight	18,000 kg	6,650 kg
drag coefficient	0.66	
rolling friction	0.013	
recuperation: absolute electrical breaking	$0.8 \, m/s^2$	
ratio between mechanical to electrical breaking system	0.5	
drivetrain efficiency	90%	
average battery efficiency	96%	

(a)

(b)

Figure 2. SEB bus (**a**) with 500 kW charging infrastructure and (**b**) Smart Wheels inner city bus.

3.1. Estimating the Energy Demand of the Bus Route

A Matlab/Simulink R2013b tool, which was developed at the institute, was used to estimate the energy consumption for a reference bus route. Detailed information on it can be found in [15]. Figure 3 shows the structure of the tool. As measurement data on the speed and height profile might not be available for all routes, synthetic profiles are generated using internet map services. Next to the accurate driving route the bus stops are also added. Depending on the actual time table, the waiting period is set to at least 15 s. The maximum inner-city speed is set to 50 km per hour. Moreover, additional stops at traffic lights or crossings are included randomly.

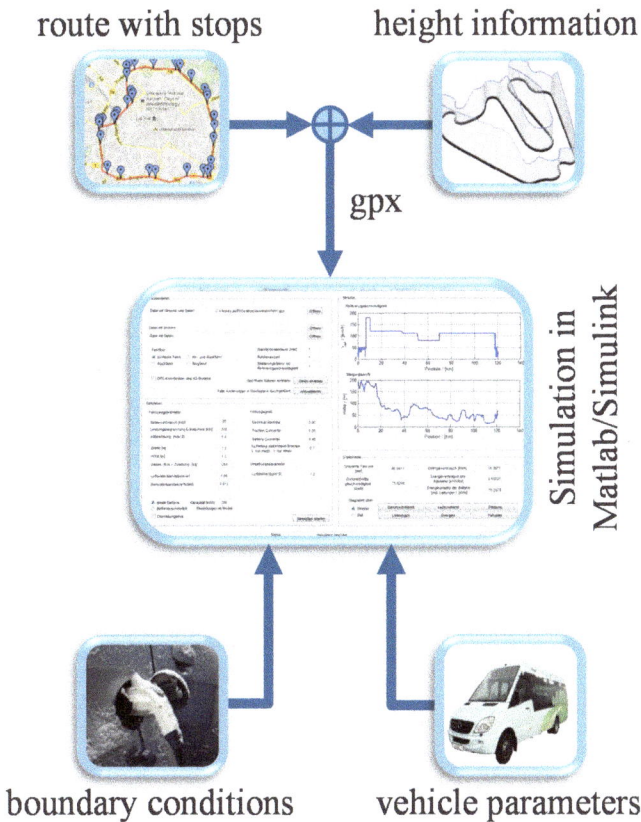

Figure 3. Matlab/Simulink tool for the demand analysis of the bus route using Google Maps for route planning based on the gpx (xml format for storing coordinate data) data.

As internet map services usually do not contain detailed height information, elevation data is obtained from the Shuttle Radar Topography Mission (SRTM) database. Additionally to the track data, vehicle parameters (Table 3) have to be

357

set and the boundary conditions such as weather and occupancy rate have to be chosen. The weather condition influences the power of the auxiliary devices whereas the occupancy rate defines the total weight of the bus. Based on the estimated profile, the vehicle parameters, and the assumed boundary conditions, a full vehicle simulation was run in Matlab/Simulink to estimate the average and maximum energy consumption during one round.

According to the explained routine, bus route 14 of the bus network in Münster, Germany was simulated for the layout of the SEB bus. Figure 4 shows the resulting speed profile as well as the correlated energy consumption of the bus route for a fully occupied bus without an electrical heating system. It shows a total consumption for traction as well as auxiliaries of over 20 kilowatt-hours per round. Due to the high amount of stops almost 30 percent can be recuperated leading to an overall energy demand of 1.52 kilowatt-hours per kilometer. The daily load can be calculated and used to choose an operating strategy based on these consumption data.

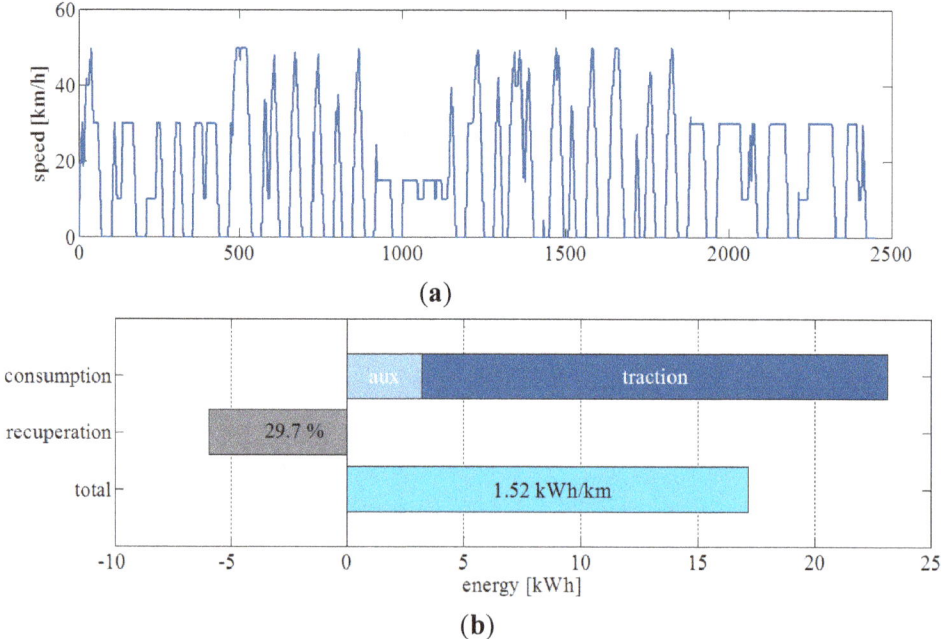

(a)

(b)

Figure 4. Speed profile and energy consumption of bus route 14 in Münster.

3.2. Operating Strategy

Knowing the energy demand of the bus route an operating strategy can be chosen. Table 4 gives an overview of the possible strategies as well as their pros and cons, which will be discussed more in detail below.

Table 4. Possible operating strategies and their pros (+) and cons (−).

Strategy	Infrastructure Effort	Charging Power	DoD	Battery Size, Weight and Costs	Flexibility	Complexity
overnight charging	+	+	−	−	−	+
fast charging at intermediate stops	−	−	+	+	−	−
fast charging at final stops	−	−	+	+	+	−

The main difference can be seen in either operating the bus without recharging in service or with an intermediate charging using a high charging power. Electric buses from China are often equipped with a battery providing enough energy for full-day operation. An example for this is the BYD K9 [16] which is also exported to Europe. The system architecture of a full day operating bus is in general easier compared to fast charging concepts as e.g., the cooling demand for the battery system is lower due to the reduced peak power. In addition to that, no expensive charging infrastructure is needed. Thus, such a concept provides the highest flexibility within scheduling. The battery can be charged overnight using only a moderate charging power, which is beneficial with regard to battery lifetime. High charging powers have negative impacts on battery aging that must not be neglected [17]. Moreover, standard charging also limits the maximum amperage at the cell level. In that case, it is determined by the highest driving power instead of possibly needing higher charging power in the fast charging strategy. This leads at the same time to a higher gravimetric and volumetric energy density. In comparison to high power (HP) cells, high energy (HE) cells have thicker layers of active material and therefore also a higher energy density [18], which is superior to other battery technologies [19,20].

However, there are also several drawbacks for the overnight charging concept that have to be taken into account. First of all the needed energy content for a full day of operation results in a very large battery system in the range of 200–400 kilowatt-hours. This limits, in a worst-case scenario, the possible payload of the buses. Currently a gravimetric energy density of around 100 watt-hours per kilogram can be achieved for state-of-the-art battery systems [21]. For the shown reference bus route with 20 trips per bus per day, an overall energy content of 307 kilowatt-hours would be needed if the bus is only charged overnight. The impact of this can also be shown by the system in the research project Smart Wheels. The used vehicle has an admissible total weight of 5650 kilograms. The demand analysis showed that a minimum energy content of 120 kilowatt-hours would be needed to fulfill the full-day operation at the desired circular route in Aachen. However, this would lead to a battery system weight of at least 1100 kilograms and therefore one fifth of the overall admissible total weight [22] which equals a reduction of the passenger load by 30%. This is, of course, unacceptable, so recharging during operation is mandatory in that case. This was achieved by recharging en route with 60 kilowatts. Compared to the average driving power of approximately 18 kilowatts, the maximum charging power is distinctively higher and therefore the key factor on system layout.

Secondly, the batteries have to be cycled with a high depth of discharge (DoD) in the overnight charging strategy which might enhance aging compared to smaller cycles. As in fast charging only smaller batteries are needed anyways, the cycle depth can be limited without a too huge influence on system weight and cost. This, however, has to be analyzed for each cell and usage condition (cycle depth full day operation *vs.* cycle depth with intermediate charging) individually. Ecker *et al.* for example found a great impact from the cycle depth for an investigated cell of Li(NiMnCo)O_2 [10] whereas Wang *et al.* on the other hand found no influence at all for an investigated LiFePO$_4$ material [23]. Hence, when designing a bus without charging en route, it is mandatory to analyze the cell's aging in detail in order to choose a cell that shows no huge dependency on cycle depth and to consider a slight over-dimensioning in order to reduce the cycle depth. Moreover, it has to be emphasized once again at that point that the higher charging power also has a negative impact on battery lifetime. Therefore, in strategy optimization, a trade-off between these factors has to be found depending on the used cells' aging characteristics.

Finally, large battery systems come at a high cost. Fast charging during operation can reduce the required battery capacity tremendously, leading to reduced vehicle investment costs. Cost estimations for Li-ion cells in 2020 range from 200 Euros per kilowatt-hour (Roland Berger) to over 140 Euros per kilowatt-hour (European Council for Automotive R&D—EUCAR) down to only 77 Euros per kilowatt-hour (United States Council for Automotive Research—USCAR) [24] and the system price scales directly with the cell costs. Tesla already nowadays claims its system price to be around 180 Euros per kilowatt-hour. A corresponding battery price drop from currently 330 Euros per kilowatt-hour to 180 Euros per kilowatt-hour on pack level in the private car sector is estimated until 2020 [25]. Therefore, system costs will be reduced a great deal compared to current systems, which can be estimated to be in the range of 600 to 800 Euro per kilowatt-hour in the public transport sector. However, in contrast to electric passenger cars, no huge economy of scale effects are expected in the public transport sector on system level. The new registrations rate is only around 33,000 buses per year in Western Europe [26]. Of these buses, around 35% account for the public transport sector whereas the rest of the buses are long distance travel [27] which are not of interest in this matter. Therefore, around 11,000 new buses are registered per year in Western Europe. This leads to an installed battery amount of approximately 220 megawatt-hours if 20 percent are fully electrified, with the constraint that a lot of different system architectures are needed to fulfill the specific route demand. In comparison a 20 percent electrification of the private car sector which is estimated by [25] (including plug-in hybrids) would mean an installed battery capacity of 36 gigawatt-hours. In this assumption, 15 million new registered cars in Western Europe per year and an estimated average battery capacity of 12 kilowatt-hours are considered. Therefore, a system price of 2.1 times the current cell price in the automotive sector is assumed as a realistic scenario for the transport sector, mainly driven by the price drop at the cell level. This would result in system costs below 400 Euros per kilowatt-hour. For intermediate charging, several different fast charging strategies are possible and shall be explained below. All of these have in common that, in addition to the impact on the drivetrain components, the impact on the grid and the energy price structure for the different charging regimes have to be considered [28,29].

Charging en route is currently realized by two different strategies. It can either be charged at the final stops of bus routes for a longer period or at several or even all intermediate stops for a very short period. Both strategies use already existing stops. Hence, no negative influence on the cycle times exists. On charging, either the fully consumed energy or only part of it can be recharged. The latter results in a depleting SoC (state of charge) during operation as can be seen in Figure 5, which depicts a strategy comparison for Smart Wheels.

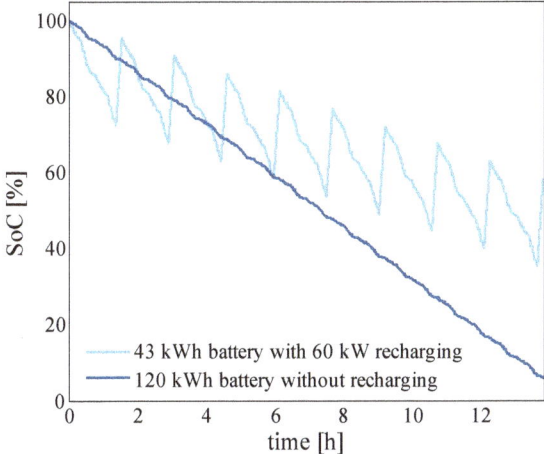

Figure 5. Comparison of operation without recharging and recharging with a depleting SoC during operation.

Recharging at each bus stop mainly involves inductive charging or charging via catenary as there is rarely enough time for automated plugging at bus stops. Hence, most of the time, the inductive charging station is directly inserted in the bus stop. The most popular example for this concept is the "primove" system from Bombardier [30]. Regarding pantograph systems especially, potentially existing electric tram or trolley bus systems within the cities can be taken into account to minimize the effort for the infrastructure tremendously [31]. As there are several common voltage levels up to 800 volts for tram catenaries, it has to be ensured that it is applicable for the bus voltage system. Examples for a catenary system are the System Electric Bus Rapid Transit from Siemens [32] and the multipole design charging pantographs of Schunk [33]. The connection to an existing DC (direct current) system results in greater flexibility as no huge infrastructure adaptations are needed, at the cost of greater complexity and weight at the bus side because chargers have to be installed in every vehicle. Therefore, this can be especially favorable for a small electric bus fleet if investment in the charging infrastructure would otherwise not pay back. Without the possibility of an existing DC feeding system, the grid connection especially is a cost-driving factor that needs to be considered carefully due to the required installation power that will not be available at a decent price at all bus stops.

4. System Design

Having decided on a charging strategy and therefore on the needed energy content of the battery, the system design process can be initiated. The left-hand side of Figure 6 shows a general overview of the components of a battery system that will be explained more in detail in the following. Moreover, in the right-hand side of Figure 6, a picture of the Smart Wheels' pack is given as a real life example. Different topologies are possible depending on the desired output voltage and the chosen cells [34]. Most of the current battery systems are built with a simple, strictly serial connection. However, there are also reasons to favor a more complex topology that connects smaller cells in parallel either on cell (parallel-series connection) or string level (serial-parallel) or consists of more than one pack.

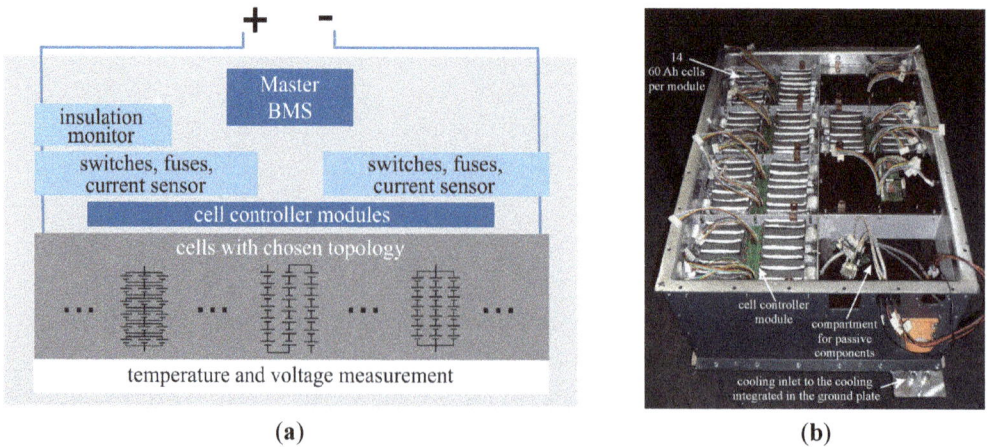

(a) (b)

Figure 6. Layout of a battery system: Schematic structure (**a**) and Smart Wheels pack (**b**).

By connecting more cells in parallel, the redundancy is enhanced as a single cell fault does not necessarily lead to a failure of the complete system. Moreover, there is always a spread in the aging of the single cell capacities as shown in [35]. This can be statistically lowered by building the mean over several cells in a parallel connection [21]. With two independent packs the operation can be continued also in case of an error such as a too high temperature in one of the systems. This can be achieved either by connecting two identical packs directly in parallel or by connecting the packs via dc-to-dc converters to a common dc-link as shown in Figure 7. The latter offers the highest flexibility as two completely different packs can be used and it is possible to have unsymmetric loads on the packs [21]. If one DC-to-DC converter per pack is used as shown in the lowest scheme in Figure 7 the DC-link voltage is adjustable. This offers further flexibility in the drivetrain design at the cost of a higher complexity due to the additional power electronics and control effort. In the research projects used as references here, the packs were connected directly in parallel to enable at least a limp-home mode but limit the additional effort at the same time.

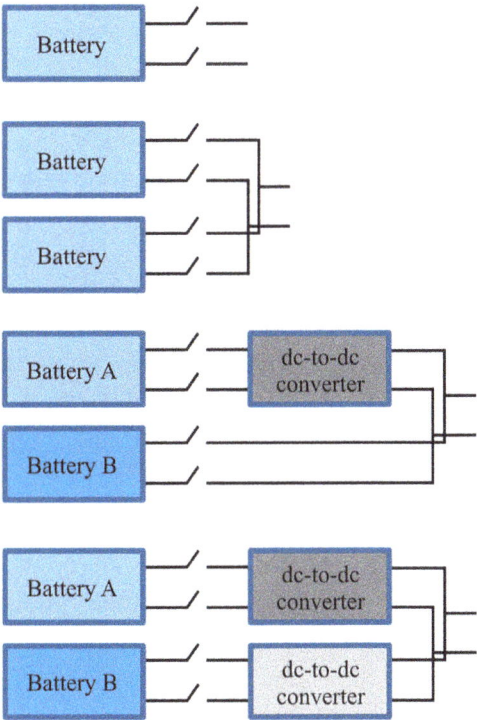

Figure 7. Topologies for batteries in automotive applications.

A further benefit of a distributed architecture is that the packs can be installed more easily outside the crash zones of the vehicles. This way, harmful deformation of the single cells can be prevented without the need for a strong housing for the batteries. If an installation outside the crash zones is not possible for a given vehicle type, further architectures can be considered that e.g., make the cell modules crash-resistant to absorb the energy in case of an accident. This was, for example, done within the research project e performance [36].

The voltage level of the system is defined by the internal connection of the cells. If a topology without DC-link is chosen, the battery voltage sets the voltage level of the traction inverter at the same time. Therefore, many of today's electric public transport buses are designed with a voltage of 700 to 800 volts as the needed driving power is also quite high and standard electronic components from the industry sector can be used in the converters. Moreover, it is the voltage already used for trolley buses and therefore further components are already on the market from this application as well.

4.1. Cell Selection

As already stated beforehand, the load profile in public buses can be predicted quite well. Hence, it is important to select a cell that best meets the specific requirements depending on the chosen operating strategy. Influencing factors are the chemistry, the estimated cycle depth, the charging strategy, and due to that, the peak power and the cooling demand. In Table 5 an overview of the most promising chemistries for public transport and their pros and cons are shown.

Table 5. Comparison of Li-ion battery chemistries with potential interest for public transport (good (+), poor (−), neutral (○)).

Chemistry	Energy Density	Admissible Charging Power	Cycle Lifetime	Voltage Curve	Deep Temperature Behavior	Safety
NMC	+	○	+	+	−	○
LFP	−	○	+	−	−	+
LTO	−	+	++	−	+	+

The different li-ion cell types are usually named by their cathode material as the anode material is classically carbon-based, such as graphite or hard carbons [37]. An exception with a new anode material is lithium titanium oxide (LTO) cells that are of potential interest in public transport due to their superior cycle lifetime, high power capability during charging [38] and good deep temperature behavior at the cost of lower voltage and therefore lower energy density compared with standard materials. The deep temperature behavior is of particular interest for public transport as recharging also has to be possible at deep temperatures. This is true for LTO as, in contrast to the conventional anode materials, lithium plating cannot occur. The lack of the risk of lithium plating is also the reason for the high admissible charging powers. Yet, this also depends on the internal structure of the cell. The four basic materials used on the cathode side are lithium cobalt oxide (LCO), lithium nickel oxide (LNO), spinel lithium manganese oxide (LMO) and lithium iron phosphate (LFP). Due to the high costs of pure LCO [39], its low thermal stability and fast capacity loss at high currents [39] as well as the lifetime issues of LNO [40], mixtures of the first three materials are often used, namely lithium nickel cobalt aluminum oxide (NCA) and especially lithium nickel manganese cobalt oxide (NMC). The ratio of the materials in NMC can be chosen in a wide range in cell design. Yet, often a 1/3-1/3-1/3 can be found which has a similar specific energy and operation voltage to LCO at significantly lower costs. As the voltage curve of NMC is comparatively steep, the management and especially SoC estimation is comparatively easy. Moreover, no hysteresis effects have to be taken into account. LFP cells offer a higher safety due to their thermal stability and a good power capability at the drawback of a relatively low average potential and therefore also energy density [39]. Summing up, it can be stated that NMC, LFP and LTO are particularly of potential interest for use in public transport. Many current systems on the market in public transport feature NMC cells due to the superior energy density in comparison to LFP and LTO. The two reference systems, SEB and Smart Wheels, were also built using NMC cells mainly due to that positive effect. However, the BYD e-bus features LFP cells, and large field tests in Munich and Vienna are also run with this chemistry. Even LTO cells can already be found, for example with the Proterra's EcoRide BE35 as well as Microvast, a manufacturer of LTO packs that is currently running buses in China, Hongkong, London and in The Netherlands.

The performance and especially the lifetime of cells with identical size and chemistry can widely differ. This risk is assessed through aging tests on a battery test bench with the corresponding application load profile as well as general parameterization and aging tests. During the cell selection process, suitable lithium-ion cells with a high charging power capability in case of a fast charging strategy have to be identified first. The resulting power profile of the energy consumption simulation can be downscaled to cell level and extended by a subsequent charging according to the chosen strategy. This procedure represents the application sufficiently and ensures realistic test conditions. Furthermore, the test results provide information about the cell performance at its end-of-life, which has to be taken into account as a worst-case scenario in the dimensioning of the cooling system.

Regarding the detailed aging analysis, calendric as well as cyclic aging have to be considered individually and added up to obtain the overall aging depending on the usage scenario [41]. Usually several cells are tested per test condition in order to take the cell spread into account [10]. In calendric aging, the influence of temperature and storage SoC are analyzed in particular. For cyclic aging, the DoD, the mean SoC and temperature and current in both, charging and discharging, have to be considered [20]. In Figure 8, an example of an aging test analyzing the DoD at a 2C cycling at 30 °C is shown. The tested cells are Kokam pouch cells such as those used in Smart Wheels and SEB. It shows the evident impact of the DoD on the aging speed. Yet even at 80 percent DoD, a sufficient cycle lifetime for public transport was achieved.

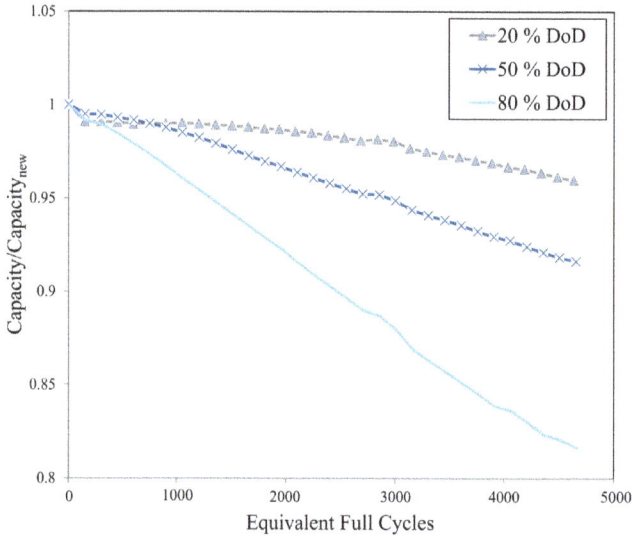

Figure 8. Reference aging analysis of a Kokam pouch cell.

Regarding the cooling demand, the cell geometry has to be taken into account. Cells types are divided into round cells, prismatic cells and pouch bag cells, which are shown in Figure 9. While pouch cells are relatively easy to cool due to their large surface area and the thin housing, the predominant direction of the thermal conduction of round cells is axial [42,43]. Moreover, a large temperature gradient can be found in some of the axial cells due to the connection to the terminals [44]. Therefore, cooling via the surface is not only more difficult due to the shape but also less effective. Regarding prismatic cells, one might assume that sufficient cooling can always be achieved via the ground plate. However, the shape of the jelly roll has to be taken into account. In some prismatic cells the jelly roll is positioned at a 90-degree angle to the tabs and therefore no good thermal linkage to the ground plate can be assumed, as the radial thermal conduction of the roll is relatively small and only a thin part of the cell is in close contact with the housing in that case, as can be seen by the internal structure in Figure 9. Yet if there is a thick enough metal cell casing it can be accounted for as a compensation for the bad connection, as in that case the housing offers good thermal conductivity to the base from all sides.

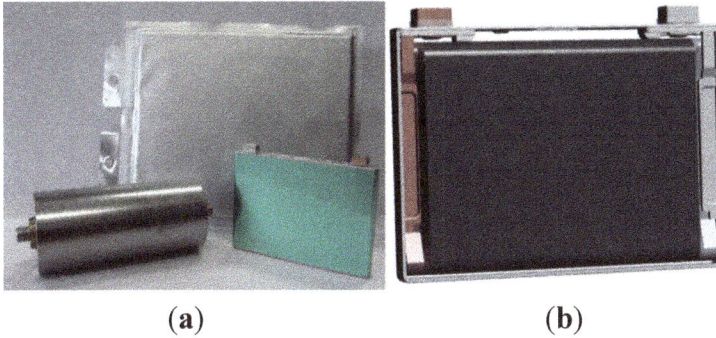

(a)   (b)

Figure 9. (a) Pouch (back), round (left) and prismatic (right) cell and (b) corresponding CAD model of the internal structure of the 5 Ah prismatic cell.

4.2. Battery Management System

The battery management system (BMS) for transport applications is mostly set up in a master-slave architecture as shown in Figure 10. Next to the measurement on the cell level, a measurement of the pack voltage and string currents (HV-interface) as well as a control of the safety devices such as the switches and precharge circuit is mandatory. Systems consisting of multiple packs are either controlled by a joint master or operated completely independently to expand the redundancy. The master, as the main intelligence of a battery, has various responsibilities [45]:

- Vehicle communication,
- Plausibility check of measurement values,
- Interconnection of packs,
- Control of the admissible charging power,
- Control of the switches as well as the precharge circuit,
- Adherence to threshold values such as end of (dis)charge voltages and temperatures,
- Monitoring of the state of health of the battery,
- Power prediction during operation, and
- Supervision of insulation-resistance.

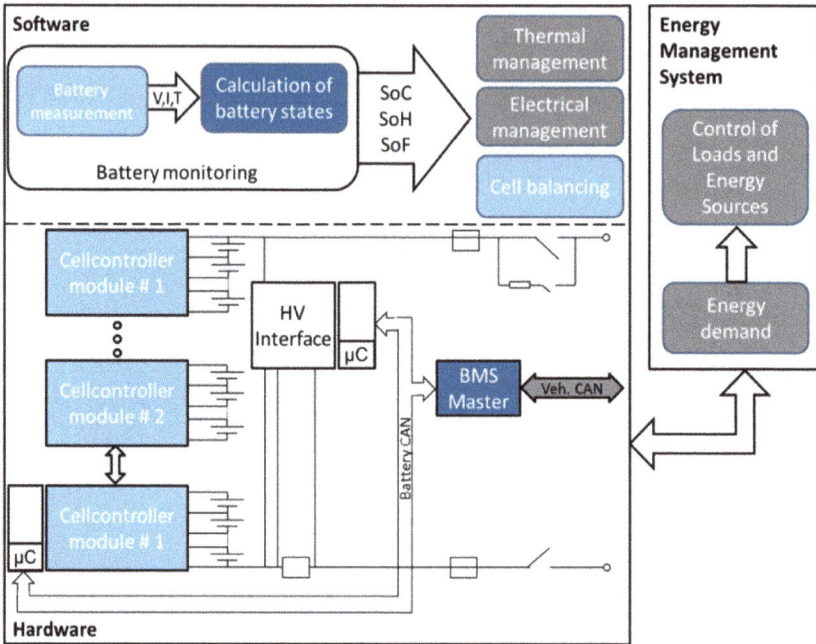

Figure 10. Structure of the battery management system on software and hardware side.

These functions on the software side of the BMS can be structured according to Figure 10. Battery states such as state of charge (SoC), state of health (SoH) and state of function (SoF) are calculated by the measured current, voltage and temperature of the single cells. They are needed for the further electrical and thermal management of the battery as well as the balancing routines which are needed to minimize the spread between the single cells in a serial connection. Regarding SoC estimation, different techniques are mostly combined, consisting of a full charge detector, an open circuit voltage module (see e.g., [46]) as well as an ampere hour balancing routine. The state of charge is a critical factor when it comes to electric vehicles, as a lot of the further management strategies are based on it. An example for this is the power prediction being directly dependent on the remaining SoC. With a faulty SoC showing high values, the voltage can drop significantly during a high power discharge pulse. By this means, a cut off voltage can be reached, resulting in a limitation of the available power which might not fulfill the actual demand anymore. Keeping an accurate SoC estimation over a long time period is challenging, especially if only rare full charge events are present, allowing for a reset of the algorithms. For electric buses this is an issue, especially if only fast charging is applied without a regular standard charge during night. Hence, sophisticated methods to improve accuracy are needed. This can be done, e.g., by Kalman filters [47]. A detailed overview and comparison of further existing methods is provided by [48].

For each cell in the series connection, a voltage measurement and balancing architecture is needed. These cell controller modules (light blue) are responsible for the preprocessing of the measured data (cell voltages and temperatures). They are usually located in close physical proximity to the actual battery modules/cells to minimize the length of the measurement cables. The importance of the low level data preprocessing has to be emphasized, as there are wide spectra of noise. In particular, high currents in combination with the short switching times of the inverter result in a strong electro-magnetic field and a wide spectrum of possible emissions [49]. Hence, suitably adapted lowpass filters for both temperature and voltage measurement are needed to limit measurement inaccuracies. As shown in [50], a low filtering has a huge impact on the measurement values and limits the usability of the data and therefore the reliability and stability of the overall system. In this context, a tradeoff between the generally desired high dynamic of the measurement and the filtering has to be found for each signal individually. Voltage and current values are needed at a high sampling rate whereas a higher time constant is admissible for the temperature values to limit the interferences due to the converters. The communication of these values between cell controller modules and BMS master is usually done by a battery internal CAN bus or alternative bus protocols such as I^2C. Regarding the temperature measurement, it is furthermore important to mention that usually the cell monitoring chips offer less temperature channels than voltage measurements. Moreover, measuring the temperature of each cell would be quite costly. Hence, in current battery systems, not all cell temperatures will be monitored individually. By thermal models of the packs the single cell temperatures can be calculated in normal operation mode by the existing temperature measurements. This, however, does not allow for a safe detection of a malfunction within a battery system leading to the possible risk of a thermal runaway. In order to prevent that, two different strategies shall be proposed. To simply detect already a temperature higher than a given threshold without measuring the actual temperature of each cell, a bimetal circuit can be used [50]. This comes at comparatively low additional effort and costs. Another more complex method is to use the already existing cell controller modules to detect the temperature by a correlation to the electrical impedance that can be measured within certain limitations, also online.

Next to the voltage measurement, the balancing is either realized by a passive or active balancing method. Passive balancing is realized by discharging the cells, having a higher voltage than a specific threshold, by a resistor and thus converting electrical energy to heat. Active balancing in contrast shifts the energy from one cell in the string to the others or another specific cell via an active converter [51,52] and is therefore more efficient than passive cell balancing. Whether an active balancing is beneficial on system level depends on the one hand on the aging spread of the cells and on the other hand on the usage scenario. The higher the aging spread the higher the benefit of the active system as it enables to even balance the cell voltages not only during the full charge but also during the discharge phase. This goes along with the dependence on the usage scenario as it will be especially of interest to balance also during discharge if full cycles are applied to the battery as this results in a higher possible driving range.

4.3. Passive Components

All parts within a battery system that do not directly add to the energy and power density are referred to as passive components. Next to the already discussed BMS, the safety devices in particular have to be considered as they are not negligible regarding weight and costs of the system. These safety systems (see Figure 10) usually consist of two main high-voltage switches (both poles of the battery) and a precharge circuit to disconnect the battery from the rest of the drive train in case of emergency and during standstill. The precharge circuit is needed on reconnecting as otherwise a high inrush current would occur due to the voltage difference between battery and DC-link. The input capacitors of the converter cannot withstand the high current, but also the switches and the cells themselves are not rated for such high currents even though it is only a pulse load [50]. The precharge resistor can be sized by the input capacitance and the time it takes until the current drops under a boundary value that is admissible for the whole circuit. Moreover, the peak and overall power that have to be dissipated by the resistor have to be considered in the design process. Additionally to the interruption of the circuit by switches, a fuse is added on one or both poles to prevent the battery from further damage in case of an external short. Mostly, the device for evaluating the insulation resistance is also fitted into the battery and a service plug is added to assure lower voltages e.g., for maintenance periods.

In Figure 11 the weight ratio of the components of the Smart Wheels' battery system (see Figure 6) is given. It can be seen that the cells account for approximately two thirds of the overall weight, which is 430 kilograms for around 44 kilowatt-hours. With a gravimetric energy density of 100 kilowatt-hours per kilogram, an equal energy density to that of the BMW i3 was achieved [53]. As the pack is mounted under the floor, the housing is also quite massive, and due to the fast charging, a liquid cooling system is needed, adding additional weight as well. The cell connectors account for almost 6 percent of the weight. However, this could be greatly reduced by welding the cells in mass production [22]. Yet the passive components account for 4 percent of the overall system, with the switches and fuses being the heaviest components. This weight greatly depends on the current rates. Moreover, it has to be kept in mind that when distributing the system in several packs especially the heavy components such as fuses and switches are needed in every pack. Hence, a tradeoff has to be found between reliability and flexibility on the one hand and system weight and achievable energy density on the other. To reduce needed space, weight and costs of the passive components, they should be directly integrated in the battery pack instead of a separate housing. This improves safety while keeping the switches in the closest possible proximity to the battery cells and ensuring there are no additional live parts within the vehicle in case of an emergency [50].

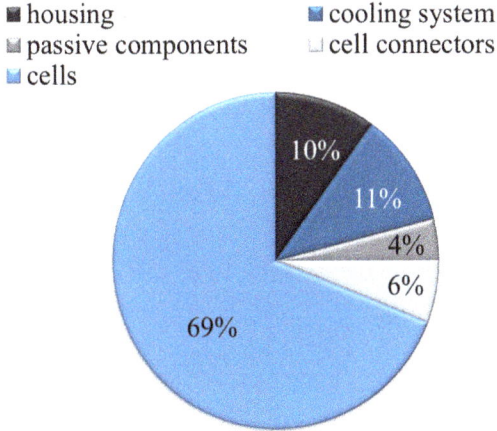

Figure 11. Weight distribution of the battery system in Smart Wheels.

Another important point regarding the passive components is their heat generation and how they might influence each other. Switches in particular, which are normally held actively shut by an electromechanic coil system, emit a significant amount of heat as can be seen in Figure 12. In the first 10 min the battery system was operated but in standby, so no load was drawn. Still, the switch temperature raised around 7 degrees Celsius as the switches were actively held shut. At the same time, fuses, which are normally melting fuses with a distinct tripping characteristic, are directly influenced by a rise in ambient air and should therefore not be sized with too narrow a margin or placed in close proximity to the switches. In the figure it can be seen that during load, the temperature of the fuse shows the strongest increase, but it cools down rapidly again in the following standstill period. In a sealed environment, care has to be taken in terms of how the heat can be dissipated by, e.g., only mounting the component on materials with good thermal conductivity.

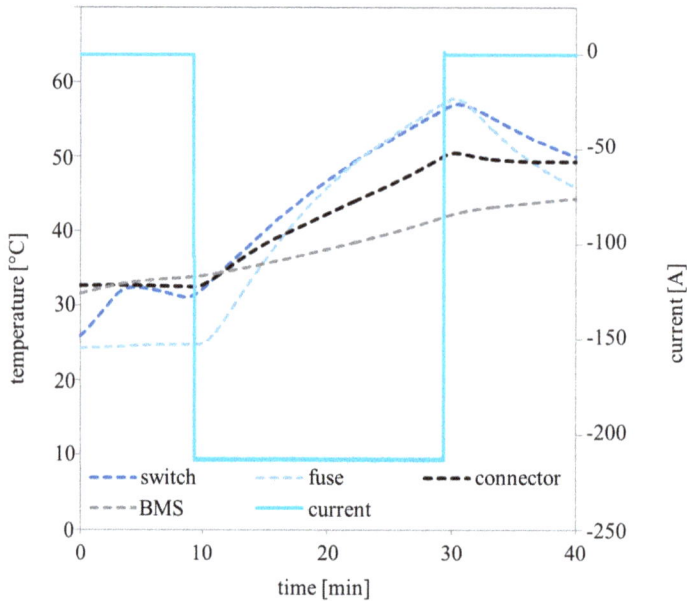

Figure 12. Heat up process of different pack components under load.

4.4. Thermal Management

Availability of the vehicle is of great importance in public transport. Excessively high or low temperatures for the battery can be potential limiting factors in that regard. At the same time, the usage scenario in a continuous full day operation in combination with fast charging is quite demanding in terms of thermal management. Therefore, an application-designed cooling system is mandatory. Moreover, as explained in Subsection 4.1 of this section, in contrast to common vehicle applications, the full life cycle of the batteries will be exploited. Hence, realizing a lower average operation temperature has a direct impact on the system lifetime.

The design goals of the cooling system have to fulfill several tasks:

- Lower the maximum temperature to reduce aging,
- Prevent deep temperature charging to avoid Lithium-plating,
- Limit the temperature gradient within the cell to prevent uneven usage of the material within the cell, and
- Limit the temperature gradient within the pack to minimize the aging diversification of the cells.

As shown in [35] the aging spread of the cells is quite high and cannot be limited by sorting the cells by their initial capacity. This effect is even more enhanced by a temperature gradient among the cells used as the Arrenhius equation [54]. According to a rule of thumb, a doubling of aging results from a ten degree temperature increase. Therefore, it is advised rather to go for the smallest possible pack gradient instead of aiming to decrease the peak temperature a bit further in consideration of the design tasks. Liquid cooling topologies favorable for this are bifilar cooling or a parallelized cooling structure with adjusted flow rates in the different channels [50].

In Figure 13 it can be seen how the gradient among the cells of the two Smart Wheels' packs changes during two hours of operation in cold ambient conditions. On the left hand side of the figure the temperature distribution in the mornings before operation is given. All cells have similar temperatures resulting in a normal distribution within the system. The slight existing differences can be explained by inner and outer cells. The center part of the system is still slightly warmer due to the huge heat capacity of the system and the coldest cells are located directly next to the housing as the packs are mounted underfloor. However, after 2 h of operation a clear separation in two groups with a delta of 1.5 to 2 degrees Celsius each can be found. This behavior can be explained by one pack being mounted at the vehicle's front axle with a higher cooling due to the air flow. In contrast, the other pack is mounted behind the rear axis and due to the shape of the vehicle under the floor mostly protected against the air flow. However, within the packs the temperature gradient could be kept small due to the bifilar cooling in the ground plate of the system.

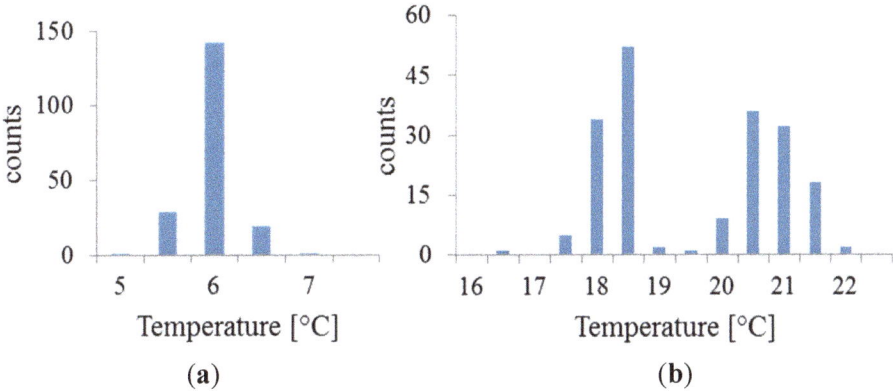

Figure 13. Temperature distribution within the two Smart Wheels' battery packs before (a) and after (b) two hours of driving.

5. Conclusions

In this work, an overview was given on how a successful electrification of the public bus sector can be achieved with battery-powered buses. It was shown that for an optimized system, not only the vehicle itself but also the charging strategy and the charging infrastructure have to be taken into account. A detailed knowledge of the expected load profile and usage scenario is mandatory. Within the battery system, several different topologies are conceivable depending on the cells used, offering different degrees of reliability and freedom. There is not a standard solution valid for all applications. Rather it is necessary to analyze and to design the battery system and the charging concept according to the local situation and demand. During system design it is important, after close management, to also take the rest of the passive components and the cooling into account to produce a system that is competitive with diesel buses and has a satisfactory lifespan.

Acknowledgments: This work has been kindly supported by the German Federal Ministry for Economic Affairs and Energy (BMWE) as part of the SmartWheels and econnect projects, as well as the German Federal Ministry for Education and Research (BMBF) within the SEB eÖPNV project.

Author Contributions: The layout of the prototype system within Smart Wheels was done by Susanne Rothgang. She also carried out the experiments concerning it and the build-up phase. Matthias Rogge simulated the energy consumption and developed tools for it. Jan Becker contributed to the multi-pack architecture. Dirk Uwe Sauer contributed to all of the parts.

Conflicts of Interest: The authors declare no conflict of interest.

References

1. Commision of the European Communities. *Results of the Review of the Community Strategy to Reduce CO_2 Emissions from Passenger Cars and Light-Commercial Vehicles*; Commision of the European Communities: Bruessels, Belgium, 2007.
2. Hartmann, N.; Oezdemir, E.D. Impact of different utilization scenarios of electric vehicles on the German grid in 2030. *J. Power Sources* **2011**, *196*, 2311–2318.
3. Vetter, J.; Novák, P.; Wagner, M.R.; Veit, C.; Möller, K.-C.; Besenhard, J.O.; Winter, M.; Wohlfahrt-Mehrens, M.; Vogler, C.; Hammouche, A. Ageing mechanisms in lithium-ion batteries. *J. Power Sources* **2005**, *147*, 269–281.
4. Käbitz, S.; Gerschler, J.B.; Ecker, M.; Yurdagel, Y.; Emmermacher, B.; André, D. Cycle and calendar life study of a graphite LiNi1/3Mn1/3Co1/3O2 Li-ion high energy system. Part A: Full cell characterization. *J. Power Sources* **2013**, *239*, 572–583.
5. Follmer, R.; Gruschwitz, D.; Jesske, B.; Quanst, S.; Lenz, B.; Nobis, C.; Köhler, K.; Mehlin, M. *Mobilität in Deutschland 2008: Ergebnisbericht; Struktur-Aufkommen-Emissionen-Trend*; Ergebnisbericht, infas, dlr: Berlin/Bonn, Germany, 2010. (In German)
6. United Kingdom Department for Transport. Collection—Transport Statistics Great Britain. Available online: http//www.gov.uk/government/collections/transport-statistics-great-britain (accessed on 14 April 2015).
7. Viswanathan, V.V.; Kintner-Meyer, M. Second use of transportation batteries: Maximizing the value of batteries for transportation and grid services. *Trans. Veh. Technol.* **2011**, *60*, 2963–2970.
8. Zhou, C.; Qian, K.; Allan, M.; Zhou, W. Modeling of the cost of EV battery wear due to V2G application in power systems. *IEEE Trans. Energy Convers.* **2011**, *26*, 1041–1050.

9. Peterson, S.B.; Apt, J.F.; Whitacre, J. The economics of using plug-in hybrid electric vehicle battery packs for grid storage. *J. Power Sources* **2010**, *195*, 2377–2384.

10. Ecker, M.; Nieto, N.; Käbitz, S.; Schmalstieg, J.; Blanke, H.; Warnecke, A.; Sauer, D.U. Calendar and cycle life study of Li(NiMnCo)O_2-based 18650 lithium-ion batteries. *J. Power Sources* **2014**, *248*, 839–851.

11. Kalinowska, D.; Kloas, J.; Kuhfeld, H.; Kunert, U. *Akualisierung und Weiterentwicklung der Berechnungsmodelle für die Fahrleistungen von Kraftfahrzeugen und für das Aufkommen und für die Verkehrsleistung im Personenverkehr (MIV)*; DIW Berlin: Berlin, Germany, 2005. (In German)

12. ISEA & EBusPlan. LCC Calculator. Available online: http://www.ebusplan.com/en/calculator/ (accessed on 1 June 2015).

13. Schwarzer, V.; Ghorbani, R. Drive cycle generation for design optimization of electric vehicles. *Trans. Veh. Technol.* **2013**, *62*, 89–97.

14. Wu, J.; Li, K.; Jiang, Y.; Lv, Q.; Shang, L.; Sun, Y. Large-scale battery system development and user-specific driving behavior analysis for emerging electric-drive vehicles. *Energies* **2011**, *4*, 758–779.

15. Sinhuber, P.; Rohlfs, W.; Sauer, D.U. Study on power and energy demand for sizing the energy storage systems for electrified local public busses. In Proceedings of the 2012 IEEE Vehicle Power and Propulsion Conference (VPPC), Seoul, Korea, 9–12 October2012; pp. 315–320.

16. BYD. BYD Electric Bus Introduction. Available online: http://www.byd.com/na/auto/ElectricBus.html (accessed on 29 June 2015).

17. Legrand, N.; Knosp, B.; Desprez, P.; Lapicque, F.; Rael, S. Physical characterization of the charging process of a Li-ion battery and prediction of Li plating by electrochemical modelling. *J. Power Sources* **2014**, *59*, 208–216.

18. Winter, M.; Brodd, R.J. What are batteries, fuell cells, and supercapacitors? *Chem. Rev.* **2004**, *104*, 4245–4269.

19. Khaligh, A.; Zihao, L. Battery, Ultracapacitor, fuel cell, and hybrid energy storage systems for electric, hybrid electric, fuell cell, and plug-in hybrid electric vehicles: State of the art. *Trans. Veh. Technol.* **2010**, *59*, 2806–2814.

20. Wang, S.; Zhao, L.; Su, X.; Ma, P. Prognostics of lithium-ion batteries based on battery performance analysis and flexible support vector regression. *Energies* **2014**, *7*, 6492–6508.

21. Rothgang, S.; Baumhöfer, T.; van Hoek, H.; Lange, T.; de Doncker, R.W.; Sauer, D.U. Modular battery design for reliable, flexible and multi-technology energy storage systems. *Appl. Energy* **2015**, *137*, 931–937.

22. Rothgang, S.; Lunz, B.; Laresgoiti, I.; Geulen, G.; Homann, J. *HV Traction Battery: From Layout to Realization*; Electric Vehicle Symposium: Los Angeles, CA, USA, 2012.

23. Wang, J.; Liu, P.; Hicks-Garner, J.; Sherman, E.; Soukiazian, S.; Verbrugge, M.; Tataria, H.; Musser, J.; Finamore, P. Cycle-life model for graphite-LiFePO$_4$ cells. *J. Power Sources* **2011**, *196*, 3942–3948.

24. Schein, H.; Holl, K. *Vorstellungen und Erwartungen aus Sicht der Zellhersteller*; Batterieforum Deutschland: Berlin, Germany, 2014. (In German)

25. Hummel, P.; Houchois, P.; Dewhurst, J.; Gandolfi, A.; Hunt, S.; Lesne, D.; Oldfield, S.; Leitch, D.; Dumoulin-Smith, J.; Gilbert, T.; *et al. Global Utilities, Autos & Chemicals. Will Solar, Batteries and Electric Cars re-Shape the Electricty Systems?*; UBS Research: Zurich, Switzerland, 2014.

26. German Association of the Automotive Industry (VDA). *Annual Report 2012*; VDA: Berlin, Germany, 2012.

27. *EU Transport in Figures 2012*; Publications Office of the European Union: Luxembourg, Luxembourg, 2012.

28. Zhang, T.; Chen, W.; Han, Z.; Cao, Z. Charging scheduling of electric vehicles with local renewable energy under uncertain electric vehicle arrival and grid power price. *Trans. Veh. Technol.* **2014**, *63*, 2600–2612.

29. Jin, C.; Tang, J.; Ghosh, P. Optimizing electric vehicle charging: A customer's perspective. *Trans. Veh. Technol.* **2013**, *62*, 2919–2927.

30. Bombadier. Bombardier Primove. Available online: http://primove.bombardier.com/products/charging.html (accessed on 29 June 2015).

31. Müller-Hellmann, A. Vom aschenbrödel zum rückgrat multimodaler E-mobilität-infrastrukturen kommunaler gleichstrombahnen beflügeln nationalen entwicklungsplan elektromobilität. *Der Nahverk.* **2012**, *12*, 15–20. (In Greman)

32. Siemens. Available online: http://www.mobility.siemens.com/mobility/global/SiteCollectionDocuments/en/road-solutions/urban/ebus-ebrt/ebus-wiener-linien-brochure-en.pdf (accessed on 29 June 2015).

33. Schunk. Charging Pantograph SLS101. Available online: http://www.schunk-sbi.com/sixcms/media.php/1973/SLS101-chargingPantograph.pdf (accessed on 14 April 2015).

34. Uddin, K.; Picarelli, A.; Lyness, C.; Taylor, N.; Marco, J. An acausal Li-ion battery pack model for automotive applications. *Energies* **2014**, *7*, 5675–5700.

35. Baumhöfer, T.; Brühl, M.; Rothgang, S.; Sauer, D.U. Production caused variation in capacity aging trend and correlation to initial cell performance. *J. Power Sources* **2014**, *247*, 332–338.

36. Schüssler, M.; Allmann, C.; Wein, M.; Landgraf, J.; Regler, C.; Schurius, M.; Gut, G.; Staats, H.; Ginsberg, S.; Hillers, T.; *et al. Research Project e Performance. From Concept to Vehicle*; Aachen Colloquium Automobile and Engine Technology: Aachen, Germany, 2012.

37. Mulder, G.; Omar, N.; Pauwels, S.; Meeus, M.; Leemans, F.; Verbrugge, B. Comparison of commercial battery cells in relation to material properties. *Electrochim. Acta* **2013**, *87*, 473–488.

38. Han, X.; Ouyang, M.; Lu, L.; Li, J. Cycle life of commercial lithium-ion batteries with lithium titanium oxide anodes in electric vehicles. *Energies* **2014**, *7*, 4895–4909.

39. Nitta, N.; Wu, F.; Lee, J.T.; Yushin, G. Li-ion battery materials: present and future. *Mater. Today* **2015**, *18*, 252–264.

40. Broussely, M.; Biensan, P.; Simon, B. Lithium insertion into host materials: the key to success for Li ion batteries. *Electrochim. Acta* **1999**, *45*, 3–22.

41. Hou, C.; Wang, H.; Ouyang, M. Battery sizing for plug-in hybrid electric vehicles in Beijing: A TCO model based analysis. *Energies* **2014**, *7*, 5374–5399.

42. Fleckenstein, M.; Bohlen, O.; Roscher, M.A.; Bäker, B. Current density and state of charge inhomogenities in Li-ion battery cells with LiFePO4 as cathode material due to temperature gradients. *J. Power Sources* **2011**, *196*, 4769–4778.

43. Maleki, H.; Hallaj, S.A.; Selman, J.R.; Dinwiddle, R.B.; Wang, H. Thermal properties of lithium-ion battery and components. *J. Electrochem. Soc.* **1999**, *146*, 947–954.

44. Robinson, J.B.; Darr, J.A.; Eastwood, D.S.; Hinds, G.; Lee, P.D.; Shearing, P.R.; Taiwo, O.O.; Brett, D.J.L. Non-uniform temperature distribution in Li-ion batteries during discharge—A combined thermal imaging, X-ray micro-tomography and electrochemical impedance approach. *J. Power Sources* **2014**, *252*, 51–57.

45. Xing, Y.; Ma, Eden W.M.; Tsui, K.L.; Pecht, M. Battery management systems in electric and hybrid vehicles. *Energies* **2011**, *4*, 1840–1857.

46. Xing, Y.; He, W.; Pecht, M.; Tsui, K.L. State of charge estimation of lithium-ion batteries using the open-circuit voltage at various ambient temperatures. *Appl. Energy* **2014**, *113*, 106–115.

47. He, W.; Willard, N.; Chen, C.; Pecht, M. State of charge estimation for electric vehicle batteries using unscented Kalman Filtering. *Microelectron. Reliab.* **2013**, *53*, 840–847.

48. Waag, W.; Fleischer, C.; Sauer, D.U. Critical review of the methods for monitoring of lithium-ion batteries in electric and hybrid vehicles. *J. Power Sources* **2014**, *258*, 114–130.

49. Maarleveld, M.; Kreisch, K.; Kellerbauer, H.; Friedrich, K. Investigation on the propagation of disturbing pulses in traction batteries of electric and hybrid electric vehicles. In Proceedings of the 2014 International Symposium on Electromagnetic Compatibility (EMC Europe), Gothenburg, Sweden, 1–4 September 2014; pp. 391–395.

50. Rothgang, S.; Nordmann, H.; Schaeper, C.; Sauer, D.U. Challenges in battery pack design. In Proceedings of the 2012 Electrical Systems for Aircraft, Railway and Ship Propulsion (ESARS), Bologna, Italy, 16–18 October 2012; pp. 1–6.

51. Gallardo-Lozano, J.; Romero-Cadaval, E.; Milanes-Montero, M.I.; Guerrero-Martinez, M.A. Battery equalization active methods. *J. Power Sources* **2014**, *246*, 934–949.

52. Einhorn, M.; Roessler, W.; Fleig, J. Improved performance of serially connected Li-ion batteries with active cell balancing in electric vehicles. *Trans. Veh. Technol.* **2011**, *60*, 2448–2457.

53. Scott, P. The New BMW i3. Available online: http//www.asymcar.com/graphics/14/i3/bmwi3b.pdf (accessed on 14 April 2015).

54. Liaw, B.Y.; Roth, E.P.; Jungst, R.G.; Nagasubramanian, G.; Case, H.L.; Doughty, D.H. Correlation of Arrhenius behaviors on power and capacity fades, impedance, and static heat generation in lithium ion cells. *J. Power Sources* **2003**, *119*, 874–886.

Improved Adaptive Droop Control Design for Optimal Power Sharing in VSC-MTDC Integrating Wind Farms

Xiaohong Ran, Shihong Miao and Yingjie Wu

Abstract: With the advance of insulated gate bipolar transistor (IGBT) converters, Multi-Terminal DC (MTDC) based on the voltage-source converter (VSC) has developed rapidly in renewable and electric power systems. To reduce the copper loss of large capacity and long distance DC transmission line, an improved droop control design based on optimal power sharing in VSC-MTDC integrating offshore wind farm is proposed. The proposed approach provided a calculation method for power-voltage droop coefficients under two different scenarios either considering local load or not. The available headroom of each converter station was considered as a converter outage, to participate in the power adjustment according to their ability. A four-terminal MTDC model system including two large scale wind farms was set up in PSCAD/EMTDC. Then, the proposed control strategy was verified through simulation under the various conditions, including wind speed variation, rectifier outage and inverter outage, and a three-phase short-circuit of the converter.

Reprinted from *Energies*. Cite as: Ran, X.; Miao, S.; Wu, Y. Improved Adaptive Droop Control Design for Optimal Power Sharing in VSC-MTDC Integrating Wind Farms. *Energies* **2015**, *8*, 7100–7121.

1. Introduction

Renewable energy especially large scale wind power generation, is rapidly becoming an alternative to traditional generation technologies due to less pollutant emission. However, large scale wind farms are often located in remote areas. Given the large capacity and long distance wind power bulk, the high voltage direct current (HVDC) based on voltage source converter (VSC) is considered to be a feasible solution to transmitting wind power because of its great advantages [1]. VSC-HVDC can control active and reactive power independently, especially as it is able to provide black-start capability [2,3], whose typical topology of two-terminal VSC-HVDC system is shown in Figure 1. Multi-terminal DC (MTDC) grids are foreseen as an alternative solution to point-to-point connections owing to their increased redundancy and higher flexibility [4]. Intensive researches have recently been conducted to resolve various technical issues in the VSC-MTDC system, such as locating and isolating of DC faults [5], operation and control of MTDC grid integrating wind farms and so on [6,7]. In order to improve the capacity and stable operation of MTDC, X. Chen *et al.* [8] have studied the control methods of hybrid MTDC grid integrating wind farms, and proposed corresponding control strategies. B. Silva and his colleagues [9] have done researches into fault ride through capacity of VSC-MTDC grid. However, there are few studies on control strategies for reducing the copper loss of MTDC grid integrating wind farms under different operation condition.

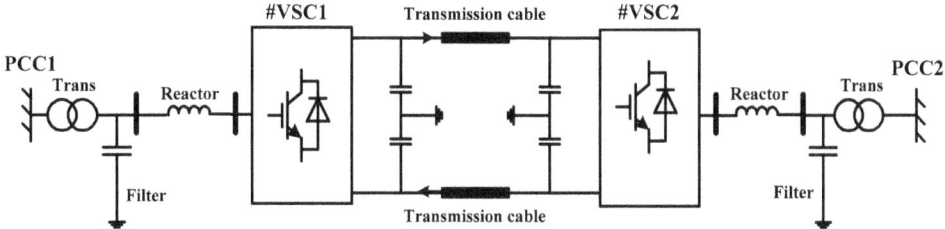

Figure 1. Configuration of voltage-source converter-high voltage direct current (VSC-HVDC) system.

The stable operation condition of VSC-MTDC is to guarantee the constant DC voltage control. DC voltage control mainly includes two types of control strategy: One is the master-slave scheme [10,11], and the other is DC voltage droop control [12,13]. Master-slave control needs communication system between different converters. One of the converters is chosen to control DC voltage of VSC-MTDC to compensate power losses, which is similar with the role of a slack bus in an AC power system, while other converters utilize the constant power control scheme. The duty of DC voltage droop control is to share imbalanced instantaneous power among two or more VSC terminals according to value of slopes [14]. As for the above control strategies, the active and/or reactive power reference or values of slopes will change the power flow of whole power system, thus affecting copper loss of VSC-MTDC system. According to Haileselassie's study [12], DC voltage droop control is considered more reliable than the master-slave scheme due to power balancing, so steady operation of MTDC grids should not depend on a main VSC-HVDC terminal.

Then there are two problems we will face: One is how to reduce copper loss as VSC-MTDC operates steadily, the other is how VSC-MTDC will react to power imbalance caused by the outage of one or more converter stations. The concept and working principle of droop control have been proposed for power sharing and frequency through VSC-MTDC grid [15–17]. Haileselassie's group [12] has discussed the impact of DC voltage drops on distribution of DC grid balancing power, and concluded that the DC voltage droop coefficient determined the degree of power sharing. Rouzbehi *et al.* [18] has studied the voltage-droop strategy based on optimal

DC power flow, all converters operate normally. However, the references above merely considered fixed droop coefficient, ignoring actual operational condition of VSC-MTDC. In another study, Chaudhuri *et al.* [19] considered the particular operating condition of each converter, presented available headroom, and proposed an adaptive droop control scheme for appropriate power sharing. In our study, to solve the two problems of VSC-MTDC above, an improved droop control would be proposed to minimize copper loss of the whole MTDC system integrating large scale wind farms under two different scenarios considering local load or not. On the basis of reducing copper loss, the available headroom of each converter station is taken into account in case of the outage, and then an adaptive improved droop control is proposed to improve the dynamic response ability of converters.

The rest of this paper is organized as follows. Modeling and control of MTDC are introduced in Section 2, followed by Sections 3 and 4 which propose the improved droop control strategy and post-contingency operation state. Then we validated this droop control of VSC-MTDC in EMTDC/PSCAD in Section 5, and drew conclusions in Section 6.

2. Modeling and Control of MTDC

In terms of the point-to-point VSC-HVDC links, one converter station is used to maintain DC link voltage which acts as a slack/swing converter station. The other stations operate in power control mode to ensure the scheduled power exchange.

2.1. Converter Modeling

The main circuit topology of converter for VSC-HVDC is composed of a conventional two-level six-bridges pulse width modulation (PWM) voltage source converter [20,21], which is shown in the Figure 2. In Figure 2, U_s refers to fundamental component of AC bus voltage, U_c is the fundamental component of converter's output voltage, δ is the angle that $\dot{U}_c$ lags to $\dot{U}_s$ parameter R and X are the inductance and resistance of phase reactor of the VSC respectively, and capacitor C_f represents the shunt filter.

Ignoring the resistance R and harmonic components, the active and reactive power absorbed by VSC-HVDC system is given in [22], because the angle δ is very small, then $\sin\delta = \delta$, $\cos\delta = 1$ active and reactive power absorbed by VSC are described as follows

$$\begin{cases} P = \frac{U_s U_c}{X}\delta \\ Q = \frac{U_s(U_s - U_c)}{X} \end{cases} \tag{1}$$

where, $X = \omega L$ is the reactance of the inverter; ω is the nominal angular frequency of AC system. Figure 2 shows a schematic diagram of VSC connected to an AC grid. The AC-side dynamics of the converter can be expressed by the following equation

$$U_{sabc} - U_{cabc} = L\frac{di_{abc}}{dt} + Ri_{abc} \tag{2}$$

where, i_{abc} represents the current flowing through interfacing reactor and coupling transformer.

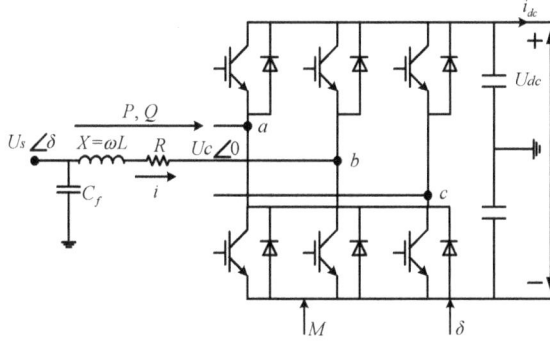

Figure 2. Main circuit topology diagram of voltage source converter.

A synchronous d-q reference approach is conventionally employed to facilitate VSC-HVDC control, and the positive-sequence three-phase voltages U_{sabc} and currents i_{abc} are transformed to d-q components U_{sdq} and i_{dq} using Park Transformation.

$$U_{sdq} = U_d + jU_q = \frac{2}{3}je^{-j\omega t}(U_{sa} + e^{-j\frac{2}{3}\pi}U_{sb} + e^{j\frac{2}{3}\pi}U_{sc}) \tag{3}$$

$$i_{dq} = i_d + ji_q = \frac{2}{3}je^{-j\omega t}(i_a + e^{-j\frac{2}{3}\pi}i_b + e^{j\frac{2}{3}\pi}i_c) \tag{4}$$

where, U_{sd} and U_{sq} are the voltage at d-axis and q-axis respectively; i_d and i_q are the current at d-axis and q-axis respectively. In the synchronous d-q reference frame, the dynamics of VSC in Equations (3) and (4) can be expressed as follows

$$U_{sd} - U_{cd} = L\frac{di_d}{dt} + Ri_d - \omega Li_q \tag{5}$$

$$U_{sq} - U_{cq} = L\frac{di_q}{dt} + Ri_q + \omega Li_d \tag{6}$$

where, $U_{cd} = MU_{dc}\sin\delta/2$; $U_{cq} = MU_{dc}\cos\delta/2$.

The instantaneous active and reactive power absorbed by VSC in *dq*-axis are given in [23], which could be described as follows

$$\begin{cases} P = 1.5U_{sd}i_d \\ Q = -1.5U_{sd}i_q \end{cases} \tag{7}$$

It can be seen from Equation (1), the angle and amplitude of ac voltage for VSC-HVDC system determine the absorbing of active and reactive power, and the following expressions could be obtained

(i)   Angle of phase shifting: $\delta = \arctan \frac{U_{cq}}{U_{cd}}$

(ii)   Modulation index: $M = \frac{\sqrt{U_{cd}^2 + U_{cq}^2}}{U_{dc}/2}$

Therefore, according to U_{cd}, U_{cq} and U_{dc} of VSC, both of the modulation index *M* and angle δ are obtained, to realize the independent control of active power and reactive power of VSC-HVDC.

2.2. Control Method of VSC-MTDC Integrating Wind Farm

As for the long distance and large transmission capacity VSC-MTDC system, the copper loss is substantial due to large resistance and imperfect power distribution of each DC power transmission line. To reduce copper loss of VSC-MTDC, we proposed an improved droop control strategy, the basic idea of which is as follows: Any one inverter converter of VSC-MTDC is chosen as a benchmark converter, and fixed droop control method is used, then the signals of power output for benchmark inverter station are transmitted to other inverter stations. Droop coefficient of other inverter stations would be adjusted according to power difference between benchmark inverter stations and other ones adaptively. The concrete implementation process will be realized in the following paragraphs.

In this paper, a four-terminal HVDC system will be considered two large-scale offshore wind farms are integrated into this four-terminal HVDC system. As is shown in Figure 3, each offshore wind farm contains its local AC network to connect individual wind farms, and each feeder is to convert the AC into DC fed to the HVDC network.

In Figure 3, there are two wind farms feeding power by cables 1 and 2 which are joined together with cable 5 at the offshore connection point. The transmitted power is then divided among cables 3 and 4 into two VSC stations. In general, the control strategies for sending and receiving feeders for VSC are different. In this study, the VSCs of sending feeder and receiving feeder are controlled by power synchronization control, and offshore VSC is connected to the wind farm. The control strategy is holding a constant magnitude of ac voltage and frequency so that power generated by wind farm can be dispatched. In addition, the control strategy of onshore VSC is holding a constant DC voltage, to realize the power balance of VSC-MTDC system.

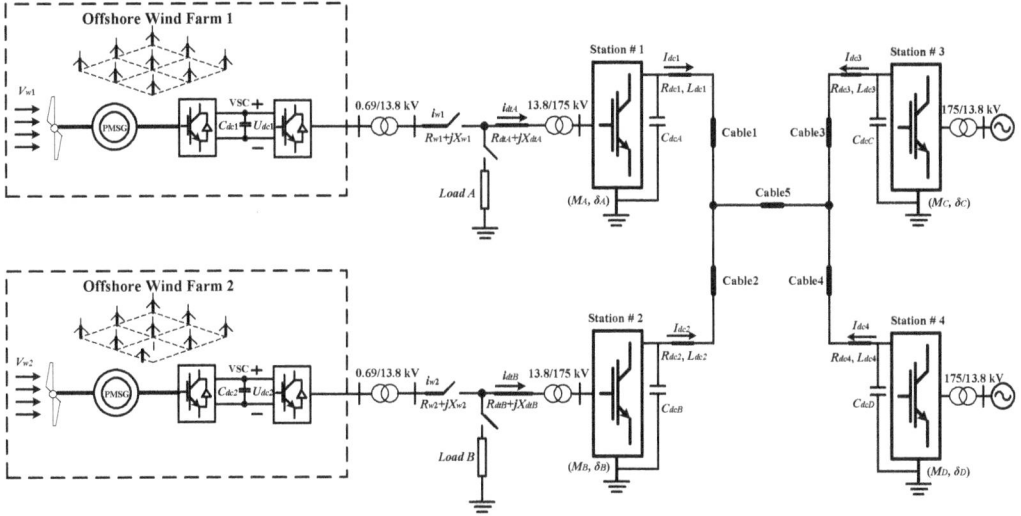

Figure 3. Wind farms integration in a four-terminal HVDC system.

2.2.1. Control of VSC for Offshore Wind Farm

The equivalent circuit of wind farm connected to VSC station is shown in Figure 4. Ignoring resistance R and harmonic components of VSC, voltage and current of ac side are described as follows

$$\begin{cases} C_1 p U_{WM} = I_{WM} - I_{CV} \\ R_1 I_{CV} + L_1 p I_{CV} = U_{WM} - U_{CV} \end{cases} \tag{8}$$

In the analysis, Equation (8) is expressed in dq-axis coordinate system, which is given by

$$\begin{cases} I_{WMd} - I_{CVd} = C_1 p U_{WMd} - \omega C_1 U_{WMq} \\ I_{WMq} - I_{CVq} = C_1 p U_{WMq} + \omega C_1 U_{WMd} \end{cases} \tag{9}$$

$$\begin{cases} U_{WMd} - U_{CVd} = (R_1 + p L_1) I_{CVd} - \omega L_1 I_{CVq} \\ U_{WMq} - U_{CVq} = (R_1 + p L_1) I_{CVq} + \omega L_1 I_{CVd} \end{cases} \tag{10}$$

where, $p = d/dt$ is differential operator.

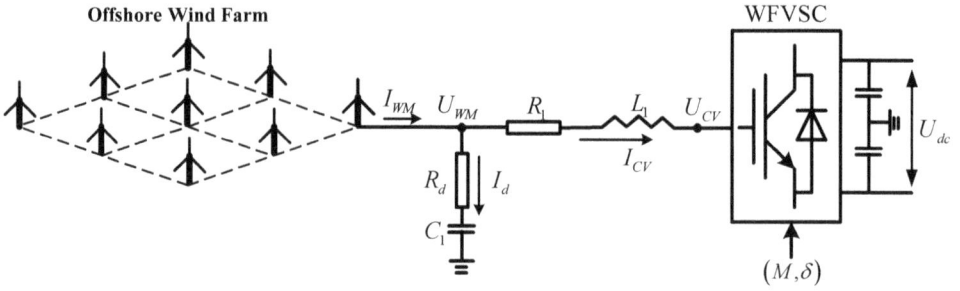

Figure 4. Equivalent circuit of wind farms connected to VSC station.

According to Equation (9), voltage loop control would be built, then reference value of current loop control can be obtained

$$
\begin{cases}
I_{CVd}^{ref} = -[k_{p1}(U_{WMd}^{ref} - U_{WMd}) + k_{i1}\int (U_{WMd}^{ref} - U_{WMd})dt] + I_{WMd} + \omega C_1 U_{WMq} \\
I_{CVq}^{ref} = -[k_{p1}(U_{WMq}^{ref} - U_{WMq}) + k_{i1}\int (U_{WMq}^{ref} - U_{WMq})dt] + I_{WMq} - \omega C_1 U_{WMd}
\end{cases}
\tag{11}
$$

Then the d-axis is coincidence with ac voltage vector, $U_{WMd}^{ref} = 1$, $U_{WMq}^{ref} = 0$. According to Equation (10), current loop control would be described as follows

$$
\begin{cases}
U_{CVd}^{ref} = -[k_{p2}(I_{CVd}^{ref} - I_{CVd}) + k_{i2}\int (I_{CVd}^{ref} - I_{CVd})dt] + U_{WMd} + \omega L_1 I_{CVq} \\
U_{CVq}^{ref} = -[k_{p2}(I_{CVd}^{ref} - I_{CVd}) + k_{i2}\int (I_{CVd}^{ref} - I_{CVd})dt] + U_{WMq} - \omega L_1 I_{CVd}
\end{cases}
\tag{12}
$$

Simplified schematic diagram of control strategy is shown in Figure 5. Back-to-back PWM converters are used for variable speed wind power generation.

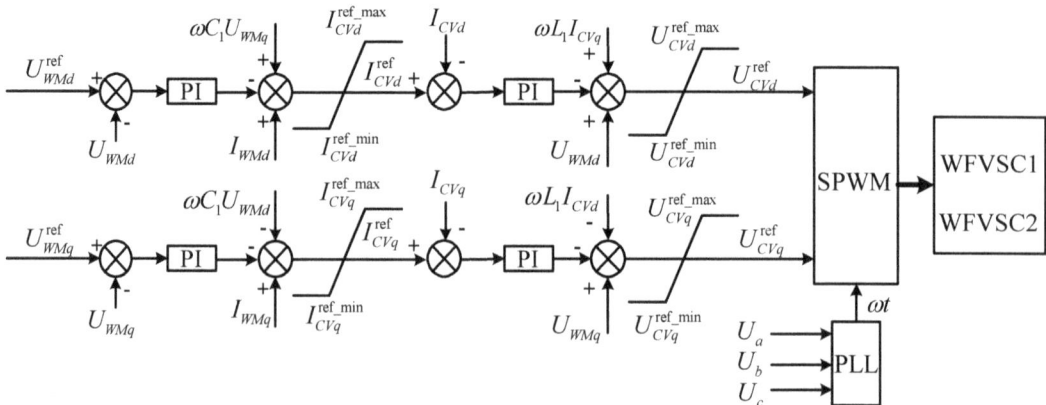

Figure 5. Control block diagram of VSC offshore wind farm.

2.2.2. Control of VSC for Onshore Stations

Onshore VSC stations would be controlled by improved adaptive droop control method. Firstly, this control strategy is holding a constant DC voltage and realizing the power balance of VSC-MTDC system. Secondly, the optimal power distribution between DC transmission lines will be realized, and copper loss of VSC-MVDC system will be reduced. The diagram of the improved adaptive droop control strategy for onshore VSC is shown in Figure 6.

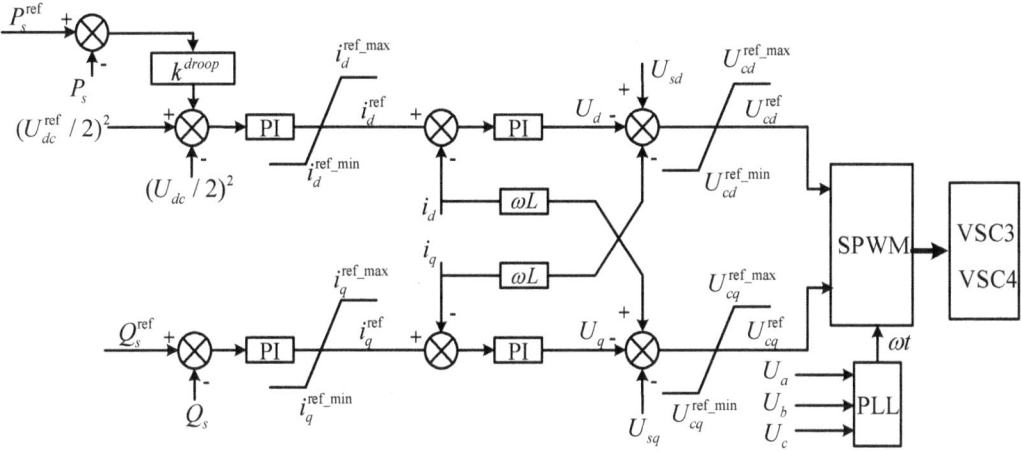

Figure 6. Control block diagram of onshore VSC.

3. Improved Adaptive Droop Control in VSC-MTDC

3.1. Fixed Droop Method

As for two terminal VSC-HVDC links, one station operates in active power control mode to ensure scheduled active power, the other operates in DC voltage control mode to provide constant DC voltage for the system and supply power loss due to line resistance, which acts as a slack VSC converter station. As for VSC-MTDC grid, if one or more converters follow an outage, remaining ones should share power gap in certain appropriate proportion. Droop coefficient is very important at different converter stations in that it determines how a power imbalance will be shared among converters, which has been discussed in [12]. If droop coefficient is identical, remaining converters are required to share power imbalance at equal ratio. On the other hand, if droop coefficient is unequal, the higher ones would have dominant contribution from active power control loop.

However, actual load condition has not been considered in a fixed droop control. In fact, if a converter follows an outage, some converters may achieve the upper limit of their capacity, though they are unable to share more power imbalance. In addition, the issue needs to be settled as to how much power ratio should be shared for each remaining station to realize minimum copper loss of VSC-MTDC. Adaptive droop control can solve these problems and will be described in the next subsection.

3.2. Design for Minimum Copper Loss

The relationship of power sharing between receiving end feeders is shown in Figure 7. It can be seen from Figure 7 that U_t and I_t are the voltage and current at sending end, U_{GSx} and U_{GSy} are the receiving end converter voltages, I_{GSx} and I_{GSy} are receiving end currents, R_{GSx} and R_{GSy} are resistance at receiving end. The voltage equations of receiving end for VSC-MTDC system are expressed as

$$
\begin{cases}
U_t = U_{GSx} + I_{GSx} R_{GSx} \\
U_t = U_{GSy} + (I_t - I_{GSx}) R_{GSy}
\end{cases}
\tag{13}
$$

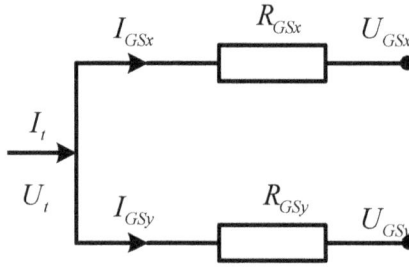

Figure 7. Power sharing between receiving end feeders.

The copper loss of VSC-HVDC between receiving end feeders is given by

$$
P_{copper}^{losss} = R_{GSx} I_{GSx}^2 + R_{GSy} (I_t - I_{GSx})^2
\tag{14}
$$

Then Equation (14) is differentiated with respect to I_{GSx}, then equating to zero. According to reference [24], the power sharing ratio for minimum copper loss between two feeders is inversely proportional to feeder resistances

$$
\frac{d P_{copper}^{losss}}{d I_{GSx}} = 2 R_{GSx} I_{GSx} - 2 R_{GSy} (I_t - I_{GSx}) = 0
\tag{15}
$$

$$
\frac{P_{GSx}}{P_{GSy}} = \frac{R_{GSy}}{R_{GSx}}
\tag{16}
$$

The power loss is minimized if the voltage magnitudes of different receiving end converters are the same [24]. If there are n cables between receiving end feeders, the power sharing ratio for minimum copper loss of VSC-MTDC system is as follows

$$P_{GS1} : P_{GS2} : \cdots : P_{GSn} = \frac{1}{R_{GS1}} : \frac{1}{R_{GS2}} : \cdots : \frac{1}{R_{GSn}} \qquad (17)$$

3.3. Improved Adaptive Droop Control

To reduce copper loss of VSC-MTDC grid, optimal strategy of power sharing based on discussions in the paragraph above is proposed, and the concrete implementation process is as follows. As for the MTDC grid, the first inverter converter is operated by fixed droop, which is given by [19]

$$\frac{U_{dc\_slack}^2 - \left(U_{dc1}^{ref}\right)^2}{4} + k_1^{droop}\left(P_{GS1}^{ref} - P_{GS1}\right) = 0 \qquad (18)$$

In this paper, the common DC reference voltage $U_{dc,slack}$ is introduced as reference voltage of MTDC system, whose value is chosen from the converter with DC link voltage control mode. The fixed droop coefficient of the first inverter converter is described as follows

$$k_1^{droop} = \frac{\left(U_{dc1,min}^{ref}\right)^2 - U_{dc\_slack}^2}{4(P_{GS1}^{ref} - P_{GS1}^{max})} \qquad (19)$$

where, $U_{dc,min}$ is minimum value of DC link voltage; P_{GSi}^{max} is the maximum value of active power for converter i. Then the improved droop control of the ith converter is given by

$$\frac{U_{dc\_slack}^2 - \left(U_{dci}^{ref}\right)^2}{4} + k_i^{droop}\left(P_{GSi}^{ref} - P_{GSi}\right) = 0 \qquad (20)$$

$$k_i^{droop} = k_1^{droop} + m_i\left(P_{GS1} - P_{GSi}\right) \qquad (21)$$

Then, m_i is defined as the power difference coefficient between the benchmark inverter station and other inverter ones, which will be deduced in the following paragraphs.

(1) Inverter converter without local load

In terms of inverter converter station, if $\delta = 0$, and $P_{GSi} = 0$; if $\delta \neq 0$, the fundamental component of converter's output voltage is described as follows

$$U_{ci} = \frac{P_{GSi}X_i}{U_s\delta} \qquad (22)$$

387

From Equations (18), (20) and (22), if $\delta = 0$, then

$$k_i^{droop}\left(P_{GSi}^{ref} - P_{GSi}\right) - k_1^{droop}\left(P_{GS1}^{ref} - P_{GS1}\right) = 0 \tag{23}$$

Substituting Equations (21) into (23), then expression of power difference coefficient of inverter stations m_i is obtained as follows

$$m_i = \left|\frac{-k_1^{droop}\left[\left(P_{GSi}^{ref} - P_{GSi}\right) + \left(P_{GS1} - P_{GS1}^{ref}\right)\right]}{\left(P_{GSi}^{ref} - P_{GSi}\right)\left(P_{GS1} - P_{GSi}\right)}\right| \tag{24}$$

Similarly, if $\delta \neq 0$, m_i is given by

$$m_i = \left|\frac{-k_1^{droop}\left[\left(P_{GSi}^{ref} - P_{GSi}\right) + \left(P_{GS1} - P_{GS1}^{ref}\right)\right] + \left(P_{GSi}^2 X_i^2 - P_{GS1}^2 X_1^2\right)/\left(4U_s^2\delta^2\right)}{\left(P_{GSi}^{ref} - P_{GSi}\right)\left(P_{GS1} - P_{GSi}\right)}\right| \tag{25}$$

(2)   Inverter converter with local load

As for the first inverter converter station, if $\delta = 0$, $P_{GSi} = 0$; if $\delta \neq 0$, then fundamental component of converter's output voltage is given by

$$U_{ci} = \frac{\left(P_{GSi} - P_{loadi}\right) X_i}{U_s\delta} \tag{26}$$

If $\delta = 0$, m_i is shown in Equation (24). Similarly, according to Equations (18), (20) and (26), if $\delta \neq 0$, m_i could be obtained as follows

$$m_i = \left|\frac{-k_1^{droop}\left[\left(P_{GSi}^{ref} - P_{GSi}\right) + \left(P_{GS1} - P_{GS1}^{ref}\right)\right] + \left[\left(P_{GSi}^{ref} - P_{loadi}\right)^2 X_i^2 - \left(P_{GS1}^{ref} - P_{load1}\right)^2 X_1^2\right]/\left(4U_s^2\delta^2\right)}{\left(P_{GSi}^{ref} - P_{GSi}\right)\left(P_{GS1} - P_{GSi}\right)}\right| \tag{27}$$

where, P_{load1} and P_{Loadi} are local load of the first and ith inverter converter respectively; and the data of m_i could be obtained according to communication device of system.

In terms of fixed droop control, if the rating of converter stations is identical, a power imbalance caused by an outage would be shared equally with other remaining converter stations. However, some converter stations may not be able to participate in power sharing equivalently due to the actual operating condition of the system. Therefore, available headroom of the ith converter is defined as [19]

$$v_i = \kappa_i - |P_i| \tag{28}$$

$$k_{ia}^{droop} = k_{i0}^{droop}\left(\frac{\kappa^{max}}{v_i}\right)^{\tau} \tag{29}$$

where, κ_i is rated capacity of the ith converter; v_i is remaining capacity of the ith converter; k_{i0}^{droop} and k_{ia}^{droop} are improved adaptive droop coefficient under pre-outage and post-outage operating conditions, respectively; $\kappa^{max} = \max(\kappa_1, \kappa_2, \cdots, \kappa_n)$ refers to maximal value of all converters' capacity; and τ is a user defined positive constant, whose value of and modal analysis of adaptive droop are discussed in [19].

4. Post-Contingency Operation

In this section, operating point of post-contingency for VSC-MTDC will be proposed. Assuming there are n converter stations in MTDC system, the active power must maintain real-time balance, which will satisfy

$$\sum_{i=1}^{n} P_i + P_{losses} = 0, \quad \sum_{i=1}^{n} P_i^{ref} = 0 \tag{30}$$

where, P_i and P_i^{ref} are the actual active power and active power reference value of the ith converter respectively. Moreover, total real power loss caused by DC line resistance is $P_{losses} = -\sum_{i=1}^{n} P_i$. According to voltage droop control strategy, improved adaptive droop control for pre-outage under steady state is described as follows

$$\left(U_{dci}^{ref}\right)^2 - U_{dc\_slack}^2 = 4\frac{P_{losses}}{\sum_i \frac{1}{k_{i0}^{droop}}} \tag{31}$$

$$P_i = P_i^{ref} - \frac{P_{losses}}{k_{i0}^{droop} \sum_i \frac{1}{k_{i0}^{droop}}} \tag{32}$$

If the converter outage of MTDC system is introduced at the nth converter, that is

$$\sum_{i=1}^{n-1} P_i^{ref} = -P_n^{ref} \sum_{i=1}^{n-1} P_i = -P_{loss}' \tag{33}$$

where, P_{loss}' is total real power loss when converter outage is introduced at the nth converter.

Following the outage of the nth converter, the steady operating point at post-contingency for VSC-MTDC system will be changed, which is given by

$$\left(U_{dci}^{ref}\right)^2 - \left(U_{dc\_slack}'\right)^2 = 4\frac{\left(P_{losses}' - P_n^{ref}\right)}{\sum_{i=1}^{n-1} \frac{1}{k_{ia}^{droop}}} \tag{34}$$

$$P'_i = P_i^{\text{ref}} - \frac{P_{losses} - P_n^{\text{ref}}}{\frac{1}{k_{ia}^{droop}} \sum\limits_{i=1}^{n-1} k_{ia}^{droop}} \tag{35}$$

Comparing the operating point between pre-outage and the post-outage, the power changes for the ith converter is described as

$$\Delta P_i = P'_i - P_i = \frac{P_{losses}}{k_{i0}^{droop} \sum\limits_i \frac{1}{k_{i0}^{droop}}} - \frac{P_{losses} - P_m^{\text{ref}}}{\frac{1}{k_{ia}^{droop}} \sum\limits_{i=1}^{m-1} k_{ia}^{droop}} \tag{36}$$

There is some differences involved improved adaptive droop coefficient between the inverter converter stations with or without local load. Figures 8 and 9 show corresponding droop strategies for the VSC-MTDC system.

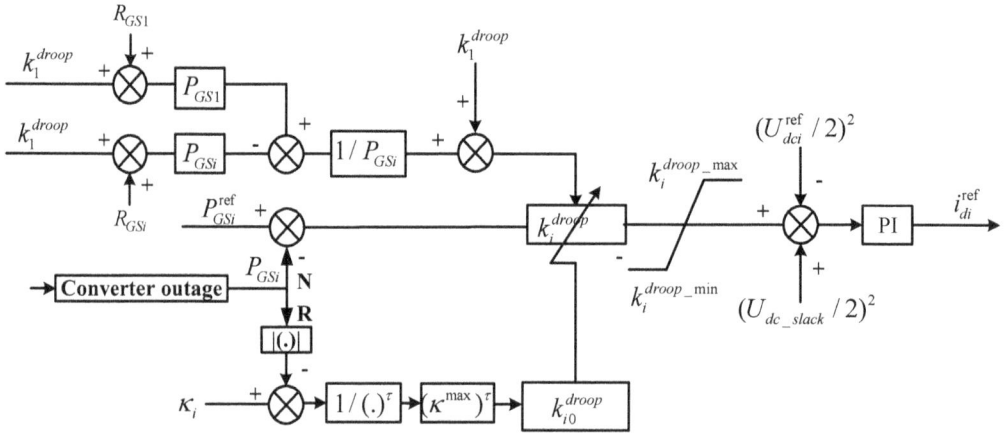

Figure 8. Block diagram of improved adaptive droop scheme without local load.

In addition, the tuning values of PI parameters for designed controller could be described from the following aspects:

In terms of the PWM converter, initial value and steady-state value of reactive current i_q are nearly equal to zero, it is not obvious change with the tuning of controller parameters. The response curves of i_d need to be studied in detail to ensure the parameters of controllers. In terms of PI parameters K_P and K_I of current inner loop, the basic principle is that does not increase overshoot amount and also ensure response speed of designed controller. The best tuning values of PI parameters for current loop are as follows

$$\tau_2 = \frac{L}{R}, \quad K_2 = \frac{L}{2T_{\text{PWM}}} \tag{37}$$

where, $T_{PWM} = T_s/2$, and T_s is the switch time period of PWM.

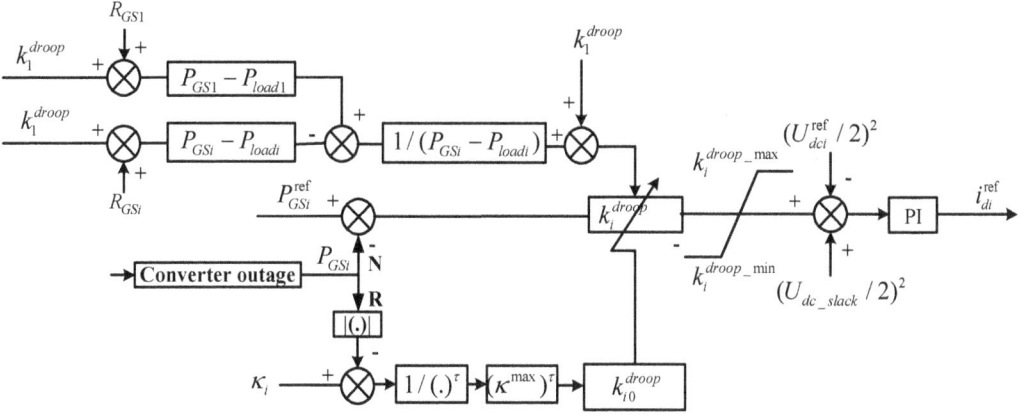

Figure 9. Block diagram of improved adaptive droop scheme with local load.

In terms of PI parameters K_P and K_I of voltage loop, if the K_P increases, the intensity of voltage rising will be improved; if the K_I increases, which can accelerate the speed of voltage into without static error state. Therefore, the best tuning values of PI parameters for voltage loop are described

$$\tau_1 = 2R_{eq}C, \quad K_1 = \frac{CU_{dcN}}{2T_{PWM}U_s} \tag{38}$$

where, the equivalent resistance $R_{eq} = U_{dcN}^2/P_N$.

In addition, the best tuning values of PI parameters for power outer loop are given by

$$\tau_1 = 10T_{PWM}, \quad K_1 = \frac{1}{2.24U_s} \tag{39}$$

Therefore, the best tuning values of PI parameters for current loop, voltage loop and power loop of designed controller could be obtained according to Equations (37)–(39).

5. Simulation and Discussion

The improved adaptive droop control of VSC-MTDC have been implemented by PSACD/EMTDC, and the configuration and control system of VSC-MTDC is shown in Figure 3. Four-terminal HVDC system is a bipolar scheme with a nominal 300 kV DC voltage. The wind farm is modeled as one aggregated PMSG driven by a single equivalent wind turbine. Parameters of DC cables and communicating transformers are displayed in Tables 1 and 2 respectively. The total rated capacity of MTDC grid is about 2180 MVA, and rated ac voltage is 175 kV, $\tau= 2.5$, $\kappa^{max}= 900$ MW.

Table 1. Parameters of DC Cables.

Number of Cable	1	2	3	4	5
Resistance/ω	0.01085	0.01085	0.008675	0.016275	0.04
Reactance/H	0.0002	0.0002	0.00015	0.003	0.015
Distance/km	20	20	50	80	100

Table 2. Parameters of Commutating Transformers.

Converters	Leakage Reactance	Capacity	Transformer Ratio
VSC1	0.15 (pu)	440 MVA	175 kV/13.8 kV
VSC2	0.15 (pu)	440 MVA	175 kV/13.8 kV
VSC3	0.15 (pu)	900 MVA	175 kV/13.8 kV
VSC4	0.15 (pu)	400 MVA	175 kV/13.8 kV

5.1. Wind Speed Variation

The performance of a VSC station integrated with wind farms is investigated in the case of variable wind speed, and the simulation results are shown in Figure 10. The gust wind speed of wind farm 1 and 2 are introduced at 1.8 s and 3.0 s with the fluctuation, slowing down from 12 to 10 m/s and ramping up from 10 to 12 m/s for wind farm 1, ramping up from 12 to 14 m/s and slowing down from 14 to 12 m/s for wind farm 2 in Figure 10a. To collect the stochastic wind power, the ac side needs to supply constant ac voltage for the wind farm network and absorb fluctuant power automatically.

With the increased or decreased wind speed at 1.8 s and 3.0 s, it can be seen from Figure 10b that active power generated by wind farm rose or fell near to a constant value, thus the control strategy of wind turbine can follow the wind speed fluctuation effectively. In addition, it can be seen from Figure 10c that the ac voltage on the wind farm side was held at 13.8 kV throughout the wind speed fluctuation as the control strategy designed. Corresponding to the power variation of the VSC rectifier side due to the wind speed change, then slack converter (VSC inverter side) would take effect to balance power flow in VSC-MTDC system.

5.2. Normal Operation of VSC-MTDC

In this section, the performance of the fixed droop control and improved adaptive droop control are compared under normal operation condition. For fixed droop, droop coefficient of station #4 is equal to 0.03418, the one at station #3 is 0.01519 according to their converter ratings. In terms of improved droop control, station #4 is a benchmark converter station whose droop coefficient is constant. The one at station #3 is 0.1424 under the condition of optimal power sharing, which is much larger than the one of reference [19]. Therefore, station #3 would share much more power transmission to reduce the copper loss of VSC-MTDC grid, and $\tau = 2.5$ was used for the improved adaptive droop control. Five different operational scenarios for inverter side converter station #3 and #4 are investigated, which mainly includes $P_{3ref}/P_{4ref} = 6, 3, 1, 1/3$ and $1/6$. Tables 3 and 4 show results of copper loss of VSC-MTDC without or with local load respectively.

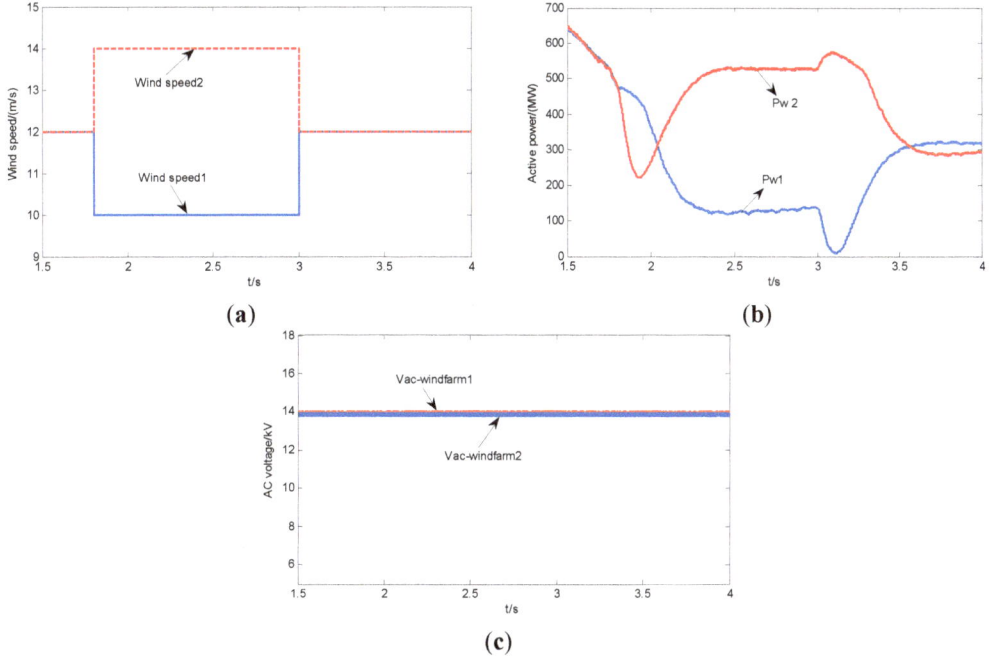

Figure 10. Simulation result with variation wind speed. (**a**) Wind speed; (**b**) AC voltage for different wind farm; (**c**) Active power on wind farm side.

Based on the above discussion, if there are two DC lines in inverter side of VSC-MTDC, the condition obtained for minimum copper loss is that $P_{3ref}/P_{4ref} = R_4/R_3 = 3.001$. From Tables 3 and 4, the minimum copper loss is about 16.7382 MW (without local load) and 16.6435 MW (with local load) respectively when $P_{3ref}/P_{4ref} = 3.0$. In addition, if the capacity ratio of P_{3ref}/P_{4ref} is much greater or smaller than the optimal ratio value 3.0, all the reduced copper loss are larger than the ones of optimal ratio.

The capacity ratio of P_{3ref}/P_{4ref} is inverse proportion to optimal ratio value of the resistance of DC line 3 and 4, and improved adaptive droop control is used for VSC-MTDC without local load and the dynamic responses of four-terminal HVDC are shown in Figure 11. It can be seen from Figure 11a,b that converter station can follow the power reference value quickly, among which stations #1 and #2 export about 300 MW. In addition, due to the obvious DC voltage loss caused by large resistance of DC line, voltage of rectifier side (stations #1 and #2) is about 306 kV and voltage of inverter side (stations #3 and #4) is about 300 kV that is hold by the introduced common DC voltage U_{dc_slack}. If DC line resistance of VSC-MTDC system increases, the voltage loss would be greater.

Table 3. Copper loss of VSC-HVDC (without local load).

Capacity Ratio	Droop Control	Proposed Method
$P_{3ref}/P_{4ref} = 6$	17.0368 MW [12]	16.8933 MW
$P_{3ref}/P_{4ref} = 3$	16.8793 MW [18]	16.7382 MW
$P_{3ref}/P_{4ref} = 1$	17.3963 MW [12]	17.1495 MW
$P_{3ref}/P_{4ref} = 1/3$	18.1844 MW [12]	17.8648 MW
$P_{3ref}/P_{4ref} = 1/6$	19.0705 MW [12]	18.7209 MW

Table 4. Copper loss of VSC-HVDC (with local load).

Capacity Ratio	Droop Control [12]	Proposed Method
$P_{3ref}/P_{4ref} = 6$	17.0156 MW [12]	16.8672 MW
$P_{3ref}/P_{4ref} = 3$	16.7841 MW [18]	16.6435 MW
$P_{3ref}/P_{4ref} = 1$	17.3718 MW [12]	17.1211 MW
$P_{3ref}/P_{4ref} = 1/3$	18.1528 MW [12]	17.8232 MW
$P_{3ref}/P_{4ref} = 1/6$	19.0115 MW [12]	18.6413 MW

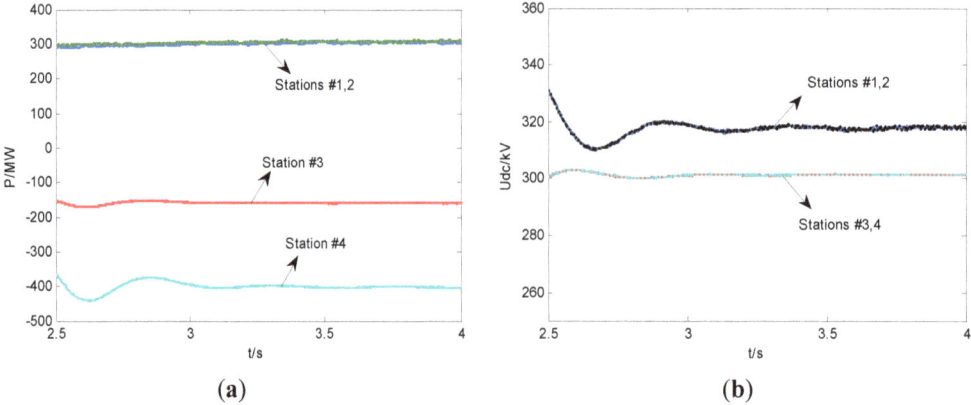

Figure 11. Dynamic response of normal operation (a) Active power (b) DC side voltage.

5.3. Converter Outage of VSC-MTDC

In this section, the performance of the fixed droop and improved adaptive droop scheme is compared in the case of a converter outage. Unequal loading conditions are considered to demonstrate the benefit of the improved adaptive droop. If a DC fault of converter appears, the differential protection will be actuated. When a certain current level is reached, IGBTs would be blocked. As DC breakers have not been used, it is necessary to block converter stations and to use AC circuit breakers to clear the fault. In this case, both rectifier and inverter converter outages are simulated which represent two different operational scenarios.

(1) *Inverter Outage:* In the first case, stations #3 and #4 imports 465 MW and 155 MW, respectively. Stations #1 and #2 export around 620 MW into the DC grid. The outage of station #4 is considered. The results of copper loss of VSC-MTDC without/with local load are shown in Table 5. It can be seen from Table 5, the amounts of copper loss without/with local load reduced from 16.7124 MW and 16.6357 MW to 16.5131 MW and 16.4265 MW as the optimal power sharing of VSC-HVDC grid (P3ref/P4ref = 3), which are more effective for improved droop control than the ones of reference [19]. The reason is that station #3 can adjust droop coefficient dynamic according to available headroom of converter capacity, and ensure the appropriate *P-U* curve, so that reduce copper loss of four-terminal HVDC system.

Table 5. Copper loss as outage of VSC4.

Type	Capacity Ratio	Droop Control [19]	Proposed Method
Without local load	$P_{3ref}/P_{4ref} = 3$	16.7124 MW	16.5131 MW
With local load	$P_{3ref}/P_{4ref} = 3$	16.6357 MW	16.4265 MW

The dynamic response of four-terminal HVDC system following inverter outage is shown in Figure 12. Figure 12a shows that in absence of station #4, slack converter station #3 operating in voltage-power droop control mode increases its power to compensate the power imbalance of the inverter. From Figure 12b, the power sharing of station #3 by improved adaptive droop is larger than that of fixed droop. The reduction of the copper loss is because improved droop control ensures a higher droop value than fixed droop. Variation of DC voltage at converter station #3 is shown in Figure 12c, which is larger than common bus voltage 300 kV with very little overshoot in its transient response for fixed droop control. Therefore, the improved adaptive droop scheme can steady the DC voltage dynamics. In addition, Figure 12d indicates that control strategy of station #3 and #4 can follow the dynamic response effectively. Then HVDC system could reach the post-outage steady-state operating point at 2.4 s. The results in this case are related to system parameters, such as the line resistances. Because the copper loss of HVDC system is square to the line resistances, if line resistances of VSC-HVDC system are much larger, much copper loss would be reduced, and then simulations will be more evident.

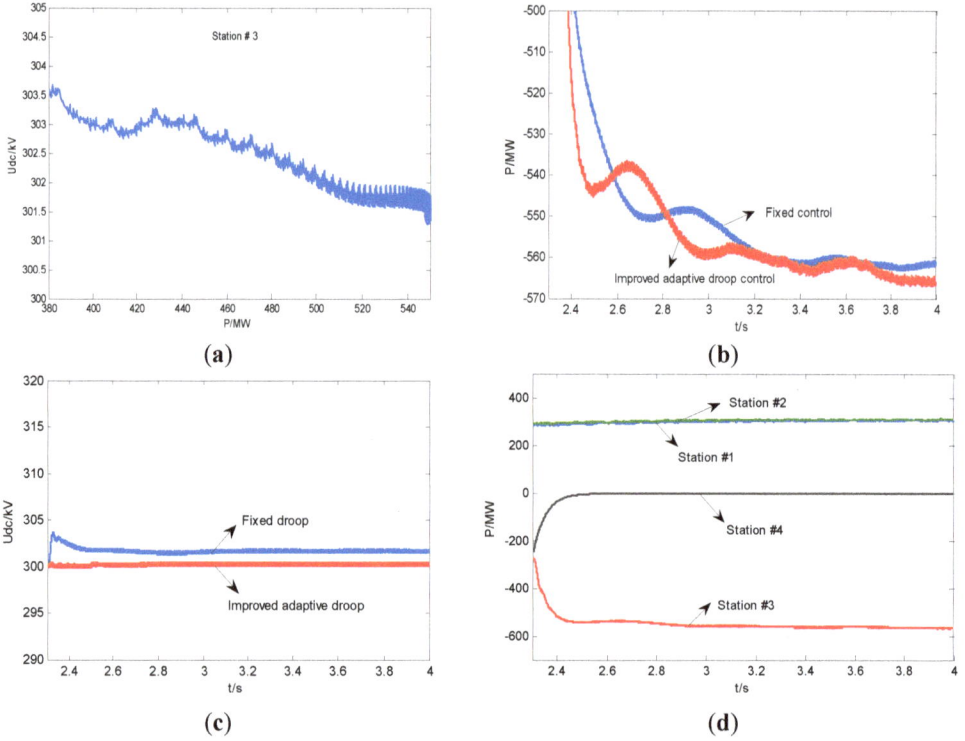

Figure 12. Dynamic responses of inverter outage (**a**) Voltage droop control of station #3; (**b**) Active power of station #3; (**c**) DC side voltage of station #3; (**d**) Active power of all stations.

(2) Rectifier Outage: In order to validate the effectiveness of proposed method, the outage of station #2 is considered, only 300MW is imported into MTDC grid. The results of copper loss of VSC-MTDC without/with local load are shown in Table 6. As we can see, if $P_{3ref}/P_{4ref} = 3.0$, improved adaptive droop control can reduce copper loss effectively. The minimum copper loss without/with local load are 4.2794 MW and 4.2481 MW respectively, which are smaller than those using methods in reference [12]. The reduced amounts of copper loss are 0.2748 MW and 0.251 MW respectively in this case.

Table 6. Copper loss of station #2.

Type	Capacity Ratio	Droop Control [12]	Proposed Method
Without local load	$P_{3ref}/P_{4ref} = 3$	4.5542 MW	4.2794 MW
With local load	$P_{3ref}/P_{4ref} = 3$	4.4991 MW	4.2481 MW

The outage of station #2 appears at 2.3 s, and dynamic response of VSC-MTDC system with fixed droop and improved adaptive droop following the contingency are shown in Figure 13. It is to be noted from Figure 13a that control strategy of station #1 and #2 can follow the power dynamics. In addition, it can be seen from Figure 13b,c that the AC voltage of station #1 can stable operate at 13.8 kV, and it has no effect on DC voltage of VSC inverter side (station #3 and 4) under condition following station #2 outage.

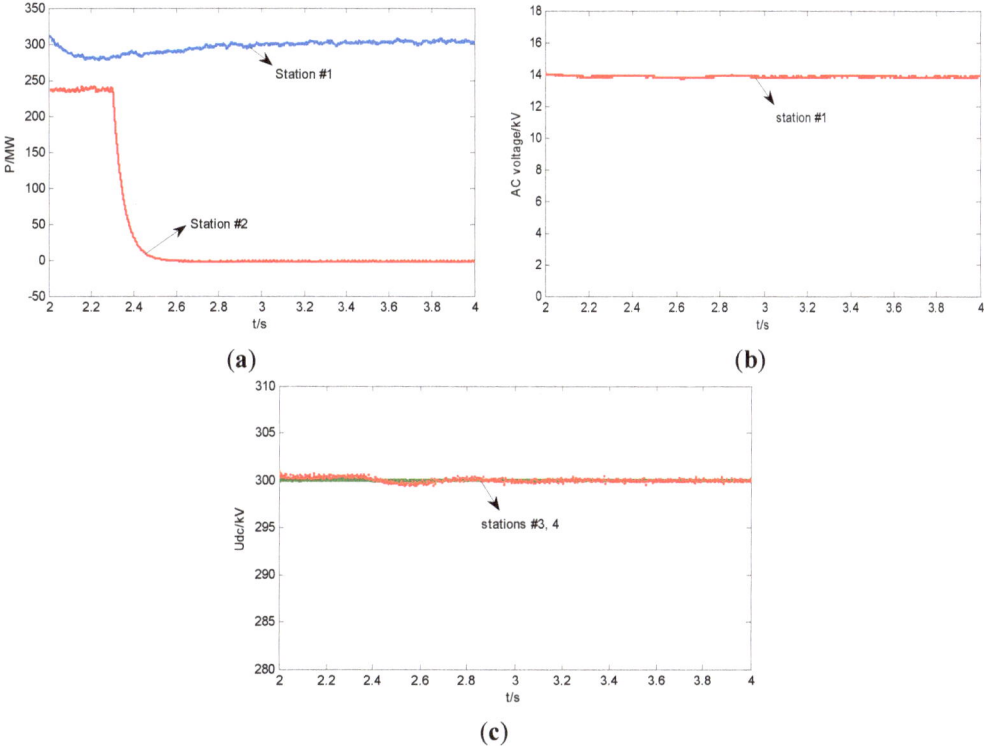

Figure 13. Dynamic response of rectifier outage (**a**) Active power of station # 1 and 2; (**b**) DC side voltage of station # 3 and 4; (**c**) AC side voltage of station # 1.

5.4. Converter Short Circuit of VSC-MTDC

In this section, the performance of the fixed droop and improved adaptive droop control is compared. If a short-circuit fault is detected in AC side close to points of common coupling (PCC), the PCC voltage will be reduced, which in turn affects voltages and currents of ac transmission grid close to PCC. To prevent destructive overcurrent, then distance protection would be actuated at this time. In this case, a three-phase short-circuit fault is introduced at 2.3 s on station #4 with a duration of 300 ms. If the short-circuit fault is detected, the AC side voltage of station #4 decreased from 13.8 kV to 7.0 kV. Dynamic responses of MTDC grid with fixed droop following the contingency are shown in Figure 14. During the disturbance, as shown in Figure 14a,b, DC voltage of station #4 decreased to 298 kV from 300 kV as using the improved droop control while the ones decreased to 293 kV from 300 kV as using fixed droop control method. Therefore, overshoot amount of improved droop control is less than the ones of fixed droop control. Since AC voltage of wind farm was controlled at constant value, its power output about 600 MW injected into VSC-MTDC network remained unchanged. In addition, improved droop control can hold DC voltage of station #3 at 300 kV while fixed droop control cannot realize this objective. The reason is that common DC voltage and available headroom of converter are considered as designing the control strategy of HVDC grid, and the proposed method could improve the stable operation of converter. From Figure 14d,e, it can be seen that the control strategy of station #3 and #4 can follow the active/reactive power dynamic response, and there is small variation for power when improved droop control is employed.

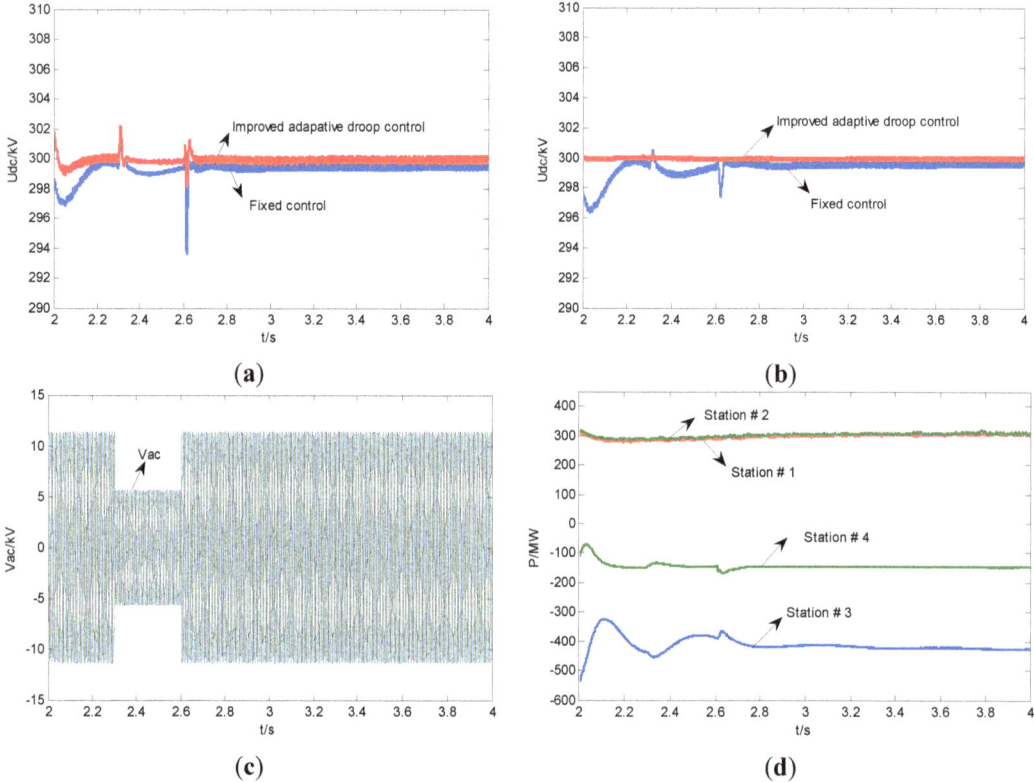

Figure 14. *Cont.*

398

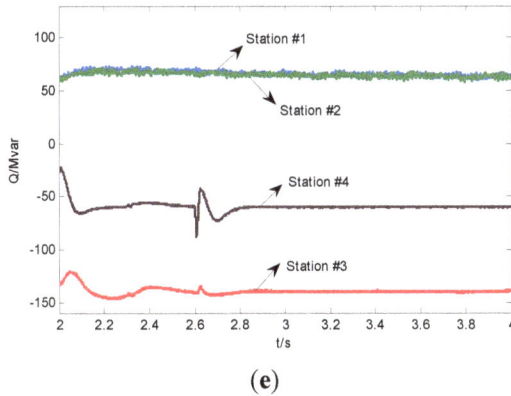

Figure 14. Dynamic response of short-circuit fault. (**a**) DC voltage of station # 4; (**b**) DC voltage of station # 3;(**c**) AC side voltage of station # 4; (**d**) Active power of VSC-MTDC; (**e**) Reactive power of VSC-MTDC.

Therefore, the improved droop control method can reduce the copper loss of system effectively, and improved stable operational performances under the condition of inverter outage, rectifier outage and a short-circuit fault of converter, and decrease dc voltage fluctuation and dc power oscillation caused by AC side fault of VSC-HVDC system. Therefore, the improved droop control scheme leads to desirable system performance.

6. Conclusions

With the development of large capacity and long distance HVDC transmission lines, the copper loss of the whole system is considerable. Therefore, the paper provides a discussion based on optimal power sharing on MTDC grids interconnecting large-scale offshore wind farms with ac mainland grids. The control strategies are proposed to reduce the copper loss of the DC transmission line.

This paper has proposed a calculation framework of the variable power-voltage droop coefficients either considering local load or not. Any inverter converter of MTDC grid can be chosen as a benchmark converter and a fixed droop control is employed. The adaptive droop coefficients of other converters are obtained by power difference between other converters and the benchmark one, the optimal power sharing of MTDC grid is realized. Compared with the fixed droop control method, the control strategy is able to reduce copper loss of MTDC grid more effectively under the condition of normal operation and following an outage of converter.

On the basis of optimal power sharing, available headroom of each converter station is considered in case of converter outage, and then the adaptive improved droop control is proposed to improve stable operation as well as reducing the copper loss of HVDC system. Transient simulations are implemented on a four-terminal MTDC grids integrated with two large capacity wind farms. The results for outages of rectifier and inverter as well as short circuit fault under three different scenarios are presented, and steady-state operating points of post-contingency are deduced in detail.

Acknowledgments: The authors acknowledge the support of projects granted by National Natural Science Foundation of China (Grant No. 51377068).

Author Contributions: The methodology followed during investigation is discussed by all authors. Simulations in PSACD/EMTDC were carried out by Xiaohong Ran and the writing of manscript was done by Xiaohong Ran, Shihong Miao provided some ideas on the paper, Yingjie Wu checked the results.

Conflicts of Interest: The authors declare no conflict of interest.

References

1. ABB Group. It's Time to Connect—Technical Description of HVDC Light Technology. 2012. Available online: http://www.abb.com/industries/ (accessed on 28 April 2015).
2. Liu, Y.; Chen, Z. A flexible power control method of VSC-HVDC link for the enhancement of effective short-circuit ratio in a hybrid multi-infeed HVDC system. *IEEE Trans. Power Syst.* **2013**, *28*, 1568–1581.
3. Flourentzou, N.; Agelidis, V.G.; Demetriades, G.D. VSC-based HVDC power transmission systems: An overview. *IEEE Trans. Power Electron.* **2009**, *24*, 592–602.
4. Cole, S.; Beerten, J.; Belmans, R. Generalized dynamic VSC MTDC model for power system stability studies. *IEEE Trans. Power Syst.* **2010**, *25*, 1655–1662.
5. Tang, L.; Ooi, B. Locating and isolating DC faults in multi-terminal DC systems. *IEEE Trans. Power Del.* **2007**, *22*, 1877–1884.
6. Ludois, D.; Venkataramanan, G. An examination of AC/HVDC power circuits for interconnecting bulk wind generation with the electric grid. *Energies* **2010**, *3*, 1263–1289.
7. Pinto, R.T.; Rodrigues, S.F.; Wiggelinkhuizen, E.; Scherrer, R.; Bauer, P.; Pierik, J. Operation and power flow control of multi-terminal DC networks for grid integration of offshore wind farms using genetic algorithms. *Energies* **2013**, *6*, 1–26.
8. Chen, X.; Sun, H.S.; Wen, J.Y.; Lee, W.; Yuan, X.; Li, N.; Yao, L. Integrating wind farm to the grid using hybrid multi-terminal HVDC technology. *IEEE Trans. Ind. Appl.* **2011**, *47*, 965–972.
9. Silva, B.; Moreira, C.L.; Leite, H.; Lopes, J.A.P. Control strategies for AC fault ride through in multi-terminal HVDC grids. *IEEE Trans. Power Del.* **2014**, *29*, 395–405.
10. Beerten, J.; Cole, S.; Belmans, R. Generalized steady-state VSC MTDC model for sequential AC/DC power flow algorithms. *IEEE Trans. Power Syst.* **2012**, *27*, 821–829.
11. Nakajima, T.; Irokawa, S. A control system for HVDC transmission by voltage sourced converter. In Proceedings of the IEEE Power Engineering Society Summer Meeting, Edmonton, AB, Canada, 18–22 July 1999; pp. 1113–1119.

12. Haileselassie, T.M.; Uhlen, K. Impact of DC line voltage drops on power flow of MTDC using droop control. *IEEE Trans. Power Syst.* **2012**, *27*, 1441–1449.

13. Yao, L.; Xu, L.; Bazargan, M.; Critchley, R. Multi-terminal HVDC grid for network interconnection and renewable energy integration. In Proceedings of the 43rd International Conference on Large High Voltage Electric Systems, CIGRE, Paris, France, 22–27 August 2010.

14. Cao, J.; Du, W.; Wang, H.; Bu, S. Minimization of transmission loss in meshed AC/DC grids with VSC-MTDC networks. *IEEE Trans. Power Syst.* **2013**, *28*, 3047–3055.

15. Lie, X.; Williams, B.W.; Yao, L. Multi-terminal dc transmission systems for connecting large offshore wind farms. In Proceedings of the IEEE Power and Energy Society General Meeting—Conversion and Delivery of Electrical Energy in the 21st Century, Pittsburgh, PA, USA, 20–24 July 2008; pp. 1–7.

16. Shu, Z.; Jun, L.; Ekanayake, J.B.; Jenkins, N. Control of multi-terminal vsc-hvdc transmission system for offshore wind power generation. In Proceedings of the 44th International Universities Power Engineering Conference (UPEC), Glasgow, Scotland, 1–4 September 2009; pp. 1–5.

17. Lu, W.; Ooi, B.T. Multi-terminal HVDC as enabling technology of premium quality power park. *IEEE Power Eng. Soc. Winter Meet.* **2002**, *2*, 719–724.

18. Rouzbehi, K.; Miranian, A.; Luna, A.; Rodriguez, P. DC voltage control and power sharing in multi terminal DC grids based on optimal DC power flow and voltage-droop strategy. *IEEE J. Emerg. Sel. Top. Power Electron.* **2014**, *2*, 1171–1179.

19. Chaudhuri, N.R.; Chaudhuri, B. Adaptive droop control for effective power sharing in multi-terminal DC (MTDC) grids. *IEEE Trans. Power Syst.* **2013**, *28*, 21–29.

20. Teodorescu, R.; Blaabjerg, F.; Pedersen, J.K.; Gengelci, E.; Enjeti, P.N. Multilevel inverter by cascading industrial VSI. *IEEE Trans. Ind. Electron.* **2002**, *49*, 832–838.

21. Su, C.W.; Jeong, I.W.; Wen, J.; Smedley, K. Drive the PMSM motor using hexagram converter. In Proceeding of the 23rd Annual IEEE Applied Power Electronics Conference Exposition, Austin, TX, USA, 24–28 February 2008; pp. 1803–1808.

22. Guerrero, J.M.; García de Vicuña, L.; Matas, J.; Castilla, M.; Miret, J. Output impedance design of parallel-connected UPS inverters with wireless load-sharing control. *IEEE Trans. Ind. Electron.* **2005**, *52*, 1126–1135.

23. Guo, C.; Zhao, C. Supply of an entirely passive AC network through a double-infeed HVDC system. *IEEE Trans. Power Electron.* **2010**, *24*, 2835–2841.

24. Abdel-Khalik, A.S.; Massoud, A.M.; Elserougi, A.A.; Ahmed, S. Optimum power transmission-based droop control design for multi-terminal HVDC of offshore wind farms. *IEEE Trans. Power Syst.* **2013**, *28*, 3401–3409.

Characteristic Analysis and Control of a Hybrid Excitation Linear Eddy Current Brake

Baoquan Kou, Yinxi Jin, Lu Zhang and He Zhang

Abstract: In this paper, a novel hybrid excitation linear eddy current brake is presented as a braking system for high-speed road and rail vehicles. The presence of the permanent magnets (PMs), whose flux lines in the primary core are oppositely directed with respect to the flux lines by the excitation windings, has the effect of mitigating the saturation of the iron in the teeth of the primary core. This allows the brake to be fed with more intense currents, improving the braking force. First, using the magnetic equivalent circuit method and the layer theory approach, the analytical model of the hybrid excitation linear eddy current brake was developed, which can account for the saturation effects occurring in the iron parts. The saturation effects make the design and control of eddy current brakes more difficult. Second, the relationship between the braking force characteristics and the design parameters were analyzed to provide useful information to the designers of eddy current brakes. Then, the controller of the hybrid excitation linear eddy current brake was designed to control the amplitude of the braking force. Finally, experimental measurements were conducted to verify the validity of the theoretical analysis.

Reprinted from *Energies*. Cite as: Kou, B.; Jin, Y.; Zhang, L.; Zhang, H. Characteristic Analysis and Control of a Hybrid Excitation Linear Eddy Current Brake. *Energies* **2015**, *8*, 7441–7464.

1. Introduction

The use of and research on eddy current brakes in high-speed road and rail vehicles have gradually increased [1–9]. Relative to the traditional mechanical friction brake, the eddy current brake has the advantages of no mechanical contact, high reliability, long working life, and lower sensitivity to environmental parameters, such as temperature and aging. Eddy current brakes have excellent braking performance, especially at high speed.

In eddy current brakes, the magnetic field can be produced by excitation winding systems or by permanent magnets. According to differences of the flux sources, eddy current brakes can be divided into three types: electric excitation eddy current brakes, permanent magnet eddy current brakes and hybrid excitation eddy current brakes. For the electric excitation eddy current brakes, the braking force can be adjustable, but an additional power supply system is required, and the braking force density is low. The permanent magnet eddy current brakes allow for elimination of the electrical supply system to simplify the device structure, but conversely, braking force modulation is not allowed, and magnet corrosion and relatively low temperature tolerance are potential hazards. The hybrid excitation eddy current brakes are a combination of the permanent magnet eddy current brakes and the electric excitation eddy current brakes and exhibit the advantages of both. Therefore, the amplitude of the braking force density is large and adjustable, and the excitation loss is low. Moreover, according to differences in the structure, eddy current brakes can be classified as radial [10–12], axial [13,14], or linear [15–17].

Analysis of and research on eddy current brakes can be found in many papers. Gay et al. [18] analyzed the relationship between the braking force and the design parameters using the analytical method and the finite element method, respectively, and experimental validation was carried out. However, the experiment results did not agree with the calculated results well. Canova et al. [19] presented an analytical model of the eddy current brake considering a 3D analytical correction. Then, the validity of the 3D analytical model was verified using 3D finite element simulations. Yazdanpanah et al. [20] developed a subdomain-based analytical model and evaluated the performance characteristics and design considerations of the device by using the analytical model. The validity of the analytical model was verified through the 3D finite element method and experimental measurement. The experimental results of the prototype brake verified the investigations and the design. In addition to the above studies, many researchers have focused on the analytical model of eddy current brakes [21–25], and the saturation effects occurring in the iron parts were omitted in most of these calculation models. Moreover, thus far, the analytical model of hybrid excitation eddy current brakes has been little studied.

In this paper, a novel hybrid excitation linear eddy current brake is presented. The remainder of this paper is organized as follows: in Section 2, the structure and working principle of the hybrid excitation linear eddy current brake are explained. The analytical model of the hybrid excitation linear eddy current brake, which can account for the saturation effects occurring in the iron parts, is deduced in Section 3. In Section 4, the parameter analysis is provided, and the controller is designed in Section 5. A three-level PWM scheme is used to reduce the switching frequency and excitation current harmonics. In Section 6, the experimental verification is carried out, and the paper is concluded in Section 7.

2. Structure and Working Principle of the Hybrid Excitation Linear Eddy Current Brake

2.1. Structure of the Hybrid Excitation Linear Eddy Current Brake

The hybrid excitation linear eddy current brake includes the primary part and the secondary part, as shown in Figure 1. The primary part consists of the excitation windings, the primary core and the permanent magnets. The permanent magnets are polarized parallel to the direction of the movement of the mover, and they are placed in notches with alternate polarity. The secondary part comprises a low resistivity conductor plate (in this paper, copper is used) on an iron backing.

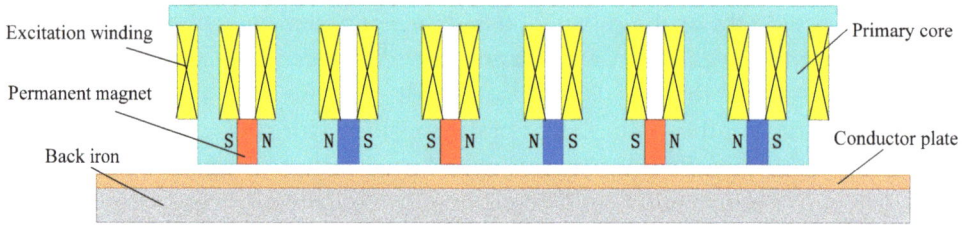

Figure 1. Structure of the hybrid excitation linear eddy current brake.

2.2. Working Principle of the Hybrid Excitation Linear Eddy Current Brake

When the excitation windings are without excitation current, the flux generated by the permanent magnets will form a magnetic short-circuit ring in the primary iron core, and it almost does not pass through the air gap, as shown in Figure 2a. When excitation current flows through the excitation windings, there are two flux loops, as shown in Figure 2b.

One is produced by the excitation windings, and the other is produced by the permanent magnets. The solid line is the flux produced by the permanent magnets, the dot and dash line is the flux produced by the excitation windings, and the dotted line is the total flux produced by the permanent magnets and the excitation windings. According to the law of electromagnetic induction, the relative movement of the primary part and the secondary conductor plate causes eddy currents in the conductor plate. Due to the interactions of the eddy currents and the magnetic flux generated by the permanent magnets and the excitation windings, the braking force between the primary part and the secondary part is produced.

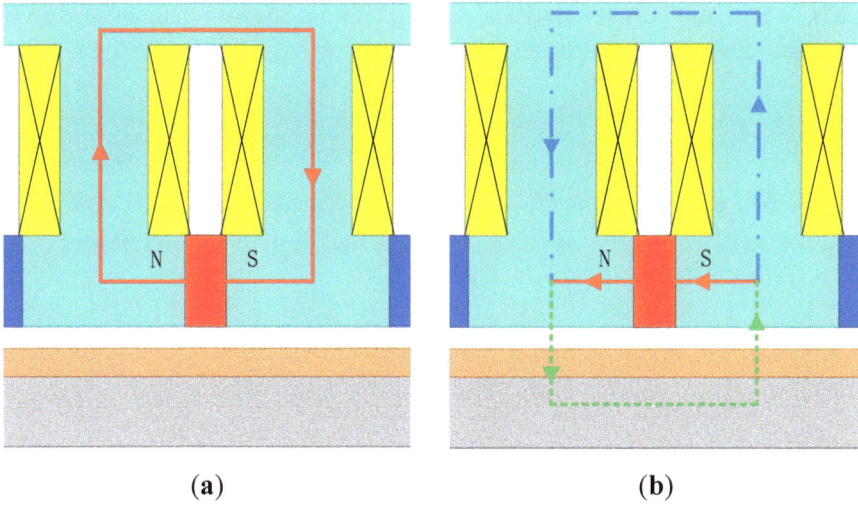

Figure 2. Flux flow: (**a**) excitation windings without excitation current and (**b**) excitation windings with excitation current.

3. Analytical Model

In this section, the analytical model of the hybrid excitation linear eddy current brake is derived. The purpose is to rapidly provide insight into the fundamental physics of the eddy current brake and preliminary design data that verify whether the performance and size are compatible with the envisioned application.

3.1. Static Field Analysis

In order to build the analytical model of the hybrid excitation linear eddy current brake, the static field is analyzed. The definition of iron part reluctances and the magnetic equivalent circuit are shown in Figure 3. All elements of the magnetic equivalent circuit are calculated in detail below. The ampere-turns of an excitation winding are as follows:

$$F_f = NI \tag{1}$$

where N and I are the turns of the excitation windings and the excitation current, respectively. The MMF (magnetic motive force) of a permanent magnet is as follows:

$$F_c = H_c h_m \tag{2}$$

where H_c and h_m are the coercivity and width of the permanent magnets, respectively. According to Figure 3a, the iron part reluctances are calculated as follows:

$$R_1 = \frac{\tau - h_m}{2\mu_0\mu_1 l_\delta b_m} \tag{3}$$

where τ is the pole pitch, μ_0 is the air permeability, μ_1 is the relative permeability of R_1, l_δ is the width of the primary core, and b_m is the height of the permanent magnets.

$$R_2 = \frac{h_s + \frac{b_m}{2} + \frac{h_j}{2}}{\mu_0\mu_2 l_\delta b_t} \tag{4}$$

where h_s is the height of slot, h_j is the height of the primary yoke, μ_2 is the relative permeability of R_2, and b_t is the tooth width.

$$R_3 = \frac{\tau}{\mu_0\mu_3 l_\delta h_j} \tag{5}$$

where μ_3 is the relative permeability of R_3.

$$R_4 = \frac{b_m}{2\mu_0\mu_4 l_\delta (\tau - h_m)} \tag{6}$$

where μ_4 is the relative permeability of R_4.

$$R_5 = \frac{h_b}{2\mu_0\mu_5 l_\delta (\tau - h_m)} \tag{7}$$

where h_b is the thickness of the back iron, and μ_5 is the relative permeability of R_5.

$$R_6 = \frac{\tau}{\mu_0\mu_6 l_\delta h_b} \tag{8}$$

where μ_6 is the relative permeability of R_6. The reluctances of the air gap and conductor plate are calculated as follows:

$$R_c = \frac{c}{\mu_0\mu_c l_\delta (\tau - h_m)} \tag{9}$$

where c is the conductor plate thickness, and μ_c is the relative permeability of the conductor plate.

$$R_\delta = \frac{\delta}{\mu_0 l_\delta (\tau - h_m)} \tag{10}$$

where δ is the air gap length. According to the magnetic equivalent circuit, the following equations are provided:

$$2F_f - F_{AB} = 2\phi_r R_2 + \frac{\phi_r}{2}R_3 \tag{11}$$

$$F_c - F_{AB} = R_m\phi_m + 2R_1\phi_m \tag{12}$$

$$F_{AB} = R_6\phi_\delta/2 + \phi_\delta(2R_4 + 2R_5 + 2R_\delta + 2R_c) \tag{13}$$

$$\phi_\delta = 2\phi_m + \phi_r \tag{14}$$

where F_{AB} is the MMF between points A and B, and ϕ_r, ϕ_m and ϕ_δ are the flux of the loops shown in Figure 3.

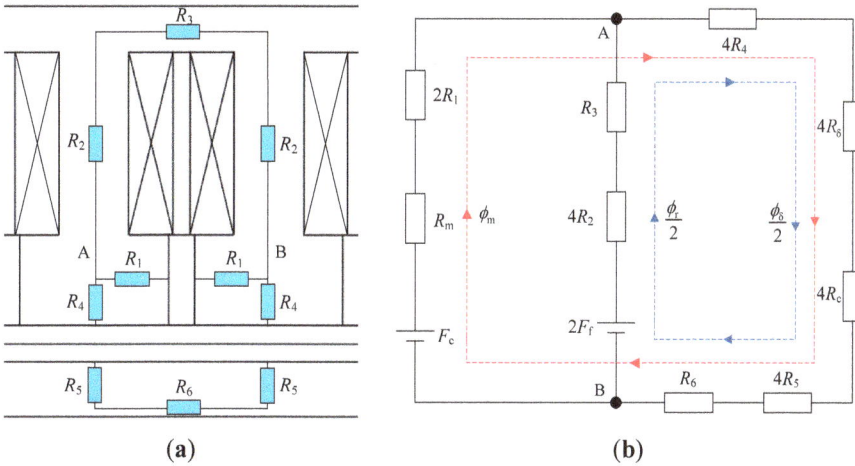

Figure 3. (a) Definition of iron part reluctances and (b) the magnetic equivalent circuit.

Thus, it is possible to calculate the amplitude of the air gap flux density:

$$B_0 = \frac{\phi_\delta}{l_\delta (\tau - h_m)} \tag{15}$$

In many researches, the air gap flux density is assumed as a rectangular wave. But the actual air gap flux density waveform is not a rectangular wave, as shown in Figure 4. Therefore in this paper, the air gap flux density waveform for different design parameters is calculated based on the finite element method (FEM), and an empirical piecewise function is proposed to fitting the actual air gap flux density waveform.

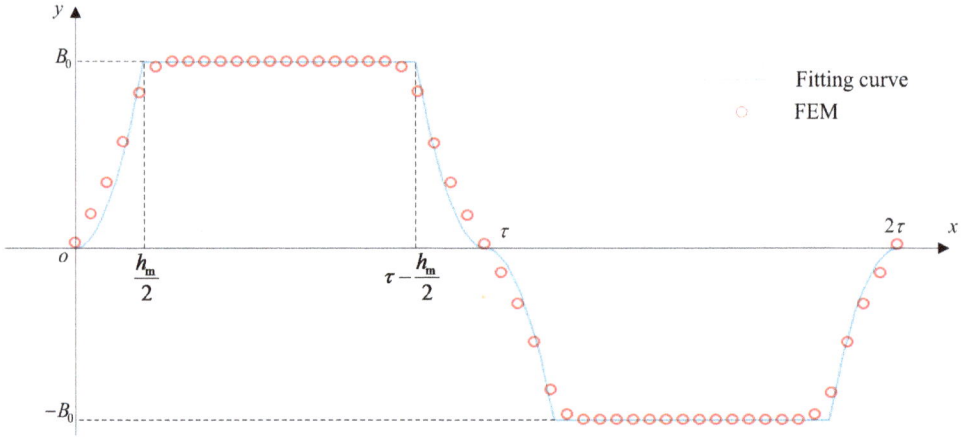

Figure 4. Waveform of the air gap flux density at the speed of 0 m/s.

The expression of the piecewise function is shown below:

$$f(x) = \begin{cases} \frac{4B_0}{h_m^2} x^2 & 0 \le x \le \frac{h_m}{2} \\ B_0 & \frac{h_m}{2} \le x \le \tau - \frac{h_m}{2} \\ \frac{4B_0}{h_m^2}(x-\tau)^2 & \tau - \frac{h_m}{2} \le x \le \tau \\ -\frac{4B_0}{h_m^2}(x-\tau)^2 & \tau \le x \le \tau + \frac{h_m}{2} \\ -B_0 & \tau + \frac{h_m}{2} \le x \le 2\tau - \frac{h_m}{2} \\ -\frac{4B_0}{h_m^2}(x-2\tau)^2 & 2\tau - \frac{h_m}{2} \le x \le 2\tau \end{cases} \tag{16}$$

Therefore, based on the Fourier decomposition, the air gap flux density is expressed as:

$$B_{\delta n} = \frac{4}{\tau} \int_0^{\frac{\tau}{2}} f(x) \cdot \sin\left(\frac{n\pi x}{\tau}\right) dx \quad n = (1,3,5\cdots) \tag{17}$$

$$B_\delta = \sum_{n}^{\infty} B_{\delta n} e^{jkx} \tag{18}$$

$$k = \frac{n\pi}{\tau} \tag{19}$$

The specific procedure is explained in below: Firstly, the amplitude of the air gap flux density B_0 is calculated using the magnetic equivalent circuit. Then, the waveform of the air gap flux density is obtained based on the empirical piecewise function. Finally, the expression of the air gap flux density is deduced using the Fourier decomposition.

3.2. Braking Force Analysis

A complete 3-dimensional analytical solution for the hybrid excitation linear eddy current brake proposed in this paper is difficult and will not be attempted. Instead, the configuration of a 2-dimensional layer model, shown in Figure 5, will be used for the analysis.

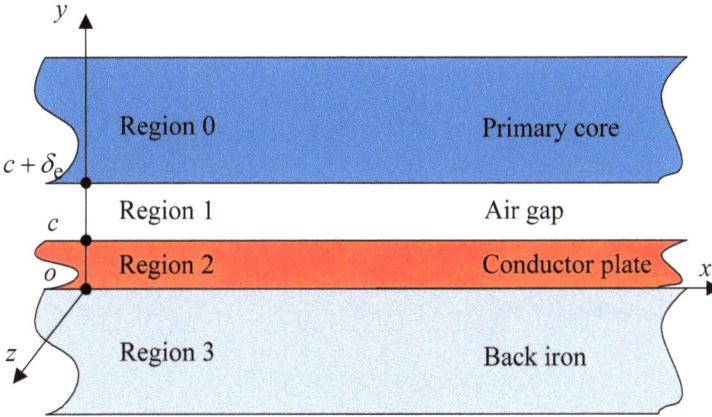

Figure 5. Two-dimensional multi-layer model.

The multi-layer model is divided into four different regions:

Region 0: primary core.
Region 1: air gap.
Region 2: secondary conductor plate.
Region 3: back iron.

In order to simplify the analysis, the following assumptions are made. The primary core, conductor plate and back iron are considered to be infinitely long in the x-direction. All currents are in the z-direction. The excitation winding, permanent magnets and salient poles are replaced by infinitely thin linear current sheets backed by smooth iron boundaries. These linear current sheets are chosen in such a way that they provide the same field in the air gap of the model having smooth structures that the original excitation windings and permanent magnets produced in the actual machine. In such a case, the actual air gap δ of the machine is replaced, using Carter's coefficient, by an effective air gap δ_e that accounts for the interpolar space and variable reluctance.

On the plane $y = c + \delta_e$, there are infinitely thin current sheets flowing in the z-direction. This current can be expressed as:

$$J_s(x) = \sum_n^\infty J_n \sin(kx + \pi) \tag{20}$$

$$\delta_e = K_\delta \delta \tag{21}$$

where J_n is the amplitude of the harmonics, and K_δ is Carter's coefficient.

The MMF that is produced by the infinitely thin linear current sheet is:

$$at = \sum_n^\infty \int_0^x J_n \sin(kx + \pi)dx = \sum_n^\infty \frac{J_n}{k}\cos(kx) \tag{22}$$

Based on the above assumptions, the air gap flux density produced by the infinitely thin linear current sheet is the same as that of Equation (18), therefore the following expression can be obtained:

$$\frac{\mu_0 \times at}{c + \delta_e} = \text{Re}(B_\delta) \tag{23}$$

$$J_n = k(c + \delta_e)\frac{B_{\delta n}}{\mu_0} \tag{24}$$

The electromagnetic equation expressed in terms of the magnetic vector potential A is:

$$\nabla^2 A = -\mu_j \mu_0 J \tag{25}$$

This equation has to be written for each region.

Region 1: in the air gap, the conductivity is zero and the field equation is:

$$\nabla^2 A_1 = 0 \tag{26}$$

In order to solve the second-order differential equation system, the variable separation method has been adopted, and, to simplify the problem, we can assume that all the electromagnetic quantities are periodic with pole pitch, which means:

$$A(x,y) = A(y)e^{jkx} \tag{27}$$

Therefore, the general solution is shown as:

$$A_{1n}(y) = C_{1n}e^{ky} + D_{1n}e^{-ky} \tag{28}$$

where C_{1n} and D_{1n} depend on the boundary conditions.

Region 2: this is the only one made of conductive material, and the field equation is:

$$\nabla^2 A_2 = -\mu_c \mu_0 J_2 \tag{29}$$

From Faraday laws:

$$J_2 = -\sigma_c v \frac{\partial A_2}{\partial x} \tag{30}$$

where σ_c is the conductivity of conductor plate and v is the speed.

Finally, the field equation becomes:

$$A''_2 (y) = \left(jk\mu_c \mu_0 \sigma_c v + k^2 \right) A_2 (y) \tag{31}$$

$$A_{2n} (y) = C_{2n} e^{\lambda y} + D_{2n} e^{-\lambda y} \tag{32}$$

where C_{2n} and D_{2n} depend on the boundary conditions and the exponential coefficient is:

$$\lambda = \sqrt[4]{k^4 + (k\mu_c \mu_0 \sigma_c v)^2} e^{j\theta} \tag{33}$$

$$\theta = \frac{1}{2} \arctan \left(\frac{\mu_c \mu_0 \sigma_c v}{k} \right) \tag{34}$$

Region 3: assuming that the conductivity in the back iron is zero, the field equation is:

$$\nabla^2 A_3 = 0 \tag{35}$$

$$A_{3n} (y) = C_{3n} e^{ky} + D_{3n} e^{-ky} \tag{36}$$

In the multi-layer model, we have considered that region 3 is extended from $y = 0$ to $y = -\infty$, thus:

$$A_3 (-\infty) = 0 \tag{37}$$

Therefore, the general solution is shown as:

$$A_{3n} (y) = C_{3n} e^{ky} \tag{38}$$

where C_{3n} depends on the boundary conditions.

The boundary conditions are shown as follows:

Between region 0 and region 1:

$$\frac{\partial A_{1n} (c + \delta_e)}{\partial y} = \mu_0 J_n \tag{39}$$

Between region 1 and region 2:

$$\left\{ \begin{array}{l} B_{1n} = B_{2n} \\ H_{1n} = H_{2n} \end{array} \right|_{y=c} \tag{40}$$

Between region 2 and region 3:

$$\left\{ \begin{array}{l} B_{2n} = B_{3n} \\ H_{2n} = H_{3n} \end{array} \right|_{y=0} \tag{41}$$

where μ_b is the relative permeability of the back iron and, in this paper, we assume that:

$$\mu_b = \mu_6 \tag{42}$$

It is possible to calculate the eddy current loss in the conductor plate:

$$P_e = \sum_{n}^{\infty} p l_\delta \sigma_c v^2 k^2 |S|^2 \tau \left(\frac{e^{2k_1c}}{2k_1} - \frac{e^{-2k_1c}}{2k_1} + \frac{2\sin(k_2c)\cos(k_2c)}{k_2} \right) \tag{43}$$

$$k_1 = \mathrm{Re}(\lambda) \tag{44}$$

$$k_2 = \mathrm{Im}(\lambda) \tag{45}$$

$$S = C_{2n} = D_{2n} \tag{46}$$

where p is the number of pole-pairs.

Finally, the braking force is calculated as:

$$F = \frac{P_e}{v} = \sum_{n}^{\infty} p l_\delta \sigma_c v k^2 |S|^2 \tau \left(\frac{e^{2k_1c}}{2k_1} - \frac{e^{-2k_1c}}{2k_1} + \frac{2\sin(k_2c)\cos(k_2c)}{k_2} \right) \tag{47}$$

3.3. Consideration of the Nonlinear Magnetization Characteristics of the Iron Core

The nonlinearity magnetization characteristics of the iron core material are a problem in the analysis of the eddy current brake because the typical operational condition of the eddy current brake requires a high supplied excitation current.

In order to address the nonlinearity of the iron core material, the iterative procedure is used. It starts by assigning an initial value to μ_1, μ_2, μ_3, μ_4, μ_5 and μ_6 to determine the reluctances of the iron core part and calculate the circuit fluxes. Next, the magnetic flux densities within R_1, R_2, R_3, R_4, R_5 and R_6 are calculated. Then, based on the B-H curve of the utilized steel, new relative permeabilities μ'_1, μ'_2, μ'_3, μ'_4, μ'_5 and μ'_6 can be obtained. The process continues until the criterion is individually satisfied for all relative permeabilities, as shown in Figure 6.

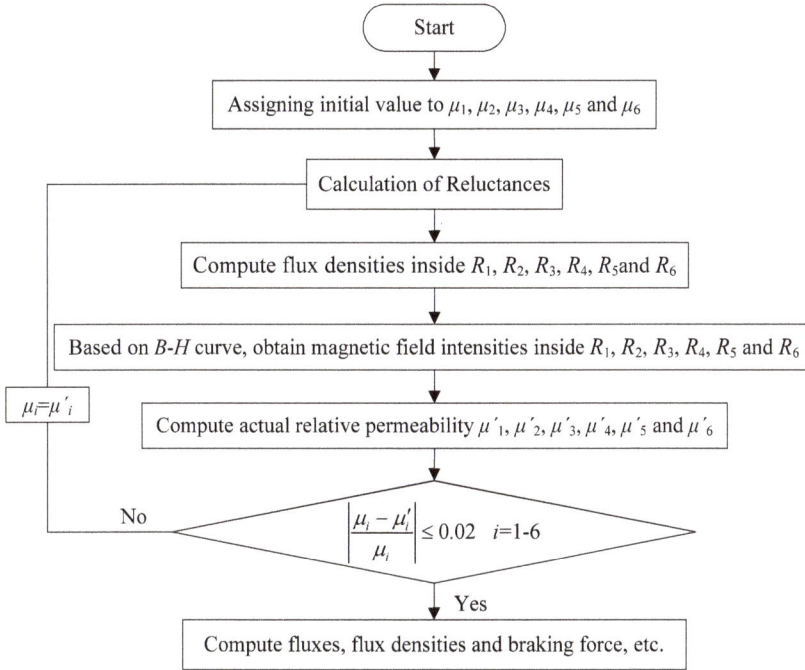

Figure 6. Iterative process of relative permeability.

Figure 7 compares the braking force obtained using the FEM with two different predicted curves for the braking force: the analytical method considering core saturation and the analytical method neglecting core saturation.

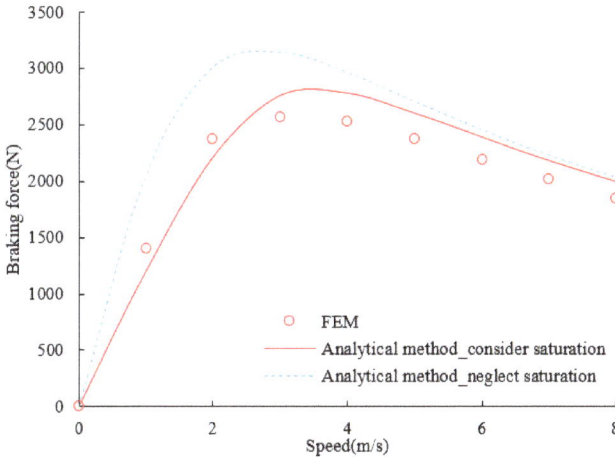

Figure 7. Braking force characteristic.

It can be seen that the difference between the analytical methods considering core saturation and neglecting core saturation is very obvious, especially in the low-speed region (in the high-speed region because of the eddy current reaction field, the saturation degree of the iron part is reduced). Therefore, in order to improve the accuracy of the analytical model, the saturation effects must be considered.

4. Parameter Analysis and Its Determination

In this section, the influence of several geometrical and physical parameters on the braking force is investigated to provide useful information to the designers of hybrid excitation eddy current brakes. The initial parameters of the hybrid excitation linear eddy current brake are shown in Table 1.

Table 1. Initial parameters of the hybrid excitation linear eddy current brake.

Symbol	Quantity	Value
h_m	width of the permanent magnet	18 mm
b_m	height of the permanent magnet	10 mm
I	excitation current	14 A
N	turns of the excitation winding	200
L	length of the primary iron core	200 mm
l_δ	width of the primary iron core	100 mm
h_j	height of the primary core yoke	11 mm
H	height of the primary iron core	49 mm
δ	air gap length	2 mm
τ	pole pitch	40 mm
c	conductor plate thickness	2 mm
h_b	back iron thickness	9 mm
b_t	tooth width	16 mm

4.1. Influence of the Air Gap Length

Figure 8 shows the braking force-speed characteristic with different air gap lengths. It can be seen that the braking force decreases gradually as the air gap increases, and this is because a large air gap length will result in a lower magnetic density. In addition, for a large air gap, the high-speed region is flatter than for a small air gap because the eddy currents are located farther from the primary part and thus have a lesser weakening influence on them. In order to obtain a larger braking force, the air gap length is chosen as 1 mm.

4.2. Influence of the Conductor Plate Thickness

The secondary conductor plate is not only an important part of the magnetic circuit but also the medium of the eddy currents. As the conductor plate thickness increases, the machine air gap effectively increases (copper has a permeability of, essentially, the air); however, the increase of the conductor plate thickness results in the increase of the conductivity of the conductor plate. Figure 9 shows the braking force-speed characteristic with different conductor plate thicknesses. It can be seen that the increasing conductor plate thickness results in a lower peak braking force and smaller critical speed. In order to obtain a larger braking force, the conductor thickness is chosen as 1 mm.

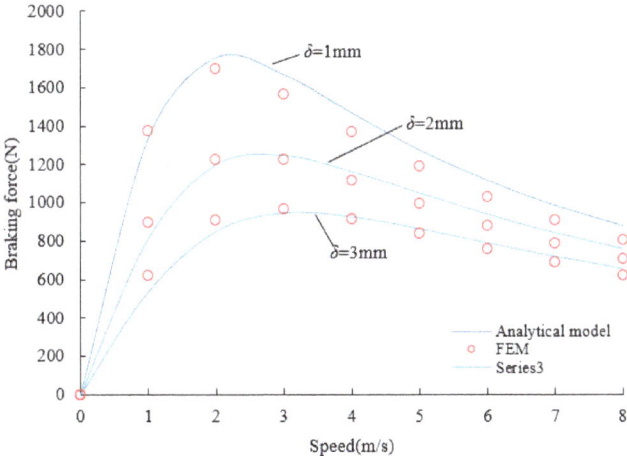

Figure 8. Braking force-speed characteristic for different air gap lengths.

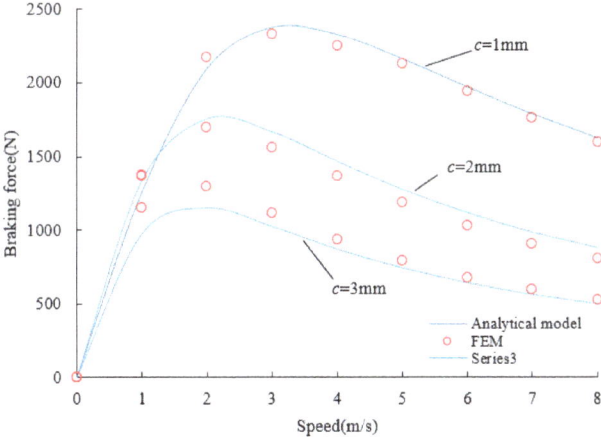

Figure 9. Braking force-speed characteristic for different conductor thicknesses.

4.3. Influence of the Conductor Material

Figure 10 shows the braking force-speed characteristic with different conductor materials. It can be seen that the slope of the braking force characteristic strongly depends on the conductivity of the conductor. The critical speed decreases as the conductivity of the conductor increases, and the conductivity has a small impact on the peak value of the braking force. Because of the limit of the experimental installation, the speed of the hybrid excitation linear eddy current brake can only take a smaller value. Therefore, the conductor material is chosen as copper.

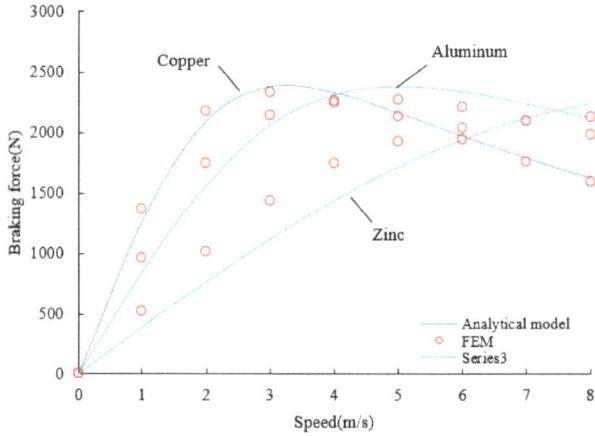

Figure 10. Braking force-speed characteristic for different conductor materials.

4.4. Influence of the Excitation Current

Figure 11 shows the braking force-speed characteristic with different excitation currents. It can be seen that the braking force increases gradually as the excitation current increases.

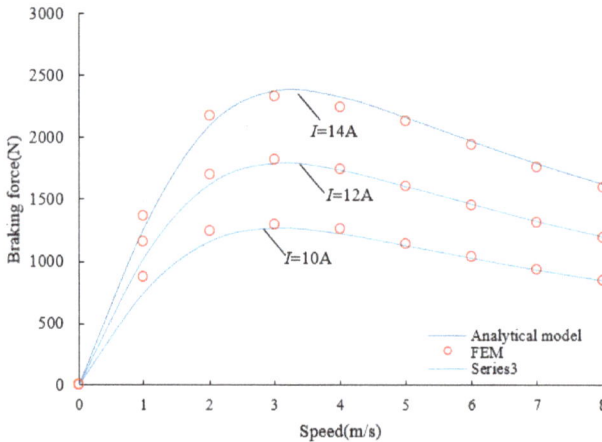

Figure 11. Braking force-speed characteristic for different excitation currents.

Moreover, the critical speed slightly increases as the excitation current increases. In order to obtain a larger braking force and guarantee the thermal stability of the hybrid excitation eddy current brake, the excitation current is chosen as 14 A.

4.5. Influence of the Magnet Dimension

The effect of the magnet dimensions (the width of the permanent magnets h_m and the height of the permanent magnets b_m) is analyzed as shown in Figure 12. The study has been performed considering a constant magnet volume. It can be seen that there are optimal magnet dimensions that will maximize the braking force density. The braking force density is the ratio of the braking force to the volume of the eddy current brake. In order to obtain a larger braking force density, the width of the permanent magnet is chosen as 13 mm and the height of the permanent magnet is chosen as 14 mm.

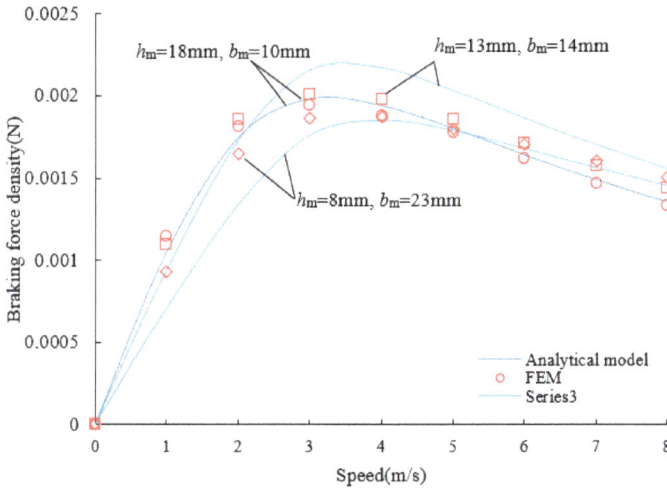

Figure 12. Braking force density-speed characteristic for different magnet dimensions.

5. Eddy Current Brake Controller

This paper presents a reliable and rugged controller of the hybrid excitation linear eddy current brake to control the amplitude of the braking force. Figure 13 shows the control scheme of the eddy current brake. There is an outer braking force controller and an inner current controller. The braking force controller outputs the reference excitation current I_{ref} according to the error between the reference braking force F_{ref} and the actual braking force F_a. The current controller outputs a reference voltage u_{ref} according to the error between the reference excitation current I_{ref} and the actual excitation current I_a. One can adjust the excitation current and move forward to adjust the braking force by controlling the reference voltage u_{ref}.

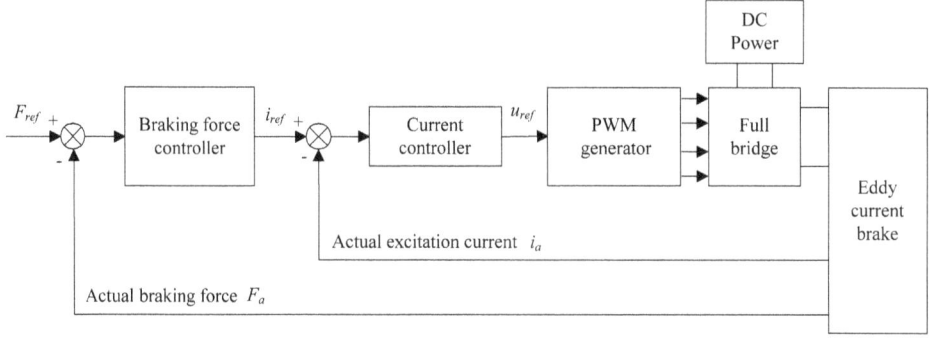

Figure 13. Control scheme of the eddy current brake.

5.1. Current Controller Design

In order to limit the amplitude of the current ripple to an acceptable extent, the three-level PWM scheme is used in this paper. For the three-level PWM scheme, the amplitude of the current ripple is independent of the direct current (DC)-link voltage, and the dynamic response characteristics of the eddy current brake system can be raised by increasing the DC-link voltage without increasing the current harmonics. Moreover, the switching frequency of an insulated gate bipolar transistor (IGBT) element is half of the output pulse frequency. This means that the switching frequency of the power elements is halved for a required dynamic performance of the eddy current brake compared with the two-level schemes.

5.2. Braking Force Controller Design

Dud to the saturation of the iron core, the relationship between the braking force and the excitation current is nonlinear. For the ease of control, the nonlinear curve is divided into several piecewise linear intervals. When the objective braking force F_{ref} is given, the corresponding interval is estimated firstly, then the corresponding excitation current I^* is calculated using the endpoints of the interval (I_1, F_1) and (I_2, F_2), as shown in Figure 14.

The expression of I^* is shown below:

$$I^* = \frac{I_2 - I_1}{F_2 - F_1}(F_{ref} - F_1) + I_1 \tag{48}$$

Finally, the reference excitation current I_{ref} is shown as:

$$I_{ref} = I^* + \Delta I \tag{49}$$

where ΔI is the output of the proportion integration (PI) controller.

418

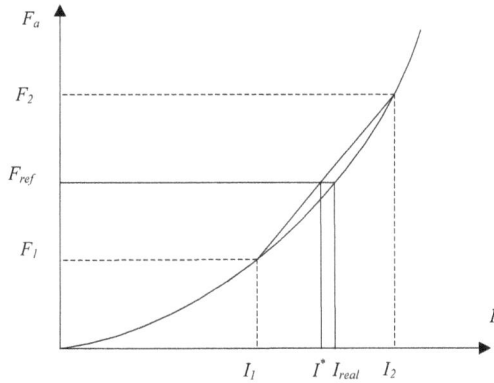

Figure 14. Schematic diagram of piecewise linearization.

A block diagram of the braking force controller is shown in Figure 15. The input of the PI controller is the error signal between the objective braking force F_{ref} and the actual braking force F_a.

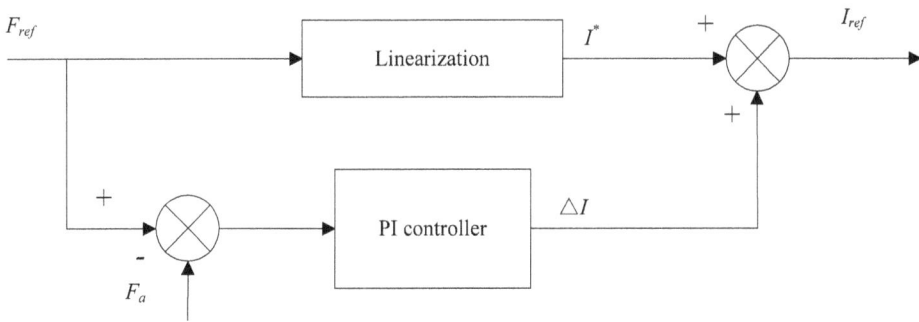

Figure 15. Block diagram of the braking force controller.

6. Numerical and Experimental Analysis

The 2-D finite element study is carried out in the same domain as the analytical problem. The field line distribution and the magnetic flux density map at a speed of 2 m/s are plotted in Figure 16a,b, respectively. It is observed that the reaction field of eddy currents tilts the field line entering the secondary conductor and shifts the epicenters of the flux density formation behind the centerlines of the magnets. Both the tilt and epicenter shift are a function of speed.

The braking force characteristics of the electric excitation eddy current brake and the hybrid excitation eddy current brake are compared at the same excitation current. The electric excitation eddy current brake has the same structure as the

419

hybrid excitation eddy current brake, except that there are not permanent magnets, as shown in Figure 17.

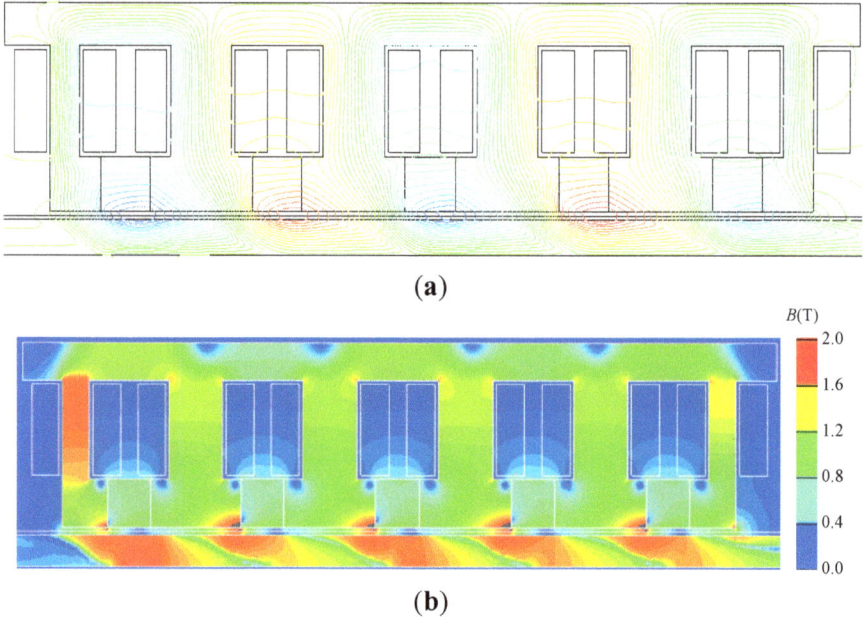

(a)

(b)

Figure 16. (a) Field line distribution and (b) the magnetic flux density map.

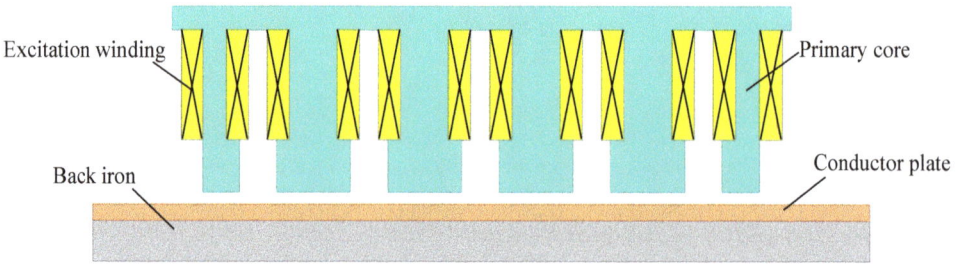

Figure 17. Structure of the electric excitation linear eddy current brake.

Figure 18a is the magnetic flux density map of the electric excitation eddy current brake at a speed of 2 m/s. The flux density of the gray area is larger than 2 T. Figure 18b is a comparison of the braking force characteristics of the electric excitation eddy current brake and the hybrid excitation eddy current brake.

As can be observed in Figure 18, the primary core of the electric excitation eddy current brake is saturated, but the primary core of the hybrid excitation eddy current brake is not saturated at the same excitation current, as shown in Figure 16b.

420

Moreover, the braking force produced by the hybrid excitation eddy current brake is larger than that produced by the electric excitation eddy current brake. In other words, the excitation loss of the hybrid excitation eddy current brake is lower than that of the electric excitation eddy current brake under the same braking force level.

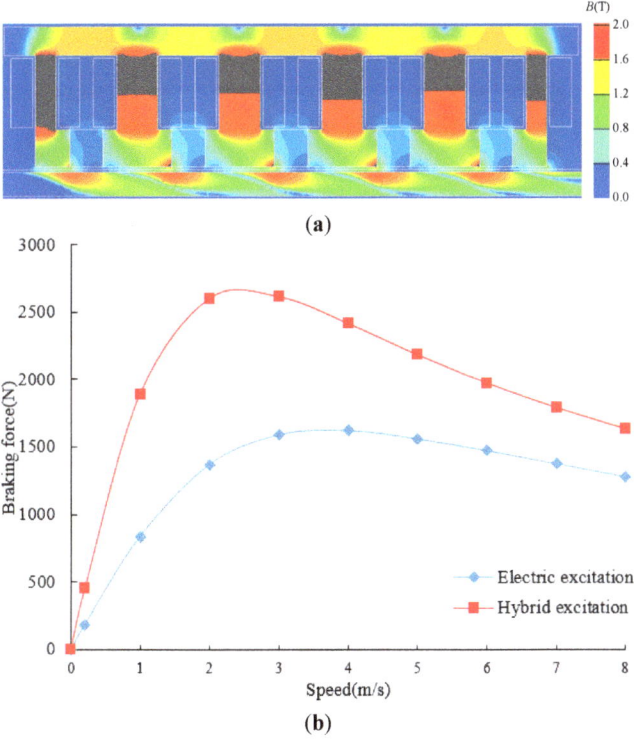

Figure 18. (**a**) Magnetic flux density map of the electric excitation eddy current brake and (**b**) a comparison between the electric excitation eddy current brake and hybrid excitation eddy current brake.

Therefore, we can conclude that the presence of the permanent magnets, whose flux lines in the primary core are oppositely directed with respect to the flux lines by the excitation windings, has the effect of mitigating the saturation of the iron in the teeth of the primary core. This allows the brake to be fed with more intense currents, improving the braking force. Through the above analysis, we can see that the hybrid excitation eddy current brake has many advantages, such as controllability, high force density and low excitation loss. The prototype of the hybrid excitation linear eddy current brake is shown in Figure 19a. The prototype includes the primary part and the secondary part. The primary part consisted of the iron core, excitation windings, permanent magnets, slides, cushion blocks, connecting plate and so on. The secondary part consisted of the conductor plate, back iron, slide rail and limited block. In this experiment, the rotation movement of a servo rotating motor is transformed into linear motion through a ball screw to drive the primary part of the eddy current brake. The test bed is shown in Figure 19b. It includes the servo rotating motor, a tension-compression sensor, the ball screw, the hybrid excitation linear eddy current brake and the controller. The measured value of the braking force can be obtained easily by measuring the output voltage of the tension-compression sensor. In this paper, the relationship between the output voltage of the tension-compression sensor and the braking force is as shown below.

$$F = \frac{V_{out}}{5mV} \tag{50}$$

In other words, when the output voltage of the tension-compression sensor is 5 mV, the braking force is 1 N.

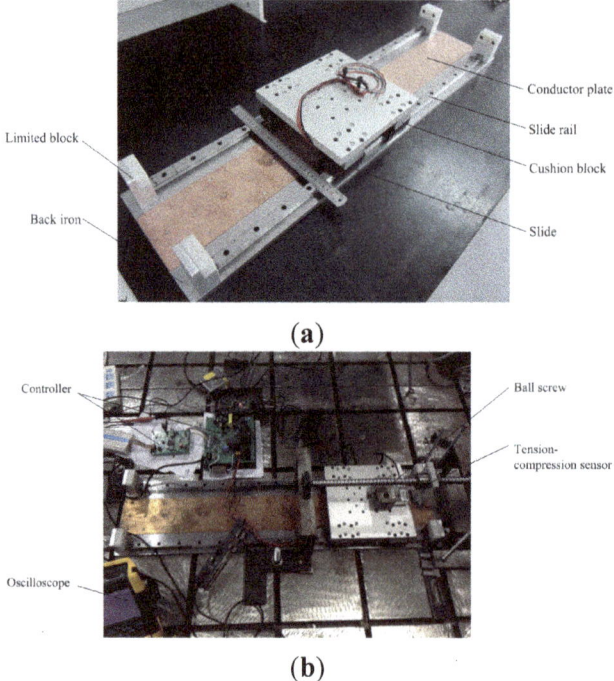

(a)

(b)

Figure 19. (a) Prototype of the hybrid excitation linear eddy current brake and (b) the test bed.

The final parameters of the hybrid excitation linear eddy current brake are shown in Table 2.

Table 2. Final parameter of the hybrid excitation linear eddy current brake.

Symbol	Quantity	Value
h_m	width of the permanent magnet	13 mm
b_m	height of the permanent magnet	14 mm
I	excitation current	14 A
N	turns of the excitation winding	200
L	length of the primary iron core	200 mm
l_δ	width of the primary iron core	100 mm
h_j	height of the primary core yoke	11 mm
H	height of the primary iron core	53 mm
δ	air gap length	1 mm
τ	pole pitch	40 mm
c	conductor plate thickness	1 mm
h_b	back iron thickness	9 mm
b_t	tooth width	16 mm

Figure 20 shows the experimental measurements of the braking force compared to the calculated values by finite element method. It can be seen that the agreement between the experimental measurements and the calculated values by the finite element method is very good. Moreover, due to the saturation of the iron core, the relationship between the braking force and the excitation current is nonlinear. As shown in Figure 19b, the length of the stator of the prototype is small, *i.e.*, the stroke of the prototype is limited. Therefore, the speed cannot reach a higher value. This problem will be solved in the future.

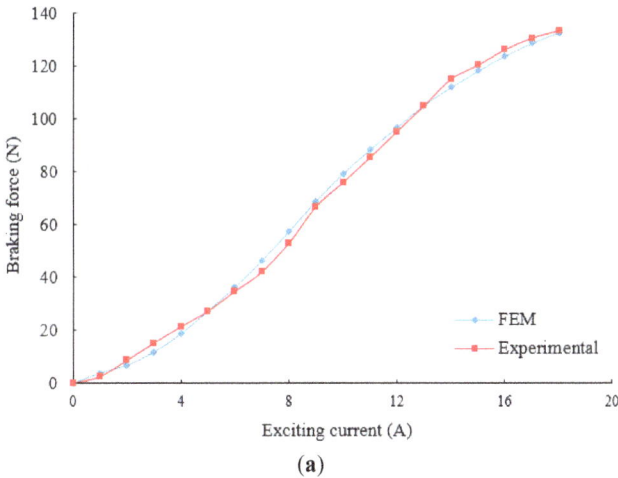

(a)

Figure 20. *Cont.*

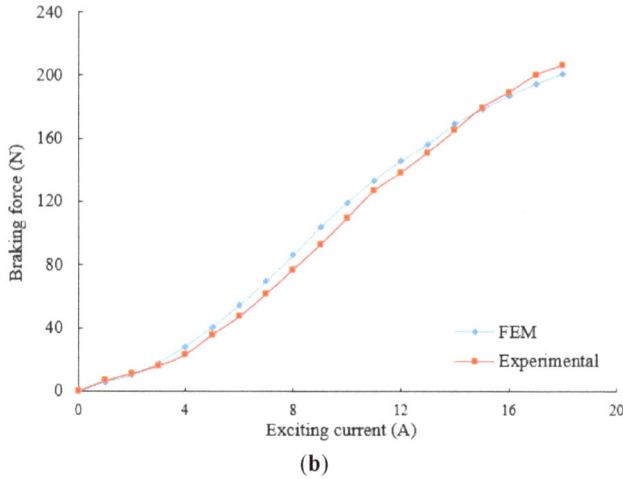

(b)

Figure 20. Comparison between the experimental measurements and calculated values via the finite element method: (**a**) $v = 0.1$ m/s and (**b**) $v = 0.15$ m/s.

The output current of the full bridge power converter with the two-level PWM schemes and the three-level PWM schemes are given in Figure 21a,b, respectively. The results show that the amplitude of the current ripple with the two-level PWM schemes is approximately 50 mA and the amplitude of the current ripple with the three-level PWM schemes is approximately 12.5 mA. That is, the amplitude of the current ripple is significantly reduced by using the three-level PWM schemes. The experimental waveforms of the braking force are shown in Figure 21c–f. It can be seen that there is no overshoot in the braking force responses.

The objective braking force, the measured value of the output voltage of the tension-compression sensor, the measured value of the braking force and the error between the objective braking force and the measured value of the braking force are shown in Table 3.

As can be observed in Table 3, the measured value of the braking force has good correlation with the objective braking force, and the availability of the eddy current brake controller is verified.

424

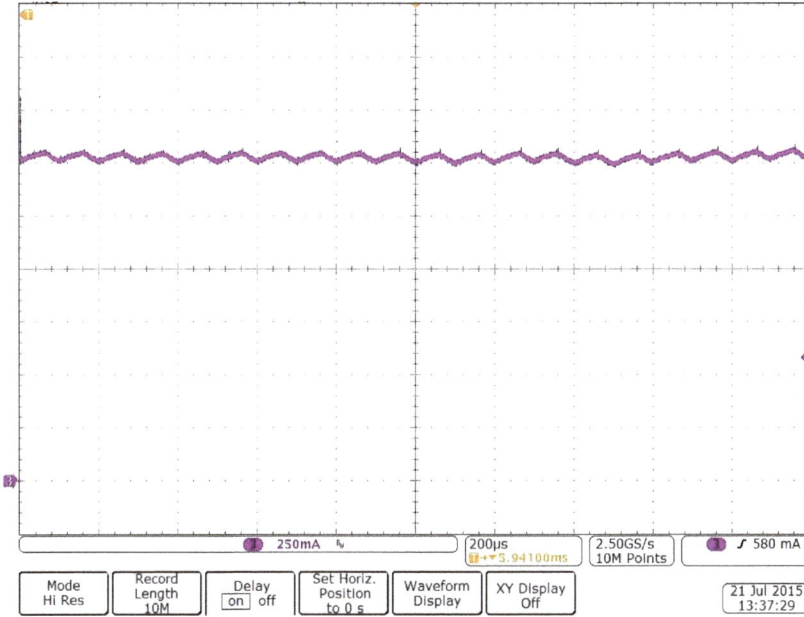

(a)

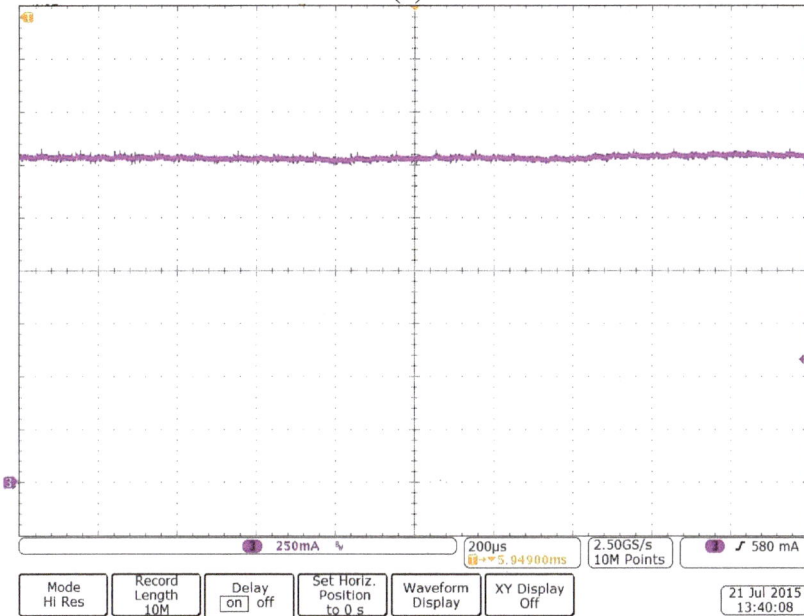

(b)

Figure 21. *Cont.*

425

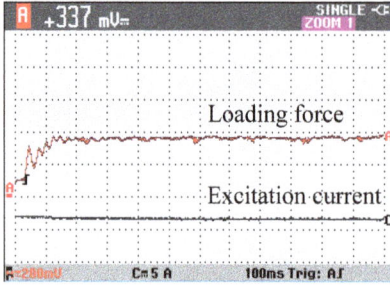

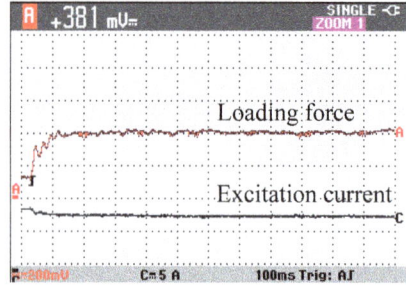

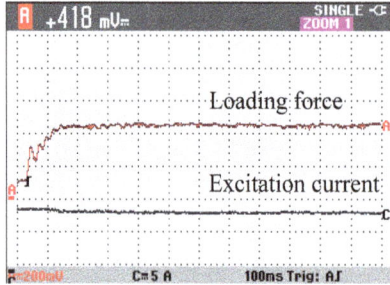

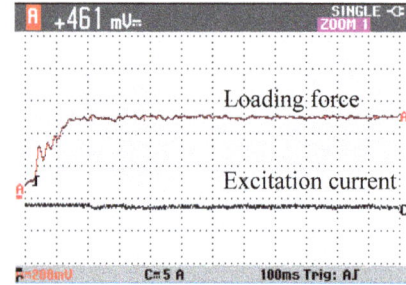

Figure 21. (**a**) Output current with the two-level PWM schemes; (**b**) output current with the three-level PWM scheme; (**c**) $F_{ref} = 70$ N; (**d**) $F_{ref} = 80$ N; (**e**) $F_{ref} = 90$ N and (**f**) $F_{ref} = 100$ N.

Table 3. Analysis of the experimental results.

Objective Braking Force	Measured Value of Output Voltage	Measured Value of Braking Force	Error
70 N	337 mV	67.4 N	3.7%
80 N	381 mV	76.2 N	4.8%
90 N	418 mV	83.6 N	7.1%
100 N	461 mV	92.2 N	7.8%

7. Conclusions

In this paper, a novel hybrid excitation linear eddy current brake was presented. The hybrid excitation linear eddy current brake has the advantages of high force density and low excitation loss compared to the electric excitation linear eddy current brakes. The validity of the analytical model was verified by the FEM and experimental tests, therefore the analytical model can be used in the preliminary design of eddy current brakes. Parametric analysis was performed to explore the influence of the design parameters on the eddy current brake performance. Moreover, the experimental results show that the eddy current brake can generate objective braking force using the controller proposed in this paper. It has been found that the proposed eddy current brake system can be used in road and rail vehicles.

Acknowledgments: This work was supported by National Science and Technology Major Projects (2012ZX04001-051).

Author Contributions: Baoquan Kou conceived the structural and experiments of the hybrid excitation eddy current brake. Lu Zhang, He Zhang and Yinxi Jin performed the experiments. Yinxi Jin analyzed the data and wrote the paper.

Conflicts of Interest: The authors declare no conflict of interest.

References

1. Jang, S.; Lee, S.; Jeong, S. Characteristic analysis of eddy-current brake system using the linear Halbach array. *IEEE Trans. Magn.* **2002**, *38*, 2994–2996.
2. Sainjargal, S.; Byun, J. Analysis and case study of permanent magnet arrays for eddy current brake systems with a new performance index. *J. Magn.* **2013**, *18*, 276–282.
3. Hecquet, M.; Brochet, P.; Lee, S.; Delsalle, P. A linear eddy current braking system defined by finite element method. *IEEE Trans. Magn.* **1999**, *35*, 1841–1844.
4. Gay, S.E. Contactless Magnetic Brake for Automotive Applications. Ph.D. Thesis, Texas A&M University, College Station, TX, USA, 2005.
5. Wang, P.J.; Chiueh, S.J. Analysis of eddy-current brakes for high speed railway. *IEEE Trans. Magn.* **1998**, *34*, 1237–1239.
6. Jang, S.; Lee, S. Comparison of three types of permanent magnet linear eddy-current brakes according to magnetization pattern. *IEEE Trans. Magn.* **2003**, *39*, 3004–3006.
7. Ha, K.; Hong, J.; Kim, G.; Lee, J.; Kang, D. A study of the design for touch free linear eddy current brake. *IEEE Trans. Magn.* **1999**, *35*, 4031–4033.
8. Ihm, H.; Lee, S.; Ham, S.; Lee, J. The influence of slit construction on the eddy current braking torque considered by 3D FEM analysis. In Proceedings of the International Conference on Electrical Machines and Systems, Wuhan, China, 17–20 October 2008; pp. 444–446.
9. Choi, J.; Shin, H.; Park, Y.; Jang, S. Torque analysis of axial flux PM type eddy current brake based on analytical field computations. In Proceedings of the International Conference on Electrical Machines and Systems, Beijing, China, 20–23 August 2011; pp. 1–5.
10. Amati, N.; Tonoli, A.; Canova, A.; Cavalli, F.; Padovani, M. Dynamic behavior of torsional eddy-current dampers: Sensitivity of the design parameters. *IEEE Trans. Magn.* **2007**, *43*, 3266–3277.
11. Lequesne, B.; Liu, B.; Nehl, T.W. Eddy-current machines with permanent magnets and solid rotors. *IEEE Trans. Ind. Appl.* **1997**, *33*, 1289–1294.
12. Shin, H.; Choi, J.; Cho, H.; Jang, S. Analytical torque calculations and experimental testing of permanent magnet axial eddy current brake. *IEEE Trans. Magn.* **2013**, *49*, 4152–4155.

13. Canova, A.; Vusini, B. Analytical modeling of rotating eddy-current couplers. *IEEE Trans. Magn.* **2005**, *41*, 24–35.
14. Mohammadi, S.; Mirsalim, M.; Vaez-Zadeh, S. Nonlinear modeling of eddy-current couplers. *IEEE Trans. Energy Convers.* **2014**, *29*, 224–231.
15. Li, P.; Ma, J.; Fang, Y. Design and analysis of hybrid excitation rail eddy current brake system of high-speed train. In Proceedings of the IEEE International Conference on Service Operations, Logistics, and Informatics, Beijing, China, 10–12 July 2011; pp. 565–569.
16. Edwards, J.D.; Jayawant, B.V.; Dawson, W.R.C.; Wright, D.T. Permanent-magnet linear eddy-current brake with a non-magnetic reaction plate. *IEE Proc. Electr. Power Appl.* **1999**, *146*, 627–631.
17. Jang, S.; Jeong, S.; Cha, S. The application of linear Halbach array to eddy current rail brake system. *IEEE Trans. Magn.* **2001**, *37*, 2627–2629.
18. Gay, S.E.; Ehsani, M. Analysis and experimental testing of a permanent magnet eddy-current brake. In Proceedings of IEEE Conference Vehicle Power and Propulsion, Chicago, IL, USA, 7–9 September 2005; pp. 756–765.
19. Canova, A.; Vusini, B. Design of axial eddy current couplers. *IEEE Trans. Ind. Appl.* **2003**, *39*, 725–733.
20. Yazdanpanah, R.; Mirsalim, M. Axial-flux wound-excitation eddy-current brakes: Analytical study and parametric modeling. *IEEE Trans. Magn.* **2014**, *50*, 1–10.
21. Choi, J.; Jang, S. Analytical magnetic torque calculations and experimental testing of radial flux permanent magnet-type eddy current brakes. *J. Appl. Phys.* **2012**, *111*.
22. Singh, A. Theory of eddy-current brakes with thick rotating discs. *Proc. IEE* **1977**, *124*, 373–376.
23. Park, M.; Choi, J.; Shin, H.; Jang, S. Torque analysis and measurements of a permanent magnet type Eddy current brake with a Halbach magnet array based on analytical magnetic field calculations. *J. Appl. Phys.* **2014**, *115*.
24. Srivastava, R.K.; Kumar, S. An alternative approach for calculation of braking force of an eddy-current brake. *IEEE Trans. Magn.* **2009**, *45*, 150–154.
25. Liu, Z.J.; Vourdas, A.; Binns, K.J. Magnetic field and eddy current losses in linear and rotating permanent magnet machines with a large number of poles. *IEE Proc.* **1991**, *138*, 289–294.

A Novel Grouping Method for Lithium Iron Phosphate Batteries Based on a Fractional Joint Kalman Filter and a New Modified K-Means Clustering Algorithm

Xiaoyu Li, Kai Song, Guo Wei, Rengui Lu and Chunbo Zhu

Abstract: This paper presents a novel grouping method for lithium iron phosphate batteries. In this method, a simplified electrochemical impedance spectroscopy (EIS) model is utilized to describe the battery characteristics. Dynamic stress test (DST) and fractional joint Kalman filter (FJKF) are used to extract battery model parameters. In order to realize equal-number grouping of batteries, a new modified K-means clustering algorithm is proposed. Two rules are designed to equalize the numbers of elements in each group and exchange samples among groups. In this paper, the principles of battery model selection, physical meaning and identification method of model parameters, data preprocessing and equal-number clustering method for battery grouping are comprehensively described. Additionally, experiments for battery grouping and method validation are designed. This method is meaningful to application involving the grouping of fresh batteries for electric vehicles (EVs) and screening of aged batteries for recycling.

Reprinted from *Energies*. Cite as: Li, X.; Song, K.; Wei, G.; Lu, R.; Zhu, C. A Novel Grouping Method for Lithium Iron Phosphate Batteries Based on a Fractional Joint Kalman Filter and a New Modified K-Means Clustering Algorithm. *Energies* **2015**, *8*, 7703–7728.

1. Introduction

With the development of EVs, battery technology has drawn more and more attention worldwide. Battery packs are core components of EVs; they are composed of hundreds or thousands of small cells joined by series-parallel connections. Owing to the subtle differences in the battery production process, the electric characteristics of the batteries are slightly different. The current flowing through the batteries will thus be inconsistent and the temperature inside the battery pack will be uneven when the inconsistent batteries are connected in parallel. In this case, battery use will accelerate their decay and some safety problems will occur [1]. In series connection, the potential of most batteries in the battery pack will not exhaust. Moreover, battery state monitoring and management, such as state of charge (SoC), state of health (SoH) estimation [2–4] and state of peak power (SoP) prediction [5,6], will become more and more difficult. Therefore, measures are necessary to guarantee the characteristics of the grouped batteries are as similar as possible.

Numerous battery grouping methods have been reported in previous studies. Kim *et al.* proposed a battery voltage and SoC consistency screening method based on matching battery DC resistance [7,8]. Schneider *et al.* used a sorting method for aged batteries [9]. In that paper, the sorting steps for these batteries are designed. Fang *et al.* proposed a method for battery classification based on the thermal behavior during the charging process [10]. In the method, differences in battery surface temperature are used to evaluate the consistency of batteries. At present, battery capacity, AC resistance, electrochemical impedance spectroscopy (EIS), voltage curve, battery model parameter, charge and discharge thermal behavior are the commonly used parameters to evaluate consistency.

In the authors' previous work [11], the reliability and effectiveness of these evaluation bases are comprehensively analyzed. The main arguments are summarized as follows: AC resistance represents only a small part of the impedance characteristics of the battery, and this parameter cannot reflect the dynamic characteristics of the battery, since the testing frequency (about 1 kHz) is much higher than the main frequency component of the driving conditions. Battery charge and discharge voltage curves can reflect the voltage consistency under the test conditions, however, the reliability of any conformance assessment will be decreased if the working conditions change. Additionally, the voltage curve is just an external manifestation of the battery, and the internal characteristics' consistency cannot be guaranteed because of the complex relationship between voltage curve and internal material properties. The surface temperature of a battery is determined by the entropy heat, resistance heat and surface heat dissipation. The parameters can reflect the battery performance, but they are susceptible to the external environment. EIS or battery model parameters can better reflect the dynamic characteristics of a battery. Capacity consistence determines that the performance of batteries in the battery pack cannot reach their full potential. However, battery capacity is obtained by constant current charge and discharge tests, EIS is measured by sweep frequency impedance tests and battery model parameters are obtained by intermittent constant current and pulse current tests. Considerable time and special testing instruments are required for these three tests.

The existing battery clustering algorithms include the self-organizing feature map algorithm (SOM) [10], fuzzy c-means (FCM) clustering algorithm, *etc.* [12–14]. These earlier works focus more on the similarity of batteries' characteristics. However, the number of elements in each group after grouping is different and not equal to the design number of batteries in each pack.

After review of the problems of the battery consistency evaluation parameter extraction methods and battery clustering algorithms, this paper presents a novel battery grouping method. This method can be divided into two parts: the first part is the battery characteristic parameter extraction, which is based on a simplified EIS model, fractional joint Kalman filter algorithm and DST—the details are described in Section 2. The second part is a new modified K-means clustering algorithm especially proposed for battery equal-number grouping—this part are presented in Section 3. The proposed method is established based on a short time charge and discharge experiment. The characteristic parameters including open circuit voltage (OCV), simplified EIS model parameters and battery capacity can be obtained. All these parameters have relatively clear physical meanings. The clustering algorithm can divide the batteries into equal-number groups according to the similarity of consistent evaluation parameters and the number of elements designed by engineers in each group.

2. Battery Parameter Extraction

2.1. Battery Model Selection

2.1.1. Comparison of Battery Models

It is essential to select a suitable model for quantifying battery characteristics. Recent battery models mainly involve equivalent circuit models based on the external electric characteristics of battery charge and discharge (ECMs for short in this paper) or electrochemical impedance spectroscopy tests (EIS models), electrochemical reaction mechanism models (ERM models), *etc.* In ECMs, resistance, capacitor and diodes are commonly used. Their structures are simple and easy to calculate. These models are widely used in battery state monitoring and management applications [5,15–19]. The typical equivalent circuit models include the R_{int} model, first order RC model, PNGV model, *etc.* Among them, the first order RC model is the most widely used because of its simple structure and high accuracy. This model is shown in Figure 1a.

EIS is usually used to study the electrochemical process of the battery electrode/electrolyte interface and in the analysis of the insertion and extraction process of the battery electrode active material. Recently, the EIS technique has had a wide range of applications in battery charging and discharging dynamic modeling. Yoon *et al.* [20] have used impedance data to evaluate the power capacity of batteries. Xu *et al.* [21] have used a lithium-ion battery EIS model and fractional order Kalman filtering algorithm to achieve accurate battery SoC estimation. Waag *et al.* [22] have used an EIS model to analyze the aging state of batteries. Compared with ECMs, EIS models not only reflect the dynamic characteristics of battery more accurately, but also the structure elements have clear physical meanings, therefore these models have attracted more and more attention. Traditional EIS tests are conducted in a laboratory environment. During the process, a small current or voltage is loaded onto the battery and the response is measured. A simple EIS model is shown in Figure 1b.

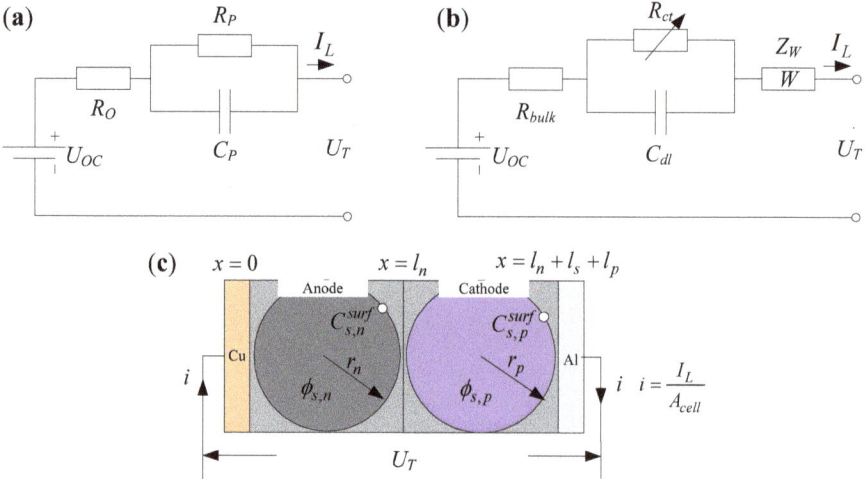

Figure 1. (a) First-order RC model; (b) a simple EIS model; (c) single particle model.

The behavior inside a battery, including the electrochemical kinetics and charge transfer processes are described in ERM models [23]. The relationship between material properties and a battery's electrical performance are established in these models. In these models, the pseudo two-dimensional model (P2D model) created by Fuller *et al.* [24] and single particle model (SP model) [25] are the most representative ones. Among these, the, SP model is most the simple ERM. The structure of a SP model is shown in Figure 1c.

For battery grouping applications, too many characteristic parameters will lead to an increase in the dimensionality of the battery characteristic vector and battery clustering will be more difficult. If the weight setting of parameters is unreasonable, the accuracy of the clustering results will be reduced. Hence, the battery model structure used for battery grouping should be simple and better reflect the battery dynamics. In the examples described above, EIS models are more suitable for battery grouping due to their fewer parameters and clear physical meaning.

2.1.2. Simplified EIS Model

The EIS test result of a lithium iron phosphate (LiFePO$_4$/graphite) battery and its fitting result based on the simple EIS model are presented in Figure 2.

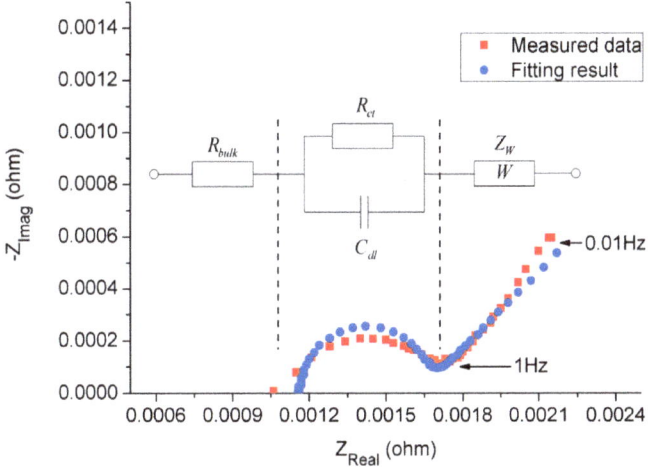

Figure 2. EIS test result and fitting result of the simple EIS model.

The values of R_{ct} and C_{dl} are 0.46 mΩ and 28.59 F, and the time constant τ is 0.01315 s. The time constant of R_{ct} and C_{dl} is very small and barely manifested on the terminal voltage in the case of electric vehicle applications. Hence, the simple EIS model can be further simplified for a LiFePO$_4$/graphite battery.

The impact of double layer capacitance on battery's external characteristics can be ignored. Moreover, R_{ct} is assumed to be constant with current at the room temperature. Thereby, the simplified EIS model shown in Figure 3 is obtained.

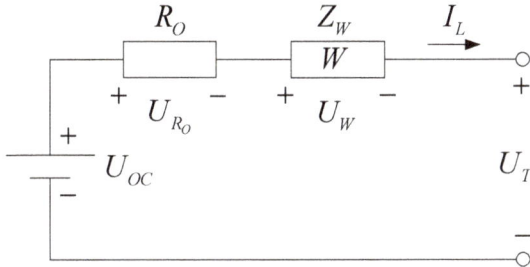

Figure 3. The simplified EIS model.

The model consists of three elements: voltage source U_{OC}, ohmic resistance R_O and Warburg impedance Z_W. U_T and I_L denote the terminal voltage and the total current flowing through the battery, respectively.

Voltage source is used to describe the open-circuit voltage characteristics of the battery. Due to the simplification of the battery model, many chemical reaction kinetics processes and the chemical reaction boundary conditions are ignored. U_{OC} represents the value of the open circuit voltage (*OCV*), and can be expressed by Equation (1) [26]:

$$U_{OC} = E_{OC,p}\left(SoC_p\right) - E_{OC,n}\left(SoC_n\right) \tag{1}$$

where $E_{OC,p}$ and $E_{OC,n}$ are open circuit potentials of positive and negative electrodes. They are functions of the *SoC* which is determined by the average concentration of ions of each electrode. U_{OC} is mainly decided by three parameters: the *SoC*'s start points of positive and negative electrodes and the cycle range of *SoC*.

R_O is a pure resistance element, which mainly reflects the medium-high frequency (typically <1 Hz) impedance characteristic of EIS; the value of this parameter is approximate the sum of the battery bulk resistance R_{bulk}, SEI film resistance R_{set} and electric charge transfer resistance R_{ct} [27], as is shown in Equation (2):

$$\begin{cases} R_O = R_{bulk} + R_{sei} + R_{ct} \\ R_{bulk} = R_{ext} + R_{O,electrolyte}^{eff} + R_{O,solid}^{eff} \\ R_{sei} = A\frac{\eta_{sei}}{I_L} \\ R_{ct} = \frac{\eta_{ct}}{I_L}, I_L = A \cdot i_0 \cdot \left(\exp\left(\frac{\alpha_a F}{RT}\right)\eta_{ct} - \exp\left(\frac{\alpha_c F}{RT}\right)\eta_{ct}\right) \end{cases} \tag{2}$$

where R_{ext} is the connecting resistance, $R_{O,electrolyte}^{eff}$ and $R_{O,solid}^{eff}$ stand for the effective part of the electrolyte ohm impedance and solid ohm impedance affecting on the terminal voltage. R_{sei} is mainly determined by the SEI film produced on the surface of negative electrode particles. R_{ct} is decided by Butler-Volmer kinetics.

Z_W is the Warburg impedance, which is usually used to describe the diffusion characteristics. It is a 0.5 order fractional element. U_W is the terminal voltage of this element. Fleischer *et al.* [28] presented the physical meaning of the parameters in Z_W; without considering the boundary conditions, Z_W can be expressed by Equation (3):

$$Z_W = \frac{RT}{cn^2F^2A} \cdot \frac{1}{\sqrt{jwD}} \tag{3}$$

where, R is the gas constant, T is the temperature, c represents the molar concentration of active particles, F is Faraday's constant, A is the active surface area, D is the diffusion coefficient of the materials. Let $X_W = \frac{RT}{cn^2F^2AD^{0.5}}$, then Equation (4) can be obtained:

$$Z_W = \frac{X_W}{(jw)^{0.5}} \tag{4}$$

The characteristics of the battery from different aspects are described by the three model parameters. Meanwhile, in the simplified EIS model, charge capacity (Q) is also a crucial parameter for the batteries, and it is meaningful to battery grouping. Four battery performance parameters, Q, U_{OC}, R_O, X_W are used for battery grouping.

2.2. Model Parameter Identification

The simplified EIS model parameter online identification method for EVs is proposed in the authors' previous work [29]. In the method, a model discrete state equation should be established first, and then, model parameters are identified based on the fractional joint Kalman filter algorithm. The details of the two parts are described as follows.

2.2.1. Model State Equation Establishment

The simplified EIS model has been established in Section 2.1. In this section, the state and observation equations are constructed as follows, taking the discharge current value $I_{L,dis}$ as the positive value and data sampling period T_s is one second. In Equation (5), Δ^r is used as a differential operator:

$$\Delta^r = \frac{d^r}{dt^r}, \ r > 0 \tag{5}$$

where, r is the differential order. When $r = 1$, Δ^1 represents a one order differential operator. As the commonly used capacitor element, if its capacitance is C, the relationship of terminal voltage U_C and the current I_C flowing through this element can be expressed by Equation (6):

$$\Delta^1 U_C = \frac{1}{C} I_C \tag{6}$$

When r is a fractional value, Δ^r represents a fractional differential operator. Similar to the capacitor, the relationship of terminal voltage U_W and the current I_L flowing through the element can be expressed as follows:

$$\Delta^{0.5} U_W = X_W I_L \tag{7}$$

The battery model parameters X_W, U_{OC}, R_O change slightly with SoC. Then, the differential equations of state and parameters can be written in the following matrix form:

$$
\Delta
\begin{bmatrix} 0.5 \\ 1 \\ 1 \\ 1 \end{bmatrix}
\begin{bmatrix} U_W \\ X_W \\ U_{OC} \\ R_O \end{bmatrix}
=
\begin{bmatrix}
0 & I_L & 0 & 0 \\
0 & 0 & 0 & 0 \\
0 & 0 & 0 & 0 \\
0 & 0 & 0 & 0
\end{bmatrix}
\begin{bmatrix} U_W \\ X_W \\ U_{OC} \\ R_O \end{bmatrix}
\tag{8}
$$

U_T is treated as the measurement observation parameter. The observation equation is written as below:

$$
U_T = U_{OC} - I_L R_O - U_W
\tag{9}
$$

Equation (8) can be discretized using a bilinear transformation method. Additionally, the system state noise w and observation noise v are considered in these equations, where these noises are assumed independent with each other. The discrete form of the state and observation equations is expressed by Equation (10):

$$
\begin{cases}
\Delta
\begin{bmatrix} 0.5 \\ 1 \\ 1 \\ 1 \end{bmatrix}
\begin{bmatrix} U_W \\ X_W \\ U_{OC} \\ R_O \end{bmatrix}_k
=
\begin{bmatrix}
0 & \frac{I_{L,k}+I_{L,k-1}}{2} & 0 & 0 \\
0 & 0 & 0 & 0 \\
0 & 0 & 0 & 0 \\
0 & 0 & 0 & 0
\end{bmatrix}
\begin{bmatrix} U_W \\ X_W \\ U_{OC} \\ R_O \end{bmatrix}_{k-1}
+ w \\[2em]
U_{t,k} = \begin{bmatrix} -1 & 0 & 1 & -I_{L,k} \end{bmatrix} \begin{bmatrix} U_W & X_W & U_{oc} & R_o \end{bmatrix}_k^T + v
\end{cases}
\tag{10}
$$

Based on the definition of the Grünwald-Letnikov fractional differential [21,30], the discrete differential form of a fractional state variable can be written as:

$$
\Delta^r x_k = \sum_{j=0}^{k} (-1)^j \binom{r}{j} x_{k-j}, \quad
\binom{r}{j} =
\begin{cases}
1 & \text{for } j = 0 \\
r(r-1)\ldots(r-j+1)/j! & \text{for } j > 0
\end{cases}
\tag{11}
$$

In this paper, the following definitions are made:

$$
\gamma_j = \text{diag} \begin{bmatrix} \binom{0.5}{j} & \binom{1}{j} & \binom{1}{j} & \binom{1}{j} \end{bmatrix}
\tag{12}
$$

$$
N = \begin{bmatrix} 0.5 & 1 & 1 & 1 \end{bmatrix}^T
\tag{13}
$$

where x in Equation (11) is $[U_W \ X_W \ U_{OC} \ R_O]^T$, then, the discrete differential form of the state and parameters of the simplified EIS model can be obtained:

$$\Delta^N x_k = x_k + \sum_{j=1}^{k} (-1)^j \gamma_j x_{k-j} \tag{14}$$

In Equation (14), the complexity of $\sum_{j=1}^{k} (-1)^j \gamma_j x_{k+1-j}$ is increased with the length of time, so this feature is not suitable for engineering applications. Additionally, the battery diffusion process does not occur infinitely, as it is constrained by the boundary conditions. Thus, k is replaced with variable length L. The value of L is chosen by applying curve fitting to the experimental data. Finally, the state and observation equations can be obtained and are shown in Equation (15):

$$\begin{cases} x_k = \underbrace{\begin{bmatrix} 0 & \frac{I_{L,k}+I_{L,k-1}}{2} & 0 & 0 \\ 0 & 0 & 0 & 0 \\ 0 & 0 & 0 & 0 \\ 0 & 0 & 0 & 0 \end{bmatrix}}_{f(x_k, I_{L,k}, I_{L,k-1})} x_{k-1} - \sum_{j=1}^{k} (-1)^j \gamma_j x_{k+1-j} + w \\[2em] y_k = \underbrace{\begin{bmatrix} -1 & 0 & 1 & -I_{L,k} \end{bmatrix}}_{g(x_k, I_{L,k})} x_k + v \\[1em] \text{where,} \begin{cases} k \le 64, L = k \\ k > 64, L = 64 \end{cases} \end{cases} \tag{15}$$

2.2.2. Model Parameter Identification

In [30], Sierociuk *et al.* put forward the fractional Kalman filter algorithm. In this method, fractional element state estimation, parameter identification and order estimation can be easily realized. Fractional model parameter can be estimated jointly with the state estimation. In order to distinguish the different applications, this method is called fractional joint Kalman filter (FJKF) in this paper. Based on this method, model parameter X_W is regarded as an implicit parameter of the state U_W, the value of X_W can be updated based on the state estimation result. Besides, the predication-correction feature of Kalman filter is effectively used. The application of this algorithm used in battery simplified EIS model parameter identification is described in Algorithm 1.

Algorithm 1 Model parameter identification based on FJKF.

Definitions:

$$A_{k-1} = \frac{\partial f(x_{k-1}, I_{L,k}, I_{L,k-1})}{\partial x_{k-1}} \bigg|_{x_{k-1}=\hat{x}_{k-1}^+} = \begin{bmatrix} 0 & \frac{I_{L,k}+I_{L,k-1}}{2} & 0 & 0 \\ 0 & 0 & 0 & 0 \\ 0 & 0 & 0 & 0 \\ 0 & 0 & 0 & 0 \end{bmatrix}$$

$$C_k = \frac{\partial g(x_k, I_{L,k})}{\partial x_k} \bigg|_{x_k=\hat{x}_k^-} = \begin{bmatrix} -1 & 0 & 1 & -I_{L,k} \end{bmatrix}$$

Step 1: Initialization, Q_k is the covariance of w_k, R_k is the covariance of noise v_k. P_k is the error covariance of the state and parameter estimated values, the initial value for each parameter is give as:

$$\hat{x}_0 = E[x], \ P_0^+ = E\left[(x-\hat{x}_0)(x-\hat{x}_0)^T\right]$$
$$R_0 = E\left[v_0 v_0^T\right], \ Q_0 = E\left[w_0 w_0^T\right]$$

Step 2: Time update:

State and parameter time update: $\hat{x}_k^- = f\left(\hat{x}_{k-1}^+, I_{L,k}, I_{L,k-1}\right)$

Error covariance time update: $P_k^- = (A_{k-1}+\gamma_1) P_{k-1}^+ (A_{k-1}+\gamma_1)^T + Q_x + \sum_{j=2}^{L} \gamma_j P_{k-j}^+ \gamma_j^T$

Step 3: Measurement update:

Kalman gain matrix update: $L_k = P_k^- (C_k)^T \left[C_k P_k^- (C_k)^T + R_k\right]^{-1}$

State and parameter measurement update: $\hat{x}_k^+ = \hat{x}_k^- + L_k^x \left[y_k - g\left(\hat{x}_k^-, I_{L,k}\right)\right]$

Error covariance measurement update: $P_k^+ = (I - L_k C_k) P_k^-$

Step 4: $k = k + 1$, repeat Step 2 and Step 3, until all data is processed.

2.3. Parameter Identification Experiment Sequence

The battery model and model parameter identification method are described in Sections 2.1 and 2.2. In order to obtain the battery model parameters and charge capacity, an experimental sequence is designed. The sequence is set up mainly based on dynamic stress tests, and there are seven steps.

The time consumption of this experiment is about eight hours, which is a little longer than the time consumption of constant current/voltage charge and discharge recommended by battery manufacturers, but shorter than the hybrid pulse power test described in the FreedomCAR battery test manual [31]. This sequence is described in Algorithm 2 and Figure 4.

438

Algorithm 2 Experimental sequence for battery parameter identification.

Step 1: Constant current discharge, until battery terminal voltage reaches the lower cut-off voltage.
Step 2: Rest for one hour.
Step 3: Charge with the battery manufacturer's recommended procedures.
Step 4: Rest for one hour.
Step 5: Apply Dynamic Stress Test on the battery. The maximum discharge current is 2C in this paper. Stop when the battery terminal voltage reaches the lower cut-off voltage. The details of DST are described in FreedomCAR battery test manual.
Step 6: Rest for several minutes.
Step 7: Discharge with a low constant current, until the battery terminal voltage reaches the lower cut-off voltage. This step causes the battery to be fully discharged, this is essential for battery grouping.

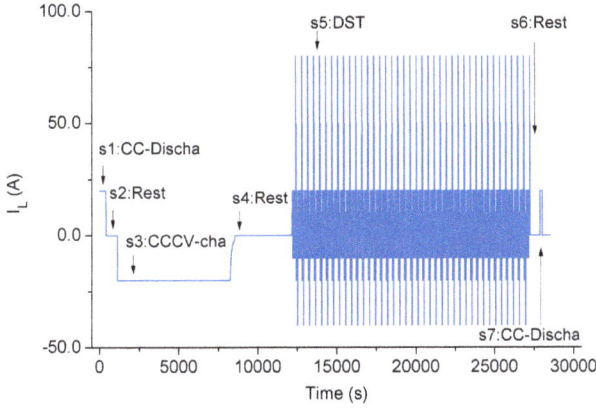

Figure 4. Battery parameter identification experimental sequence.

3. Battery Feature Clustering

3.1. Data Preprocess

3.1.1. Data Down-Sampling

Battery model parameters U_{OC}, R_O, X_W can be obtained based on the method proposed in Section 2, and they are functions of test time. It is necessary to translate the function into the depth of discharge (DoD) of the battery. DoD of a battery i at time k is defined by Equation (16):

$$DoD_i(k) = \frac{\int_{t=0}^{t=k} I_L dt}{Q_i} \times 100\% \qquad (16)$$

Here, Q_i is the charge capacity of battery i, its value can be calculated by the Ah-counting method, and this process is expressed by Equation (17):

$$Q_i = \int_{step5}^{step7} I_L dt \qquad (17)$$

In order to reduce the amount of calculations, a down-sampling process is executed on the parameter identification results. Additionally, model parameter identification results of $DoD = 0\%$–10% and 90%–100% are discarded from further analysis.

3.1.2. Crude Data Exclusion

Typically, operational errors are unavoidable in the experimental process. For example, test cables may connect unreliably, equipment channels fail and so on. Apart from these errors, the failures in the battery production process can also lead to the existance of crude values in the model parameter identification results. In this paper, the crude values are discarded based on the Laiyite criterion (3σ criterion). Additionally, since the number of samples remaining should not larger than the designed numbers n_0 for battery grouping, another step is added to the crude data discarding process. The details of this process are shown in Algorithm 3.

Algorithm 3 Crude data excluding process.

Step 1: Calculate the average curve of each model parameters of the samples, additionally, calculate the average value of battery capacities.
$$X_{avg}(DoD) = \frac{1}{n}\sum_{i=1}^{n} X(DoD)$$
Where, X represents U_{OC}, R_O or X_W.
Step 2: Calculate the average distance between the parameter curve and the average curve, that is the distance between battery capacities and the average capacity.
$$D_{X,i} = mean\left(X_i(DoD) - X_{avg}(DoD)\right), i = 1, ..., n$$
Step 3: Calculate the standard deviation of each distance parameter.
$$\sigma_X = \sqrt{\frac{1}{n}\sum_{i=1}^{n} D_{X,i}^2}$$
Step 4: Discard the samples that do not meet the requirement of Laiyite criterion. The Laiyite criterion is: $|DX_i| < 3\sigma_X$
Step 5: Regard the remaining samples as the evolution objects. Reset the number of evolution objects n, and repeat Steps 1–4, until all the remaining parameters are eligible with the requirement of Laiyite criterion.
Step 6: If $n > n_0$, randomly discard the samples which have maximum value of $|DX_i|$ until the number of remaining samples is equal to n_0. Ultimately, the samples used for battery grouping are obtained.

3.2. Consistency Evaluation Parameter Generation

In Section 3.1, D_Q, $D_{U_{OC}}$, D_{R_O} and D_{X_W} of each remaining battery are obtaine; these parameters have different units and should be normalized for further analysis, the results are represented by D_Q^*, $D_{U_{OC}}^*$, $D_{R_O}^*$ and $D_{X_W}^*$. The four parameters describe the performance of batteries, but cannot be used directly for battery consistency evaluation, because they have different effects on battery consistency. It is necessary to weight the parameters according to defined rules. In this paper, the weights are set up based on the parameters' inconsistent influence on the terminal voltage. Battery terminal voltage can be expressed as:

$$U_T = U_{OC} - U_{R_O} - U_W \tag{18}$$

The relationship between Q's inconsistent value ΔQ and ΔU_{OC} is described in Figure 5. In order to simplify the calculation, the OCV curve can be regard as a straight line, and Equation (19) can be obtained.

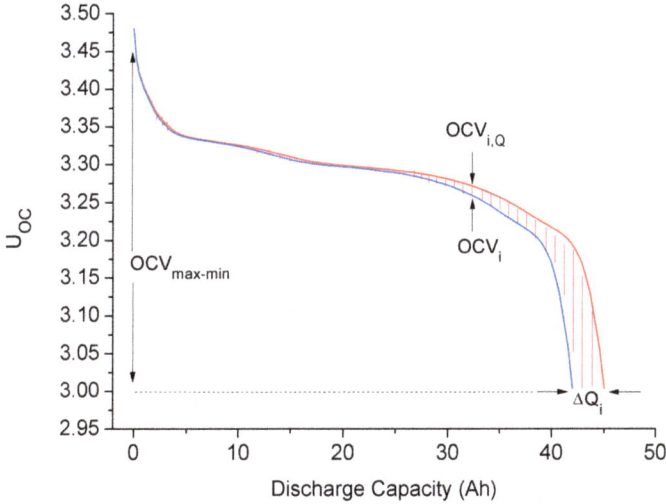

Figure 5. Effect of capacity inconsistency on U_{OC} (U_T).

$$\Delta U_{T,Q} = \frac{\frac{1}{2}\Delta OCV_{max-min}\Delta Q}{Q_{avg}} \tag{19}$$

where, $\Delta OCV_{max-min}$ is about 0.48 V.

The relationship between ΔU_T and ΔU_{OC}, ΔR_O can be easily obtained; this are expressed as:

$$\Delta U_{T,U_{OC}} = \Delta U_{OC} \tag{20}$$

$$\Delta U_{T,R_O} = I_L \Delta R_O \tag{21}$$

However, the influence of ΔX_W on ΔU_T cannot be obtained directly. Thus, the following analysis is carried out. Assuming the initial state of U_W is zero, when a step current excitation I_L is applied to Z_W, the voltage response after k s can be calculated. When $k = 1$, the voltage response is given as:

$$U_{W,1} = X_W I_L - (-1)^1 \binom{0.5}{1} U_{W,0} \tag{22}$$

The voltage response of Z_W when $k \geq 2$ is given as:

$$U_{W,k+1} = X_W I_L - \sum_{j=1}^{L} (-1)^j \binom{0.5}{j} U_{W,k+1-j} \tag{23}$$

Finally, the voltage response of Z_W at time k can be given as:

$$U_{W,T} = a(k) X_W I_L \tag{24}$$

where, $a(k)$ is coefficient variable of $X_W I_L$, and is determined by L in the state discrete equations. I_L and X_W are constant, the coefficient a can be expressed in Figure 6. The maximum value of a is $a_{\max}(k) = 14$.

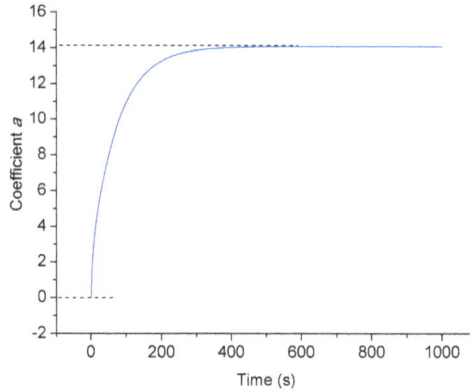

Figure 6. Changes of the coefficient a with time.

During charge or discharge, the working time is much longer than the length of time presented in the figure. $a_{\max} = 14$, the value of a changes mainly near the maximum value. The relationship between ΔU_T and ΔX_W can be described as:

$$\Delta U_W = a_{\max} I_L \Delta X_W \tag{25}$$

The median working current 1C (40 A) is used as the reference current $I_{L,ref}$ in the Equations (24) and (25), the weight proportion of each parameter can be described as:

$$w_Q : w_{U_{OC}} : w_{R_O} : w_{X_W} = \frac{\frac{1}{2}OCV_{max-min}\Delta Q}{Q_0} : \Delta U_{OC} : I_{L,ref}\Delta R_O : a_{max}I_{L,ref}\Delta X_W \qquad (26)$$

After the parameters are weighted, the feature vector $Cell\left(D_{Q,w}^*, D_{U_{OC},w}^*, D_{R_O,w}^*, D_{X_W,w}^*\right)$ which consists of four consistency evaluation parameters for each cell is obtained.

3.3. Battery Equal-Number Clustering

Battery grouping is a special kind of clustering problem. In the traditional clustering method, elements are grouped based on the similarity of feature data, and there is no pre-set category features. The whole clustering process works without supervision. However, battery grouping is a little different from the traditional method, because the amount of elements in each group should meet the requirements of the battery pack. Hence, a special equal-number clustering method is needed for battery grouping.

3.3.1. K-Means Clustering Method

K-means clustering method is a popular partitioning clustering method [32]. In this method, the initial cluster centers are randomly selected from the elements to be clustered, and the remaining elements are clustered into N groups based on the similarity (distance), and the cluster center of each cluster is recalculated after that. The variance function shown in Equation (27) is generally used as the standard measurement function in this method. The main steps of K-means are shown in Algorithm 4.

$$E = \sum_{n=1}^{N} \sum_{x \in c_n} ||x - \mu_n||^2 \qquad (27)$$

3.3.2. Kd-Tree Cluster Center Initialization

Unfortunately, K-means clustering method is local optimization strategy, and it is sensitive to the initial cluster center and can easily fall into a local optimal solution. A good initial cluster center can avoid the local optima issue. Redmond, et al. proposed a cluster center initial method based on kd-tree. kd-tree is established for data density estimation, and the initial cluster centers are selected from kd-tree's leave boxes based on an improved max-min distance method. This cluster center initial method is much better than the random cluster center initial method. The calculation is relatively simple. Readers can refer to [33] for further details of the algorithm.

Algorithm 4 Main steps of K-means clustering method.

Step 1: Initialize the cluster centers. Select N cluster centers randomly. $\mu_1, \mu_2, ..., \mu_k$.
Step 2: Calculate the distance between each sample and each cluster center. Clustering the samples to the nearest cluster based on the minimum distance principle. For sample p, it is belong to the cluster which is obtained by:

$$c^{(p)} = \text{argmin} \, || x^{(p)} - \mu_n ||^2$$

Step 3: Recalculate the cluster centers. The mean value of each cluster is used as the new cluster center:

$$\mu_n = \frac{\sum_{p=1}^{M} 1\{c^{(p)}=n\} x^{(p)}}{\sum_{p=1}^{M} 1\{c^{(p)}=n\}}$$

Step 4: Repeat Step 2 and Step 3, until the cluster centers are no longer changing or only changes minimal.

3.3.3. The New Modified K-Means Clustering Method

Sections 3.3.1 and 3.3.2 describe the K-means clustering method using kd-tree cluster center initialization. The battery equal-number clustering method proposed in this paper is based on this method and the clustering rules are redesigned. The target of the method is partitioning the samples into N clusters and the amount of elements in each cluster is equal to M. The objective function of equal-number clustering is the same with the traditional K-means method, as shown in Equation (27). An adjusting rule is developed to equalize the numbers as below:

Rule 1: Element number equalization

The cluster which has the largest number of elements should be determined first. Then, the elements of this cluster are sorted based on the distances between them and their cluster center. The element with the farthest distance is assigned to another cluster whose number of elements is less than the pre-set value M and the distance of this element to the new cluster should be as near as possible. This assigning rule takes distance and element number into account at the same time. We repeat this process until the number of elements in each cluster is equal to the pre-set value. Equal-number clusters can be accomplished based on Rule 1. However, this process cannot guarantee the accuracy of the element's attribution. Thus, an element exchange criterion between two clusters is set up.

Rule 2: Element exchange criterion

This rule is made to regulate the element exchanging process between two selected clusters. In order to explain this rule, several definitions are made:

Master cluster C_m: C_m is the cluster who initiates the element exchange process. The cluster center and element i in this cluster is expressed by μ_m and $x_{m,i}$.

Slave cluster C_s: C_s is the cluster waiting for element exchange with C_m. The cluster center and element j is expressed μ_s and $x_{s,j}$.

Master (Slave) center vector $\overrightarrow{\mu_m\mu_s}$ ($\overrightarrow{\mu_s\mu_m}$): The vector's starting point is μ_m (μ_s) and ending point is μ_s (μ_m).

Master (Slave) proximal element $x_{m,ns}$ ($x_{s,nm}$): The element of C_m (C_s) near μ_s (μ_m). If $\overrightarrow{\mu_m x_{m,i}} \cdot \overrightarrow{\mu_m\mu_s} > 0$, element i is a Master proximal element Similarly, If $\overrightarrow{\mu_s x_{s,j}} \cdot \overrightarrow{\mu_s\mu_m} > 0$, element j is a slave proximal element.

In these two clusters, the exchange process is only allowed in the proximal elements because the exchange possibility between two proximal elements is much higher than between non-proximal elements. This limitation is an effective approach to reduce calculation complexity and improve the convergence rate of the clustering algorithm.

In order to decrease the value of the measurement target of the new modified K-means method, exchange process between C_m and C_s occurs when the requirements of Equation (28) is satisfied:

$$||x_{m,i} - \mu_m||^2 + ||x_{s,j} - \mu_s||^2 > ||x_{s,j} - \mu_m||^2 + ||x_{m,i} - \mu_s||^2 \qquad (28)$$

Based on the analysis and two new rules for cluster elements attribution, the new modified K-means method is established. This equal-number clustering steps are described in the Algorithm 5 and some of the main steps are shown in Figure 7.

Algorithm 5 Steps of equal-number clustering method.

Step 1: Cluster centers initialization. Cluster center is calculated using the kd-tree cluster center initialization method.

Step 2: Cluster initialization. The elements are clustered to each cluster according to the minimum distance criterion described in K-means clustering method. Step1 and Step 2 are shown in Figure 7a.

Step 3: Element number equalization. Equalize the number of each cluster according to rule 1. This step is shown in Figure 7b.

Step 4: Distance data storage. Two arrays are set up for each cluster, named main storage array and auxiliary storage array. Calculate the distances from each element to all centers. The results are stored in the main storage arrays, but auxiliary storage array is not used here.

Step 5: Elements exchange among the clusters.

Step 5.1: Set one of the clusters as a master cluster C_m, another one is regard as a slave cluster C_s. Find the proximal elements in these two clusters. This step is shown in Figure 7c.

Step 5.2: Select a proximal element in C_m and a proximal element in C_s as the quasi-exchange elements, If the two elements can meet the requirements of Equation (28), take out the two elements from their pervious clusters, exchange the belonging of elements and store their feature parameters in the auxiliary storage array. If not, select another proximal element in C_s as the quasi-exchange element and repeat the calculation of Equation (28) and determine whether to exchange elements or not, until all $x_{s,nm}$s are traversed. This step is shown in Figure 7d.

Step 5.3: Select another proximal element in C_m as the quasi-exchange element, repeat Step 5.2, until all $x_{m,ns}$s are traversed.

Step 5.4: Set another cluster as a new C_s, repeat Step 5.2 and Step 5.3, until all clusters apart from the master cluster are treated as a slave cluster one time for element exchange.

Step 5.5: Set another cluster as a new C_m, repeat Step 5.2, Step 5.3 and Step 5.4, until all the clusters are treated as a master cluster one time.

Step 6: Take out the data stored in the auxiliary storage array, and push them onto the main storage array of each cluster. Step 6: Recalculate the center of each cluster. This step is shown in Figure 7e.

$$\mu_n = \frac{\sum\limits_{p=1}^{M} 1\{c^{(p)}=n\}x^{(p)}}{\sum\limits_{p=1}^{M} 1\{c^{(p)}=n\}}$$

Step 7: Repeat Steps 4–6, until cluster centers do not change or only change a little.

$$\sum ||\Delta\mu_n|| < \varepsilon$$

Where, ε is the threshold value of process stopping, the equal-number clusters are obtained.

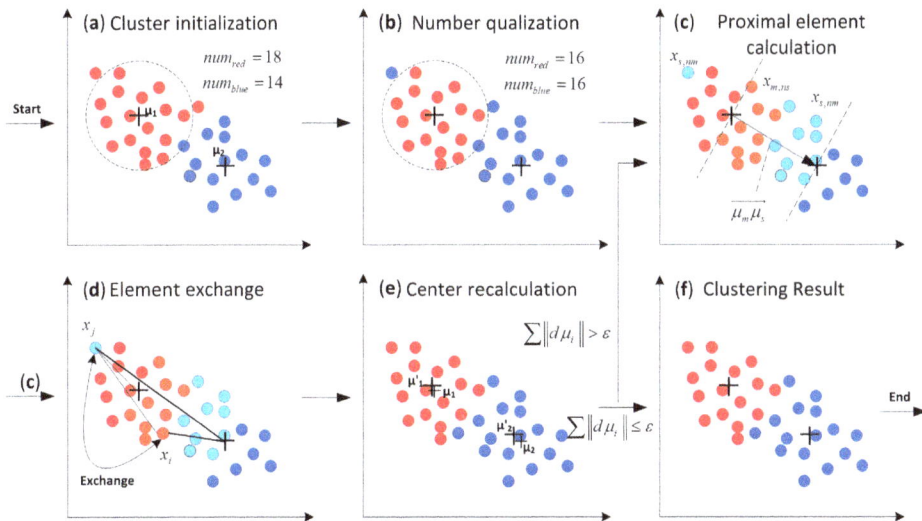

Figure 7. Main steps of equal-number clustering (**a**) Cluster initialization; (**b**) Element number equalization; (**c**) Proximal element determination; (**d**) Elements exchange; (**e**) Cluster center recalculation; (**f**) Result of the clustering method.

4. Experimental Details

Battery parameter identification and battery grouping verification experiments were implemented at CALB Co. Ltd. (Luoyang, China). A series of battery test systems were used for battery testing. Model and parameter identification verification experiments were carried out with an Arbin BT2000 battery test system in the Harbin Institute of Technology. The devices under test are 95 energy-type LiFePO$_4$/graphite batteries produced by CALB. The typical capacity (Q_0) is 42 Ah, cut-off voltages are 3.65 V and 2.5 V, maximum constant discharge current is 2C. The experiments were conducted at room temperature and standard atmospheric pressure.

4.1. Parameter Identification Experimental Details

The sequence of parameter identification experiment has been described in Section 2.3, this sequence is applied to the 95 batteries. The terminal voltage and load current value are recorded in the experiment.

4.2. Verification Experimental Details

There are two verification experiments in this paper. One is a model and parameter identification validation experiment, the other is a battery grouping verification experiment.

447

4.2.1. Model and Parameter Identification Verification Experiments

In order to validate the accuracy of the simplified EIS model and parameter identification method, a load current sequence combining the DST profile with constant current discharge profile is established. The change of DoD in each segment is 10%, the interval of every segment is 1 h. This sequence is shown in Figure 8.

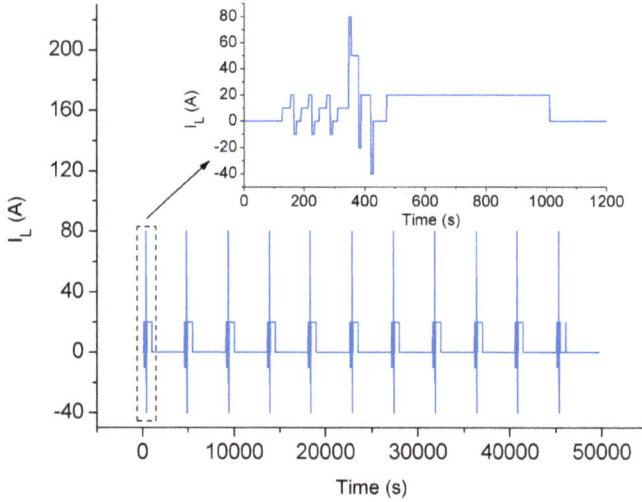

Figure 8. Model and parameter identification verification experimental sequence.

A1: The segment of the test data in $DoD = 50\%$ is selected for model accuracy verification. U_{OC} only changes slightly in this segment, and the value of U_{OC} can be regarded as constant. The simulation results of the first-order RC model and the simplified EIS model are compared. Genetic algorithm (GA) is used for model parameter identification in this process.

A2: The terminal voltage of each segment's first point is regarded as the value of U_{OC} at the initial time. The value of U_{OC} at the end point of this segment is determined by the terminal voltage of the next segment's first point. The change of U_{OC} is considered to be linear with the change of DoD in the center of this segment. GA and FJKF are used to get the impedance parameters in the model. The accuracy of FJKF is verified by comparing the parameter identification results of the two methods.

4.2.2. Battery Grouping Verification Experiments

In order to verify the accuracy of the battery grouping method, typically, batteries should be connected in parallel or series and experiments performed. In this paper, the batteries are not connected in parallel or in series because the analysis is too complex and meaningful results may not be obtained. Instead of that, five simple experiments are implemented on each battery respectively. Battery grouping accuracy is evaluated by the aggregation degree of the voltage curves in each group. The experiments include: 0.5 C-rate charge and 0.5 C-rate discharge, 1 C-rate charge and 1 C-rate discharge and dynamic stress test.

5. Results and Discussion

5.1. Battery Grouping Result Analysis

5.1.1. Battery Parameter Identification Result Analysis

Battery model parameters are extracted from the experiment using the FJKF algorithm described in Section 2.3, and battery capacity is obtained using Ah-counting. Battery capacity and its distribution are shown in Figure 9, where the capacity values are located between 40 Ah and 44 Ah, and the distribution is approximately normal. Battery model parameters and terminal voltage estimation errors are shown in Figure 10.

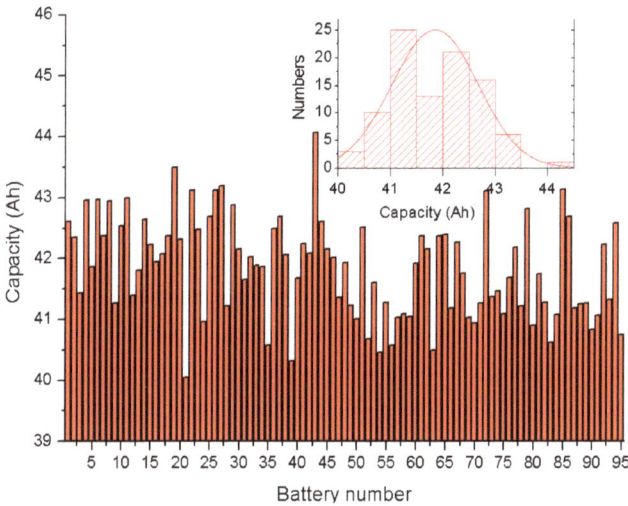

Figure 9. Battery capacities and distribution.

In Figure 10a,b, there exist apparent crude values in the parameter curves. In Figure 10d, the terminal voltage estimation errors in most regions (DoD = 10%–90%) are smaller than 5 mV. This phenomenon can partly verify the accuracy of the simplified EIS model and FJKF algorithm. The parameters in the regions of DoD = 0%–10% and DoD = 90%–100% is unreliable because of the large simulation error existing in these regions. The data are discarded.

449

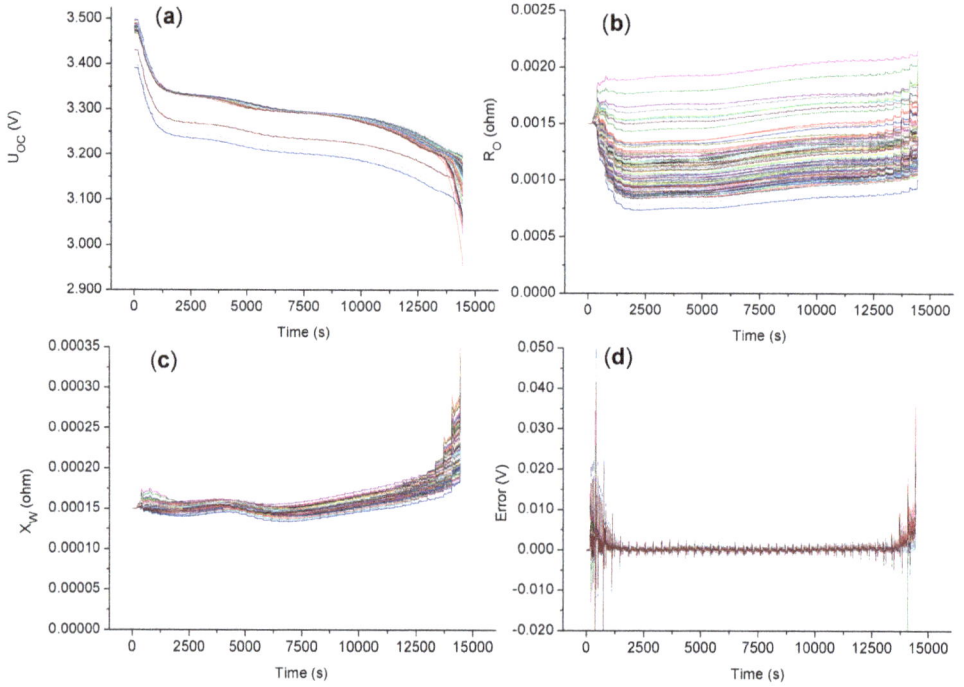

Figure 10. Model parameter identification result (a) U_{OC}; (b) R_O; (c) X_W; (d) U_T error.

5.1.2. Battery Clustering Result Analysis

Model parameters are down sampled and the crude data is discarded according to the data preprocessing process described in Section 3.1. The result is shown in Figure 11. Then, the weights of the battery performance parameters are calculated based on Equation (26). The values of $OCV_{max-min}$, $I_{L,ref}$ and a_{max} are obtained in Section 3.2 and the other coefficients in this equation can be obtained from the preprocess result shown in Figure 11:

$$\begin{cases} \Delta Q = 3.5 \text{ Ah} \\ \Delta U_{OC} = 8.0 \times 10^{-3} \text{ V} \\ \Delta R_O = 5.0 \times 10^{-4} \text{ } \Omega \\ \Delta X_W = 1.2 \times 10^{-5} \text{ } \Omega \end{cases} \tag{29}$$

Then, the weights can be obtained. After normalizing the minimum weight and rounding the value of each weight, Equation (30) can be obtained:

$$w_Q : w_{U_{OC}} : w_{R_O} : w_{X_W} \approx 3 : 1 : 3 : 1 \tag{30}$$

450

Battery clustering result are shown in Figure 12. The samples are clustered into four groups. In each group, the consistency evolution parameters are very similar with each other.

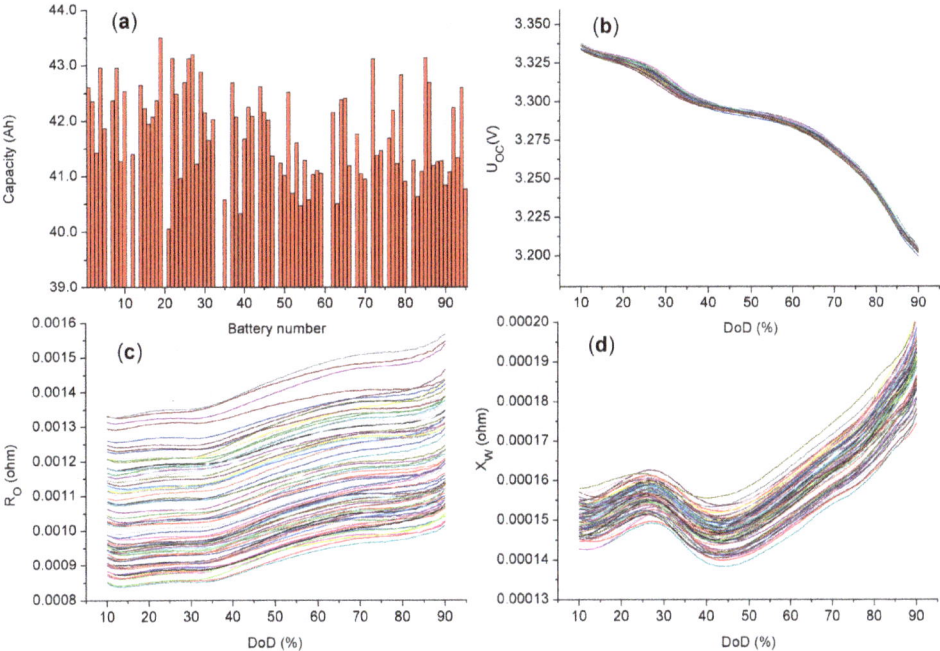

Figure 11. The characteristic parameters after preprocess (**a**) Q; (**b**) U_{OC}; (**c**) R_O; (**d**) X_W.

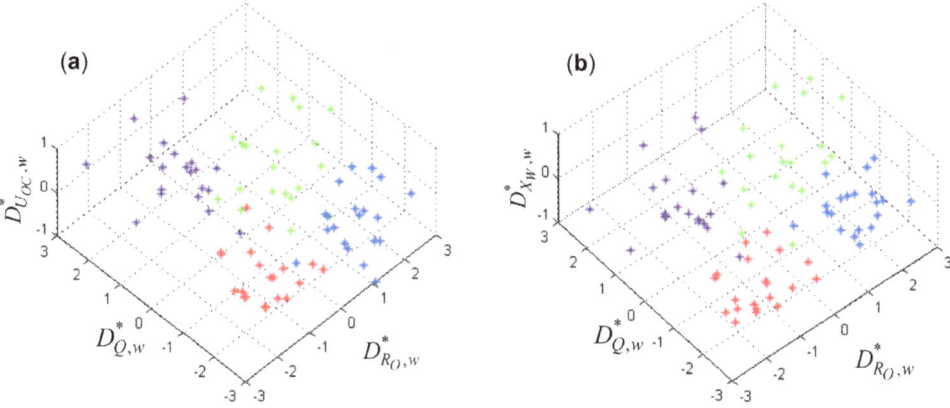

Figure 12. Battery feature vector clustering result (**a**) Clustering result from the view of $D^*_{Q,w}$, $D^*_{R_O,w}$ and $D^*_{U_{OC},w}$; (**b**) Clustering result from the view of $D^*_{Q,w}$, $D^*_{R_O,w}$ and $D^*_{U_{OC},w}$.

5.2. Verification of the Battery Grouping Method

5.2.1. Model Accuracy Verification

In this section, the first order RC model is compared with the simplified EIS model. In order to obtain the model parameters, a genetic algorithm (GA) is used to find the optimal value of the parameters except for U_{OC}. Equation (31) is used as the objective function of GA.

$$f(X) = \frac{1}{N} \sum_{k=1}^{N} \left(U_{T,k} - \hat{U_{T,k}} \right)^2 \tag{31}$$

The measured terminal voltages of DST in $DoD = 50\%$ are used as the true values. The optimization results of model parameters are used for terminal voltage estimation and predication. Results are shown in Figure 13.

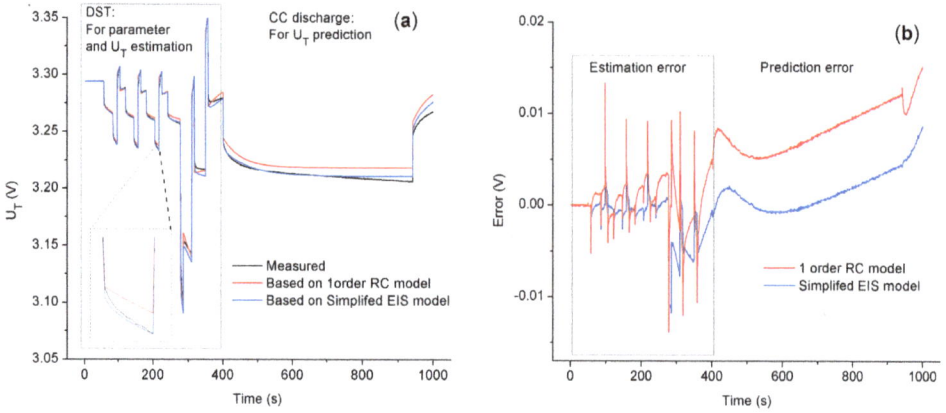

Figure 13. Comparison of battery model simulation results (**a**) simulation results and measured values of terminal voltage; (**b**) simulation errors.

In the test results, the root mean square (RMS) voltage error of simulation results of the first-order RC model and the simplified EIS model are 6.9 mV and 2.8 mV. The estimation errors in DST region are 2.7 mV and 2.8 mV, and prediction errors are 8.7 mV and 2.7 mV. The simulation accuracy of the simplified EIS model is higher than the accuracy of the first-order RC model. Compared to the first-order RC model, the simplified EIS model can represent the impedance characteristics of the battery much better. Furthermore, the number of parameters is less than that of the first-order RC model.

5.2.2. Model Parameter Identification Accuracy Verification

The RMS error of terminal voltage is 1.03 mV in the result which is shown in Figure 9. In this section, FJKF and GA are applied to the same experimental data for model parameter identification. The identification results of U_{OC}, R_O and X_W are shown in Figure 14. The results of the two methods are very similar. The accuracy and effectiveness of FJKF can be verified from the results.

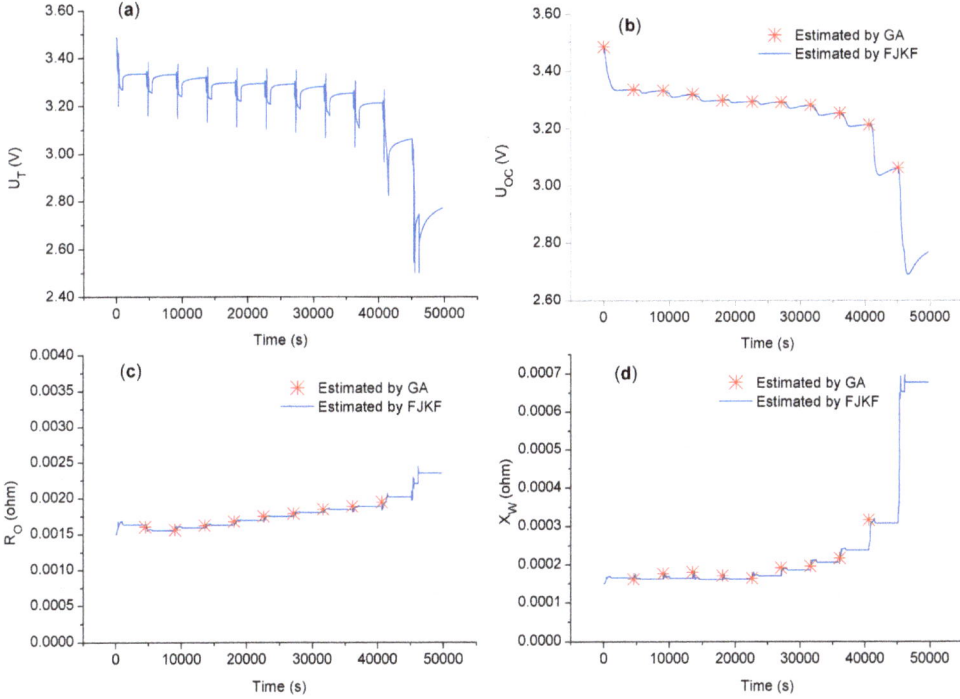

Figure 14. Comparison of the model parameter identification methods: (**a**) terminal voltage; (**b**) U_{OC}; (**c**) R_O; (**d**) X_W.

5.2.3. Battery Grouping Accuracy Verification

In order to verify the validity of the proposed method for battery grouping, three other battery grouping methods—capacity matching, R_O matching and the voltage curve matching method—are used to compare with this method. The first two methods are easy to implement. The voltage curve matching method, proposed in [34], is based on the similarity of average distance, Euclidean distance and correlation coefficient, but the clustering method for three parameters was not described. In this paper, the voltage curve matching method is based on the average distance between the voltage curves and the mean curve. Additionally, 1C-rate discharge curves are used for battery grouping in this method. The clustering effective evaluation criterion is dominated based on the average Euclidean distance of the battery voltage curves of each cluster. The evaluation equation is shown as:

$$E\left(group_i\right) = \frac{1}{N} \sum_{n=1}^{N} \sqrt{\frac{1}{K} \sum_{k=1}^{K} \left(U_{T,i}(k) - U_{T,gn}^{avg}(k)\right)^2} \tag{32}$$

The battery grouping evaluation results of these methods are shown in Table 1. As is seen from the results, the voltage dispersion in each group after adopting the proposed clustering method is much smaller than the other ones under most working conditions, so the new grouping method has relatively higher accuracy when compared with the other methods.

Table 1. Comparison of battery grouping results.

Method	Capacity Matching ($\times 10^{-3}$ V)			R_O Matching ($\times 10^{-3}$ V)			Curve Matching ($\times 10^{-3}$ V)			Proposed Method ($\times 10^{-3}$ V)		
Dispersion	Max	Min	Avg.	Max	Min	Avg.	Max	Min	Avg.	Max	Min	Avg.
DST	21.56	9.483	15.40	38.43	19.18	25.94	16.50	6.910	13.41	15.80	5.682	10.78
0.5 C charge	26.02	9.406	17.00	43.61	27.31	32.71	22.78	7.803	18.22	19.60	6.425	13.98
0.5 C discharge	21.53	9.391	15.31	41.90	26.13	31.23	19.40	7.776	15.82	19.17	7.548	14.21
1 C charge	30.35	15.88	24.40	42.65	28.98	33.67	22.74	17.89	20.32	22.93	16.85	19.64
1 C discharge	24.40	16.13	20.46	42.92	27.97	32.32	23.32	9.052	17.03	21.80	9.939	16.71

6. Conclusions

A grouping method for LiFePO$_4$/graphite batteries is proposed. The method is mainly based on the fractional joint Kalman filter and a new modified K-means algorithm. Batteries can be divided into groups with same number of elements and similar battery characteristics in each group. The innovations and contributions of the proposed method are as follows:

1. A simplified EIS battery model is adopted for battery characteristic description. The accuracy of this model is relatively higher and the number of parameters is less than in the first order RC model under the tested conditions. The model is suitable for battery grouping.
2. A fractional joint Kalman filter is used for parameter identification. Fractional parameters X_W can be estimated jointly with the state U_W of the fractional element Z_W. This algorithm is appropriate and effective for fractional model parameter identification.

3. A new modified K-means clustering method is proposed for battery equal-number grouping. In this method, a feature vector with four parameters is used to describe the performance of the battery from different angles. Two rules are designed to equalize the numbers of elements in each group and exchange samples among groups. The rules are added to the K-means clustering method and replace the criterion of element attribution calculation.

This battery grouping method is effective for LiFePO$_4$/graphite battery grouping. Furthermore, this method has the potential to be utilized in battery reuse applications. The simplified EIS model and FJKF algorithm can also be used for battery online state estimation or prediction. The equal-number clustering method can be applied to other clustering applications which require not only similarity, but also equal-numbers in each cluster. Future works will include evaluating the algorithm for the abovementioned applications.

Acknowledgments: This research was supported by the Research and Development of Application Technology Plan Project in Heilongjiang Province of China (GA13A202), the NSFC-EPSRC Collaborative Research Initiative in Smart Grids and the Integration of Electric Vehicles (51361130153), and the Science and Technology Project of State Grid Corporation of China. The authors thank the support of China Aviation Lithium Battery (CALB) Co., LTD, Teng Wang and Yanyan Zhao's help on this research. The author would also like to acknowledge the reviewers' corrections and helpful suggestions.

Author Contributions: Xiaoyu Li designed the algorithm, some parts of the experiment and wrote the main parts of the manuscript. Kai Song checked the whole manuscript and gave some suggestions. Guo Wei and Rengui Lu designed some parts of experiments. Chunbo Zhu checked the results and the whole manuscript.

Conflicts of Interest: The authors declare no conflict of interest.

Nomenclature

Symbols

U_T	terminal voltage of a battery, V
I_L	load current, A
OCV	open circuit voltage, V
U_{OC}	open circuit voltage, V
E_{OC}	open circuit potential of electrode, V
R_O	ohmic resistance, Ω
R_{bulk}	bulk resistance, Ω
R_{ext}	connect resistance, Ω
R_{sei}	SEI resistance, Ω
R_{ct}	charge transfer resistance, Ω
C_{dl}	double layer capacitance, F
Z_W	Warburg impedance, $\Omega s^{-0.5}$

X_W	Warburg resistance, Ω
U_W	terminal voltage of Z_W, V
Q	nominal capacity, Ah
Q_0	typical capacity of the battery, Ah
w	weight of a parameter, 1
σ	standard deviation, V or Ω
D	distance, V or Ω
E	evaluation parameter, V or Ω
C	cluster
N	total number of clusters
n	cluster n
M	total number of samples
p	sample p
μ	cluster center

Subscripts, superscript

eff	effective value
avg	average value
k	time step index
m	master
s	slave
(p)	sample p
$\hat{}$	estimation value
$\tilde{}$	prior estimation value
$\hat{+}$	posteriori estimation value
$\rightarrow$	vector
$*$	normalized value

Abbreviations

EV	electric vehicle
EIS	electrochemical impedance spectroscopy
FJKF	fractional joint Kalman filter
DST	dynamic stress test
SoC	state of charge
DoD	depth of discharge
SEI	solid electrolyte interface
RMS	root mean square

References

1. Wang, T.; Tseng, K.J.; Zhao, J.; Wei, Z. Thermal investigation of lithium-ion battery module with different cell arrangement structures and forced air-cooling strategies. *Appl. Energy* **2014**, *134*, 229–238.

2. Xiong, B.; Zhao, J.; Wei, Z.; Skyllas-Kazacos, M. Extended Kalman filter method for state of charge estimation of vanadium redox flow battery using thermal-dependent electrical model. *J. Power Sources* **2014**, *262*, 50–61.

3. Wang, T.; Zhu, C.; Pei, L.; Lu, R.; Xu, B. The State of Arts and Development Trend of SOH Estimation for Lithium-Ion Batteries. In Proceedings of the 2013 IEEE Vehicle Power and Propulsion Conference (VPPC), Beijing, China, 15–18 October 2013; pp. 1–6.

4. Andre, D.; Appel, C.; Soczka-Guth, T.; Sauer, D.U. Advanced mathematical methods of SOC and SOH estimation for lithium-ion batteries. *J. Power Sources* **2013**, *224*, 20–27.

5. Pei, L.; Zhu, C.; Wang, T.; Lu, R.; Chan, C.C. Online peak power prediction based on a parameter and state estimator for lithium-ion batteries in electric vehicles. *Energy* **2014**, *66*, 766–778.

6. Xiong, R.; He, H.; Sun, F.; Liu, X.; Liu, Z. Model-based state of charge and peak power capability joint estimation of lithium-ion battery in plug-in hybrid electric vehicles. *J. Power Sources* **2013**, *229*, 159–169.

7. Kim, J.; Shin, J.; Chun, C.; Cho, B.H. Stable Configuration of a Li-Ion Series Battery Pack Based on a Screening Process for Improved Voltage/SOC Balancing. *IEEE Trans. Power Electron.* **2012**, *27*, 411–424.

8. Kim, J.; Cho, B.H. Screening process-based modeling of the multi-cell battery string in series and parallel connections for high accuracy state-of-charge estimation. *Energy* **2013**, *57*, 581–599.

9. Schneider, E.L.; Oliveira, C.T.; Brito, R.M.; Malfatti, C.F. Classification of discarded NiMH and Li-Ion batteries and reuse of the cells still in operational conditions in prototypes. *J. Power Sources* **2014**, *262*, 1–9.

10. Fang, K.; Chen, S.; Mu, D.; Wu, B.; Wu, F. Investigation of nickel–metal hydride battery sorting based on charging thermal behavior. *J. Power Sources* **2013**, *224*, 120–124.

11. Li, X.; Wang, T.; Pei, L.; Zhu, C.; Xu, B. A comparative study of sorting methods for lithium-ion batteries. In Proceedings of the 2014 IEEE Conference and Expo Transportation Electrification Asia-Pacific (ITEC Asia-Pacific), Beijing, China, 31 August–3 September 2014; pp. 1–6.

12. Diamond, R.A.; Wang, H.; Chen, F.; Wilke-Douglas, M. Cell Preparation and Enrichment for FCM Analysis and Cell Sorting. In *In Living Color*; Springer Lab Manuals; Springer: Berlin/Heidelberg, Germany, 2000; pp. 111–141.

13. Guo, L.; Liu, G.W. Research of Lithium-Ion Battery Sorting Method Based on Fuzzy C-Means Algorithm. *Adv. Mater. Res.* **2012**, *354–355*, 983–988.

14. Yun, W.U.; Ge-Chen, L.I.; Chen, S.L. The Application of BP Nerual Network in Battery-sorting. *J. Harbin Univ. Sci. Technol.* **2001**, *5*, 54–57.

15. Feng, F.; Lu, R.; Zhu, C. A Combined State of Charge Estimation Method for Lithium-Ion Batteries Used in a Wide Ambient Temperature Range. *Energies* **2014**, *7*, 3004–3032.

16. Plett, G.L. Extended Kalman filtering for battery management systems of LiPB-based HEV battery packs Part 1. Background. *J. Power Sources* **2004**, *134*, 262–276.

17. Plett, G.L. Extended Kalman filtering for battery management systems of LiPB-based HEV battery packs Part 2. Modeling and identification. *J. Power Sources* **2004**, *134*, 252–261.

18. Plett, G.L. Extended Kalman filtering for battery management systems of LiPB-based HEV battery packs Part 3. State and parameter estimation. *J. Power Sources* **2004**, *134*, 277–292.

19. Xiong, R.; Sun, F.; He, H.; Nguyen, T.D. A data-driven adaptive state of charge and power capability joint estimator of lithium-ion polymer battery used in electric vehicles. *Energy* **2013**, *63*, 295–308.

20. Yoon, S.; Hwang, I.; Lee, C.W.; Ko, H.S.; Han, K.H. Power capability analysis in lithium ion batteries using electrochemical impedance spectroscopy. *J. Electroanal. Chem.* **2011**, *655*, 32–38.

21. Xu, J.; Mi, C.C.; Cao, B.; Cao, J. A new method to estimate the state of charge of lithium-ion batteries based on the battery impedance model. *J. Power Sources* **2013**, *233*, 277–284.

22. Waag, W.; Käbitz, S.; Sauer, D.U. Experimental investigation of the lithium-ion battery impedance characteristic at various conditions and aging states and its influence on the application. *Appl. Energy* **2013**, *102*, 885–897.

23. Luo, W.; Lyu, C.; Wang, L.; Zhang, L. A new extension of physics-based single particle model for higher charge–discharge rates. *J. Power Sources* **2013**, *241*, 295–310.

24. Fuller, T.F.; Doyle, M.; Newman, J. Relaxation phenomena in lithium-ion-insertion cells. *J. Electrochem. Soc.* **1994**, *141*, 982–990.

25. Zhang, D. Modeling Lithium Intercalation of a Single Spinel Particle under Potentiodynamic Control. *J. Electrochem. Soc.* **1999**, *147*, 831–838.

26. Zhang, L.; Lyu, C.; Hinds, G.; Wang, L.; Luo, W.; Zheng, J.; Ma, K. Parameter Sensitivity Analysis of Cylindrical LiFePO$_4$ Battery Performance Using Multi-Physics Modeling. *J. Electrochem. Soc.* **2014**, *161*, A762–A776.

27. Fleischer, C.; Waag, W.; Heyn, H.; Sauer, D.U. On-line adaptive battery impedance parameter and state estimation considering physical principles in reduced order equivalent circuit battery models part 2. Parameter and state estimation. *J. Power Sources* **2014**, *262*, 457–482.

28. Fleischer, C.; Waag, W.; Heyn, H.; Sauer, D.U. On-line adaptive battery impedance parameter and state estimation considering physical principles in reduced order equivalent circuit battery models : Part 1. Requirements, critical review of methods and modeling. *J. Power Sources* **2014**, *260*, 276–291.

29. Li, X.; Zhu, C.; Wei, G.; Lu, R. Online Parameter Estimation of LiFePO$_4$ Battery Simplified Impedance Spectroscopy Model Based on Fractional Joint Kalman Filter. *Trans. Chin. Electrotechnical Soc.* **2015**, in press.

30. Sierociuk, D.; DzielińSki, A. Fractional Kalman filter algorithm for the states, parameters and order of fractional system estimation. *Int. J. Appl. Math. Comput. Sci.* **2006**, *16*, 129–140.

31. *FreedomCAR Battery Test Manual for Power-Assist Hybrid Electric Vehicles*; U.S. Department of Energy: Washington, DC, USA, 2003.
32. Mcqueen, J. Some methods for classification and analysis of ultivariate observations. In Proceedings of the Berkeley Symposium Mathematical Statistics Probability; University of California Press: Berkeley, CA, USA, 1967; pp. 281–297.
33. Redmond, S.J.; Heneghan, C. A Method for Initialising the K-Means Clustering Algorithm Using Kd-Trees. *Pattern Recognit. Lett.* **2007**, *28*, 965–973.
34. Hanze, D.Z.L.G. Cell classification system based on automatic curve-recognition. *Chin. J. Power Sources* **2000**, *2*, 65–69.

Stability Analysis for Li-Ion Battery Model Parameters and State of Charge Estimation by Measurement Uncertainty Consideration

Shifei Yuan, Hongjie Wu, Xuerui Ma and Chengliang Yin

Abstract: Accurate estimation of model parameters and state of charge (SoC) is crucial for the lithium-ion battery management system (BMS). In this paper, the stability of the model parameters and SoC estimation under measurement uncertainty is evaluated by three different factors: (i) sampling periods of 1/0.5/0.1 s; (ii) current sensor precisions of $\pm 5/\pm 50/\pm 500$ mA; and (iii) voltage sensor precisions of $\pm 1/\pm 2.5/\pm 5$ mV. Firstly, the numerical model stability analysis and parametric sensitivity analysis for battery model parameters are conducted under sampling frequency of 1–50 Hz. The perturbation analysis is theoretically performed of current/voltage measurement uncertainty on model parameter variation. Secondly, the impact of three different factors on the model parameters and SoC estimation was evaluated with the federal urban driving sequence (FUDS) profile. The bias correction recursive least square (CRLS) and adaptive extended Kalman filter (AEKF) algorithm were adopted to estimate the model parameters and SoC jointly. Finally, the simulation results were compared and some insightful findings were concluded. For the given battery model and parameter estimation algorithm, the sampling period, and current/voltage sampling accuracy presented a non-negligible effect on the estimation results of model parameters. This research revealed the influence of the measurement uncertainty on the model parameter estimation, which will provide the guidelines to select a reasonable sampling period and the current/voltage sensor sampling precisions in engineering applications.

Reprinted from *Energies*. Cite as: Yuan, S.; Wu, H.; Ma, X.; Yin, C. Stability Analysis for Li-Ion Battery Model Parameters and State of Charge Estimation by Measurement Uncertainty Consideration. *Energies* **2015**, *8*, 7729–7751.

1. Introduction

The lithium-ion battery has been widely utilized as a promising power source of hybrid-electric vehicles (HEVs) and pure electric vehicles (EVs) for its high energy and power density, no memory effect, and slow rate of self-discharge. Reduced safety hazards and an efficient Li-ion battery system can be achieved by developing an advanced battery management system (BMS). The model parameters and state of charge (SoC) are two critical indicators for an efficient BMS to operate the battery system safely and extend the cell life longevity.

460

Many studies have been performed in literature about the battery model parameters and SoC estimation. According to their highlights, these works can be classified around four categories: (I) operation environments; (II) varied aging states; (III) modeling and algorithm error; and (IV) measurement uncertainty.

In Class I, He *et al.* [1] realized the influence different charging and discharging rates had on the cell capacity. They formulated the coulombic efficiency that related with the current rates (1/3C–3C) and flowing direction to achieve a more accurate SoC estimation. To improve the accuracy of the battery model further, the authors He *et al.* [2] and Xing *et al.* [3] considered the temperature effect with the range of −20–60 °C and 0–40 °C. In their research, the coulombic efficiency was expressed as the polynomial equation of temperature. Wang *et al.* [4] considered the current rates and temperature effect together to correct the coulombic efficiency by creating a table to achieve more precise estimation of SoC and energy. In contrast, Liu *et al.* [5] addressed this issue by using the back-propagation neural network (BPNN) model to manage the current rate and temperature effect, and their simulation results also presented great performance. In Class II, the authors [6,7] analyzed the aging effect in the SoC estimation for lithium-ion polymer battery (LiPB) by two methods: first, the cell nominal capacity was re-modified according to the actual health environments; second, the cell's open circuit voltage (OCV) was re-expressed as a second order polynomial equation, which is related with the SoC and modified cell capacity. Based on the recursive least square (RLS) and adaptive Kalman filter algorithm, the SoC estimation could be maintained with high accuracy. Dai *et al.* [8] applied the SoC estimation on a Li-ion battery pack of multiple inconsistent cells with the averaged cell model. Then they attempted to estimate the SoC for each individual cell, and the simulation results indicated good performance for the model and algorithm. Zhong *et al.* [9] analyzed the difference between the cells and the impact of balance control to minimize the cell capacity variation. The battery pack in serial and parallel connections and the passive balance control effect were considered to evaluate the impact of cell uniformity and inconsistency.

In Class III of the modeling and algorithm error, the battery model should first be established. In general, the battery model consists of model parameters and OCV in two parts. The two major factors of parameters uncertainty in varying aging and temperature effects have been reviewed previously. This review will address the OCV variation and SoC estimation algorithm uncertainty under different environments. Liu *et al.* [10] conducted the OCV test from 0 °C to 45 °C, and compared the OCV variation with temperature effect by creating a table. Hu *et al.* [11] provided the OCV value for lithium nickel-manganese-cobalt oxide (LiNMC) and lithium iron phosphate (LiFePO$_4$) cell by a series of polynomial functions with variables of temperature and SoC. Xiong *et al.* [12] presented the OCV functions for four

461

different kinds of chemistry cells, such as carbon/lithium manganese oxide (C/LMO), lithium titanate oxide/nickel-manganese-cobalt-oxide (LTO/NMC), carbon/NMC (C/NMC) and carbon/lithium iron phosphate (C/LFP). Their simulation showed that the adaptive extended Kalman filter (AEKF) based SoC estimation method is suitable for multiple kinds of cells and obtained good estimation results with a maximum error being less than 3%. About the adaptive algorithm for SoC estimation, the extended Kalman filter (EKF) algorithm is firstly applied to the lithium ion polymer battery by Plett [13–15] in the work. Afterward, more extensive studies were conducted to assess the effect of the initial SoC creation, process covariance R_k, and measurement covariance Q_k [16,17]. To overcome the uncertainty of measurement noise covariance and initial SoC creation, the unscented Kalman filter [2,18,19], the particle filter [9,20,21], the AEKF [1,7,12], and the adaptive observer [22,23] techniques are proposed for SoC estimation. In Class IV, the measurement uncertainty remains an emerging research field according to our knowledge. Liu et al. [10] analyzed the current measurement with drift noise effect on SoC estimation by a dual-particle-filter estimator. In their study, the drift current was considered as an undetermined static parameter in the battery model to eliminate the drift current effect. Xia et al. [24] conducted the SoC estimation under the measurement error of both 2.5% voltage noise and 5% current noise. The results showed the SoC estimation error would not exceed 4.5%. However, as stated in [25], the current sensors have a certain grade of measurement accuracy and resolution. The amplifier circuit also exhibited some accuracy issues with the adaptation of resistors, capacitors and power lines. The micro-control-unit (MCU) involved the rounding errors in the process of analogue to digital (A/D) conversion and calculation. The current signal with harmonics also caused measurement errors during the sampling process with the MCU. All these noises and uncertainties in the measurement loop will generate errors for model parameters and SoC estimation. The detailed information is summarized and listed in the following Table 1.

Table 1. Factors influencing on the model parameters and state of charge (SoC) estimation.

Attribution	Item in detail	Reference
I. Operation environments	I.1. Current direction and rates	[1–5]
	I.2. Temperature effect	
II. Varied aging states	II.1. Battery capacity loss during aging cannot be counted in the SoC calculation	[6,7]
	II.2. Cell inconsistency	[8,9]
	II.3. Self-discharging cannot be counted for by the SoC algorithm	unknown
III. Modeling and algorithm error	III.1. Parameters uncertainty due to varying aging	[6,7]
	III.2. Parameters uncertainty due to varying temperature	[2–5]
	III.3. OCV uncertainty due to varying aging	[6]
	III.4. OCV uncertainty due to varying temperature	[10]
	III.5. OCV uncertainty due to battery types	[11]
	III.6. Different sampling rate T_s	unknown
	III.7. Battery initial SoC remains unknown	[16]
	III.8. Different noise covariance R_k, Q_k	
	III.9. Different model accuracy and adaptive algorithms	[1]
IV. Measurement uncertainty	IV.1. Current measurement with drift noise	[10]
	IV.2. Different sampling rate T_s	unknown
	IV.3. Current sensor with different resolution	unknown
	IV.4. Current sensor with different precision	[24]
	IV.5. Voltage sensor with different resolution	unknown
	IV.6. Voltage sensor with different precision	[24]
	IV.7. Loading/excitation profile dependence	unknown

1.2. Motivations and Contributions

The measuring system will inevitably bring in uncertainty and errors between measured signal and true signal from sampling aliases, conversion loss, and rounding calculation solutions. In this paper, the key innovation and contribution was to evaluate the effects of the measurement uncertainty on battery model parameter and SoC estimation both in theory and experiments. Different from other research, the measurement uncertainty is assumed from three different factors: (a) sampling periods of 1/0.5/0.1 s; (b) current sensor precisions of $\pm 5/\pm 50/\pm 500$ mA; and (c) voltage sensor precisions of $\pm 1/\pm 2.5/\pm 5$ mV. Afterward, the perturbation analysis of current and voltage measurement uncertainties, model stability analysis and parametric sensitivity analysis of model parameters were conducted respectively.

The analytical result and conclusion provided guidelines that an engineer could use to choose the optimal sampling periods and current/voltage sensor precisions for improved estimation accuracy of model parameters and SoC.

1.3. Organization of the Paper

This paper will be organized as follows. In Section 2, the discrete battery model is given and the perturbation analysis of current/voltage measurement uncertainty, model stability analysis and parametric sensitivity analysis are conducted. In Section 3, the AEKF algorithm was adopted for the SoC estimation based on the auto regressive exogenous (ARX) battery model. In Section 4, the experiment setup is established, and the measured OCV-SoC curve is presented. In Section 5, the simulations were implemented upon three different factors. Then, the comparison is performed to analyze the weighted importance of factors on parameter and SoC estimation. Finally, the paper ends with a concluding remark.

2. Battery Model Analysis and Parameter Estimation

2.1. Auto Regressive Exogenous (ARX) Battery Model

As shown in Figure 1, the equivalent circuit model is employed to simulate the cell dynamic performance. It consists of an ohmic resistor R_o, open circuit voltage V_{oc} (SoC) and one R_pC_p network connected in series. The resistor R_o represents the cell internal resistance. The R_pC_p network describes the electrochemical polarization dynamics. The electrical behavior of the lithium-ion battery can be expressed by the following equation:

$$\begin{cases} \dot{U}_p = -\frac{1}{C_pR_p}U_p + \frac{1}{C_p}I_L \\ U_t = V_{oc}+U_p+R_oI_L \end{cases} \tag{1}$$

where U_p indicates the polarization voltage across the RC network. I_L stands for the applied current and U_t stands for the terminal voltage. SoC is defined by current integration as:

$$SoC(k+1) = SoC(k) + \frac{\eta_i T_s}{Q_n} \cdot I_L(k) \tag{2}$$

where η_i is the coulombic efficiency, which is related with the current flow direction and magnitude, temperature, and the degradation/aging status. Herein, as the coulombic efficiency is not the priority in this research, the value of η_i is assumed to be 1. T_s is the sampling interval, and Q_n is the nominal capacity.

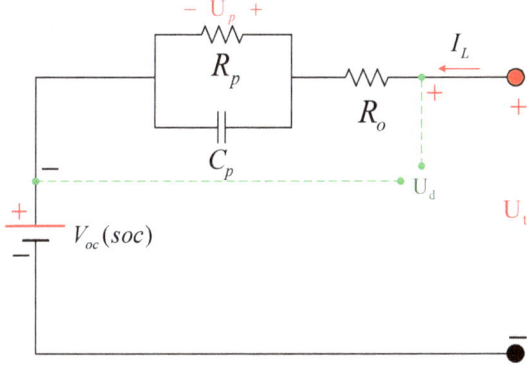

Figure 1. The schematic diagram of the equivalent circuit model for lithium-ion battery.

A bilinear transformation method $s = \frac{2}{T_s}\frac{1-z^{-1}}{1+z^{-1}}$ is employed to discrete the battery model into z domain for a given sampling interval, and the discrete transfer function is given as Equation (3):

$$G(z^{-1}) = \frac{b_0 + b_1 z^{-1}}{1 + a_1 z^{-1}} \tag{3}$$

With the model parameters $[a_1, b_0, b_1]$ equated as:

$$a_1 = -\frac{T_s - 2R_p C_p}{T_s + 2R_p C_p}; b_0 = \frac{R_o T_s + R_p T_s + 2R_o R_p C_p}{T_s + 2R_p C_p}; b_1 = \frac{R_o T_s + R_p T_s - 2R_o R_p C_p}{T_s + 2R_p C_p} \tag{4}$$

It can be found the model parameters $[a_1, b_0, b_1]$ are not only subjected to R_o, R_p, C_p, but also related to the sampling period T_s. R_o, R_p, and C_p can be solved based on the inverse equations of a_1, b_0 and b_1:

$$R_o = \frac{b_0 - b_1}{1 + a_1}; R_p = \frac{2(a_1 b_0 + b_1)}{1 - a_1^2}; C_p = \frac{T(1 + a_1)^2}{4(a_1 b_0 + b_1)} \tag{5}$$

After discretization, Equation (3) can be rewritten in the form of ARX as follows:

$$U_d(k) = a_1 U_d(k-1) + b_0 I_L(k) + b_1 I_L(k-1) \tag{6}$$

where k is the time point, $k = 1,2,3, \ldots . N$. Since $U_d = U_t - V_{oc}$:

$$
\begin{aligned}
U_t(k) &= [V_{oc}(k) - a_1 V_{oc}(k-1)] + a_1 U_t(k-1) + b_0 I_L(k) + b_1 I_L(k-1) \\
&\approx (1 - a_1) V_{oc}(k-1) + a_1 U_t(k-1) + b_0 I_L(k) + b_1 I_L(k-1)
\end{aligned} \tag{7}
$$

The information vector is defined as the following $\Phi(k) =$ $\begin{bmatrix} U_L(k-1) & I_L(k) & I_L(k-1) & 1 \end{bmatrix}$ and the parameter vector is defined as $\Theta(k) = \begin{bmatrix} a_1 & b_0 & b_1 & (1-a_1)V_{oc} \end{bmatrix}^T$, then they can be combined as the following:

$$Y(k) = \Phi(k) \cdot \Theta(k) \tag{8}$$

The RLS algorithm [26] is an effective method for online parameter identification. As an improved algorithm, the bias correction recursive least square (CRLS) technique [27] is used in this paper. The basic idea of CRLS is to eliminate the estimation bias by adding a correction term in the RLS estimation algorithm. Therefore, the performance and numerical convergence of the algorithm can be prompted to a higher level. The detailed implementation of CRLS algorithm could refer to our previous work [28].

2.2. Perturbation Analysis of Measurement Uncertainty

In the literature, there are many adaptive methods for parameter identification, such as RLS, EKF, and adaptive observers. All these methods can give the recursive-identified value for model parameters. However, they cannot tell the parameter variation under current and voltage measurement uncertainty. In this section, the parameter variation is theoretically analyzed under the perturbation of current/voltage measurement uncertainty. Based on above Equations (4) and (7), the formula of parameters (R_o, R_p, C_p) can be deduced as follows:

$$R_o = \frac{(T_s+2R_pC_p)U_t(k)+(T_s-2R_pC_p)U_t(k-1)-2T_s \cdot V_{oc}(k-1)-(R_p \cdot T_s)(I_L(k)+I_L(k-1))}{(T_s+2R_pC_p)I_L(k)+(T_s-2R_pC_p)I_L(k-1)} \tag{9}$$

$$R_p = \frac{T_s \cdot (U_t(k)+U_t(k-1)-2V_{oc}(k-1))-R_oT_s \cdot (I_L(k)+I_L(k-1))}{-2C_p \cdot (U_t(k)-U_t(k-1))+(T_s+2R_pC_p)I_L(k)+(T_s-2R_pC_p)I_L(k-1)} \tag{10}$$

$$C_p = \frac{-T_s \cdot (U_t(k)+U_t(k-1)-2V_{oc}(k-1))+(R_o+R_p)T_s \cdot (I_L(k)+I_L(k-1))}{2R_p \cdot (U_t(k)-U_t(k-1))-2R_oR_p \cdot (I_L(k)-I_L(k-1))} \tag{11}$$

These expressions can be used to analyze the impact of measurement errors $(\delta I_L, \delta U_t)$ on the model parameters (R_o, R_p, C_p). Let's take the ohmic resistance R_o, for an example, to illustrate how the current sensor error δI_L affects the estimated results of R_o. By incorporating the current error δI_L, the new modified current $(I_L + \delta I_L)$ will replace the I_L in Equation (9) to achieve the new estimated R_o'. Then, the relative error of R_o can be defined as:

$$\varepsilon = \left| \frac{R_o' - R_o}{R_o} \right| \times 100\% \tag{12}$$

By this method, the effect of current sensor error δI_L on the parameter variation can be analyzed. Since the battery system is time-varying, the pulse current excitation (1.0C, 30 s) is chosen as a typical profile. In this research, the current sensor error δI_L is assumed as $\pm 5/\pm 50/\pm 500$ mA. The calculation result is listed in Figure 2a–c. Similarly, the voltage sensor error δU_t ($\pm 1/\pm 2.5/\pm 5$ mV) effect on the parameter variation can be assessed in Figure 2d–f).

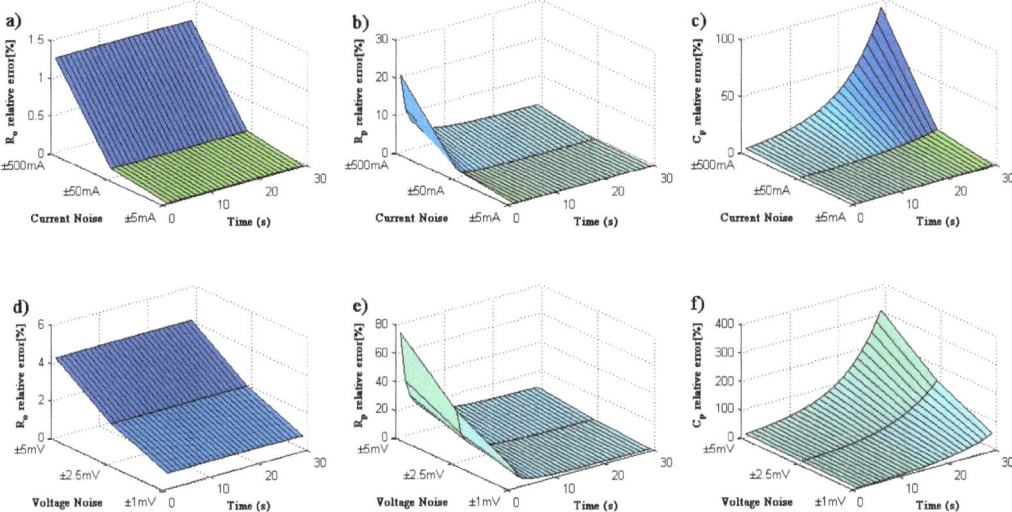

Figure 2. Relative errors of model parameters: (**a**) R_o, (**b**) R_p, (**c**) C_p under current noises; and (**d**) R_o, (**e**) R_p, (**f**) C_p under voltage noises.

Seen from Figure 2, the ohmic resistance R_o is more robust than the R_p and C_p. As the current noise and voltage noise enhanced, the relative error of parameters enlarged. The R_p and C_p are used to express the battery dynamic performance, and at the beginning of pulse current excitation, the load current flows mostly upon the capacitor. Therefore, the relative error of capacitor C_p is smaller at the beginning. In contrast, the relative error of resistance R_p is much larger at the beginning. The minimum values of relative error of model parameters (R_o, R_p, C_p) under current/voltage noises are summarized in Table 2.

Table 2. Minimum values of relative error of model parameters (R_o, R_p, C_p) under current/voltage noises.

Item	Magnitude of noise	Parameter relative error (%) @ 0.2C			Parameter relative error (%) @ 1.0C		
		R_o	R_p	C_p	R_o	R_p	C_p
Current noise	$\delta I_L = \pm 5$ mA	0.0625	0.1513	0.1411	0.0125	0.0303	0.0282
	$\delta I_L = \pm 50$ mA	0.6224	1.5058	1.4111	0.1249	0.3024	0.2822
	$\delta I_L = \pm 500$ mA	6.000	14.407	14.111	1.2397	2.9966	2.8223
Voltage noise	$\delta U_t = \pm 1$ mV	4.1667	8.5613	9.2046	0.8333	1.7123	1.8815
	$\delta U_t = \pm 2.5$ mV	10.416	21.403	23.518	2.0833	4.2806	4.7038
	$\delta U_t = \pm 5$ mV	20.833	42.806	47.037	4.1667	8.5613	9.4076

2.3. Model Stability Analysis

Model stability is critical to indicate the model stable and robust level under external perturbation. As the model stability is increased, the model output will be much stable; meanwhile the identification of model parameters will be much easier. The two key factors (poles and zeros) will be used to reveal the model stability level in a quantifiable form. A detailed process for poles and zeros calculation is in the author's previous work [28].

To assess the model stability, the battery parameters are assumed as $R_o = 0.002\ \Omega$; $R_p = 0.001\ \Omega$; $C_p = 8000$ F; $T_s = 1$ s, 0.5 s, 0.2 s, 0.1 s, 0.02 s, and the parameter sets $[a_1, b_0, b_1]$ of the ARX model is calculated using Equation (4). Meanwhile, the model poles and zeros can be computed to present the model stability level and the result is listed in Table 3.

From Table 3, the poles and zeros of the system increased as T_s decreased, which reveals the model stability is degraded. According to the Lyapunov's first stability criterion, the model stability will become much poorer as the eigenvalues of the ARX model get close to one. That is, perturbations caused by noise and unmodeled dynamics could significantly influence the accuracy of model parameter identification. To this point, the sampling rate should be lower (*i.e.*, sampling period should be higher), to improve the model stability and the robustness of parameter identification. From an engineering viewpoint, it is recommended to restrict the eigenvalues within a range of 0–0.95. In other words, T_s should be larger than one threshold, such as $T_s \geq 0.5$ s.

In another way, the sampling period T_s should be chosen modestly enough to capture the significant variation or critical events of Li-ion cell dynamics. In the viewpoint of hardware runtime, the sampling period T_s should be sufficient for the SoC calculation on the ECU platform with the discrete battery model, CRLS and AEKF algorithms. Therefore, the optimized time sampling period must be selected in a tradeoff way by considering the model stability, parametric sensitivity, system-sampling precision and the hardware runtime.

Table 3. Parameter variation and stability analysis for the auto regressive exogenous (ARX) model.

Sampling T_s (s)	Parameter variation			Stability analysis	
	a_1	b_0	b_1	Pole	Zero
1	0.88235	0.00205	−0.00170	0.88235	0.82857
0.5	0.93939	0.00203	−0.00184	0.93939	0.91044
0.2	0.97530	0.00201	−0.00193	0.97530	0.96319
0.1	0.98757	0.00200	−0.00196	0.98757	0.98142
0.02	0.99750	0.00200	−0.00199	0.99750	0.99625

2.4. Parametric Sensitivity Analysis

In this research, the model parameters are identified online by CRLS. In other words, the variations of a_0, b_0, or b_1 will affect the model parameter set P of R_o, R_p, and C_p. The sensitivity of the model parameters to the changes in variable α (such as a_0, b_0, or b_1) is given by the partial differentiation of $P(s)$ with respect to α and is denoted as:

$$S_\alpha^P = \frac{\alpha}{P} \cdot \frac{\partial P}{\partial \alpha} \tag{13}$$

where P is the parameter set of R_o, R_p, and C_p as defined in Equation (5), and α is a_0, b_0, or b_1. To be specific [28]:

$$S_\alpha^P = \begin{bmatrix} S_{a_1}^{R_o} & S_{b_0}^{R_o} & S_{b_1}^{R_o} \\ S_{a_1}^{R_p} & S_{b_0}^{R_p} & S_{b_1}^{R_p} \\ S_{a_1}^{C_p} & S_{b_0}^{C_p} & S_{b_1}^{C_p} \end{bmatrix} \tag{14}$$

Based on the sensitivity equation, the parametric sensitivity has been calculated at four sampling periods (T_s = 1 s, 0.5 s, 0.1 s and 0.02 s), and the results are listed in Table 4.

First, the sensitivity of R_o with respect to parameters $[a_0, b_0, b_1]$ was not great and always retained within $[-0.45–0.54]$. As the growth of T_s, the sensitivity of R_o on a_0, b_0, and b_1 became more uniform. This illustrated that R_o was much more robust and more easily identifiable regardless of the sampling period T_s. This characteristic was confirmed in our later simulation. Second, the sensitivity of R_p and C_p increased quasi-linearly as T_s decreased. Specifically, a small disturbance in a_0, b_0, and b_1 may cause large fluctuations in R_p and C_p when T_s is much smaller; therefore, this feature will also increase the difficulty for real-time parameter identification. In practice, it is suggested to limit T_s to be greater than a certain level to maintain good stability for parameter identification.

Table 4. Parametric sensitivity analysis of the ARX model at four sample periods ($T_s = 1/0.5/0.1/0.02$ s).

Item	Sensitivity	a_1	b_0	b_1
Sample @ $T_s = 1$ s	R_o	−0.4688	0.5469	0.4531
	R_p	23.4374	16.4062	−15.4062
	C_p	−15.4688	−16.4063	15.4063
Sample @ $T_s = 0.5$ s	R_o	−0.4844	0.5234	0.4766
	R_p	47.4687	32.4531	−31.4531
	C_p	−31.4844	−32.4531	31.4531
Sample @ $T_s = 0.1$ s	R_o	−0.4969	0.5047	0.4953
	R_p	239.4938	160.4906	−159.4906
	C_p	−159.4969	−160.4906	159.4906
Sample @ $T_s = 0.02$ s	R_o	−0.4994	0.5009	0.4991
	R_p	1199.5	800.4981	−799.4981
	C_p	−799.4994	−800.4981	799.4981

3. Adaptive Extended Kalman Filter Algorithm

In this section, the model states and parameters are estimated jointly based on real measurements of the current, voltage and temperature. In theory, the joint state and parameter estimation algorithm could provide a more accurate estimation result for battery model parameters and SoC. The general working principle of the joint estimation algorithm is displayed as Figure 3.

The AEKF [1,17,29] is an advanced method for system state estimation, especially when the system process and measurement noise are unknown. This algorithm can avoid the estimation error divergence effectively due to its robustness property. Furthermore, it can enhance the performance in the SoC estimation enormously.

Firstly, the general form of state space representation is presented as:

$$
\begin{aligned}
X_{k+1} &= f(X_k, u_k) + w_k \\
Y_k &= g(X_k, u_k) + v_k \\
w_k &\sim N(0, Q_k), \quad v_k \sim N(0, R_k)
\end{aligned}
\tag{15}
$$

where X_k is the model state, u_k is the model input, k is the time index, $f(\cdot)$ and $g(\cdot)$ indicate the process equation and output equation of the battery model, respectively, w_k is a discrete time process white noise with a covariance matrix Q_k, whose initial value can be chosen by the state X_k properties. Similarly, v_k is a discrete time measurement white noise with covariance matrix R_k, whose initial value can be determined according to the voltage sensor precision.

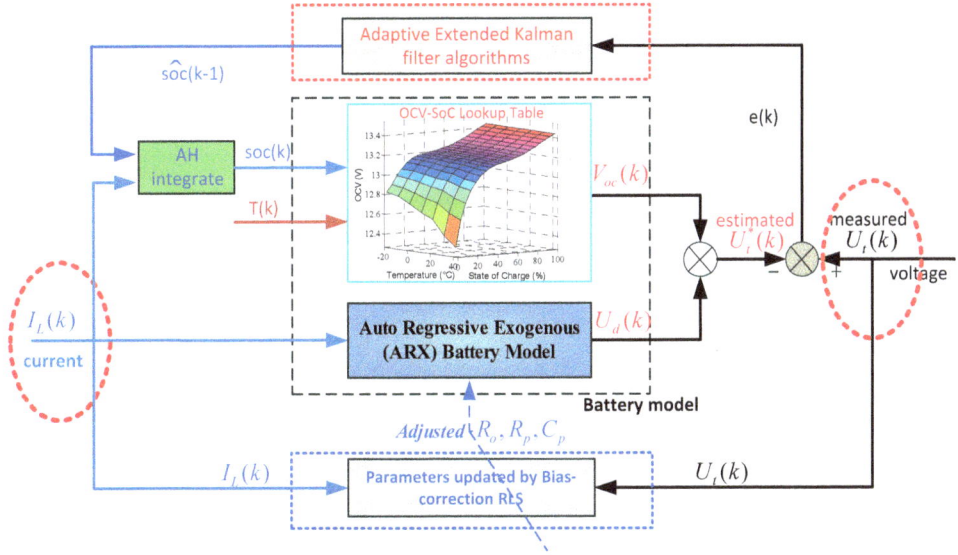

Figure 3. General diagram of the battery model parameters and SoC joint estimation with adaptive extended Kalman filter (AEKF) and correction recursive least square (CRLS) algorithms.

The battery model Equations (1) and (2) can be transformed as state space form:

$$[X_{k+1}] = \begin{bmatrix} SoC(k+1) \\ U_p(k+1) \end{bmatrix} = \underbrace{\begin{bmatrix} 1 & 0 \\ 0 & \exp(\dfrac{-T_s}{R_p C_p}) \end{bmatrix}}_{A_k} \underbrace{\begin{bmatrix} SoC(k) \\ U_p(k) \end{bmatrix}}_{state: X_k} + \underbrace{\begin{bmatrix} \dfrac{\eta_i T_s}{Q_n} \\ R_p \left[1 - \exp(\dfrac{-T_s}{R_p C_p})\right] \end{bmatrix} [I_L(k)]}_{input: u_k} \qquad (16)$$

$$[Y_k] = [U_t(k)] = \underbrace{\begin{bmatrix} \dfrac{dV_{oc}(SoC)}{dSoC}\Big|_{SoC=SoC(k)} & -1 \end{bmatrix}}_{C_k} \underbrace{\begin{bmatrix} SoC(k) \\ U_p(k) \end{bmatrix}}_{state: X_k} + \underbrace{[-R_o][I_L(k)]}_{input: u_k} \qquad (17)$$

After several iterations, the estimated model voltage will converge to the truly measured value; meanwhile, the estimated SoC will converge to the true or optimal value.

4. Experimental Setup

The test bench setup is shown as Figure 4. It consists of a battery cycler, a thermal chamber for temperature control, and a computer for script programming and data storage. The battery testing system is responsible for loading the battery module with maximum charging/discharging current of ± 500 A. The measurement error of the current/voltage transducer inside the cycler is within 0.25%. The key specification of the LiFePO$_4$ cell is listed in Table 5.

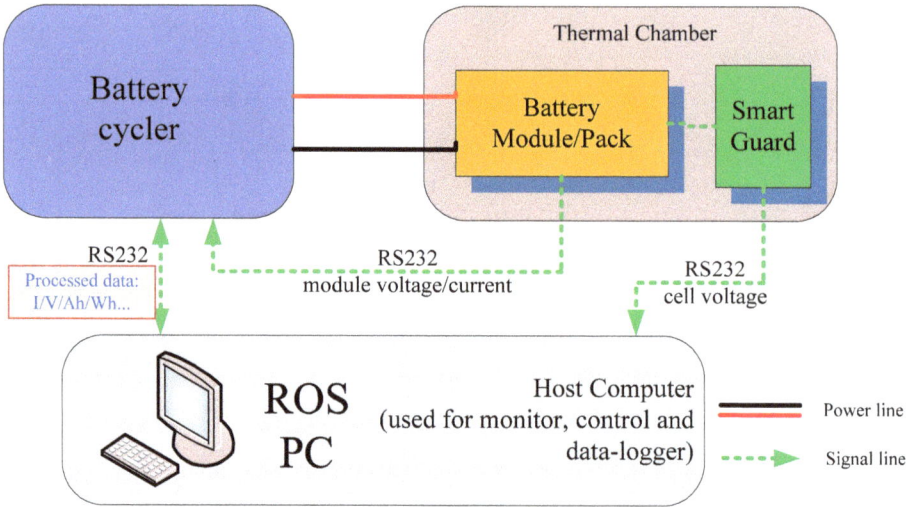

Figure 4. Configuration of the battery test bench.

Table 5. Main specifications of the test cell.

Type	Rated capacity	Rated energy	Maximum current	Rated voltage	Upper voltage	Lower voltage
LiFePO$_4$	60 A h	192 W h	3C, 180 A	3.20 V	3.65 V	2.10 V

The LiFePO$_4$ cell is cycled with the OCV test as stated in [16]. According to the emphasis and priority in this research, the averaged OCV is employed to simplify the hysteresis phenomena of the OCV under charging and discharging process. The computed result of averaged OCV is listed in Figure 5. The open circuit voltage $V_{oc}(SoC)$ can be expressed by a polynomial function as:

$$V_{oc}(SoC) = K_0 + K_1 \cdot SoC + K_2 \cdot SoC^2 + K_3 \cdot SoC^3 + K_4 \cdot SoC^4 \tag{18}$$

where K_i ($i = 0, 1, \ldots, 4$) are the polynomial coefficients to fit the averaged OCV with respect to different SoC based on the least square techniques, and the specific values are 3.1292, 0.00025, 0.00085, 0.0421, 0.0076, respectively. By incremental capacity analysis (ICA), the $dSoC/dOCV$ reaches its upper limit of 0.0145 at the voltage of about 3.280 V, which means the 1 mV estimated error of OCV will result in about 1.45% for SoC estimation bias.

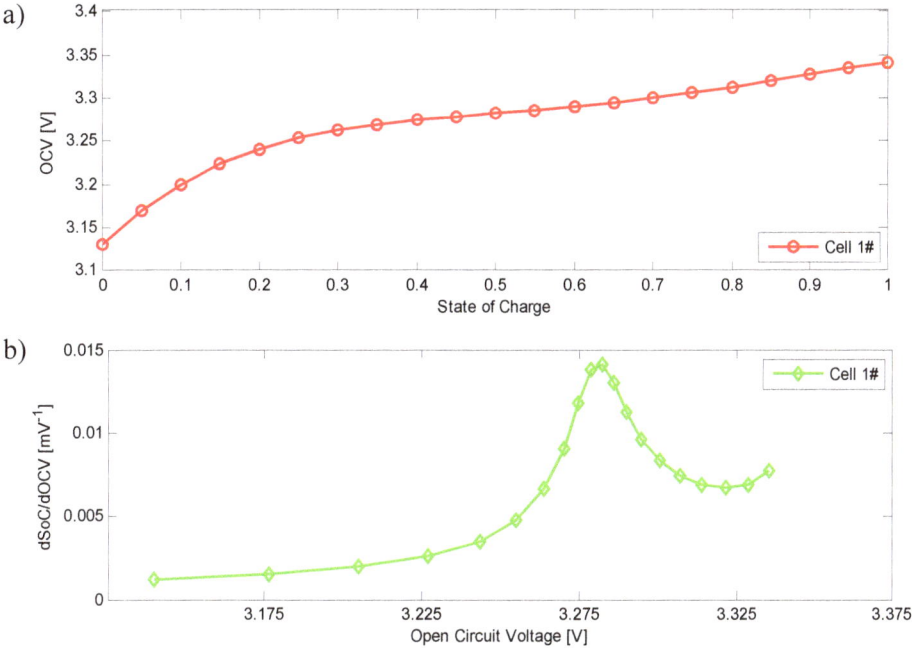

Figure 5. (**a**) The open circuit voltage (OCV) test and (**b**) incremental capacity analysis (ICA) result at temperature 25 °C.

5. Simulation and Discussion

In this section, the effects of measurement uncertainty on battery model parameters and SoC estimation is evaluated in the following three aspects: (i) sampling periods of 1/0.5/0.1 s; (ii) current sensor accuracy of ±5/±50/±500 mA; and (iii) voltage sensor accuracy of ±1/±2.5/±5 mV. In Section 5.4, the simulations of these three different scenarios are compared to evaluate the impact of each factor on the model parameters and SoC estimation. The federal urban driving sequence (FUDS) profile is a typical experiment cycle to assess the model and algorithm performance. In this research, the parameter sets $\Theta = [R_o, R_p, C_p] = [0.002, 0.001, 8000]$ is adopted as a baseline.

5.1. Sampling Period Effect

To evaluate the effect of the sampling periods on the battery model parameters and SoC estimation, three different sampling periods of 1/0.5/0.1 s are selected in the simulation with the CRLS and AEKF algorithm. The estimation results of the model parameters under the FUDS loading profiles are shown in Figure 6.

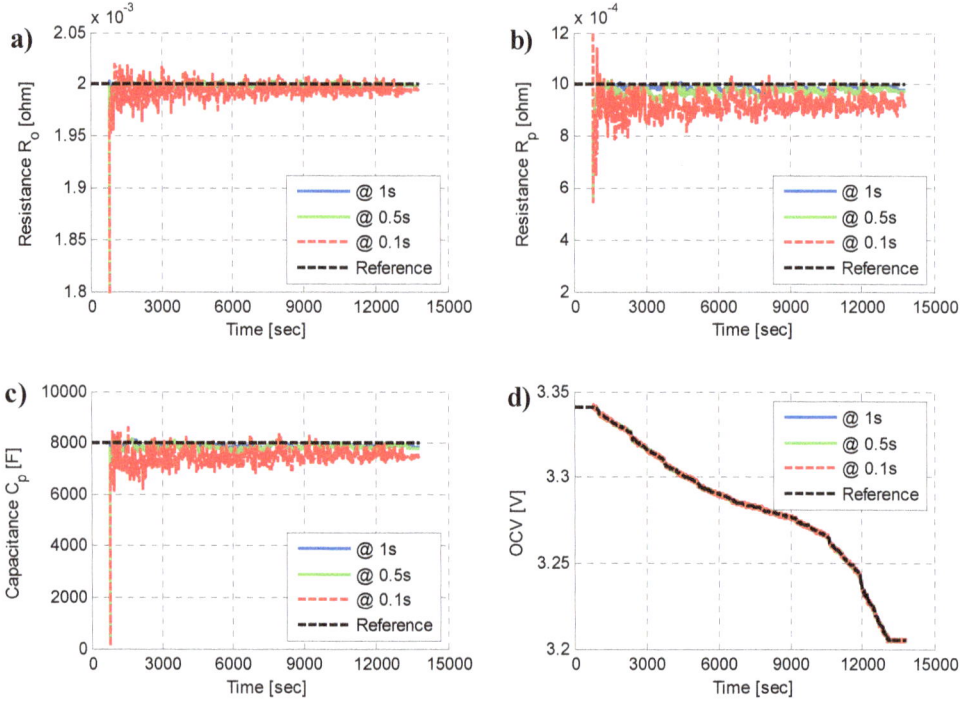

Figure 6. Estimated model parameters: (a) R_o, (b) R_p, (c) C_p and (d) OCV at the sampling periods of 1/0.5/0.1 s.

Figure 6a shows the estimated ohmic resistance R_o and the reference value. The maximum estimation error of R_o increases from 0.2408% to 1.9072% when the sampling rate increased from 1.0 s to 0.1 s. Figure 6b,c lists the estimated polarization resistance R_p and polarization capacitance C_p respectively. When the sampling period is 1 s, the maximum estimation errors of R_p and C_p are 3.3859% and 2.6605%. As the sampling period decreased to 0.1s, the maximum estimation errors of R_p and C_p increased hugely to 21.9172% and 23.81%. If three parameters (R_o, R_p and C_p) are compared together, it can be found the pair of R_p and C_p is much more sensitive to the noise. This conclusion can be verified by the previous model stability and parameter sensitivity analysis. As the sampling period declines from 1 s to 0.1 s, the model stability will degrade, and the parameter sensitivity will be intensified, which means the perturbation to the model parameters will be enhanced under the same noise excitation. Finally, the estimated OCV is plotted in Figure 6d. It reveals that the maximum estimation error of OCV remains nearly the same level. In other words, the sample period exhibits slighter effect on the OCV estimation.

Figure 7 is the estimation results of SoC and SoC error for three different sampling rates under the FUDS loading profiles. From Figure 7, we find that the SoC estimation errors for three different sampling rates are 1.432%, 1.536% and 1.729%. These results reveal that the effect of sampling rate on the SoC estimation accuracy is not significant. Through the comparison, it can infer that the sampling rate has more influence on the model parameters estimation than the SoC estimation. The statistical error analysis of model parameters and SoC, such as maximum error and root-mean-square error (RMSE) is shown in Table 6.

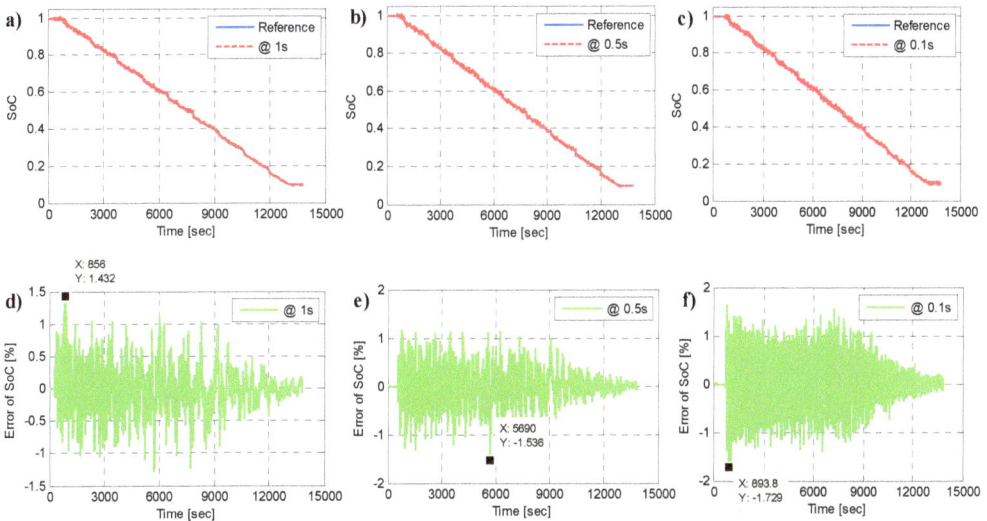

Figure 7. SoC estimation results at the sampling periods of (**a**) 1 s, (**b**) 0.5 s, (**c**) 0.1 s; and the SoC error at the sampling periods of (**d**) 1 s, (**e**) 0.5 s, (**f**) 0.1 s.

Table 6. The statistical error analysis of model parameters and SoC estimation error. RMSE: root-mean-square error.

Item	Sample @ 1.0s				Sample @ 0.5s				Sample @ 0.1s			
Parameters	R_o	R_p	C_p	SoC	R_o	R_p	C_p	SoC	R_o	R_p	C_p	SoC
Maximum error (%)	0.24	3.38	2.66	1.43	0.38	7.33	5.09	1.53	1.90	21.9	23.8	1.72
RMSE (%)	0.06	1.34	1.23	0.33	0.11	2.55	2.10	0.34	0.38	8.39	7.31	0.36

5.2. Current Sensor Accuracy Effect

In the engineering application, the accuracy of the current sensor/transformer is divided into six grades of 0.1, 0.2, 0.5, 1, 3, 5, according to China's national standards GB 1208-1997. The number listed here indicates that the accuracy of the current sensor with the unit of the percentage. To be instinctive, some typical current transducers with detailed specification [30,31] are collected in the following Table 7. From the table, it can be found that the sensor accuracy is varied from ± 50 mA to ± 500 mA for different kinds of current transducers.

Table 7. The accuracy information comparison of three typical current sensors.

Company and product name	LEM DHAB S/25 (Geneva, Switzerland)	LEM LA100-P (Geneva, Switzerland)	Allegro, ACS758 LCB-100B-PSF-T (Worcester, MA, USA)
Transducer type	Open loop	Closed loop	Closed loop
Supply voltage	5 V	5 V	3.3 V or 5 V
Primary current I_p	± 25 A for ch1; ± 200 A for ch2	± 100 A	± 100 A
Output voltage V_{sn}	0.25–4.75 V	-	Vref ± 2 V
Overall accuracy @I_p, $T = 25\,°C$	$\pm 4\%, \pm 500$ mA	$\pm 0.45\%, \pm 50$ mA	$\pm 2.4\%, \pm 150$ mA
Linearity error	$<\pm 1\%$	$<\pm 0.15\%$	$<\pm 1.25\%$
Operation temperance	$-40.125\,°C$	$-40.85\,°C$	$-40.150\,°C$
Response time	<25 ms	<1 µs	<4 µs

In this research, three current precisions of $\pm 5 / \pm 50 / \pm 500$ mA are chosen to evaluate the effect of current sensor accuracy on the model parameters and SoC estimation. Figure 8 is the estimation results of the model parameters under the FUDS loading profiles. Figure 8a shows the estimated ohmic resistance R_o and the reference value. It reveals that the maximum estimation error of R_o increases from 0.0263% to 0.3921% with the current sensor accuracy increased from ± 5 mA to ± 500 mA. Figure 8b,c lists the estimated polarization resistance R_p and polarization capacitance C_p respectively. When the current sensor accuracy is ± 5 mA, the maximum estimation errors of R_p and C_p are 0.7180% and 0.5015%. As the current sensor accuracy increased to ± 500 mA, the maximum estimation errors of R_p and C_p also increased greatly to 3.5399% and 2.4398%. Finally, the estimated OCV is plotted in Figure 8d. It reveals that the maximum estimation error of OCV varies from 0.0433 mV to 1.9094 mV as the current accuracy increases from ± 5 mA to ± 500 mA. The statistical error analysis of model parameters, such as maximum error and RMSE is shown in Table 8.

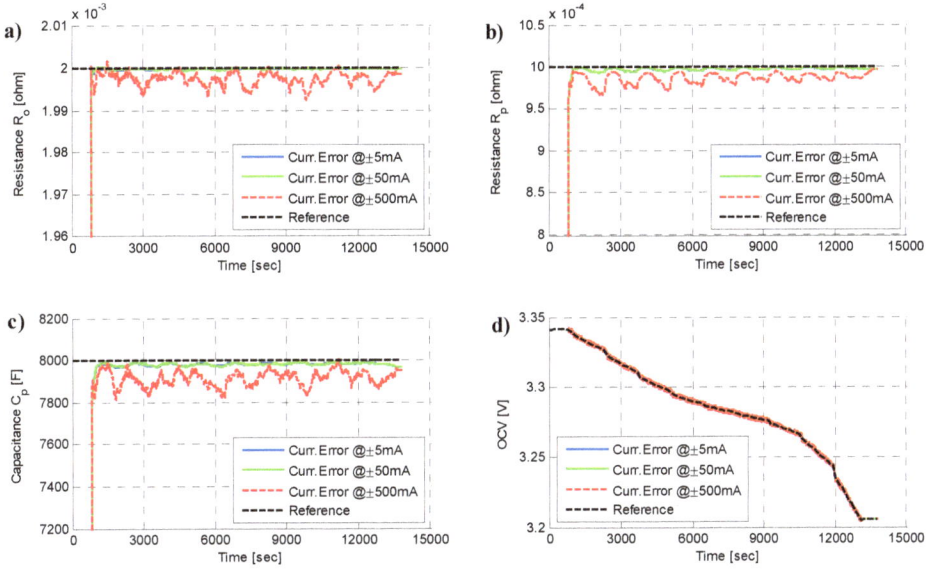

Figure 8. Estimated model parameters: (**a**) R_o, (**b**) R_p, (**c**) C_p, and (**d**) OCV at the current precisions of $\pm 5/\pm 50/\pm 500$ mA.

Table 8. The statistical error analysis of model parameters and SoC at the current sensor precisions of $\pm 5/\pm 50/\pm 500$ mA.

Item	Current accuracy: ± 5 mA				Current accuracy: ± 50 mA				Current accuracy: ± 500 mA			
Parameters	R_o	R_p	C_p	SoC	R_o	R_p	C_p	SoC	R_o	R_p	C_p	SoC
Maximum error (%)	0.02	0.71	0.50	0.06	0.04	0.69	0.45	0.26	0.39	3.53	2.43	2.76
RMSE (%)	0.01	0.33	0.25	0.01	0.01	0.33	0.25	0.06	0.15	1.50	1.15	0.54

Figure 9 is the estimation results of SoC and SoC error for three different current sensor precisions under the FUDS loading profiles. From Figure 9, we find that the SoC estimation errors for three different precisions are 0.0628%, 0.2607% and 2.7671%. These results reveal that the effect of current precisions on the SoC estimation accuracy is evident.

5.3. Voltage Sensor Accuracy Effect

As for the voltage sensor, the accuracy issue will occur with the adaptation of analogue and digital elements, the rounding error of MCU, the sampling alias and harmonics or even electro-magnetic interference, *etc.* Some typical voltage sensor/transducers with detailed specification [32–34] are collected in the following Table 9. From the table, it can be found that the sensor accuracy is varied from ± 1.2 mV to ± 12.5 mV for different kinds of voltage transducers.

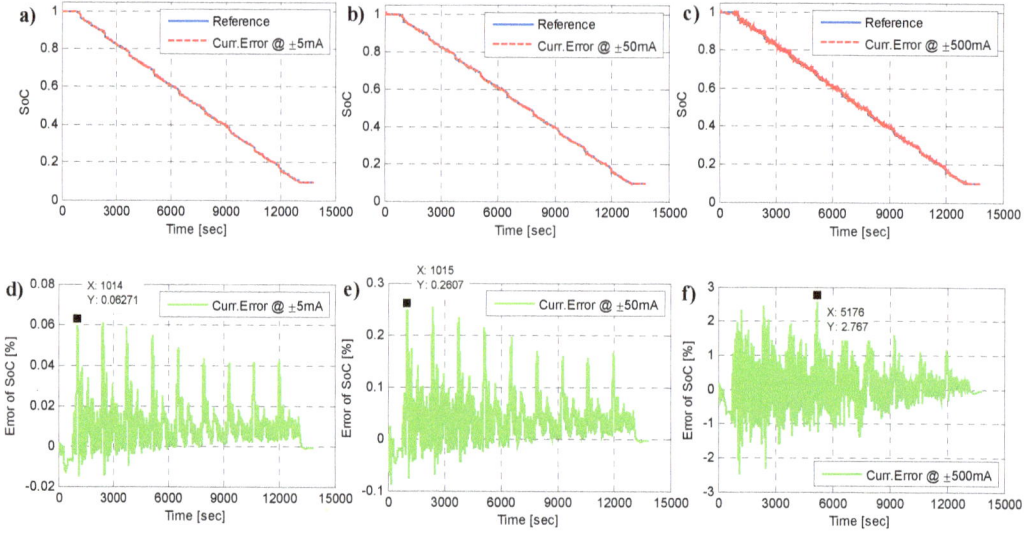

Figure 9. SoC estimation results at the current sensor precisions of (**a**) ±5 mA, (**b**) ±50 mA, (**c**) ±500 mA; and the SoC error at the current sensor precisions of (**d**) ±5 mA, (**e**) ±50 mA, (**f**) ±500 mA.

Table 9. The accuracy information comparison of three typical voltage sensors.

Company and product name	Linear Technology Co. LTC6802 (Milpitas, CA, USA)	Texas Instrument Co. bq76PL536 (Dallas, TX, USA)	Maxim Co. MAX11068 (San Jose, CA, USA)
Voltage meas. channels	12	6	12
AD resolution (Bit)	16	14	12
AD conversion time	57 µs	6 µs	10 µs
Typical voltage accuracy	±1.2 mV	±3.0 mV	±12.5 mV
Maximum voltage accuracy	±8.3 mV	±18.0 mV	±50.0 mV
Operating temp. range	−40 °C to +85 °C	−40 °C to +85 °C	−40 °C to +105 °C
Cell balancing	12 channels	6 channels	12 channels
Input voltage range (V)	0–75	0–30	6–72

To evaluate the effect of voltage precisions on model parameters and SoC estimation, three voltage accuracies of ±1/±2.5/±5 mV are adopted in the simulation. Figure 10 is the estimation results of the model parameters under the FUDS loading profiles. Figure 10a shows the estimated ohmic resistance R_o and its reference value. It indicates that the maximum estimation error of R_o increases from 0.3351% to 1.7986% with the voltage sensor accuracy increased from ±1 mV to ±5 mV. Figure 10b,c lists the estimated polarization resistance R_p and polarization capacitance C_p respectively. When the voltage sensor precision is ±1 mV, the maximum estimation errors of R_p and C_p are 2.5103% and 2.3691%.

As the voltage sensor precision rises to ± 5 mV, the maximum estimation errors of R_p and C_p ascends abundantly to 20.7418% and 15.9086%. Finally, the estimated OCV is plotted in Figure 10d. It reveals that the maximum estimation error of OCV varies from 0.7227 mV to 3.4698 mV, which is nearly in accordance with voltage precisions. The statistical error analysis of model parameters and SoC, such as maximum error and RMSE is shown in Table 10.

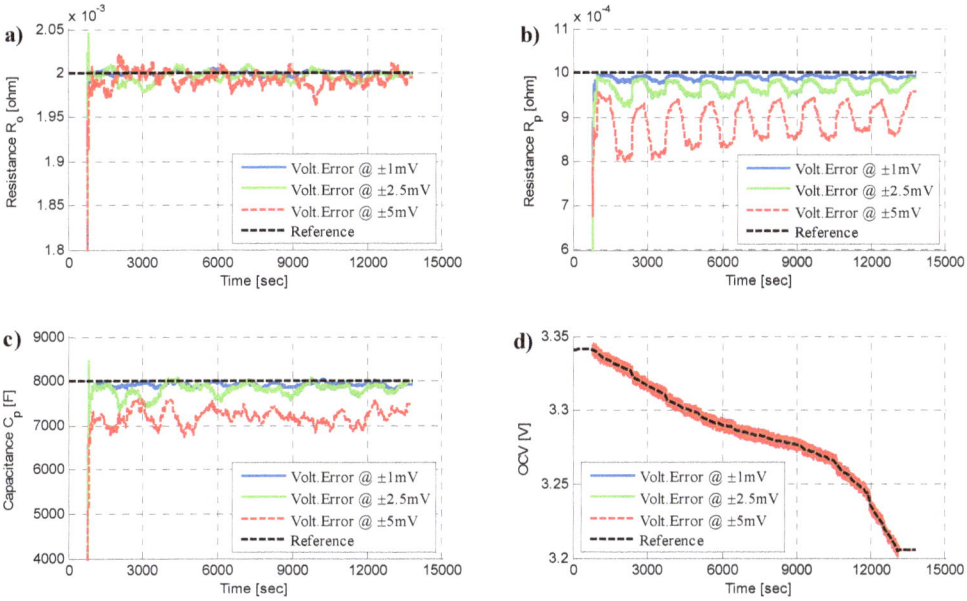

Figure 10. Estimated parameters: (a) R_o, (b) R_p, (c) C_p, and (d) OCV at the voltage sensor precisions of $\pm 1/\pm 2.5/\pm 5$ mV.

Table 10. The statistical error analysis of model parameters and SoC at the voltage sensor precisions of $\pm 1/\pm 2.5/\pm 5$ mV.

Item	Voltage accuracy: ± 1 mV				Voltage accuracy: ± 2.5 mV				Voltage accuracy: ± 5 mV			
Parameters	R_o	R_p	C_p	SoC	R_o	R_p	C_p	SoC	R_o	R_p	C_p	SoC
Maximum error (%)	0.33	2.51	2.36	2.73	1.08	7.79	8.52	6.86	1.79	20.74	15.9	14.42
RMSE (%)	0.11	1.08	1.07	0.54	0.34	3.64	2.88	1.64	0.51	12.01	10.11	3.54

Figure 11 is the estimation results of SoC and SoC errors for three different voltage sensor precisions under the FUDS loading profiles. It shows that the maximum SoC estimation errors for three voltage accuracies are 2.7387%, 6.8638% and 14.4207%. This simulation reveals that there is an effect of current sensor precisions on the SoC estimation accuracy. In the engineering application, to achieve the higher precise estimation for SoC (<5%), the voltage sensor accuracy should be limited to less than ±2 mV.

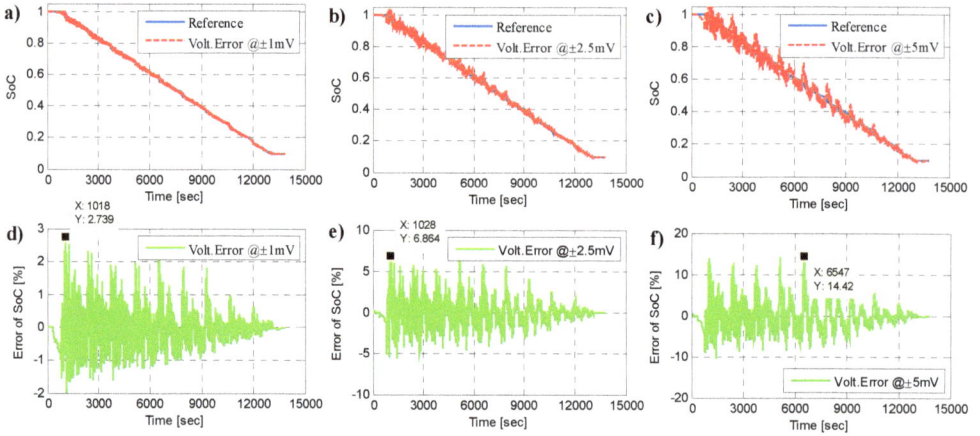

Figure 11. SoC estimation results at the voltage sensor precisions of (**a**) ±1 mV, (**b**) ±2.5 mV, (**c**) ±5 mV; and the SoC error at the voltage sensor precisions of (**d**) ±1 mV, (**e**) ±2.5 mV, (**f**) ±5 mV.

5.4. Results Comparison and Discussion

Through the comparison of the simulation results, as shown in Figure 12, four meaningful results can be drawn:

(1)   The variation of sampling periods (0.1–1 s) has a significant impact on parameter estimation accuracy. Specifically, the sampling time has a relatively small effect to estimate the ohmic resistance R_o, with the maximum error of 2%. However, the sampling time presents the significant influence for polarization resistance R_p and polarization capacitance C_p with the maximum error of 23%. This result is verification of the previous parameter sensitivity analysis, which means the raised parameter sensitivity of R_p and C_p will be more sensitive to external perturbations. Therefore, an accurate estimate of the R_p and C_p will encounter greater difficulty. On the other hand, the variation of sampling rate has a less significant effect on the SoC estimation error. Therefore, changing the sampling time is not the optimal choice to obtain improved estimation accuracy of the SoC.

(2)   The variation of current sensor precisions (±5/±50/±500 mA) shows little influence for model parameters estimation. For instance, when the current sensor accuracy is ±500 mA, the maximum error of R_p and C_p is about 3.5% and the maximum error of SoC is about 2.76%. Therefore, to restrict the estimation

accuracy of the model parameters and SoC, the current sensor accuracy is recommended to be lesser than ±50 mA.

(3) The variation of voltage sensor precisions (±1/±2.5/±5 mV) has significant impact both on model parameter estimation and on SoC estimation. When the voltage accuracy is ±5 mV, the maximum estimation error of R_o, R_p and C_p is 1.79%, 20.74% and 15.90%, respectively. It reveals that the error of R_o is acceptable, while the error of R_p and C_p is hardly acceptable. As the voltage accuracy decreases to ±1 mV, the maximum estimation error of R_o, R_p and C_p is in the acceptable range of 0.33%, 2.51% and 2.36%. For the SoC estimation, the maximum SoC error increases from 2.73%, 6.86% to 14.42%, as the voltage sensor accuracy ascends from ±1 mV, ±2.5 mV to ±5 mV. Therefore, to ensure an accurate SoC estimation (<5%), the voltage sensor precision should be less than ±2 mV. This conclusion can also be drawn from the ICA result.

(4) The weighted importance of factors on parameters and SoC estimation can be sorted as (by descending order): voltage sensor accuracy > sampling period > current sensor accuracy based on the above comparison.

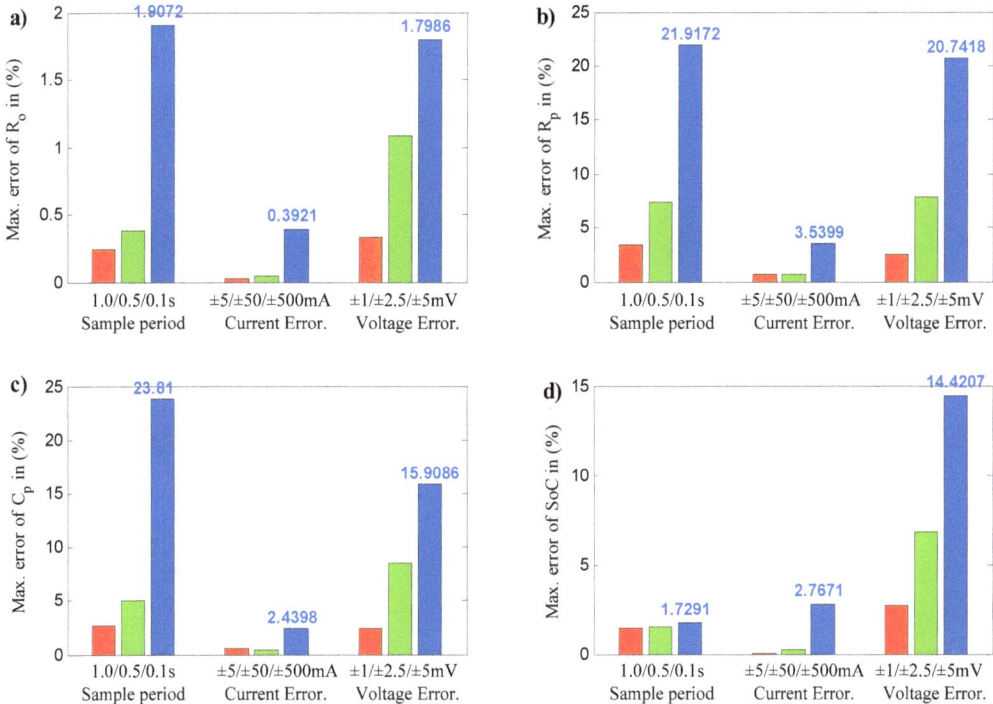

Figure 12. Comparison of three factors on the stability of model parameters and SoC estimation: (**a**) R_o, (**b**) R_p, (**c**) C_p, and (**d**) SoC.

In general, the optimal selection approach for time sampling period T_s, current/voltage sensor precisions, as shown in Figure 13, could be summarized as:

(a) Firstly assess the parameters of battery system, such as R_o, R_p, C_p and capacity, then conduct the perturbation analysis with the Equations (9)–(11) according to the precision requirement of model parameters. The optimized current/voltage sensor precision could be computed.

(b) About the given SoC precision requirement, the user can calculate the voltage precision with the ICA. Take this research for example: If the minimal SoC estimation precision is limited as 1.5%, our voltage sensor precision could be calculated as 1.034 mV with the ICA result in Figure 5.

(c) Compare the voltage precision in Steps (a) and (b), then choose the minimized result. If the minimized result is from Step (a), re-compute Step (a) again to update the current sensor precision.

(d) About the given model parameters, the user can conduct the model stability analysis and parametric sensitivity analysis. Then the threshold of time sampling period T_s could be gain. The optimized time sampling period T_s could be selected in a tradeoff way by considering the model stability, parametric sensitivity, system-sampling precision and the hardware runtime.

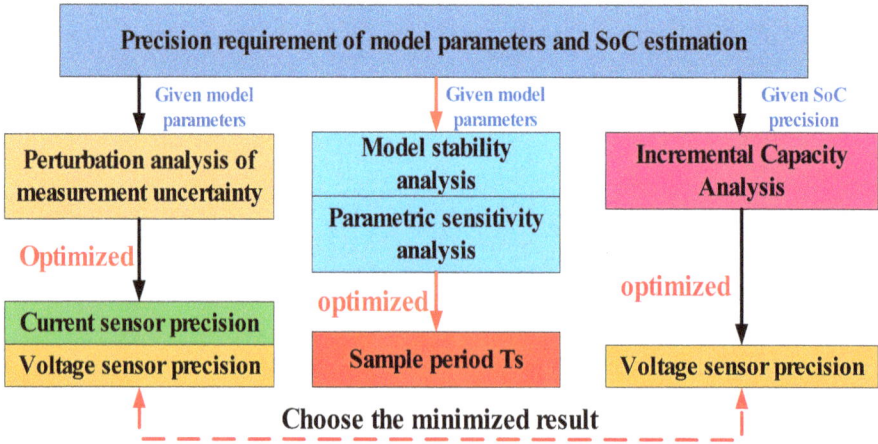

Figure 13. The selection approach for sample period T_s, current/voltage sensor precisions.

6. Conclusions

In this paper, the stability of model parameters and SoC estimation have been analyzed and simulated with FUDS profiles. In summary, the main concluding remarks are given as follows:

(1) The model stability and parametric sensitivity have been analyzed under different sampling periods T_s (0.02, 0.1, 0.2, 0.5 and 1 s). The results reveal that the increase of sampling period T_s will be beneficial to the model stability and parameter identifiability. From an engineering viewpoint, it is recommended to restrict the eigenvalues of the ARX model within a range of 0–0.95. That is, T_s should be larger than one threshold, such as $T_s \geq 0.5$ s.

(2) The variation of sampling periods (0.1–1 s), has a significant impact on parameter estimation accuracy but a less significant effect on the SoC estimation error. Therefore, to improve the estimation accuracy of the SoC, it is not optimal to change the sampling time.

(3) The variation of current sensor precision ($\pm 5/\pm 50/\pm 500$ mA) shows little influence for model parameters and SoC estimation. To restrict the estimation accuracy of the model parameters and SoC, the current sensor accuracy is recommended to be less than ± 50 mA.

(4) The variation of voltage sensor precision ($\pm 1/\pm 2.5/\pm 5$ mV) has significant impact on both the model parameter estimation and SoC estimation. To ensure the SoC estimation accuracy (<5%), the voltage sensor accuracy should be less than ± 2 mV.

(5) According to the parameter variation analysis under the perturbation of current/voltage measurement uncertainty, the weighted importance of factors on parameter and SoC estimation can be sorted as (by descending order): voltage sensor accuracy > sampling period > current sensor accuracy.

Acknowledgments: This research work is supported by a grant from the National High Technology Research and Development Program of China (863 Program) (No.2011AA11A229), the Specialized Research Fund for the Doctoral Program (SRFDP) of Higher Education (No.20090073120051), and the U.S.-China Clean Energy Research Center Clean Vehicles Consortium (CERC-CVC) (No. 2010DFA72760-305).

Author Contributions: Shifei Yuan and Hongjie Wu conducted the main experiment/simulation and drafted the main part of the paper. Xuerui Ma checked out and revised the paper. Chengliang Yin provided insightful suggestions on the research and experiment.

Conflicts of Interest: The authors declare no conflict of interest.

References

1. He, H.; Xiong, R.; Zhang, X.; Sun, F.; Fan, J. State-of-charge estimation of the lithium-ion battery using an adaptive extended Kalman filter based on an improved thevenin model. *IEEE Trans. Veh. Technol.* **2011**, *60*, 1461–1469.

2. He, Z.; Gao, M.; Wang, C.; Wang, L.; Liu, Y. Adaptive state of charge estimation for Li-ion batteries based on an unscented Kalman filter with an enhanced battery model. *Energies* **2013**, *6*, 4134–4151.

3. Xing, Y.; He, W.; Pecht, M.; Tsui, K.L. State of charge estimation of lithium-ion batteries using the open-circuit voltage at various ambient temperatures. *Appl. Energy* **2014**, *113*, 106–115.

4. Wang, Y.; Zhang, C.; Chen, Z. A method for joint estimation of state-of-charge and available energy of LiFePO$_4$ batteries. *Appl. Energy* **2014**, *135*, 81–87.

5. Liu, X.; Wu, J.; Zhang, C.; Chen, Z. A method for state of energy estimation of lithium-ion batteries at dynamic currents and temperatures. *J. Power Sources* **2014**, *270*, 151–157.

6. Xiong, R.; Sun, F.; Chen, Z.; He, H. A data-driven multi-scale extended Kalman filtering based parameter and state estimation approach of lithium-ion olymer battery in electric vehicles. *Appl. Energy* **2014**, *113*, 463–476.

7. Xiong, R.; Sun, F.; Gong, X.; Gao, C. A data-driven based adaptive state of charge estimator of lithium-ion polymer battery used in electric vehicles. *Appl. Energy* **2014**, *113*, 1421–1433.

8. Dai, H.; Wei, X.; Sun, Z.; Wang, J.; Gu, W. Online cell SOC estimation of Li-ion battery packs using a dual time-scale Kalman filtering for EV applications. *Appl. Energy* **2012**, *95*, 227–237.

9. Zhong, L.; Zhang, C.; He, Y.; Chen, Z. A method for the estimation of the battery pack state of charge based on in-pack cells uniformity analysis. *Appl. Energy* **2014**, *113*, 558–564.

10. Liu, X.; Chen, Z.; Zhang, C.; Wu, J. A novel temperature-compensated model for power Li-ion batteries with dual-particle-filter state of charge estimation. *Appl. Energy* **2014**, *123*, 263–272.

11. Hu, X.; Li, S.; Peng, H.; Sun, F. Charging time and loss optimization for linmc and LiFePO$_4$ batteries based on equivalent circuit models. *J. Power Sources* **2013**, *239*, 449–457.

12. Xiong, R.; Gong, X.; Mi, C.C.; Sun, F. A robust state-of-charge estimator for multiple types of lithium-ion batteries using adaptive extended Kalman filter. *J. Power Sources* **2013**, *243*, 805–816.

13. Plett, G.L. Extended Kalman filtering for battery management systems of LiPB-based HEV battery packs: Part 1. Background. *J. Power Sources* **2004**, *134*, 252–261.

14. Plett, G.L. Extended Kalman filtering for battery management systems of LiPB-based HEV battery packs: Part 2. Modeling and identification. *J. Power Sources* **2004**, *134*, 262–276.

15. Plett, G.L. Extended Kalman filtering for battery management systems of LiPB-based HEV battery packs: Part 3. State and parameter estimation. *J. Power Sources* **2004**, *134*, 277–292.

16. Yuan, S.; Wu, H.; Yin, C. State of charge estimation using the extended Kalman filter for battery management systems based on the arx battery model. *Energies* **2013**, *6*, 444–470.

17. Han, J.; Kim, D.; Sunwoo, M. State-of-charge estimation of lead-acid batteries using an adaptive extended Kalman filter. *J. Power Sources* **2009**, *188*, 606–612.

18. Sun, F.; Hu, X.; Zou, Y.; Li, S. Adaptive unscented Kalman filtering for state of charge estimation of a lithium-ion battery for electric vehicles. *Energy* **2011**, *36*, 3531–3540.

19. Santhanagopalan, S.; White, R.E. State of charge estimation using an unscented filter for high power lithium ion cells. *Int. J. Energy Res.* **2010**, *34*, 152–163.

20. Shao, S.; Bi, J.; Yang, F.; Guan, W. On-line estimation of state-of-charge of Li-ion batteries in electric vehicle using the resampling particle filter. *Transp. Res. Part D* **2014**, *32*, 207–217.

21. Gao, M.; Liu, Y.; He, Z. Battery State of Charge Online Estimation Based on Particle Filter. In Preoceeeings of the 2011 4th International Congress on Image and Signal Processing (CISP), Shanghai, China, 15–17 October 2011; IEEE: New York, NY, USA, 2011; pp. 2233–2236.

22. Hu, X.; Sun, F.; Zou, Y. Estimation of state of charge of a lithium-ion battery pack for electric vehicles using an adaptive luenberger observer. *Energies* **2010**, *3*, 1586–1603.

23. Kim, I.-S. The novel state of charge estimation method for lithium battery using sliding mode observer. *J. Power Sources* **2006**, *163*, 584–590.

24. Xia, B.; Chen, C.; Tian, Y.; Sun, W.; Xu, Z.; Zheng, W. A novel method for state of charge estimation of lithium-ion batteries using a nonlinear observer. *J. Power Sources* **2014**, *270*, 359–366.

25. Zhang, X.; Mi, C. *Vehicle Power Management: Modeling, Control and Optimization*; Springer: Berlin/Heidelberg, Germany, 2011.

26. Hu, X.; Sun, F.; Zou, Y.; Peng, H. Online Estimation of an Electric Vehicle Lithium-Ion Battery Using Recursive Least Squares with Forgetting. In Preoceeeings of the American Control Conference (ACC), San Francisco, CA, USA, 29 June–1 July 2011; IEEE: New York, NY, USA, 2011; pp. 935–940.

27. Feng, D.; Tongwen, C.; Li, Q. Bias compensation based recursive least-squares identification algorithm for MISO systems. *IEEE Trans. Circuits Syst. II Express Briefs* **2006**, *53*, 349–353.

28. Wu, H.; Yuan, S.; Zhang, X.; Yin, C.; Ma, X. Model parameter estimation approach based on incremental analysis for lithium-ion batteries without using open circuit voltage. *J. Power Sources* **2015**, *287*, 108–118.

29. Xiong, R.; He, H.; Sun, F.; Liu, X.; Liu, Z. Model-based state of charge and peak power capability joint estimation of lithium-ion battery in plug-in hybrid electric vehicles. *J. Power Sources* **2013**, *229*, 159–169.

30. Lem—Current Tranducer, Voltage Transducer, Sensor, Power Measurement. Available online: http://www.lem.com/ (accessed on 26 June 2015).

31. Allegro MicroSystems LLC. Available online: http://www.allegromicro.com/ (accessed on 26 June 2015).

32. Linear Technology—Home Page. Available online: http://www.linear.com/index.php (accessed on 26 June 2015).

33. Analog, Embedded Processing, Semiconductor Company, Texas Instruments—TI.Com. Available online: http://www.ti.com/ (accessed on 26 June 2015).

34. Maxim Integrated. Analog, Linear, and Mixed-Signal Devices from Maxim. Available online: http://www.maximintegrated.com/en.html (accessed on 26 June 2015).

Numerical Analysis on Combustion Characteristic of Leaf Spring Rotary Engine

Yan Zhang, Zhengxing Zuo and Jinxiang Liu

Abstract: The purpose of this paper is to investigate combustion characteristics for rotary engine via numerical studies. A 3D numerical model was developed to study the influence of several operative parameters on combustion characteristics. A novel rotary engine called, "Leaf Spring Rotary Engine", was used to illustrate the structure and principle of the engine. The aims are to (1) improve the understanding of combustion process, and (2) quantify the influence of rotational speed, excess air ratio, initial pressure and temperature on combustion characteristics. The chamber space changed with crankshaft rotation. Due to the complexity of chamber volume, an equivalent modeling method was presented to simulate the chamber space variation. The numerical simulations were performed by solving the incompressible, multiphase Unsteady Reynolds-Averaged Navier–Stokes Equations via the commercial code FLUENT using a transport equation-based combustion model; a realizable $\kappa - \omega$ turbulence model and finite-rate/eddy-dissipation model were used to account for the effect of local factors on the combustion characteristics.

Reprinted from *Energies*. Cite as: Zhang, Y.; Zuo, Z.; Liu, J. Numerical Analysis on Combustion Characteristic of Leaf Spring Rotary Engine. *Energies* **2015**, *8*, 8086–8109.

1. Introduction

This paper introduces a novel rotary engine, "Leaf Spring Rotary Engine". The leaf spring rotary engine is a pistonless design based on the Wankel engine. A pistonless rotary engine is an internal combustion engine that does not use pistons in the way a reciprocating engine does, but instead uses one or more rotors, sometimes called rotary pistons. An example of a pistonless rotary engine is the Wankel engine. The basic concept of a (pistonless) rotary engine avoids the reciprocating motion of the piston with its inherent vibration and rotational-speed-related mechanical stress. As of 2006, the Wankel engine is the only successful pistonless rotary engine, but many similar concepts have been proposed and are under various stages of development. A Hamilton Walker prototype engine was construction in 1968 [1,2]. Walker's engine is believed to be the second working rotary engine after the Wankel engine. The Quasitubine, or Qurbine, engine is a proposed pistonless rotary engine using a rhomboidal rotor whose sides are hinged at the vertices in 1996. As well as an internal combustion engine, the Quasitubine has been proposed as a possible pump design, and a possible stirling engine [3]. It has been demonstrated as a pneumatic engine using stored compressed air, and as a steam engine [4]. The essential concept, the Ramgen engine, is the incorporation of one or more ramjets onto the rim of a rotor such that the thrust from the ramjets acts tangentially, causing it to rotate at supersonic ram speeds. The engine is based on ramjet technology, which produces propulsive force by increasing the momentum of the working fluid via combustion and expansion through a supersonic nozzle [5]. The Rotary Piston Machine (RKM) is a proposed form of machine. It can be used either to transform pressure into rotational motion or the converse rotational motion into pressure. It is still in development and offers very high potential in pump market [6]. One application for RKM is miniaturization, like Wankel engine. The Sarich orbital engine is a type of internal combustion engine, invented in 1972 by Ralph Sarich. The theoretical advantage is that there is no high-speed contact area with the engine walls, unlike in the Wankel engine in which edge wear is a problem. However, the combustion chambers are divided by blades, which do have contact with both the walls and the rotor, and are said to have been difficult to seal due to the perpendicular intersection with the moving impeller. A wave disk engine is a type of pistonless rotary engine being developed at Michigan State University and Warsaw Institute of Technology. The engine has a spinning disk with curved blades. Once fuel and air enter the engine, the rotation of the disk creates shockwaves that compress the mixture. The Wave Disk Micro-Engine concept has also been developed [7]. The Jonova engine is a type of pistonless rotary engine designed by Nowakowski at the University of Arizona. Consisting of only four moving parts; the newer Jonova engine loses no energy through vibration and has a wide power stroke over 240 degrees of rotation, as opposed to four-stroke engines, which have a power stroke of 180 degrees on alternating rotations or two-stroke engines with a power stroke of 180 degrees on each rotation.

In this paper, the designed displacement of the engine is 1.77 cm^3 and the theoretical compression ratio is 7.3. The size of the combustion chamber is very small so that the engine can be classified as micro- or meso-scale engine. With the rapid development of MEMS (Micro Electro Mechanic system) technology, various micro- and meso-scale devices and systems, including micro turbines, robots, satellites and portable electric devices, are continuously emerging. Because electrochemical batteries have some disadvantages such as short life spans, long recharging periods and low energy densities, combustion based on micro-power-generation devices are supposed to be potential alternatives due to the much higher energy densities of hydrocarbon fuels compared to batteries [8–10]. The micro-combustor is an important component in which the chemical energy of hydrocarbon is converted into thermal energy via combustion. Therefore, the development of a micro-combustor with a wide operation range has attracted increasing attention over the past few years. Epdtein $et\ al.$ firstly designed a 2-mm-height, 66-mm^3-chamber-volume combustor reported by Mehra and Waitz *et al.* [11] with three pieces of silicon wafers. Mehra *et al.* [12] designed a 195-mm^3-chamber-volume new combustor with six layer of silicon wafers based on previous design. Researchers [13,14] in MIT have started studying micro gas turbines with 21-mm-length and 3-mm-height. Kelvin *et al.* [15] proposed three different displacement micro gas turbines. Ochoa *et al.* [16,17] constructed Swiss-roll combustor with the material of Bi$_2$Te$_3$. Researchers [18] at university of Michigan developed micro internal combustion swing engine (MICSE) with the dimensions 61 (H) × 61 (W) × 34 (D) mm. Georgia Honeywell Technology Center and Air Force Research Lab (AFRL) [19] studied micro technology-based energy and chemical systems (MECS) free-piston knock engine. Minotti and Sciubba [20] designed a cylindrical combustor of 29 cm^3 to provide 2 KW of thermal power.

Among various design considerations in developing combustion-based micro power generation systems, the core is to guarantee stable burning in micro combustors during operations. Therefore, investigations into microflames to provide fundamental data for designing stable burning micro combustors have been reported [21,22]. However, there are some challenges to investigating stable combustion in micro combustors. Firstly, the increased heat losses and wall radical capture due to large surface area-to-volume ratio, which makes it difficult to sustain a stable flame under small scales [23–25]. Another critical problem is the shortened residence time of fuel/oxidant mixture in the combustor. For most hydrocarbon fuels, the quenching distance is about a few millimeters, which is of the same order of magnitude of the flame thickness. Previous experimental studies demonstrated that stable flames of CH$_4$–O$_2$ mixture can be achieved in a 0.5 mm diameter tube. Meanwhile, the numerical method, which is able to yield detailed information of parameters within the small space, has been widely used to investigate micro-combustion processes. Raimondeau *et al.* [26] used two-dimensional parabolic simulations to model flame propagation in microchannels. It was found that in very small reactors, radial gradients and temperature discontinuity at the wall were negligible but became significant as the diameter increased. Karagiannidis *et al.* [27] investigated numerically the hetero-/homogeneous steady combustion and the stability limits of methane-fueled catalytic micro reactors in a 1-mm-gap channel at pressures of 1 and 5 bar, respectively. A full-elliptic two-dimensional model was used to investigate coupling of hetero-/homogeneous combustion, heat transfer mechanisms, influence of solid thermal conductivity, surface radiation and flow confinement. Norton *et al.* [28] studied the effects of microburner wall conductivity, external heat losses, burner dimensions, and operating conditions on combustion characteristics and the steady-state, self-sustained flame stability of propane/air mixtures by solving computational fluid dynamics (CFD) model of microburner. Norton *et al.* [29] studied the effects of microburner dimensions, conductivity and thickness of wall materials, external heat losses, and operating conditions on combustion characteristics and flame stability using computational fluid dynamics (CFD) model of microburner. CFD based on numerical simulation [30,31] has been proven to be an effective approach to analyze the performance of the micro-combustor under various conditions. It was found that chemical kinetics and heat transfer mechanism within micro-combustors are essential for the development of combustion-based power MEMS devices. Kaisare *et al.* [32] used a one-dimensional (1D) model to investigate the flame characteristics and stability of homogeneous combustion in micro-scale (1 mm) channels, and the roles of heat recirculation and heat loss on the mechanisms of flame extinction and blowout. Li *et al.* [33] performed a numerical study of H_2–air premixed combustion in the micro channels with a detailed chemical reaction mechanism by solving the two-dimensional fully elliptic governing equations of continuity, momentum, energy and species, coupled with the energy equation in the solid wall. The results indicated that various boundary and physical conditions, including the combustor size and geometry, inlet velocity profile, axial heat conduction in the solid wall and slip-wall and temperature jump at the gas–solid interface have effects on the flame temperature to different extents. Li *et al.* [34] also investigated effect of combustor size, geometry and boundary conditions on flame temperature for CH_4–air mixture by a numerical model. Lee *et al.* [35] used a two-dimensional CFD simulation with a reduced kinetic mechanism and a detailed transport model to predict the distributions of temperature, fuel and radicals for single microflames near the stability limits and in the stable region. Shih *et al.* [36] studied computationally the combustion of hydrogen/methane-blended fuels for micro gas turbines. The simulations were performed with three-dimensional compressible $\kappa - \varepsilon$ turbulent flow model and presumed probability density function for chemical reaction. Wan *et al.* [37] investigated numerically combustion characteristics of H_2/air mixture in a micro-combustor with wall cavities. The effects of inlet velocity, equivalence ratio, and the length–depth ratio of the cavity were studied.

Based on previous methods, it is known that numerical simulation provides a convenient, credible and cost-effective approach to investigate micro-combustion phenomena and underlying mechanisms. Therefore, in the present investigation, the influence of rotational speed, air–fuel ratio, initial pressure and temperature on combustion characteristics of leaf spring rotary engine was investigated using a 3D CFD model. In this paper, the concentration of the gasoline/air mixture is specified at the inlet. The reaction mechanism is a one-step global reaction ($O_2 = 21\%$, $N_2 = 78\%$).

2. Structure and Principle, Application Mode

The leaf spring rotary engine is similar to the Wankel rotary engine, which is a 4-stroke cycle engine in structure and working principle. The energy released by the air-fuel mixture drives the rotor and output shaft to move. The rotor centerline aligns with the housing centerline, which is different from the Wankel rotary engines mentioned above. The end of the spring is fixed to the rotor, and the other end depends on its elasticity to contact with combustion wall. The chemical energy is converted to mechanical energy through compression, combustion, and exhaust processes. Compared with a traditional piston engine, the leaf spring rotary engine does not require a crank and complex connecting rod mechanism. The engine operates smoothly and impact force is small when leaf springs are added to the chamber. Compared with the traditional Wankel engine, the leaf spring rotary engine removes eccentric parts so that the effect of noise and vibration due to unbalance can be eliminated. Taking flexible leaf spring gives the chamber a better sealing property than the Wankel engine. Apex seal is guaranteed by close fitting between the cylinder and the spring because of its flexibility. Leakage mainly occurs at the area between the piston and the housing. The simpler structure of the leaf spring rotary engine leads to less mass leakage than the Wankel engine between the rotor and the cylinder. Furthermore, an appropriate compression ratio is selected, which also decreases the mass of the leaf spring rotary engine [38]. The leaf spring rotary engine consists of front cylinder cover, back cylinder cover, rotor, leaf spring, cylinder, output shaft, flywheel, and so on (Figure 1a).

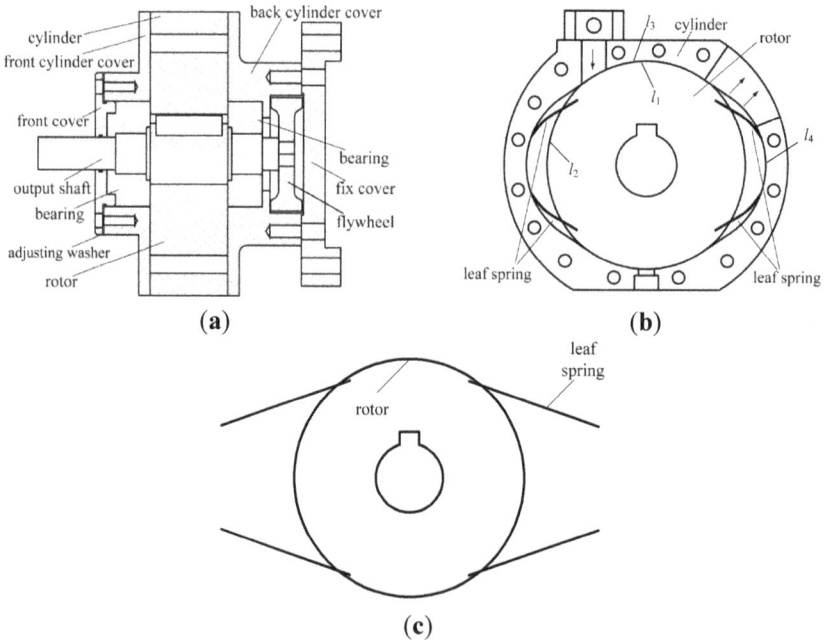

Figure 1. *Cont.*

490

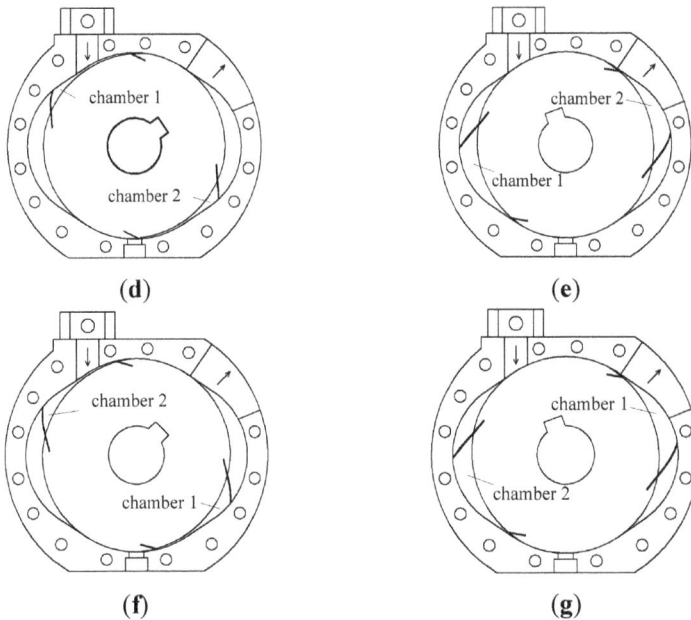

Figure 1. (**a**) Sketch of leaf spring rotary engine; (**b**) sketch of the rotor–spring–cylinder assembly; (**c**) sketch of the rotor; (**d**) intake procession of the engine; (**e**) compression procession of the engine; (**f**) combustion procession of the engine; and (**g**) exhaust procession of the engine.

The combustion chamber is enclosed by the rotor, front and rear springs, and front and back cylinder covers. One end of the leaf spring is fixed on the rotor groove, and the other end contacts the cylinder, relying on its elasticity (Figure 1b). The leaf spring is deformed by the force between cylinder and leaf spring to change the volume of combustion chamber, which is the impetus of intake stroke, compression stroke, combustion stroke, and exhaust stroke.

The working process of the leaf spring rotary engine is shown in Figure 1d–g. The designed rotary engine contains symmetric double combustion chambers. In this work, taking chamber 1 as an example to describe working process: (1) In the intake phase, releasing the stored energy from spring and gas expansion power in chamber 2 together drives the rotor to rotate. The volume of combustion chamber 1 increases, then negative pressure is formed, thus fresh air flows into combustion chamber 1 because of the pressure difference between the inside and outside. (2) In the compression phase, under the action of fly wheel inertia, the fresh charge within combustion chamber 1 is compressed and waste gas is exhausted. Flywheel inertia converts into compression energy. (3) In the combustion phase, the electric heater plug ignites the mixture, causing gas expansion and the release of stored energy together to promote the reed rotor and to inhale gas to chamber 2. (4) In exhaust phase, the spring in chamber 1 is compressed by inertia force to drive gas to discharge.

Based on the structure of the leaf spring rotary engine, the rotor is enclosed within two circles. The original point of the absolute coordinate is located at the cross-section center of the output shaft. The center of one circle, marked with l_1, is placed at the original point and the radius is 25 mm, whereas the center of the other circle, marked with l_2, is situated at (2.2, 0) and the radius is 26.6 mm. Taking the leaf part of the coordinate system as an example, the outline equation of the rotor is shown as Equation (1).

$$\begin{cases} (x-2.2)^2 + y^2 = 26.6^2 & x_1 < x < x_2 \\ x^2 + y^2 = 25^2 & x_2 < x < 0 \end{cases} \tag{1}$$

The outline of the cylinder is symmetrical about the original point. Taking the segment of the cylinder in the second quadrant of the coordinate system as an example, it is surrounded by $arcl_1$, $arcl_2$, and the tangent line of the two arcs, l_4. The center of circular $arcl_3$ is situated at $(-10, 0)$ and the radius is 19.5 mm. The molded line of the cylinder is shown as Equation (2).

$$\begin{cases} (x+10)^2 + y^2 = 19.5^2 & x_3 < x < x_4 \\ ax + by + c = 0 & x_4 < x < x_5 \\ x^2 + y^2 = 25^2 & x_5 < x < 0 \end{cases} \tag{2}$$

where, x_3 is the cross point coordinate between l_3 and abscissa axis, x_4 is the cross point abscissa of l_3 and l_4, and x_5 is the cross point abscissa between l_1 and l_4.

The intake port positions the upper side of the arc l_1 and is adjacent to the cross point of l_1 and l_4. The intake port is completely closed by the rotor during compression in order to prevent combustible mixture in the non-combustion space. In addition, there is no clearance to allow the air fleeing in the cylinder because the cylinder design matches the rotor suitably.

The present state of battery technology severely limits the development of pure electric devices. Until significant advances in battery technology are made, hybrid propulsion systems provide an interim solution [39]. The leaf spring rotary engine is a potential power source to power generation due to high density, which makes it convenient for transportation. The goal of the MEMS Rotary Engine Power System (REPS) project is to develop an autonomous, portable power system capable of producing electrical power on the order of watts with an energy density better than conventional batteries [40]. Such system leverages the specific energy advantage of liquid hydrocarbon fuels over current portable power sources. The MEMS REPS is an interdisciplinary project divided in research areas such as engine fabrication, engine sealing, fuel delivery, packaging and testing. The MEMS REPS unit will be designed to produce power from the leaf spring rotary engine that will be integrated with an external small electrical generator. The leaf spring rotary engine is like a Wankel engine that operates on a 4-stroke cycle. A fuel and air mixture is drawn into the engine through the carburetor. The electrical generator extracts mechanical power from engine with a coupled shaft. In a series hybrid configuration, the devices, including portable electronics and equipment used in remote settings, derive instantaneous performance from energy storage system, but its sustained operation relies on an auxiliary power-generating unit (APU). The engine operates in a gasline-fueled spark-ignited mode. In this engine, the shaft will be used to convert chemical energy into mechanical power. A coupled shaft will be needed to extract mechanical power to drive electrical generator used to generate electrical energy. The output shaft of the rotary engine is centered relative to the rotor housing, and can be easily coupled with a generator to produce electricity. Electric energy will be stored in batteries, which can be used to provide energy for small devices like small aircrafts, satellites, robots, transport devices and so on. At the same time, devices will provide energy feedback to APU. A novel generator may need to be designed to match the leaf spring rotary engine due to its characteristics. Previously, linear generators to match free-piston engines have been designed and investigated [41]. Therefore, the leaf spring rotary engine is a very efficient and reasonable power plant for electric equipment. The leaf spring rotary engine may be a potential power source for future application in small electric devices.

When used as a range extender for a series PHEV, the rotary combustion engine showed better performance than the reciprocating engine. The rotary engine has the advantage of a high power-to-weight ratio, more compact size and packaging, and reduced noise, vibration and harshness (NVH) compared to the reciprocating engine. Reduced NVH is especially critical when extending the range of electric vehicles, as the occupants of the vehicle will be accustomed to the smooth and silent operation of electric traction system, and may dislike NVH produced by a conventional power train unit with a reciprocating engine range extender. These benefits come at the expense of lower fuel economy [42]. In a series hybrid, the internal combustion engine can be operated at its most efficient point by maintaining a specific constant speed. An electronic engine control unit (ECU) is also required to control the fueling and ignition timing on a cycle-by-cycle basis to control the engine speed and stroke.

3. Volume Calculation

The flexibility of the leaf spring determines the volume variation of the leaf spring rotary engine. Thus the spring deformation calculation becomes an important part of the volume calculation. The leaf spring rotor is assembled as a cantilever structure, and remains perfectly elastic in the work process. The spring deformation, which is caused by concentrated load imposed on the free end, is regarded as static behavior without considering complex dynamics characteristic. The geometric nonlinear deforming of the leaf spring satisfies the Euler–Bernoulli Equation (Equation (3)).

$$\frac{y''}{\left[1 + (y')^2\right]^{\frac{3}{2}}} = -\frac{M}{EI} \tag{3}$$

It yields Equations (4) and (5).

$$y\,(x) = \int_0^x \frac{W(\lambda)}{\sqrt{1 - [W(\lambda)]^2}} d\lambda \tag{4}$$

$$W\,(x) = \frac{pl}{2EI}\left[x^2 - (L - l)^2\right] \tag{5}$$

Moreover, the length of the leaf spring can be calculated using arc length integral Equation (6).

$$L = \int_0^{L_0} \left[1 + (y')^{\frac{1}{2}}\right] dx \tag{6}$$

$W(x)$ is the function of horizontal displacement l of the free end. Using trial and error method [43,44], Matlab calculation procedures is performed to calculate the horizontal displacement l in each load step applied to the free end of the leaf spring. The flowchart of the Matlab program is illuminated in Figure 2a [45]. The deflection curve of the leaf spring can be obtained by fitting a constant length to the leaf spring, and then getting the volume of combustion chamber by mathematical integration.

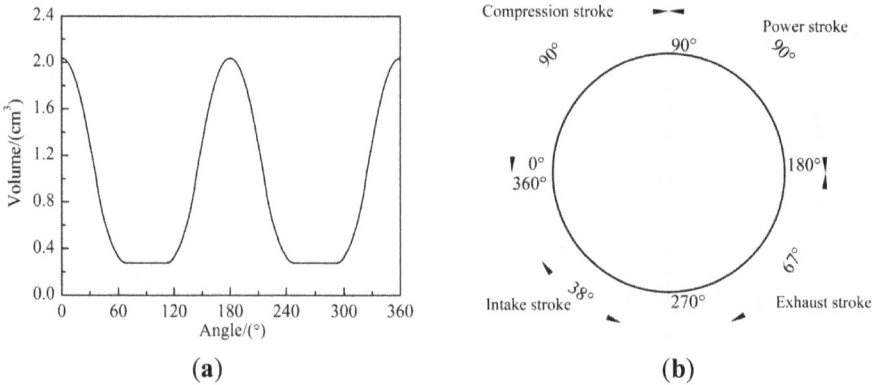

Figure 2. (a) Variation profile of combustion chamber volume; (b) The phase chart of the leaf spring rotary engine.

494

Variation of combustion chamber volume corresponds to rotary angle during the engine work process. The relationship of them is shown in Figure 2a. This finds a compression ratio of about 7.3 and the working volume of the combustion chamber is 1.77 cm^3.

With regard to the change rule of the combustion chamber volume, and based on the locating standard point, which is the midpoint of the rotor molded line facing the combustion chamber, the phase chart of the leaf spring rotary engine is shown in Figure 2b.

4. Methods

4.1. Numerical Methods

The mixture in the chamber was simulated using a commercially available CFD code, FLUENT 14.5. The time-dependent ordinary sets of the continuity equation, the three-dimensional Navier–Stokes (N-S) equation, the energy conservation equation and the species conservation equations were solved with the finite volume method. The code allows for multi-component diffusion, thermal diffusion, variable thermochemical properties and variable transport properties. A realizable $\kappa - \varepsilon$ turbulence model and finite-rate/eddy-dissipation model are used to account for the effect of factors on the combustion characteristics. Turbulence energy κ and ε are introduced to N-S equation by Re-Normalisation Group (RNG) $\kappa - \varepsilon$ method. Then, new turbulence model can be written as follows:

$$\frac{\partial\,(\rho k)}{\partial t} + \frac{\partial\,(\rho k u_i)}{\partial x_i} = \frac{\partial}{\partial x_j}\left(\alpha_k \mu_{eff}\frac{\partial k}{\partial x_j}\right) + G_k + \rho\varepsilon \tag{7}$$

$$\frac{\partial(\rho\varepsilon)}{\partial t} + \frac{\partial(\rho\varepsilon u_i)}{\partial x_i} = \frac{\partial}{\partial x_j}(\alpha_\varepsilon \mu_{eff}\frac{\partial\varepsilon}{\partial x_j}) + \frac{C_{1\varepsilon}{}^*\varepsilon}{k}G_k - C_{2s}\rho\frac{\varepsilon^2}{k} \tag{8}$$

where $\mu_{eff} = \mu + \mu_t$, $\mu_i = \rho C_\mu k^2/\varepsilon$, $C_{1\varepsilon}{}^* = C_{1s} - \eta(1 - \eta/\eta_0)/(1 + \beta\eta^3)$, $\eta = (2E_{ij} \cdot E_{ij})^{1/2} \cdot k/\varepsilon$, $E_{ij} = \frac{1}{2}(\frac{\partial u_i}{\partial x_j} + \frac{\partial u_j}{\partial x_i})$, $C_\mu = 0.0845$, $\alpha_k = \alpha_s = 1.39$, $C_{1s} = 1.42$, $C_{2s} = 1.68$, $\eta_0 = 4.337$, $\beta = 0.012$.

The turbulent model can apply to computer compressible fluid, which needs to reflect the change of density. Adding new source term to κ equation and ε equation can be written as follows:

$$S_k = -\frac{2}{3}\rho k\nabla \cdot \bar{u} - \frac{2}{3}\mu_l\,[\nabla \cdot \bar{u}]^2 \tag{9}$$

$$S_\varepsilon = -\left[\frac{2}{3}C_1 - C_2\right]\rho\varepsilon\nabla \cdot \bar{u} - \frac{1}{3}C_1\rho\,[\nabla \cdot \bar{u}]^2 \tag{10}$$

Centrifugal forces of rotary engine affect fluid dynamics. When centrifugal acceleration is parallel to the solid wall of the increasing boundary layer, secondary flow happens in field parallel to the boundary layer. When centrifugal acceleration is perpendicular to the solid wall, the structure of turbulence is affected; meanwhile, eddy-dissipation rate increases and turbulence energy reduces. Therefore, based on modified method of Buoyancy flow, eddy-dissipation rate can be modified as follows [46–48]:

$$S_\varepsilon' = -\frac{\varepsilon}{k}\left[C_1 G_k\left(1 + C_3 R_f\right) - C_2 \rho \varepsilon\right] \tag{11}$$

where $R_f = 0.5\rho v(rw)^2/G_k$, $G_k = 2\mu_l \overline{S_{ij}S_{ij}}$, $\mu_l = \rho C_\mu k^2/\varepsilon$, $C_\mu = 0.0845$, $C_1 = 1.44$, $C_2 = 1.92$, $C_3 = -1$.

4.2. Equivalent Modeling Method

It is known from engine principles that the combustion model of chamber space excites three dynamic boundaries. In this paper, the rotor was set as a fixed reference system. The cylinder revolves around the center point of the rotor, and the leaf spring is compressed or extended by force. This method can save on simulation calculation costs. A sharp corner excites between leaf spring and rotor under ideal condition due to a narrow chamber. A sharp corner is not helpful for CFD mesh generation and combustion computation. The contact area for the spring and cylinder is not constant and deflection value near the fixed end is less than one the value near the free end. Thus it is able to consider the leaf as two parts. The one part is incorporated into the cylinder, and the other part is considered the linear segment. The two lines represent reeds; the upper line is cylinder profile, while the lower line is the thermal profile of the rotor, as shown in Figure 3. Length of the epitrochoidal increase or decrease depends on change of chamber volume. The angle between spring and rotor changes from sharp to large acute angle. Therefore, ensuring the accuracy of computation and reducing the difficulty of meshing reduce the calculation costs.

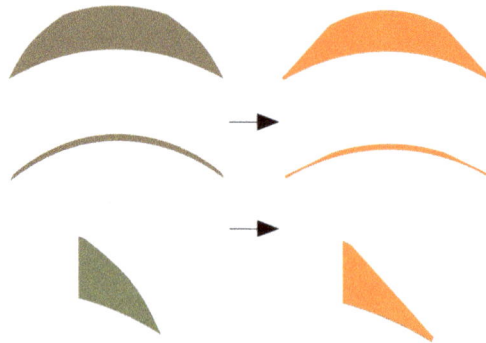

Figure 3. Equivalent model of chamber.

The mesh density was finalized at 42,640 grids and 46,848 hexahedral elements (Figure 4a). Both of the springs and the cylinder profile constitute dynamic boundaries; the rotor profile constitutes a static boundary. The contrast diagram shown in Figure 4b is given for CFD simulation and theoretical calculation. It can be seen from Figure 4b that equivalent modeling method is valid and reasonable. Chamber volume using mathematical method is very accurate and credible because the structure of the chamber is relatively fixed without considering the effects of temperature on material distortion.

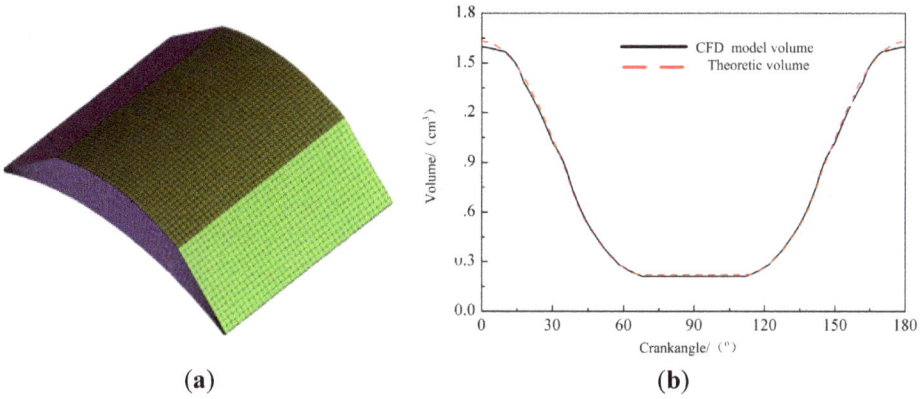

(a)   (b)

Figure 4. (a) Chamber mesh model; (b) Contrast diagram for CFD simulation and theoretical calculation.

5. Analysis of Combustion Process

5.1. Initial Conditions and Boundary Conditions

In present work, intake parameters were imposed on the entire computational domain as initial conditions (shown in Table 1) for numerical simulation. In order to take into account combustion characteristics, the effect of heat transfer between the outer wall and air is not considered. Thus, wall boundary is in adiabatic condition. Parts temperatures are estimated by the iteration method, which is applied on the boundary of fluid–solid coupling [49]. Crankangle is at 90°, corresponding to the ignition start.

Table 1. Initial conditions and boundary conditions.

Parameter	Value	Unit
Combustion Duration	3.2	ms
Inlet pressure	0.101325	MPa
Inlet Temperature	300	K
Excess air ratio	1	—
Cylinder Wall Temperature	380	K
Cover Surface Temperature	360	K
Rotor Surface Temperature	450	K
Spring Surface Temperature	400	K
Ignition Time	90	°
Fuel LHV	44200	KJ/Kg

5.2. Numerical Simulation of Combustion Process

Numerical simulation of combustion process is helpful to further master combustion mechanism. Temperature is one of the most important parameters to characterize a combustion process. Therefore, choosing the temperature field (shown in Figure 5) represents the combustion process of the engine.

At the primary stage, constant-volume combustion happens. Additionally, flame kernel spins with engine rotation. When counter-clockwise is set as a reference system, unburned mixture ahead of the flame flows in the same direction of the flame and unburned mixture behind the flame diffuses in another direction. In others words, there are more mixtures into flame front at the same moment. It is found that flame backwards mixture propagation speed is higher than anterior mixture at 0.8 ms. Propagation speed of flame with calabash shape was much smaller than rotational speed (shown in Figure 5b). Before 1 ms, flame in the region between the rotor and fire surface cannot propagate to the reed. Thus, heat load on the spring is relatively small. After 1 ms, flame propagation speed started to increase and temperature of mixture closed to pre- and post-springs keeps rising. Elastic energy of the front reed was released, which led to volume increasing close to the front reed. Therefore, flame propagation speed started to increase at anticlockwise direction under the action of pressure difference and a new flame kernel was formed at the center of the front reed (shown in Figure 5d). After that, combustible mixture was full combustion so that chamber temperature continuously increased up to 3058 K (Figure 5e,f). The region of maximum temperature concentrated in the middle of the cylinder and the rotor.

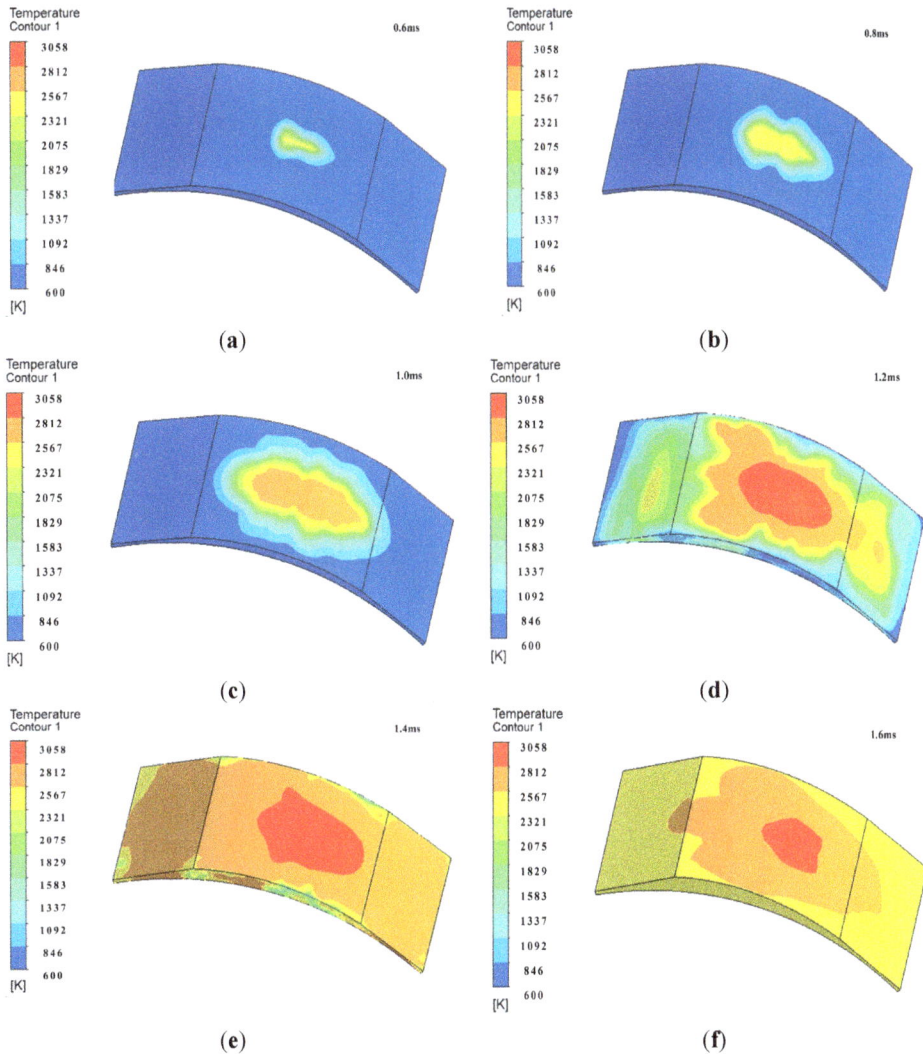

Figure 5. (**a**) Temperature field at 0.6 ms; (**b**) Temperature field at 0.8 ms; (**c**) Temperature field at 1.0 ms; (**d**) Temperature field at 1.2 ms; (**e**) Temperature field at 1.4 ms; (**f**) Temperature field at 1.6 ms.

5.3. Verification of Heat Release Rate

Based on CFD simulation of engine combustion, thermodynamic release model was able to verify by CFD simulation results. Therefore, the results of thermodynamic simulation were feasible and reliable. Heat release model adopted the wiebe release model [50] in this paper. Thermodynamic analysis had been made in a previous paper [51]. In order to better contrast, the same initial conditions were employed in

499

adiabatic CFD model and thermodynamic release model. That is, engine speed was 3000 r/min, compression beginning pressure was 0.1013 MPa, initial temperature was 305 K, excess air coefficient was 1, wall was adiabatic, and ignition timing was 90° crankangle. Heat release model was checked by modifying weight coefficient m and empirical coefficient a. In the present work, m and a were 3.9 and 3, respectively (shown in Figure 6).

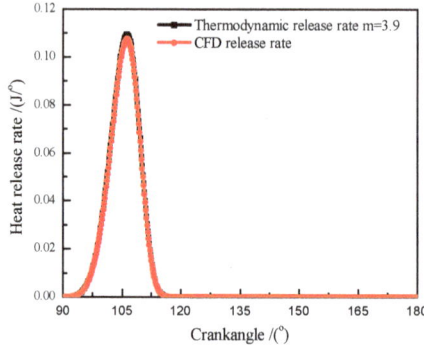

Figure 6. Verification of the thermodynamic model.

6. Results and Discussion

Combustion characteristic of leaf spring rotary engine can be investigated by mixture pressure, heat release rate, burning duration, maximum pressure rise rate and so on. Parameters for engine speed, excess air coefficient, initial pressure and initial temperature are important to combustion simulation. In this paper, the combustion characteristic and combustion processes at different initial conditions were investigated by CFD model. Combustion initial angle is compression end angle while simulating combustion process.

6.1. Speed

The effect of engine speed on chamber pressure, maximum pressure rise rate, heat release rate and burning duration seems to be negligible within the range it has been simulated. For air–fuel mixture, it is shown that four different speeds were simulated and the lower the speed, the higher the maximum pressure (Figure 7a). Lines connecting symbols are only for the sake of visualization. The increase of the engine speed results in pressure peak moving toward the larger crankangle. When the crankangle is higher than 120° and engine is at low rotational speed, the curve of pressure decreased to the same level due to the timing increase of constant volume combustion. In order to understand this result, maximum pressure rise rate can be examined (Figure 7b). Maximum pressure rise rate decreases with engine speed increase. The value changes from 0.47 Mpa/°CA at 1200 r/min to 0.34 Mpa/°CA at 3000 r/min, decreasing by 27.7%.

With the increase of the engine speed, the curve of heat release rate moved toward the right and covers a wider range of crankangle (Figure 7c). The peak value decreases by 65% when engine speed changes from 1200 r/min to 3000 r/min. With the same excess air ratio condition, the increase of engine speed is helpful for reducing combustion duration (Figure 7d), $i.e.$, flame propagation speed is proportional to engine speed. The reason is that turbulent flow in chamber results in the increase of flame wrinkle due to increasing speed so that unburned mixture into flame speeds up. Furthermore, it can be seen from Figure 7d that combustion duration increases with the rising of excess air ratio for the same engine speed and the time interval becomes large.

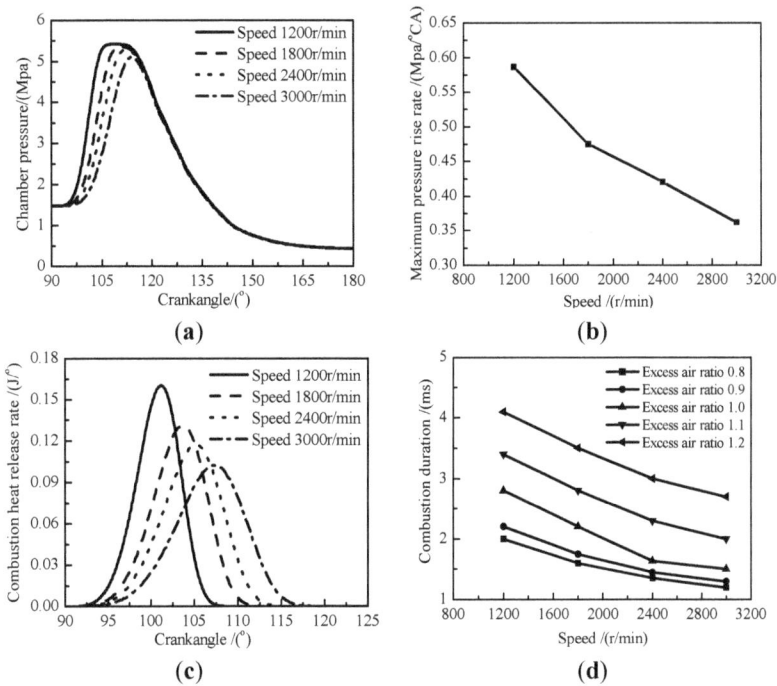

Figure 7. (**a**) Chamber pressure for different speed; (**b**) Maximum pressure rise rate for speed. (**c**) Combustion heat release rate for different speed; (**d**) Combustion duration for different excess air ratio and speed.

6.2. Excess Air Ratio

With the same engine speed for 3000 r/min and fully-opened throttle, peak pressure in chamber decreases with excess air coefficient increase (Figure 8a).

Based on an excess air coefficient of 1, the proportion of peak pressure toward the two sides is increasing. The peak pressure at excess air ratio 1.2 is obviously lower than the one at excess air ratio 1.1. With the increase of the excess air ratio, peak pressure moves toward the right and maximum pressure rise rate decreases from 0.63 MPa/°CA to 0.15 MPa/°CA (Figure 8b). The whole curve is on the decline and the trend becomes slow near an excess air ratio 1.0. Figure 8c shows that as crankangle increases, the difference among several excess air ratios becomes larger. The excess air ratio 1.2 takes much longer distance to reach its peak heat release rate. In addition to the significant difference in terms of chemical kinetics, mass ratio also plays an important role in yielding this result. Combustion duration and excess air ratio profiles are plotted in Figure 8d. It can be seen that burning duration is gradually increasing among the range.

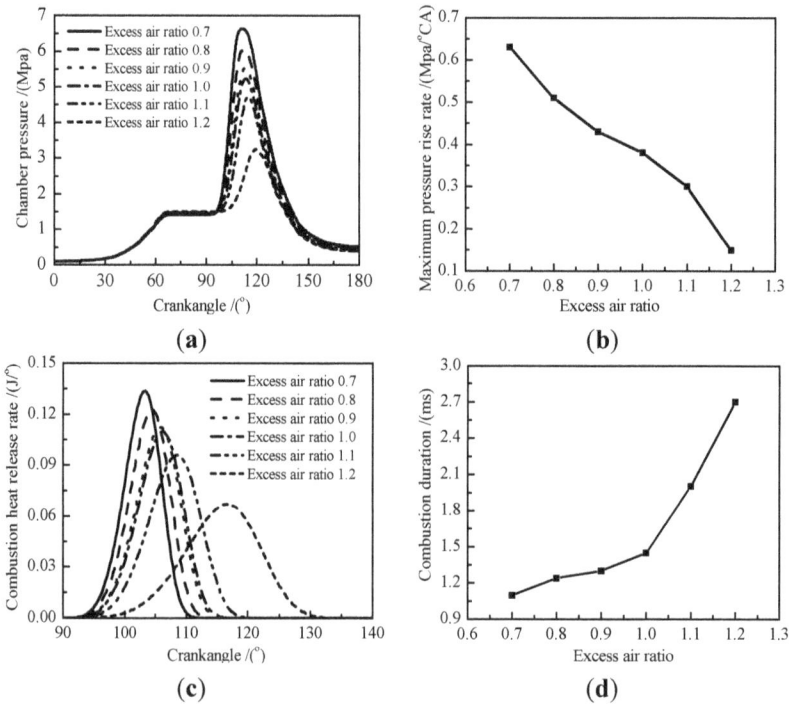

Figure 8. (a) Chamber pressure for excess air ratio; (b) Maximum pressure rise rate for excess air ratio; (c) Combustion heat release rate for excess air ratio; (d) Combustion duration for different excess air ratio.

Figure 9 illustrates the reaction process of the chamber for different excess air ratios. Under the conditions of same excess air ratio, flame varied from laminar to turbulence with the development of combustion. In Figure 9a, flame showed obvious laminar characteristic at 1.1 ms. However, turbulence flame happened at 1.4 ms until the end of combustion due to the changes of chamber volume and gases temperature. Decreased excess air ratio increased the variation tendency and reduced the time from laminar to turbulence. It is well established that propagation speed of flame increases with decreasing excess air ratio, but increasing degree becomes smaller and smaller. Figure 9c,d shows that the reaction process is nearly the same.

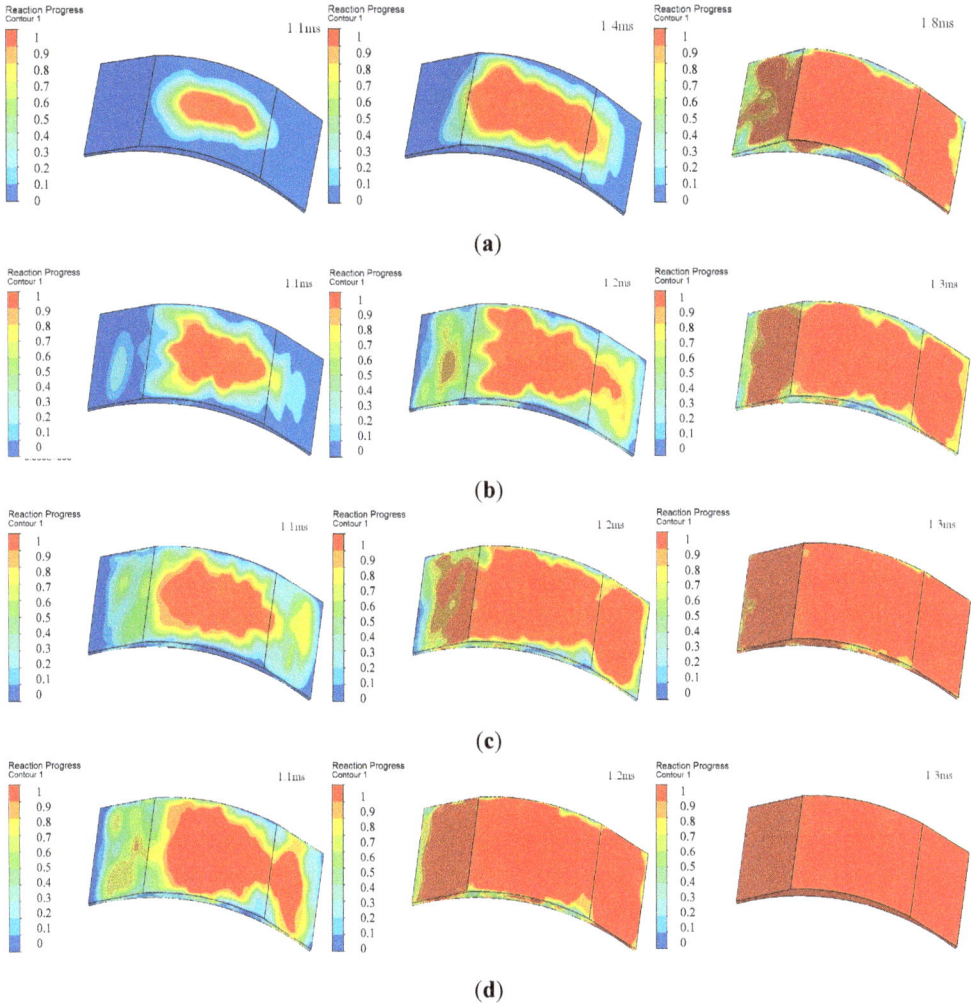

(a)

(b)

(c)

(d)

Figure 9. (a) Reaction process at excess air ratio 1.1; (b) Reaction process at excess air ratio 1.0; (c) Reaction process at excess air ratio 0.9; (d) Reaction process at excess air ratio 0.8.

6.3. Initial Pressure

Initial pressure is another factor to affect engine combustion. No significant difference in terms of chamber pressure is found over the entire crankangle from 90° to 180° (Figure 10a). However, it can be seen that initial pressure affects the position where the pressure is up to peak. The difference increases with the increasing pressure. The pressure position is essentially an overall result of gas quality and heat transfer loss. Gas quality increase leads to the increase of initial pressure. Although heat was lost in the chamber, the effect of heat loss due to higher increase of pressure caused by gas quality can be negligible. Figure 10b shows that maximum pressure rise rate against initial pressure. It can be seen that it increases with the increasing initial pressure. The results attribute to increasing flame propagation speed caused by temperature rising. Maximum pressure rise rate increased from 0.345 Mpa/°CA to 0.435 Mpa/°CA. the difference for several initial pressures cannot be clearly identified (Figure 10c). This can be seen as the heat release rate slightly increasing as initial pressures increases and the peak shifts to the leaf. Meanwhile, combustion duration decreases from 1.62 ms to 1.51 ms and the trend slows down, as shown in Figure 10d, which indicates that flame propagation speed is enhanced.

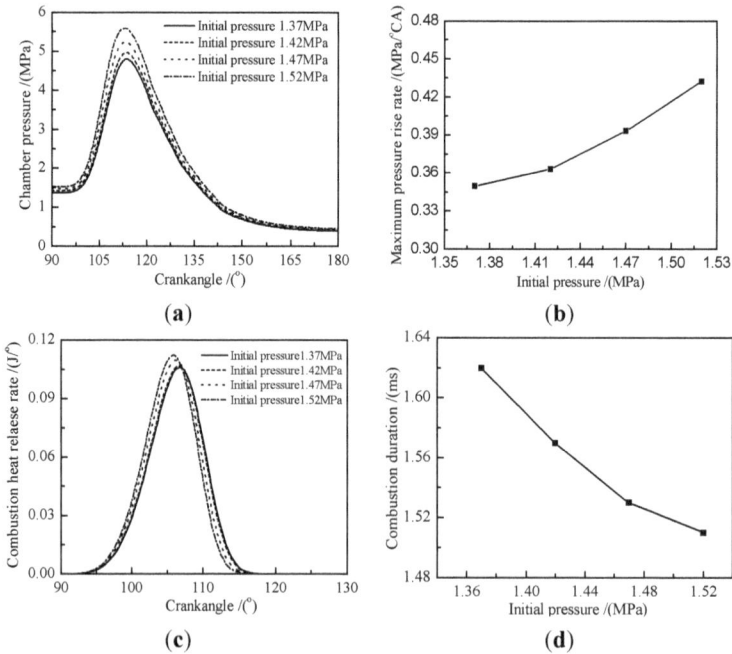

Figure 10. (a) Chamber pressure for initial pressure; (b) Maximum pressure rise rate for initial pressure; (c) Combustion heat release rate for initial pressure; (d) Combustion duration for different initial pressure.

Figure 11 illustrates the reaction process of chamber for different initial pressures. Under the conditions of same initial pressure, combustion characteristic relies mainly on laminar with the development of subsidiary wrinkled turbulence. After 1.1 ms, turbulence gradually increased. The reason is that increased pressure leads to the increase of flame propagation speed. Another reason is variation of chamber volume. Reaction process was nearly the same, at 1.1 ms. After that, turbulence characteristic for different pressure was more obvious, and the more a combustion center started to appear. When comparing the reaction process for initial pressures of 1.47 MPa and 1.52 MPa, combustion completeness is mostly the same.

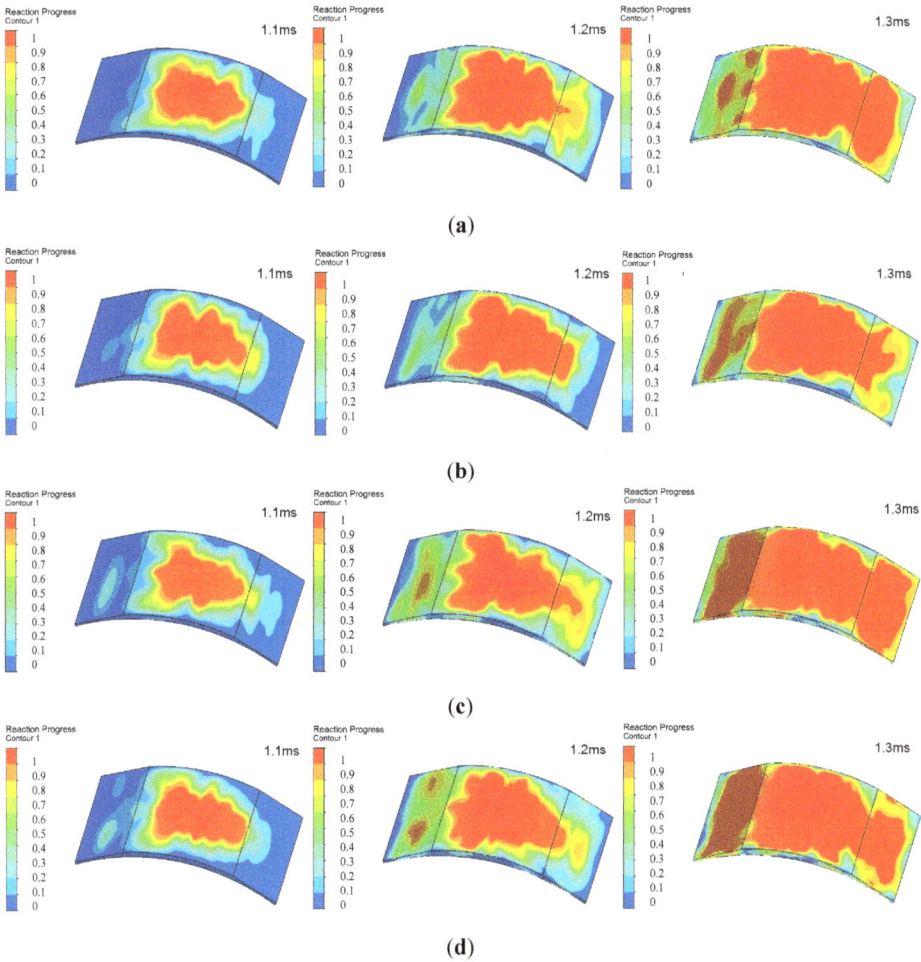

Figure 11. (**a**) Reaction process at pressure 1.37 MPa; (**b**) Reaction process at pressure 1.42 MPa; (**c**) Reaction process at pressure 1.47 MPa; (**d**) Reaction process at pressure 1.52 MPa.

6.4. Initial Temperature

The effect of initial temperature on combustion characteristics seems to be non-negligible with the range of combustion process. Taking into account the fact that the initial density of fuel/air mixture decreases as initial temperature increases, it is clear that mixture mass is inversely proportional to initial temperature under the same conditions, including engine speed 3000 r/min, excess air ratio 1 and initial pressure 1.47 Mpa. Figure 12a,b shows that chamber pressure and pressure rise rate increase as initial temperature increases. This is due to mass of fuel/air mixture decreases as initial temperature increases without other influence factors. Figure 12c,d shows that the difference for chamber pressure and heat release rate become unclear as initial temperature increases. At the same time, the peak moves toward the left. Heat release rate increases from 0.08 J/°CA at 580 K to 0.105 J/°CA at 620 K. Figure 12d indicates that flame propagation speed increases as initial temperature increases.

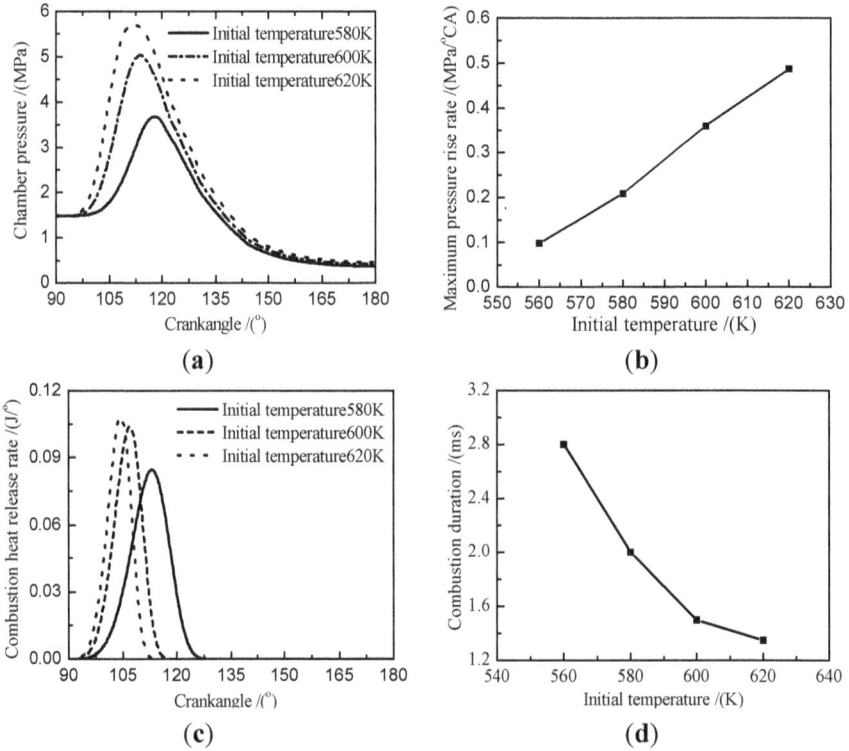

Figure 12. (a) Chamber pressure for initial temperature; (b) Maximum pressure rise rate for initial temperature; (c) Combustion heat release rate for initial temperature; (d) Combustion duration for different initial temperature.

Figure 13 illustrated reaction process of chamber for different initial temperature. Under the conditions of the initial temperature 580 K, combustion characteristic was laminar at 1.1 ms, after that combustion characteristic relied mainly on laminar with the development of subsidiary turbulence, turbulence continues to increase, but flame propagation speed is slower. When initial temperature was up to 600 K, combustion completeness was 0.4 ms earlier than 580 K, but was 0.1 ms longer than 620 K.

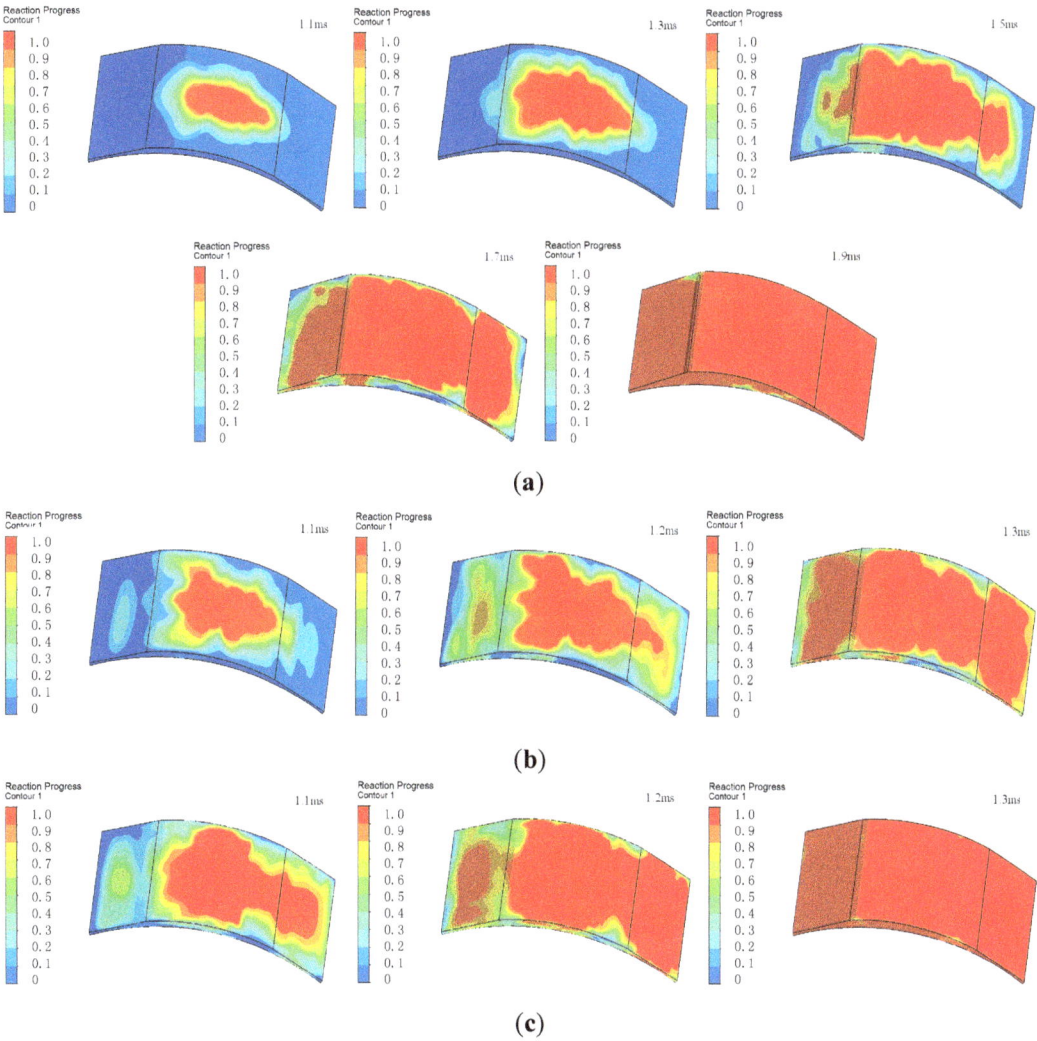

(a)

(b)

(c)

Figure 13. (**a**) Reaction process at temperature 580 K; (**b**) Reaction process at temperature 600 K; (**c**) Reaction process at temperature 620 K.

7. Conclusions

In this work, a novel leaf spring rotary engine as a new research object was used to investigate the combustion properties inside a small-sized engine. A 3D numerical model is developed to study the influence of several operative parameters on combustion characteristics. An equivalent modeling method was presented to simulate real boundary conditions. Engine speed, excess air ratio, initial pressure and initial temperature were controlled to simulate the engine operating conditions.

(1) In the process of compression, fluid motion is mainly affected by deformation of the reed. When deforming, swirl is formed near to the spring. After constant volume burning, fluid motion mainly manifested as laminar characteristics.

(2) In the process of combustion, the increase of speed makes reduction of burning duration, sustainable heat release angle increase, peak pressure and release rate decrease; when excess air ratio decreases, burning duration reduces, the curve of heat release moves to the direction decrease of crankshaft, peak pressure and release rate increases.

(3) With the same temperature condition, when initial pressure increases, burning duration reduces but the amplitude is smaller, peak pressure and release rate increases. Under different temperature conditions, the trend of burning duration is the same as the one for different pressure conditions. However, as the heat release curve and pressure shift to near the top dead center, the pressure increases at the same time.

Author Contributions: Yan Zhang, Zhengxing Zuo and Jinxiang Liu checked and discussed the simulation results. Yan Zhang confirmed the series of simulation parameters and arranged and organized the entire simulation process. Zhengxing Zuo participated in establishing the simulation model. Jinxiang Liu revised the paper. Zhengxing Zuo and Jinxiang Liu made many useful comments and simulation suggestions. In addition, all authors reviewed the manuscript.

Conflicts of Interest: The authors declare no conflict of interest.

Nomenclature

x_1	Cross point coordinate of l_2 and abscissa axis
x_2	Cross point abscissa of l_2 and l_1
x_3	Cross point coordinate between l_3 and abscissa axis
x_4	Cross point abscissa of l_3 and l_4
x_5	Cross point abscissa between l_1 and l_4
M	Flexural torque at the free end [N·m]
p'	Concentrated load at the free end [N]
EI	Bending rigidity of the leaf spring
l	Horizontal displacement of the free end [m]

L	Integration length of leaf spring [m]
L_0	Actual length of leaf spring [m]
y	Flexibility of leaf spring [m]
$W(x)$	Function of horizontal displacement l of the free end
ρ	Density [kg/m^3]
u_i	Velocity of i direction [m/s]
C_μ	Specific heat
k	Kinetic energy of turbulent fluctuation [J]
ε	Dissipation ratio of kinetic energy for turbulent fluctuation [%]
G_k	Kinetic energy of turbulent fluctuation for buoyant force [J]
$\alpha_k,$	Prandtl number
$\overline{\alpha}_\varepsilon$	Mean velocity [m/s]
μ_{eff}	Effective viscosity coefficient
R_f	Richardson number
x_1	Cross point coordinate of l_2 and abscissa axis

References

1. Hamilton Walker Rotary Engine. *Weekly News* **1968**, 7.
2. Walker, D.F.; Hamilton, P.H. A new class of rotary piston suitable for compressors, pumps, and internal combustion engines. *Proc. Inst. Mech. Eng.* **1972**, *186*, 743–753.
3. Quasiturbine Stirling engine, Stirling engine idea on the Quasiturbine website. Available online: https://en.wikipedia.org/wiki/Quasiturbine (accessed on 1 April 2015).
4. Gilles, S.H.; Roxan, S.H.; Ylian, S.H. Quasiturbine: Low RPM high torque driven turbine for top efficiency power modulation. *ASME Turbo Expo* **2007**, *3*, 17–26.
5. Heiser, W.H.; Pratt, D.T. *Hypersonic Airbreathing Propulsion*; American Institute of Aeronautics and Astronautics (AIAA): Reston, VA, USA, 1994.
6. Schapiro, B.; Terlitesky, L. The RKM (RKM) rotary piston machines with the jumping momentously Axis. In Proceeding of the International Conference on Sustainable Automotive Technology 2008, Melbourne, Australia, 4–9 November 2008.
7. Janusz, P.; Dawid, D. Numerical investigation of the wave disk micro-engine concept. *Int. J. Gas Turbine Propuls. Power Syst.* **2008**, *2*, 1–8.
8. Fernandez-Pello, A.C. Micropower generation using combustion: Issues and approaches. *Proc. Combust. Inst.* **2002**, *29*, 883–899.
9. Sirignano, W.A.; Pham, T.K.; Dunn-Rankin, D. Miniature-scale liquid-fuel-film combustor. *Proc. Combust. Inst.* **2002**, *29*, 925–931.
10. Zamaschikov, V.V. Combustion of gases in thin-walled small diameter tubes. *Combust. Explos. Shock Waves* **1995**, *131*, 10–16.

11. Mehra, A.; Waitz, I.A. Development of a hydrogen combustor for a microfabricated gas turbine engine. In Proceeding of the Solid-State Sensor and Actuator Workshop, Hilton Head, SC, USA, 8–11 June 1998.

12. Mehra, A.; Zhang, X.; Ayon, A.A.; Waitz, I.A.; Schmit, M.A.; Spadaccini, C.M. A six-wafer combustion system for a silicon micro gas turbine engine. *J. Microelectromech. Syst.* **2000**, *9*, 517–527.

13. Epstein, A.H.; Senturia, S.D.; Al-Midani, O.; Anathasuresh, G.; Ayon, A.; Breuer, K.; Chen, K.-S.; Ehrich, F.E.; Esteve, L.; Frechette, G.; *et al.* Micro-heat engines, gas turbines and rocket engines—The MIT micro engine project. In Proceedings of the AIAA 28th Fluid Dynamics Conference, Snowmass, CO, USA, 29 June–2 July 1997.

14. Epstein, A.H.; Senturia, S.D.; Ayon, A.; Breuer, K.; Chen, K.-S.; Ghodssi, R.; Jacobson, S.; Lang, J.; Nagle, S.; Orr, D.; *et al.* Power MEMS and Microengines. In Proceeding of the International Conference on Solid State Sensors and Actuators, Chicago, IL, USA, 16–19 June 1997.

15. Fu, K.; Knobloch, A.J.; Martinez, F.C. Design and experimental results of small-scale rotary engine. In Proceeding of the ASME International Mechanical Engineering Congress and Exposition, New York, NY, USA, 11–16 November 2001.

16. Ochoa, F.; Eastwood, C.; Ronney, P.D.; Dunn, B. Thermal transpiration based microscale propulsion and power generation devices. In Proceeding of the 7th International Microgravity Combustion Workshop, Cleveland, OH, USA, 4 June 2003.

17. Maruta, K.; Takeda, K.; Ahn, J.; Borer, K.; Ronney, P.D.; Deutschmann, O. Extinction limits of combustion in microchannels. In Proceeding of the 29th International Syposium on Combustion, Sapporo, Japan, 21–26 July 2002.

18. Dahn, W.J.; Ni, J.; Mijit, K.; Mayor, J.R.; Qiao, G.; Benajmin, A.; Gu, Y.; Lei, Y.; Papke, M. Micro internal combustion swing enigne (MICSE) for portable power generation system. In Proceedings of the 40th AIAA Aerospace Sciences Meeting, Reno, NV, USA, 14–17 January 2002.

19. Aichlmayr, H.T.; Kittelson, D.B.; Zachariah, M.R. Design Consideration, Modeling, and Analysis of Micro-Homogeneous Charge Compression Ignition Combustion Free-Piston Engine. Ph.D. Thesis, The University of Minnesota, St. Paul, MN, USA, 2002.

20. Minotti, K.; Sciubba, E. LES of a Mesocombustion chamber with a detailed chemistry model: Comparison between the Flamelet and EDC Models. *Energies* **2010**, *3*, 1943–1959.

21. Yuasa, S.; Oshimi, K.; Nose, H.; Tennichi, Y. Concept and combustion characteristics of ultra-micro combustors with premixed flame. *Proc. Combust. Inst.* **2004**, *30*, 2455–2462.

22. Lee, K.H.; Hong, Y.T.; Kim, K.B.; Kwon, O.C. Stability limits of premixed microflames at elevated temperatures for portable fuel processing devices. *Int. J. Hydrog. Energy* **2008**, *33*, 232–239.

23. Maruta, K. Micro and mesoscale combustion. *Proc. Combust. Inst.* **2011**, *33*, 125–150.

24. Wang, H.O.; Luo, K.; Lu, S.Q.; Fan, J.R. Direct numerical simulation and analysis of a hydrogen/air swirling premixed flame in a micro combustor. *Int. J. Hydrog. Energy* **2011**, *36*, 3838–3849.

25. Bedr, A.L.; Rutigliano, M.; Balat-Pichelin, M.; Cacciatore, M. Atomic oxygen recombination on quartz at high temperature: Experiments and molecular dynamics simulation. *Langmuir* **2006**, *22*, 7208–7216.

26. Raimondeau, S.; Norton, D.; Vlachos, D.G.; Masel, R.I. Modeling of high-temperature microburners. *Proc. Combust. Inst.* **2002**, *29*, 901–907.

27. Karagiannidis, S.; Mantzaras, J.; Jackson, G.; Boulouchos, K. Hetero-/Homogeneous combustion and stability maps in methane-fueled catalytic microreactors. *Proc. Combust. Inst.* **2007**, *31*, 3309–3317.

28. Norton, D.G.; Vlachos, D.G. A CFD study of propane/air microflameStability. *Combust. Flame* **2004**, *138*, 97–107.

29. Norton, D.G.; Vlachos, D.G. Combustion characteristics and flame stability at the microscale: A CFD study of premixed methane/air mixtures. *Chem. Eng. Sci.* **2003**, *58*, 4871–4882.

30. Hua, J.S.; Wu, M.; Kumar, K. Numerical simulation of the combustion of hydrogen-air mixture in micro-scaled chambers. Part I: Fundamental study. *Chem. Eng. Sci.* **2005**, *60*, 3497–3506.

31. Hua, J.S.; Wu, M.; Kumar, K. Numerical simulation of the combustion of hydrogen-air mixture in micro-scaled chambers Part II: CFD analysis for a micro-combustor. *Chem. Eng. Sci.* **2005**, *60*, 3497–3506.

32. Kaisare, N.S.; Vlachos, D.G. Optimal reactor dimensions for homogeneous combustion in small channels. *Catal. Today* **2007**, *120*, 96–106.

33. Li, J.; Chou, S.K.; Li, Z.W.; Yang, W.M. A comparative study of H_2-air premixed flame in micro combustors with different physical and boundary conditions. *Combust. Theory Model.* **2008**, *12*, 325–347.

34. Li, J.; Chou, S.K.; Yang, W.M.; Li, Z.W. A numerical study on premixed micro combustion of CH_4-air mixture: Effect of combustor size, geometry and boundary conditions on flame temperature. *Chem. Eng. J.* **2009**, *150*, 213–222.

35. Lee, K.H.; Kwon, O.C. A numerical study on structure of premixed methane-air microflames for micropower generation. *Chem. Eng. Sci.* **2007**, *62*, 3710–3719.

36. Shih, H.Y.; Liu, C.R. A computational study on the combustion of hydrogen/methane blended fuels for a micro gas turbines. *Int. J. Hydrog. Energy* **2014**, *39*, 15103–15115.

37. Wan, J.L.; Yang, W.; Fan, A.; Liu, Y.; Yao, H.; Liu, W.; Du, Y.Q.; Zhao, D.Q. A numerical investigation on combustion characteristics of H_2/air mixture in a micro-combustor with wall cavities. *Int. J. Hydrog. Energy* **2014**, *39*, 8138–8146.

38. Sprague, S.B.; Walther, D.C.; Park, S.W.; Pisano, A.; Fernandez, A. Effect of leakage on optimal compression ratio for small-scale rotary engine. In Proceedings of the 45th AIAA, Reno, NV, USA, 8–11 January 2007.

39. Cawthorne, W.R.; Famouri, P.; Chen, J.D.; Clark, N.N.; McDaniel, T.I.; Atkinson, R.J.; Nandkumar, S.; Atkinson, C.M.; Petreanu, S. Development of a linear alternator-engine for hybrid electric vehicle applications. *IEEE Trans. Veh. Technol.* **1999**, *48*, 1797–1802.

40. Fu, K.; Knobloch, A.J.; Martinez, F.C.; Walther, D.C.; Fernandz-Pello, C.; Pisano, A.P.; Liepmann, D. Design and fabrication of a silicon-based MEMS Rotary Engine. In Proceeding of the 2001 ASME International Mechanical Engineering Congress and Exposition, New York, NY, USA, 11–16 November 2001.

41. Feng, H.H.; Song, Y.; Zuo, Z.Z.; Shang, J.; Wang, Y.D.; Anthony, P.R. Stable operation and electricity generating characteristic of a single-cylinder free piston engine linear generator: simulation and experiments. *Energies* **2015**, *8*, 765–785.

42. Scott, V.; Adam, S.; Jesse, R.; Jae, W.P. A numerical investigation on the efficiency of range extending systems using advanced vehicle simulator. *J. Power Sources* **2011**, *196*, 3360–3370.

43. Liu, J.Y.; Yuan, R. Rigid-flexible coupling dynamics with consideration of geometric nonlinearity and thermal effect. *Chin. J. Solid Mech.* **2008**, *29*, 72–76.

44. Wu, H.H.; Liu, H.; Yang, Z.Y.; Liu, P. Application of imitate linearity equal system in nonlinear problem of geometric large deformation. *J. Wuhan Univ. Technol.* **2008**, *32*, 963–966.

45. Wang, D.J.; Zuo, Z.X. Design and numerical simulation of leaf spring rotary engine. In Proceedings of the 2012 Asia-Pacific Power and Energy Engineering Conference, Shanghai, China, 27–29 March 2012; pp. 46–50.

46. Bradshaw, P. *Effects of Streamline Curvature on Turbulent Flow*; Advisory Group for Aerospace Research and Development: Neuilly sur Seine, France, 1973; pp. 169–173.

47. Moore, J. *Effects of Coriolis Forces on Turbulent Flow in Ro-tating Channels*; Gas Turbine Laboratory: Cambridge, MA, USA, 1967; pp. 89–90.

48. Sloan, D.G. Modeling swirl in turbulent flow. *Energy Combust. Sci.* **1986**, *11*, 163–260.

49. Li, Y. Simulation and Application of Solid-liquid Coupled Heat Transfer Internal Combustion Engines. Ph.D. Thesis, Zhejiang University, Hangzhou, China, 2006.

50. Shah, R.K. Advances in heat exchanger design. *Am. Soc. Mech. Eng.* **1986**, *66*, 112–122.

51. Zhang, Y.; Zuo, Z.X.; Yuan, C.H.; Wang, D.J. Analysis on performance of leaf spring rotary engine. *Energy Proced.* **2014**, *61*, 984–989.

Space Charge Behavior in Paper Insulation Induced by Copper Sulfide in High-Voltage Direct Current Power Transformers

Ruijin Liao, Ende Hu, Lijun Yang and Yuan Yuan

Abstract: The main insulation system in high-voltage direct current (HVDC) transformer consists of oil-paper insulation. The formation of space charge in insulation paper is crucial for the dielectric strength. Unfortunately, space charge behavior changes because of the corrosive sulfur substance in oil. This paper presents the space charge behavior in insulation paper induced by copper sulfide generated by corrosive sulfur in insulation oil. Thermal aging tests of paper-wrapped copper strip called the pigtail model were conducted at 130 °C in laboratory. Scanning electron microscopy (SEM) was used to observe the surface of copper and paper. Pulse electroacoustic (PEA) and thermally stimulated current (TSC) methods were used to obtain the space charge behavior in paper. Results showed that both maximum and total amount of space charge increased for the insulation paper contaminated by semi-conductor chemical substance copper sulfide. The space charge decay rate of contaminated paper was significantly enhanced after the polarization voltage was removed. The TSC results revealed that copper sulfide increased the trap density and lowered the shallow trap energy levels. These results contributed to charge transportation by de-trapping and trapping processes. This improved charge transportation could be the main reason for the decreased breakdown voltage of paper insulation material.

Reprinted from *Energies*. Cite as: Liao, R.; Hu, E.; Yang, L.; Yuan, Y. Space Charge Behavior in Paper Insulation Induced by Copper Sulfide in High-Voltage Direct Current Power Transformers. *Energies* **2015**, *8*, 8110–8120.

1. Introduction

Oil-paper is widely used as an insulating medium in various high voltage apparatuses, such as power transformers, capacitors, and HVDC devices, which play a vital role in electrical power networks [1,2]. Aging of oil-paper can eventually threaten the safe operation of electrical devices. The space charge will accumulate within the paper materials under the action of DC high voltage, leading to the distortion of internal local electrical field distribution [3,4]. Therefore, the aging speed of paper material is accelerated. Unfortunately, researchers have recently faced a new problem, namely, corrosive sulfur in insulating oil. Corrosive sulfur can form the semi-conductor substance copper sulfide, which will be deposited on the paper surface [5,6]. Insulating resistance and dielectric property of oil-paper insulation have decreased because of emergence of this substance. The understanding of space charge behavior in paper contaminated by copper sulfide is of significant interests.

The sources of corrosive sulfur in oil have not yet been completely identified, but the indications suggest that they are the residuals of the refining process [7]. The emergence of copper sulfide by-product caused by the chemical reaction between corrosive sulfur (mainly dibenzyl disulfide DBDS) in the insulation oil and copper conductor greatly change the dielectric characteristic of paper. The deposition of copper sulfide on the paper surface distorts the electric field, decreases the insulating resistance, and accelerates the aging of insulating materials or even causes direct breakdown [5,8].

The space charge behavior of aged oil is complicated because of the formation of polar substances such as water, acid, and organic molecules. The processes of charge injection, migration, and decay in paper material are directly influenced by these substances. However, to date, knowledge about the relationship between the copper sulfide deposition on paper and space charge behavior is still lacking. On the other hand, distorted electric field caused by the copper sulfide deposition plays a vital role in the decreasing breakdown voltage of insulation paper [9]. However, to the authors' knowledge, space charge behavior is likely to be one of the most fundamental reasons for the characteristic change in paper material. The insulation risk caused by copper sulfide from the perspective of micros should be determined. Thus, in this paper, we aimed to understand the effect of copper sulfide on the space charge behavior of accelerated thermal aging paper material, and analyze the failure of paper material caused by copper sulfide deposition.

2. Experiments

Selected corrosive sulfur DBDS was purchased from a local company. The insulation oil used in this experiment was 25# mineral oil without any corrosive sulfur or DBDS substance. The corrosivity of the oil was based on the procedures stipulated by IEC62535. Table 1 shows the typical properties of this 25# commercially available insulation oil.

Three oil samples were prepared: one was base oil (called sample I), which had no corrosive sulfur or DBDS. The second oil sample was called sample II, in which 500 ppm DBDS was added. The third oil sample was called sample III, in which 1000 ppm DBDS was added. Three oil samples with different DBDS concentrations were prepared. Paper-wrapped copper windings and these oil samples were dried at 90 °C/50 Pa and then finished the oil impregnation [10]. Windings and oil samples were thermally aged for 24 days at 130 °C to obtain three paper samples. One aged paper was uncontaminated by copper sulfide deposition, whereas the other two were contaminated by copper sulfide deposition with different levels of severity. The detailed information is listed in Table 2.

Table 1. Properties of tested mineral base insulation oil.

Characteristics	Items	Properties
Function	Density (20 °C), kg/m^3	883.2
	Viscosity (40 °C), mm^2/s	9.67
	Viscosity(−20 °C), mm^2/s	488.9
	Breakdown voltage, kV	62
Stability	Acidity, mg KOH/g	0.008
	Corrosivity	None

Table 2. Testing samples and condition of thermal aging tests.

Specimen	Description	Dibenzyl disulfide (DBDS) concentration (mg/kg)	Temperature (°C)
Sample 1	Oil + winding	0	130
Sample 2	Oil + winding + DBDS	500	130
Sample 3	Oil + winding + DBDS	1000	130

The paper-wrapped copper windings were provided by Chongqing ABB Transformer Co., Ltd. (Chongqing, China). Figure 1a shows the winding, and copper strip is wrapped by insulation paper. The windings were processed to pig-tail model and each model consists of two windings, the dimension of this model is shown in Figure 1b. The length of the winding is 75.00 mm and the tail at the end of winding is 15.00 mm.

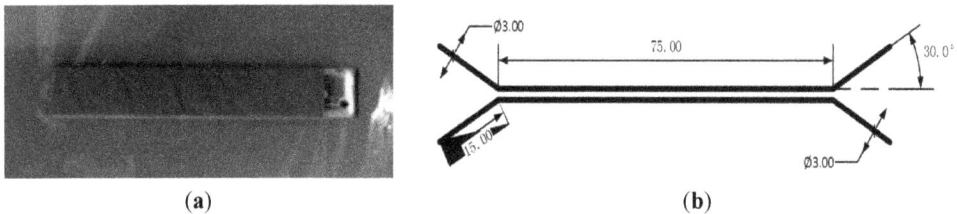

(a)　　　　　　　　　　　　　　　　(b)

Figure 1. Paper-wrapped copper winding and dimension pig-tail model. (a) winding; (b) dimension of pig-tail model.

The alternating current (AC) breakdown voltage of oil samples was obtained according to the IEC 60156 standard. Partial discharge inception voltages (PDIV) of oil-impregnated papers were measured based on the pig-tail model shown in Figure 1b. One of the windings in the pig-tail model was connected with high voltage supply (AC) through the hole, which was located in the end of the winding, and another winding was connected with the ground electrode. The pig-tail model was immersed in insulation oil during the PD test.

Scanning electron microscopy (SEM) was used to analyze the deposition on the paper surface. The space charge behaviors in paper materials were measured by PEA method based on previous studies [11,12]. Trap characteristics of paper samples were tested using TSC method. The setup and operating procedure of PEA and TSC are described in detail in previous publications [13].

3. Results

3.1. Appearance of Copper Strip and Paper

The comparison of three samples, including the appearance of copper and paper surface, is shown in Figure 2. No deposition was formed on the paper surface in sample I, as shown in Figure 2a. The paper surface was clear without any contamination. However, some silvery white deposition could be clearly observed in Figure 2b. The copper surface became darker compared with that in Figure 2a, and some parts of the paper surface were covered by deposition. For sample III (Figure 2c), the surface copper became darker than the former two copper samples. The whole surface of paper was covered by deposition.

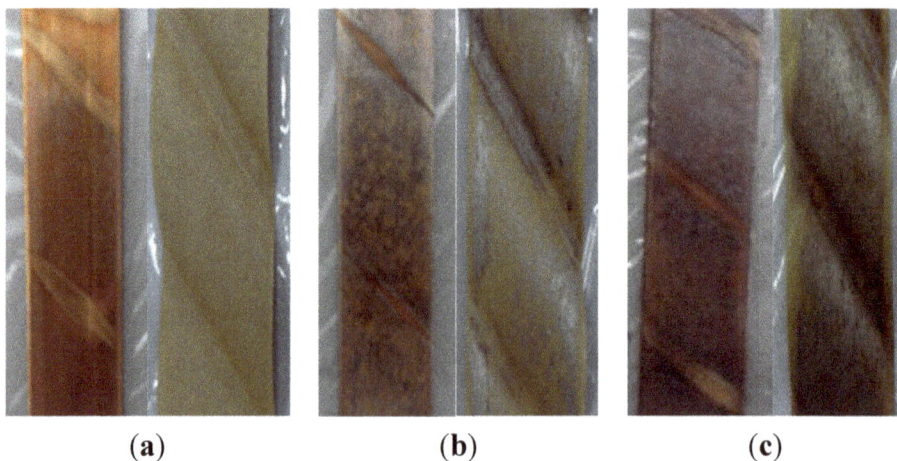

(a)   **(b)**   **(c)**

Figure 2. Appearance of copper and paper surface of samples (a) sample I; (b) sample II; and (c) sample III.

3.2. Scanning Electron Microscopy (SEM) of Copper Strips and Paper

SEM images of the three copper strip samples are shown in Figure 3. Lines were clear on the copper surface in sample I, as shown in Figure 3. The copper strip had a flat surface without bulges. The surface of the copper strip of sample II, which was contaminated by corrosive sulfur, became rough and numerous granular sediments were present on the copper surface. White particles appeared and lines were covered by particles. Nevertheless, the contamination condition in sample III was the most serious one. The SEM images of the three paper samples at a magnification of hundreds of times (10 μm) are shown in Figure 3. The fiber texture of paper was clearly visible in paper sample I. By contrast, the fiber texture of paper sample II was not easy to visualize, because deposition appeared on the paper surface. This appearance became fuzzy and more serious in paper sample III. The gaps between fibers were filled with white particles and some fibers were covered by particles.

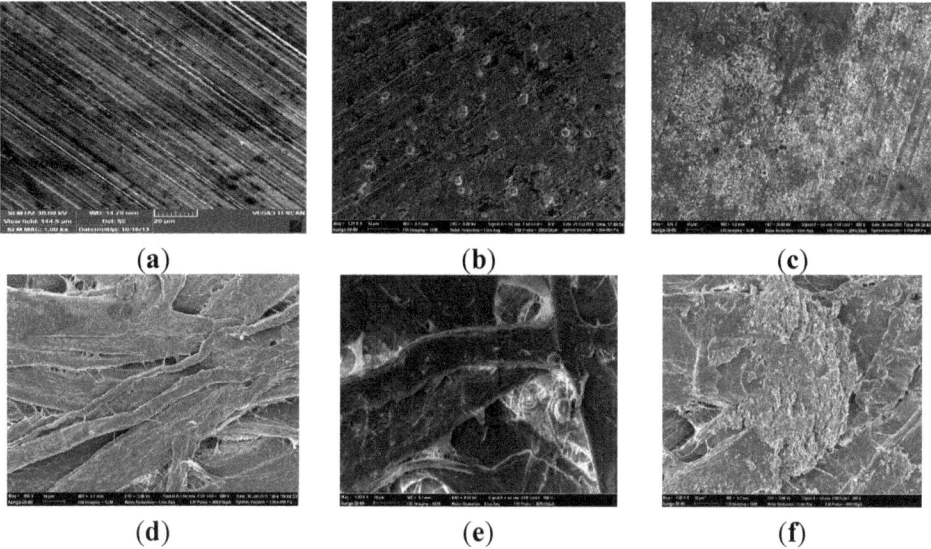

| (a) | (b) | (c) |

| (d) | (e) | (f) |

Figure 3. Scanning electron microscopy (SEM) of copper strip surface: (**a**) Sample I; (**b**) Sample II; (**c**) Sample III and paper surface: (**d**) Sample I; (**e**) Sample II; (**f**) Sample III.

3.3. Electrical Properties of Insulation Paper

Most studies have proven that the deposition on the copper surface and paper surface is copper sulfide substance (mainly Cu_2S) [14]. The electrical properties of the paper material changed because of the emergence of this kind of substance as result of its semi-conductive nature [15]. SEM and energy dispersive X-ray (EDX) spectrum were used to analyze this substance, and the results are shown in Figure 4 (working distance was kept at 1 μm, magnification = 4.66 K×). As shown in Figure 4a, the copper surface was covered by a large number of granules, the element composition of which was confirmed by EDX (Figure 4b). Figure 4c shows that the atomic number between Cu and S was 39.70/11.22, which is close to 4:1 (Cu_2S:4:1). Another study [16] also pointed out that the deposition is mainly Cu_2S, which contains small amounts of CuO and CuS.

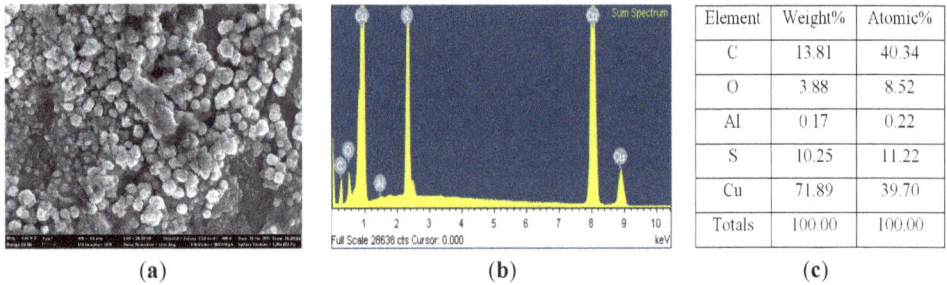

Element	Weight%	Atomic%
C	13.81	40.34
O	3.88	8.52
Al	0.17	0.22
S	10.25	11.22
Cu	71.89	39.70
Totals	100.00	100.00

(a) (b) (c)

Figure 4. (a) SEM image; (b) energy dispersive X-ray (EDX) analysis; and (c) element composition of copper surface.

The breakdown strength of oil-impregnated paper wrapped copper winding with and without copper sulfide deposition is shown in Table 3. Compared with paper-wrapped copper winding of sample I, the breakdown voltage of oil and PDIV of the winding of sample II decreased by 4.6% and 21.73%, respectively. Nevertheless, the degree of decrease for sample III was more obvious. The breakdown and partial discharge inception voltage of sample III decreased by 14.45% and 42.86%, respectively. The variation in breakdown voltage and PDIV indicated that the presence of copper sulfide on paper contributed to the decrease in electrical properties, and this contribution may be caused by the semi-conductor properties of copper sulfide.

Table 3. Deposition effects on electrical strength of oils and paper-wrapped copper winding. PDIV: partial discharge inception voltages.

Items	Parameter	Sample I	Sample II	Sample III
Oil	Breakdown voltage (kV)	49.8	47.5	42.6
	Standard deviation (kV)	3.14	3.62	3.29
Paper	Partial discharge (PD) inception voltage (kV)	16.1	12.6	9.2
	PD pico-coulomb of PDIV (pC)	209	610	725

3.4. Space Charge Behavior of Insulation Paper

The charge transport characteristic in insulation paper will change with the introduction of copper sulfide. The PEA method was used to obtain the space charge behavior in paper. After samples were subjected to a DC electric field, the space charge distribution in paper samples is shown in Figure 5.

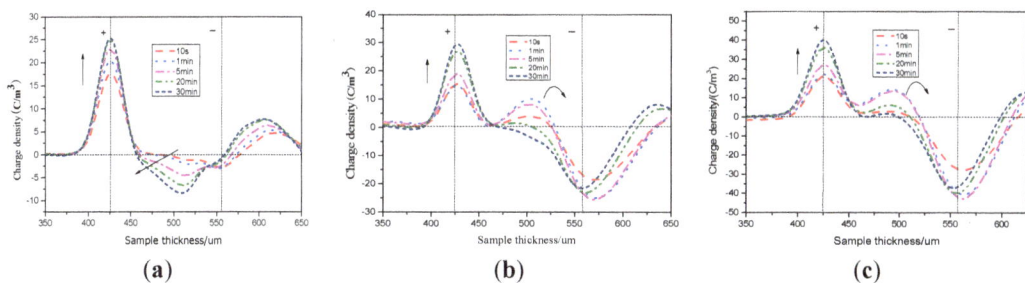

Figure 5. Volt-on space charge distribution of paper samples: (**a**) sample I; (**b**) sample II; and (**c**) sample III.

The anode peak was sharp and evident. As shown in Figure 5b, the peak at the anode electrode increased compared with that in Figure 5a. Copper sulfide deposition had a significant effect on the space charge behavior. The maximum space charge in paper sample II (Figure 5b: 29 C/m^3) was higher than that of sample I (Figure 5a: 24 C/m^3) when the same DC voltage was applied, implying that more charge carriers were formed in the paper contaminated by copper sulfide. The maximum space charge reached the highest value in sample III (Figure 5c: 40 C/m^3). Notably, for oil-paper contaminated by copper sulfide deposition, the positive charge injection behavior was different from that of paper without contamination. For the oil/paper sample without contamination, negative charges were injected into the inside of paper (around 500 μm) immediately once an electric field was applied. The amount of negative charge from this injection increased as the electric field applied. Therefore, most charges inside paper (around 500 μm) were negative, as shown in Figure 5a. For sample II, not only negative but also positive charge can be found inside of paper. A positive charge observed at the same position (500 μm) initially increased and then decreased after 1 min, as shown in Figure 5b. The positive charge bump increased, indicating that the positive charge injection was strengthened when paper was contaminated by copper sulfide. For sample III, a positive charge bump was found in the beginning of voltage application. However, the maximum positive charge density inside paper reached 15 C/m^3, which was higher than that in sample II (10 C/m^3). Results in Figure 5 demonstrated that higher deposition in paper, resulting in a more obvious positive charge injection.

Figure 6 shows the space charge distribution measured after the removal of the applied voltage for oil-paper samples with different contents of copper sulfide deposition. The maximum space charge in contaminated paper around the anode (about −5.5 C/m^3 as shown in Figure 6b) was higher than that of uncontaminated paper (about −2.5 C/m^3 as shown in Figure 6a). Nevertheless, the maximum space charge in sample III was the highest (about −11 C/m^3 as shown in Figure 6c). For all paper samples, a majority of space charge in the bulk disappeared after

5 min. The space charge decay of sample II was 2.85-fold of that in sample I after the external electrical field was removed for 5 min. Thus, the deposition on the paper could enhance the dissipation mobility of the charges.

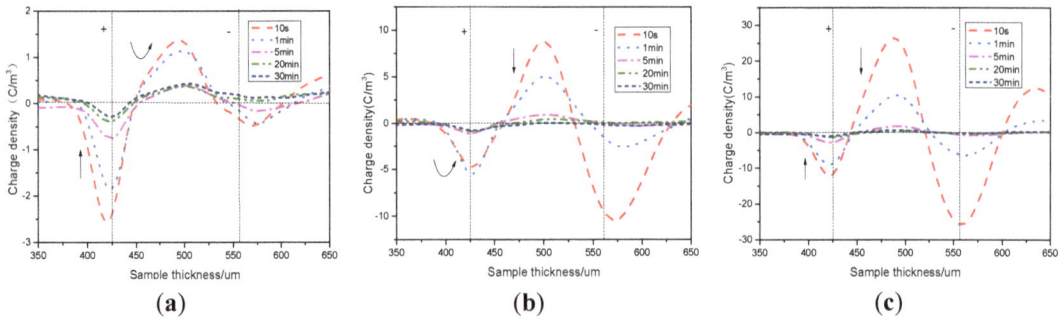

Figure 6. Volt-off space charge distribution of paper samples: (a) without contamination; (b) with contamination; and (c) with more contamination.

The decay speed of the three samples was determined using the following math model:

$$Q(t) = Q(0) + Ae^{-\frac{t}{\tau}} \tag{1}$$

where $Q(t)$ is space charge amount at time t, $Q(0)$ is the residual space charge amount when the decay is moving toward stability, and τ is the decay time constant. The relationship is shown in Table 4.

Table 4. Relationship between space charge amount and time for oil impregnated papers.

Sample	Formula	R-squared
I	$Q(t) = 72.3\exp(-t/4.1) + 27.5$	0.989
II	$Q(t) = 522.8\exp(-t/2.18) + 73.1$	0.985
III	$Q(t) = 1217.6\exp(-t/1.7) + 132.9$	0.971

The τ represents the decay speed of space charge. A low value τ results in higher decay speed of space charge. Thus, the space charge in contaminated paper had higher decay speed than that of uncontaminated paper.

4. Discussion

Copper sulfide deposition on paper exerted great effects on the electrical properties and space charge dynamics in oil-paper samples. The emergence of copper sulfide deposition on paper not only enhanced the charge accumulation but also improved the mobility of the charge in the samples. Oil-paper degradation caused by thermal aging is irreversible. The aging by-products deposited on the insulation paper cannot be removed. The decreased breakdown performance of sample II and sample III was largely attributed to the by-product copper sulfide (Cu_2S) because of the physical/chemical interaction between DBDS and copper, as shown in Figure 7 [14].

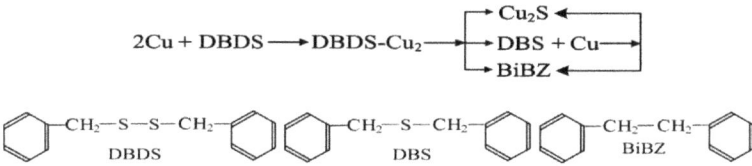

$$2Cu + DBDS \longrightarrow DBDS\text{-}Cu_2 \longrightarrow \begin{cases} Cu_2S \\ DBS + Cu \\ BiBZ \end{cases}$$

CH$_2$—S—S—CH$_2$ DBDS   CH$_2$—S—CH$_2$ DBS   CH$_2$—CH$_2$ BiBZ

Figure 7. Chemical structure of dibenzyl disulfide (DBDS) and the reaction between DBDS and copper conductor.

DBDS, is a corrosive sulfur that can attack copper windings under certain conditions. The DBDS-Cu complex, which is soluble in oil, formed. Part of the DBDS-Cu complex remained on the copper surface and then decomposed into Cu_2S. Portions of the complex transferred from the copper surface to the paper surface through bulk oil. Copper sulfide eventually formed on the paper surface.

Copper sulfide deposition can not only deposit on the paper surface but also permeate into paper [17]. The distribution of the electric field was distorted because of this deposition and permeation, leading to the downtrend of PDIV and breakdown voltage [9]. The decreased PDIV and breakdown voltage was related to space charge behavior. However, space charge distribution and decay are associated with trapping/de-trapping processes and charge transport in insulation paper materials [18,19]. It was reported that high conductivity leads to the high mobility of the charges trapped in the samples [20]. The Cu_2S could enhance the dissipation mobility of the space charges and mobility of charges in insulation paper because of its conductive characteristic. The higher this substance content in insulation paper, the faster the trapped charge moves. Thus, the trap characteristics of two samples (uncontaminated paper and contaminated paper) were tested to further investigate the charge behavior, and the results are shown in Figure 8.

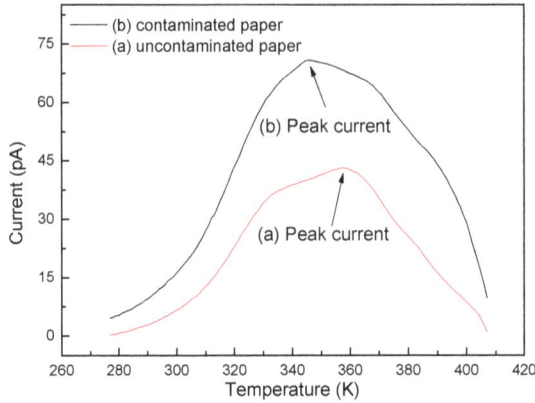

Figure 8. Thermally stimulated current (TSC) test results of two insulation paper samples: (**a**) uncontaminated paper; (**b**) contaminated paper.

The current peak related to trap energy level in sample II (contaminated by copper sulfide) shifted to a lower temperature compared with that in sample I (from 356 K to 344 K). This phenomenon indicated that shallower trap levels were generated in contaminated paper. Trapped charges and trap levels were calculated to analyze the trap characteristics. The trap charge amounts Q_{tsc} and the trap level E could be calculated according to the TSC test results using the following equations:

$$Q_{tsc} = \int_{t_0}^{t_1} I(t)dt = \frac{60}{\beta} \int_{T_0}^{T_1} I(T)dT \qquad (2)$$

$$E = \frac{2.47 T_m{}^2 k}{\Delta T} \qquad (3)$$

where T_m is the corresponding temperature to the peak current (K); β is the heating rate (K/min); ΔT is the temperature difference from the peak current to the half peak current (K); and k is the Boltzmann constant. Table 5 shows the TSC parameters of two paper samples.

Table 5. Thermally stimulated current (TSC) test results.

Parameter	Sample I	Sample II
Peak current (pA)	42.891	68.995
Temperature of peak current (K)	356	344
Trapped charges (nC)	0.546	0.912
Trap levels (eV)	0.819	0.476

As shown in Table 5, the trap level of contaminated paper was 0.476 eV, which is lower than that of uncontaminated paper (0.819 eV). Nevertheless, the trap charge amount of contaminated paper showed opposite characteristics (higher than that of uncontaminated paper). Lower trap level and higher trap charge amounts indicated that numerous shallower traps were present in contaminated paper. Furthermore, the shallower trap level could accelerate the de-trapping velocity of trapped electrons, leading to a more frequently charged transport. Therefore, electrons could easily accumulate energy and become thermion because of the decrease in trap levels. Electric tree and discharge channels formed easily. Excessive amounts of shallower traps also had an adverse effect on the anti-aging properties of insulation paper. However, the current peak, which was connected to the maximum trap density for contaminated paper was higher than that of uncontaminated paper, indicating that contaminated paper possessed higher trap density. These findings resulted in high-speed charge transportation via de-trapping and trapping processes in shallower traps.

5. Conclusions

This paper presents the space charge behavior in insulation paper induced by copper sulfide in HVDC power transformers. SEM, PEA, and TSC methods were used to conduct corresponding experiments and analyses. The results of the reported work and analyses are summarized as follows:

(1) The electrical insulating characteristics of paper material contaminated by copper sulfide decreased largely because of its semi-conductor properties.

(2) Copper sulfide on paper not only enhanced charge accumulation but also improved the mobility of the charge in the paper materials.

(3) The space charge behaviors of the thermally aged paper materials before and after contamination demonstrated that shallower trap energy level and higher trap density were generated because of the introduction of copper sulfide. These two reasons directly led to the rapid transfer of charge carriers via de-trapping and trapping processes. This phenomenon may explain why dielectric strength and breakdown voltage of aged paper materials contaminated by copper sulfide decreased compared with uncontaminated paper.

Acknowledgments: The authors acknowledge the National Natural Science Foundation of China (51277187) with supporting this research.

Author Contributions: All authors contributed extensively to the work presented in this paper. Ruijin Liao supervised the project. Ende Hu conducted the detail experiments and measurements. Lijun Yang and Yuan Yuan performed profiles and results analysis. All authors discussed the results and implications and commented on the manuscript at all stages.

Conflicts of Interest: The authors declare no conflict of interest.

References

1. Okabe, S.; Kohtoh, M.; Amimoto, T. Investigation of electrostatic charging mechanism in aged oil-immersed transformers. *IEEE Trans. Dielectr. Electr. Insul.* **2011**, *17*, 287–293.
2. Saha, T.K. Review of time-domain polarization measurements for assessing insulation condition in aged transformers. *IEEE Trans. Dielectr. Electr. Insul.* **2003**, *18*, 1293–1301.
3. Blaise, G. Space charge physics and the breakdown process. *J. Appl. Phys.* **1995**, *77*, 2916–2927.
4. Wang, X.; Yoshimura, N.; Murata, K.; Tanaka, Y.; Tanaka, T. Space charge characteristics in polyethylene. *J. Appl. Phys.* **1998**, *84*, 1546–1550.
5. Scatiggio, F.; Tumiatti, V.; Maina, R.; Tumiatti, M.; Pompili, M.; Bartnikas, R. Corrosive sulfur induced failured in oil-filled electrical power transformers and shunt reactors. *IEEE Trans. Power. Deliv.* **2009**, *24*, 1240–1248.
6. De Carlo, R.M.; Bruzzoniti, M.C.; Sarzanini, C.; Maina, R.; Tumiatti, V. Copper contaminated insulating mineral oils-testing and investigation. *IEEE Trans. Dielectr. Electr. Insul.* **2013**, *20*, 557–563.
7. Martins, M.A.G.; Gomes, A.R.; Pahlavanpour, B. Experimental study of passivated oil corrosiveness after depletion of the passivator. *IEEE Electr. Insul. Mag.* **2009**, *25*, 23–27.
8. Bolliger, D.; Pilania, G.; Boggs, S. The effect of aromatic and sulfur compounds on partial discharge characteristics of hexadecane. *IEEE Trans. Dielectr. Electr. Insul.* **2013**, *20*, 801–813.
9. Rajan, J.S.; Rudranna, N. Electric stress distribution in paper oil insulation due to sulphur corrosion of copper conductors. *J. Electrost.* **2013**, *71*, 429–434.
10. Liao, R.; Guo, C.; Wang, K.; Yang, L.; Grzybowski, S.; Sun, H. Investigation on thermal aging characteristics of vegetable oil-paper insulation with flowing dry air. *IEEE Trans. Dielectr. Electr. Insul.* **2013**, *20*, 1649–1658.
11. Ahmed, N.H.; Srinivas, N.N. Review of space charge measurements in dielectrics. *IEEE Trans. Dielectr. Electr. Insul.* **1997**, *4*, 644–656.
12. Tang, C.; Chen, G.; Fu, M.; Liao, R.J. Space charge behavior in multi-layer oil-paper insulation under different DC voltage and temperatures. *IEEE Trans. Dielectr. Electr. Insul.* **2010**, *17*, 775–780.
13. Du, Y.; Lv, Y.; Li, C.; Chen, M.; Zhou, J.; Li, X.; Tu, Y. Effect of electron shallow trap on breakdown performance of transformer oil-base nanofluids. *J. Appl. Phys.* **2011**, *110*.
14. Toyama, S.; Tanimura, J.; Yamada, N.; Nagao, E.; Amimoto, T. Highly sensitive detection method of dibenzyl disulfide and the elucidation of the mechanism. *IEEE Trans. Dielectr. Electr. Insul.* **2009**, *16*, 509–515.
15. Rudranna, N.; Rajan, J.S. Modeling of copper sulphide migration in paper oil insulation of transformers. *IEEE Trans. Dielectr. Electr. Insul.* **2012**, *19*, 1642–1649.
16. Kawarai, H.; Uehara, Y.; Mizuno, K.; Toyama, S.; Nagao, E.; Hosokawa, N.; Amimoto, T. Influences of oxygen and 2, 6-di-tert-butyl-p-cresol on copper sulfide deposition on insulating paper in oil-immersed transformer insulation. *IEEE Trans. Dielectr. Electr. Insul.* **2012**, *19*, 1884–1890.

17. Maina, R.; Tumiatti, V.; Pompili, M.; Bartnikas, R. Corrosive sulfur effect in transformer oil and remedial procedures. *IEEE Trans. Dielectr. Electr. Insul.* **2009**, *16*, 1655–1663.
18. Chong, Y.L.; Chen, G.; Miyake, H.; Matsui, K.; Tanaka, Y.; Takada, T. Space charge and charge trapping characteristics of cross-linked polyethylene subjected to ac electric stresses. *J. Phys. D Appl. Phys.* **2006**, *39*, 1658.
19. Huang, X.; Jiang, P.; Yin, Y. Nanoparticle surface modification induced space charge suppression in linear low density polyethylene. *Appl. Phys. Lett.* **2009**, *95*.
20. Hao, J.; Chen, G.; Liao, R.; Yang, L.; Tang, C. Influence of moisture on space charge dynamics in multilayer oil-paper insulation. *IEEE Trans. Dielectr. Electr. Insul.* **2012**, *19*, 1456–1464.

Comparative Study of Surface Temperature Behavior of Commercial Li-Ion Pouch Cells of Different Chemistries and Capacities by Infrared Thermography

Shovon Goutam, Jean-Marc Timmermans, Noshin Omar, Peter Van den Bossche and Joeri Van Mierlo

Abstract: The non-uniform surface temperature distribution of a battery cell results from complex reactions inside the cell and makes efficient thermal management a challenging task. This experimental work attempts to determine the evolution of surface temperature distribution of three pouch type commercial cells: Nickel Manganese Cobalt oxide (NMC)-based 20 Ah cell, Lithium Iron Phosphate (LFP) 14 Ah, and Lithium Titanate Oxide (LTO) 5 Ah battery cell by using contact thermistor and infrared (IR) thermography. High current (up to 100 A) continuous charge/discharge and high current (80 A) micro pulse cycling profile were applied on the cells. It was found that thermistor based temperature profile varied cell to cell, especially the LTO cell. Among the investigated cells, the NMC cell shows highest temperature rise and the LTO cell the lowest rise. IR (Infrared) images revealed the spatial distribution of surface temperature, in particular the location of the hottest region varies depending not only on the geometrical and material properties of the cell, but also the type of loads applied on the cells. Finally, a modeling perspective of the cell temperature non-uniformity is also discussed.

Reprinted from *Energies*. Cite as: Goutam, S.; Timmermans, J.-M.; Omar, N.; Van den Bossche, P.; Van Mierlo, J. Comparative Study of Surface Temperature Behavior of Commercial Li-Ion Pouch Cells of Different Chemistries and Capacities by Infrared Thermography. *Energies* **2015**, *8*, 8175–8192.

1. Introduction

As an alternative to gasoline fuel, rechargeable energy storage systems (RESS) have garnered the attention of automotive industries since the last decade. A growing number of electric vehicles (EV) and hybrid electric vehicles (HEV) in the present market depicts the rapid growing demand [1–3]. Being of superior performance in terms of high power and energy density, EV and HEV manufacturers largely depends on lithium-ion batteries as the rechargeable energy source nowadays. However, automotive application requires batteries with high performance as well as with a high level of safety. For instance, heat generated within the cell is one of the major concerns for safety and performance of the battery cell. Total heat generation within the cell can generally be attributed to the contributions of reversible heat (entropic heat components related to the electrochemical reactions) and irreversible heat (ohmic and polarization resistance heat components). On the one hand, high cell temperature, even locally, can have a detrimental influence on the cell performance. For instance, cycle life of the cell can be heavily influenced leading to accelerated ageing of the cell [4,5]. On the other hand, excessive and uncontrolled heat can lead to thermal runway and ultimately may lead to explosion of the cell [6]. Therefore, monitoring and optimizing the operating temperature of the cell during operation are of crucial importance.

It has been well established that modeling of the thermal behavior of battery cells can play a vital role in cell temperature monitoring and can also provide scopes for the development of battery cells for better thermal management and improved thermal safety of, not only a single battery cell, but also a battery pack consisting several cells [7,8]. In reality, a precise validation process of these thermal models can be complex and extremely difficult. Validation through comparing cell surface temperature obtained from model and physical measurements is, comparatively, a less complex and faster method of model validation [9]. However, due to the complexity of electrochemical reactions (e.g., effect of side reactions) and uneven current density distribution inside the cell, the spatial distribution of cell surface temperature can be non-uniform [10–15], in particular, for large format pouch cells for automotive applications. Moreover, the pattern of the heat distribution, e.g., location of maximum and average temperature, can be different depending on the type of applied load profile [15]. This particular behavior of temperature non-uniformity makes the thermal management of a battery pack more challenging in order to maintain a uniform temperature inside the battery pack during operation. Improvements have to be made in battery cells design, and the location oriented active/passive cooling system of battery pack has to be incorporated to overcome this challenge [16–18].

Now, among the different methods of measuring cell surface temperature, the single point (minute area) measurement (e.g., contact thermocouple, thermistor, *etc.*) is not adequate to measure the spatially non-uniform temperature distribution. In this case, thermal infrared (IR) imaging is a potential tool that can be used to observe and measure this spatially non-uniform temperature distribution with reasonable high accuracy. Additionally, battery cell surface temperature distribution information obtained as such can be used for monitoring surface temperature locations of interest (maximum and/or average temperature) through single point measurements for a large number of cells during performance-characterization testing. Despite the high importance of non-uniform battery cell surface temperature measurement, to the authors' knowledge, very few works have been published dedicated to this topic using IR thermography [12,14,19]. Veth *et al.* have analyzed surface temperature evolution through IR thermography in their work [14]. However, the work was performed on an NMC cell only and the thermal behavior was studied only under a continuous charge and discharge load profile.

In this experimental work, non-uniform surface temperature evolution of three commercial pouch cells, of different chemistries and capacities, were studied through both infrared thermography and contact thermistor measurements. Continuous charge and discharge current, ranging from 10 A up to 100 A, and pulse current up to 80 A, were applied to determine the evolution of surface temperature distribution.

2. Experimental Procedures

The chemistries of three lithium-ion commercial pouch type battery cells that were used are: Lithium Nickel Manganese Cobalt oxide (NMC), Lithium Iron Phosphate (LFP), and Lithium Titanate Oxide (LTO). The nominal capacities of these cells are 20 Ah, 14 Ah, and 5 Ah, respectively. Two types of load profiles were applied on the cells. One type consisted of several consecutive continuous complete charge (constant current, CC) and discharge (CC) cycles with currents of 10 A, 20 A, 40 A, 60A, 80 A, up to 100 A. Sufficient rest time (1–3 h depending on the current rate) was incorporated between successive operations (charge/discharge) and cycles in order to allow the cell surface temperature to return to ambient temperature. The complete charge and discharge were performed according to the conditions presented in Table 1. On the other type of load profile, high-current micro-pulse cycling was incorporated. The micro-pulse cycling load profile consisted of ~1000 cycles of charge and discharge current pulses. Each of these cycles sequentially consisted of a charge pulse for 2 s at 80 A, 2 s rest (zero current), a discharge pulse for 2 s at 80 A, and 2 s rest. All the tests were performed by using an ACT 0550 (80 channels) battery tester (PEC®, Leuven, Belgium). Surface temperature was logged by using an NTC 5K thermistor (EPCOS®, Munich, Germany) with a measuring range of 0–100 °C and a tolerance of 1%. Simultaneously, IR images were captured by a Ti25 thermal imager (FLUKE®, Everett, WA, USA) at regular time intervals. In order to achieve accurate results from the IR thermography, cells were placed in a semi-closed (to observe behavior at natural heat transfer) and dark environment (to avoid visible light interference). Moreover, the cell surfaces were painted uniformly with a dull black paint in order to minimize the effect of the visible light reflections. The IR camera was calibrated within the range 0–120 °C, with a maximum error of 2% by the manufacturer. The thermal sensitivity, noise-equivalent temperature difference (NETD) was $\leq$0.09 °C at a 30 °C target temperature ($\leq$90 mK). For the calibration of the experiment, emissivity correction was performed by measuring a known uniform surface temperature painted with the same dull black paint as used on the cell surface. Emissivity was set to ~0.99. A comparative test was carried out to compare the measurement taken by the thermistor and the IR camera during a continuous discharge at 100 A of a NMC cell. The cell surface was divided into 16 equal rectangular portions and two points were selected at two points of intersection of the dividing lines, as shown in Figure 1a,b, which shows the temperature profile of the corresponding two positions on the cell surface. It was found that, throughout the process, the maximum difference between the measurements made by the thermistor and the IR camera was ~1 °C. By means of this comparative test measurement, the accuracy of the IR temperature measurement has been quantified and validated over the complete measurement range.

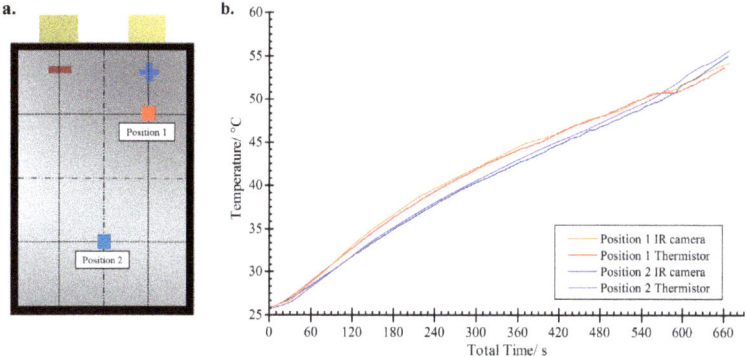

Figure 1. (a) Two locations on the cell surface subjected to measurement comparison; **(b)** Temperature profile measured by the IR camera and the thermistor at the corresponding locations of the cell surface during a discharge at 100 A of an NMC, 20 Ah cell.

Table 1. Mechanical and electrical properties of the NMC, LFP and LTO Cells.

Properties		NMC	LFP	LTO
Mechanical				
Body	Length, mm	217	216	235
	Width, mm	130	130	173
	Thickness, mm	7.1	7.1	4
Tab	Length, mm	40	40	40
	Width, mm	30	45	85
	Weight, g	428	380	262
Electrical				
Nominal Voltage, V		3.65	3.2	2.2
Nominal Capacity, Ah		20	14	5
End of Charge Voltage, V		4.2	3.65	2.8
End of Discharge Voltage, V		3	2	1.5
Ac Impedance (1 KHz), mΩ		<3	<5	0.7
Specific energy, Wh/kg		174	120	42
Energy Density, Wh/L		370	230	90
Specific Power (DoD 50%, 10 S), W/kg		2300	2500	2250
Power Density (DoD 50%, 10 S), W/L		4600	4500	4400
Maximum Charge Current, A		–	–	150
Maximum Discharge Current, A		100	140	150

3. Cell Characteristics

Internal resistance values of the three cells have been measured by means of the standardized hybrid pulse power capability (HPPC) test. The pulse currents for this HPPC test were adapted to be the same as the currents used for surface temperature measurements, in particular 20, 40, 60, 80, and 100 A. The pulses were performed at three different SoC levels, namely at an 80%, 50%, and 20% SoC level. Figure 2 shows the internal resistance trend for all three cells.

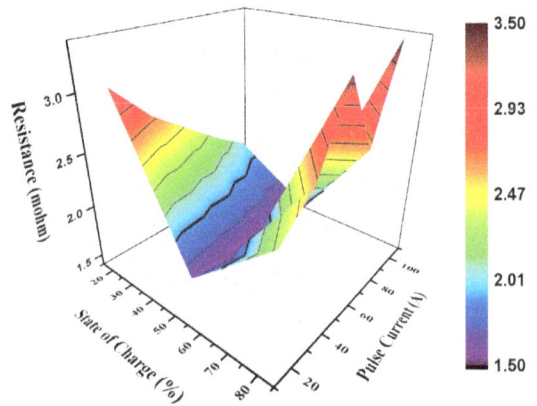

Figure 2. Internal resistances measured by HPPC test at an 80%, 50% and 20% SoC level with pulse currents of 20, 40, 60, 80, 100 A of (**a**) NMC discharge pulses; (**b**) LFP discharge pulses; (**c**) LTO discharge pulses; (**d**) LTO charge pulses.

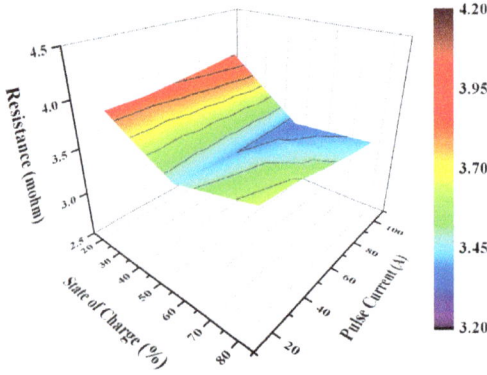

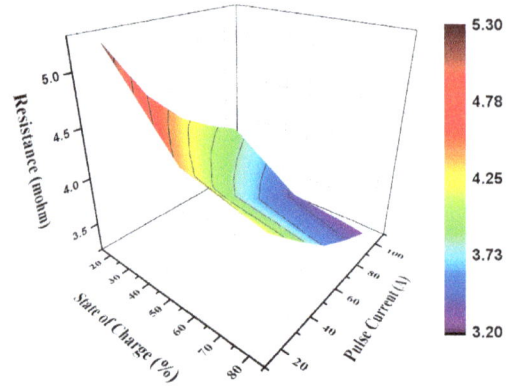

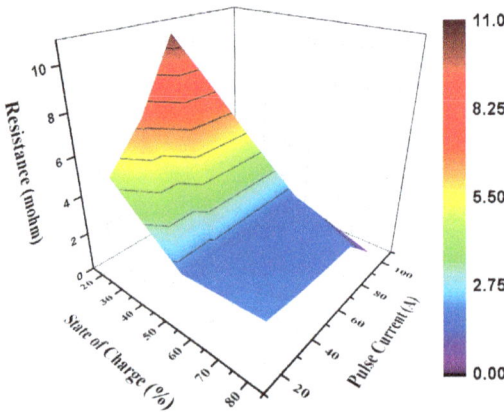

From these graphs, we can see that in the case of discharging, and for all three chemistries, higher internal resistance is observed at a low state of charge of the cell. Further, the lowest internal resistance is observed at a high state of charge, in combination with the higher current values. In the case of the LTO chemistry charging pulses, the lowest internal resistance is found at the mid value of the state of charge.

4. Results and Discussion

4.1. Continuous Charge and Discharge

Figure 3 shows the surface temperature profile of the 20 Ah NMC cell obtained by a thermistor placed near the positive tab during complete charge at 20 A and complete discharge at 10 A, 20 A, 40 A, 60 A, 80 A and 100 A respectively. The temperature profile according to this thermistor measurement during charging showed an initial rise until the state of charge (SoC) level reaches *ca.* 50%. Following, the temperature remains fairly steady until it reaches a 100% SoC level (4.2 V). Discharge at different rates showed, comparatively, a different trend in the temperature profile. For instance, during discharge at 10 A, the temperature rose until a *ca.* 70% SoC level was reached and was followed by a drop until it rose sharply again after it reached a *ca.* 30% SoC level.

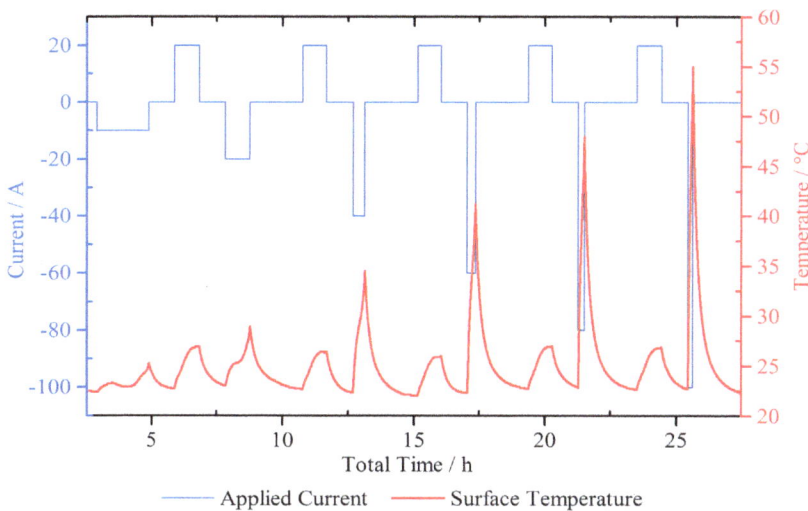

Figure 3. Thermistor (placed near the positive tab of the cell) based temperature profile (red line) of NMC cell and corresponding load profile (blue line) during continuous charge at 20 A and discharge at 10 A, 20 A, 40 A, 60 A, 80 A, and 100 A. Discharge is denoted by negative current.

In order to explain the phenomenon of temperature drop, one may consider the relative dominance of reversible and irreversible heat contributions. It was found that endothermic entropy change is the result of phase change in the electrode material at a certain SoC level range [20]. Both of the electrodes may undergo volume and phase changes. At the cathode, these changes are based on the ratio of lithium and other elements (e.g., Cobalt) at the cathode. Additionally, at the anode, these changes are based on the ratio of lithium and carbon during intercalation/deintercalation of lithium [20–22]. However, at a higher current rate, the contribution of irreversible heat (*i.e.*, polarization resistance heat and ohmic resistance heat) becomes dominant. Therefore, temperature drop due to entropy change became comparatively less significant at 20 A discharge. At higher current of 40 A and 60 A, this effect became trivial, shown by a slight change in the steepness (slope) of the temperature profile. At 80 A and 100 A the effect is negligible.

The IR images of Figure 4 depict the spatial distribution of the cell surface temperature during a discharge at 100 A. Figure 4a shows the schematic of the relative size and orientation of the tabs of the NMC cell. The black border represents the extruded pouch of the cell. The following IR images were represented according to the orientation of the schematic. Now, it is clear from the temperature distribution pattern of the IR images that, initially, the most heated regions were at the adjacent areas of the tabs of the cell and slightly higher near the positive tab. This can be attributed to the comparatively higher resistance of the aluminum positive tab and current collector (compared with the copper negative tab). However, this finding is inconsistent with the findings of Veth *et al.*, who observed that the maximum temperature was initially near the negative tab [14]. Apparently, this inconsistency appeared because of the variation in commercial battery cell design (e.g., surface area of the tabs and current collectors, *etc.*) [16]. Nonetheless, with the progression of the discharge towards ending, the temperature distribution became more spatially uniform over the whole surface of the cell (Figure 4c,d), with the most heated region located in the center region of the cell. This observation of the shifting of hottest point is in congruence with the observation of Li *et al.* [23], who investigated cell temperature by inserting sensors inside the cell. This observation can be attributed to the faster depletion of the active species near the positive area due to the higher current density. Similar patterns were observed with charge and discharge at other current rates. It is important to mention here that individual IR images have different temperature scales. Here, the principal objective was to observe the spatial non-uniform temperature distribution of a cell at a certain state. By incorporating a different temperature scale, it was possible to observe and show the distinguishable non-uniformity of surface temperature distribution at a particular stage of operation. Figure 4d shows the temperature distribution of heat dissipation at rest (no applied current). This pattern suggests that the heat dissipation rate was higher at the upper half region of the cell (the half that contains the tabs), apparently because of higher heat transfer through the tabs.

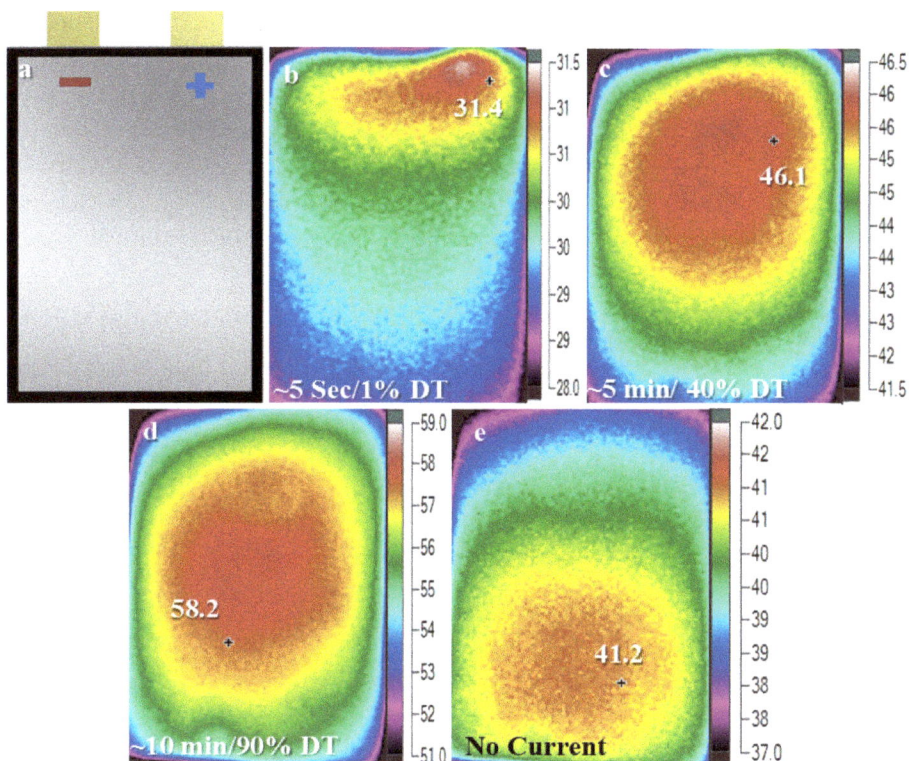

Figure 4. (a) Schematic of the NMC 20 Ah cell. IR images during discharge at 100 A along with the location of respective maximum temperature; (b) after ~5 s of discharge or ~1% of total discharge time (DT); (c) after 5 min or ~40% DT; (d) after 10 min or ~ 90% DT; (e) during rest (zero current). Different temperature scales were used here to show distinguishable non-uniformity of surface temperature distribution.

The LFP cell consisted of a similar graphite based anode, therefore, the thermistor (placed at adjacent area of the positive tab) based surface temperature profile shows similar trend as the trend of the NMC cell (see Figure 5), particularly during discharge at all current rates. However, the temperature profile during charge showed a slight variation. At the beginning of the charge process, a small drop in temperature was observed, followed by a sharp rise until the SoC reached a level of *ca.* 50%–60%. After that, again, a small drop appeared, followed by a rise until the end of charge.

In order to explain this variation from the NMC cell, the following factors can be considered. (i) It was found that the entropy of an LFP cathode does not change significantly during the charge/discharge process [21], therefore, a major portion of the entropy related heat is contributed by the graphite anode; (ii) an LFP cell operates within a lower voltage window compared to NMC; (iii) LFP shows a two phase system of LiFePO$_4$ and FePO$_4$ during Li$^+$ intercalation/deintercalation [24]. Additionally, it was also shown that the distribution of these two phases, which vary across the electrode, are at a certain SoC level [25]. Additionally, the geometrical attributes of the cell also had to be considered. The LFP cell is slightly smaller than the NMC cell with comparatively wider tabs (see Figure 6a). Although, despite of this geometrical variation, spatial temperature distribution showed similar evolution of the contour, as shown in Figure 6a–c.

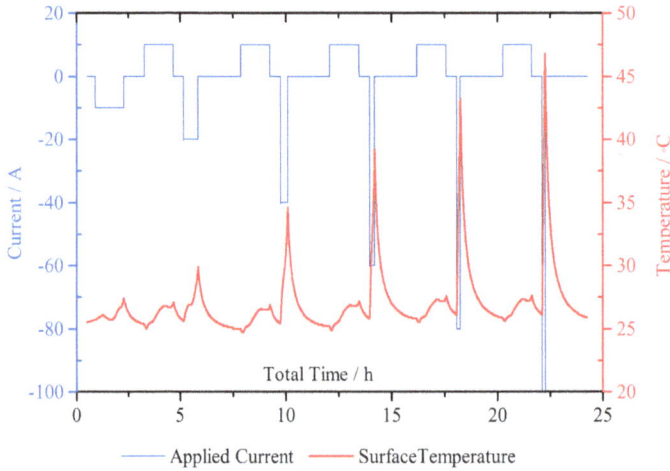

Figure 5. Thermistor (placed near the positive tab of the cell) based temperature profile (red line) of an LFP cell and the corresponding load profile (blue line) during a continuous charge at 10 A and discharge at 10 A, 20 A, 40 A, 60 A, 80 A and 100 A. Discharge is denoted by negative current.

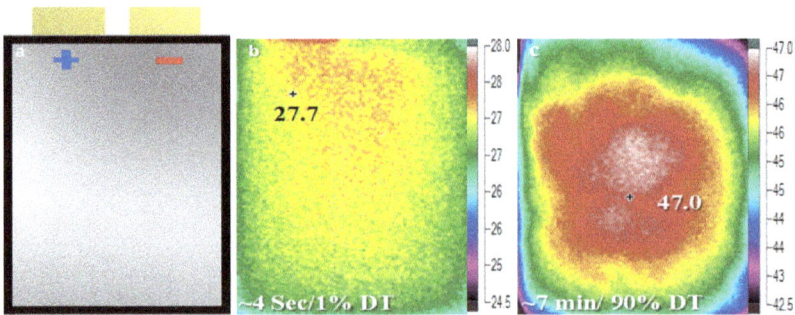

Figure 6. (a) Schematic of the LFP 14 Ah cell. IR images during discharge at 100 A along with the location of respective maximum temperature; **(b)** after ~4 s of discharge or ~1% of discharge time (DT); **(c)** after 7 min or ~90% DT.

534

The LTO anode based cells have a distinct difference compared to the other two cells (*i.e.*, the NMC and LFP cells). Titanate based anode has a higher electrical conductivity and a bigger surface area, thus, it shows lower impedance compared to graphite anode (see Figure 2 and Table 1). One of the reasons behind this variation is the distinction of solid-electrolyte interface (SEI) layer formation and the thickness characteristics between graphite and tianate anodes [3,26]. Thus, the net heat generation in the LTO cells is much lower, which makes the LTO cell a potential candidate for a cell with higher thermal safety, even during very fast charging (within less than 30 min). It was also found that the entropy change is insignificant in an LTO based commercial full cell [22]. These properties were reflected in the surface temperature profile of the LTO cell (see Figure 7). Discharge at different rates showed a sharp rise without any noticeable drop in the surface temperature (Figure 7a). However, charge at a low current (10 A) showed a change of slope in the rising temperature profile, approximately at the middle of the charge process, which further diminished gradually during charging at high current rates (20–100 A) (Figure 7b). Additionally, the LTO battery cell had different geometrical attributes. For instance, much wider tabs were located at the opposite side of the cell (see Figure 8a). Despite of the difference in material and geometrical properties, the spatial temperature distribution of the cell surface showed similar contours to the NMC and LFP cells (see Figure 8b,c).

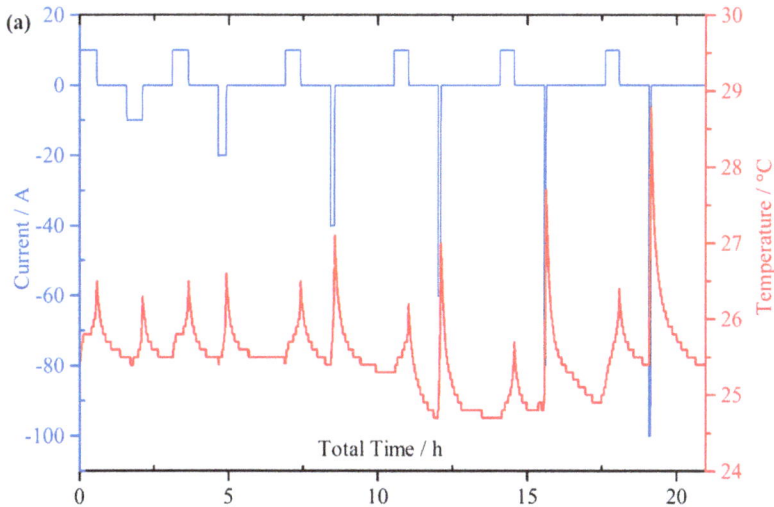

Figure 7. *Cont.*

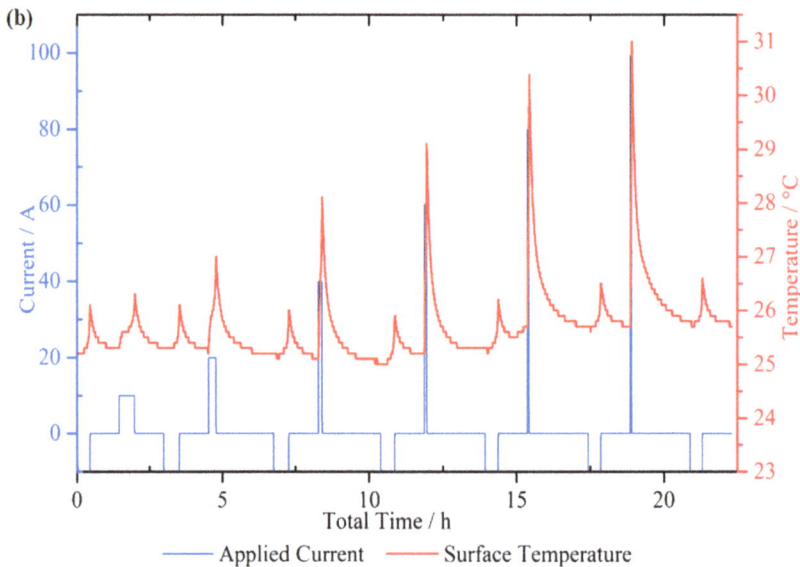

Figure 7. Thermistor (placed near the positive tab of the cell) based temperature profile (red line) of LTO cell and corresponding load profile (blue line) during (**a**) continuous charge at 10 A and discharge at 10 A, 20 A, 40 A, 60 A, 80 A and 100 A, respectively; (**b**) continuous charge at 10 A, 20 A, 40 A, 60 A, 80 A and 100 A and discharge at 10 A, respectively. Discharge is denoted by negative current.

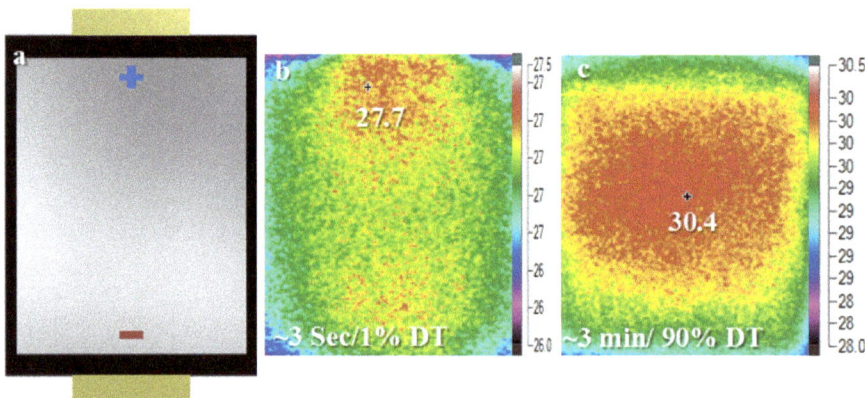

Figure 8. (**a**) Schematic of the LTO 5 Ah cell. IR images during discharge at 100 A along with the location of respective maximum temperature; (**b**) after ~3 s of discharge or ~1% of discharge time (DT); (**c**) after 3 min or ~90% DT.

536

A summary of surface temperature rise of all the three cells is presented in Figure 9. Here, ΔT represent the temperature rise for each process (charge/discharge) based on the measurement by the thermistor. In addition, it was calculated as follows:

$$T = T_{end} - T_{start} \tag{1}$$

where, T_{end} = Temperature at the end of the process (charge/discharge), T_{start} = Temperature at the start of the process (charge/discharge).

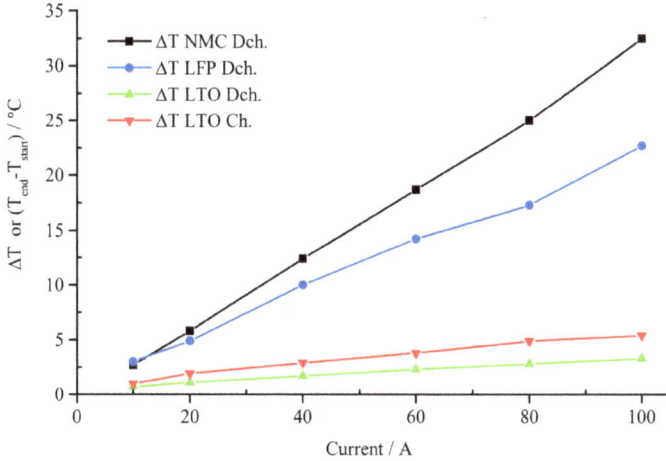

Figure 9. Temperature rise evolution in the NMC cell (during discharge), the LFP cell (during discharge) and the LTO cell (during both charge and discharge) at different current rates.

4.2. Micro Pulse Cycling

Thermistor (placed at the adjacent area of the positive tab) based surface temperature profile of the three cells showed a similar trend during high current (80 A) micro pulse cycling (see Figure 10). The inset of Figure 10 shows one complete cycle of ~1000 micro pulses of charge and discharge cycles. At the beginning, the temperature followed a sharp rise. Approximately after ~500 cycles or ~1 h, the temperature reached a steady state condition (*i.e.*, variation <1 °C per 5 min). Micro pulse cycling was performed on all the cells, which were at an 80% SoC level. In previous work, we showed that micro pulse cycling performed at different SoC levels showed similar surface temperature profiles. Additionally, the temperature rise did not change significantly at different SoC levels [15].

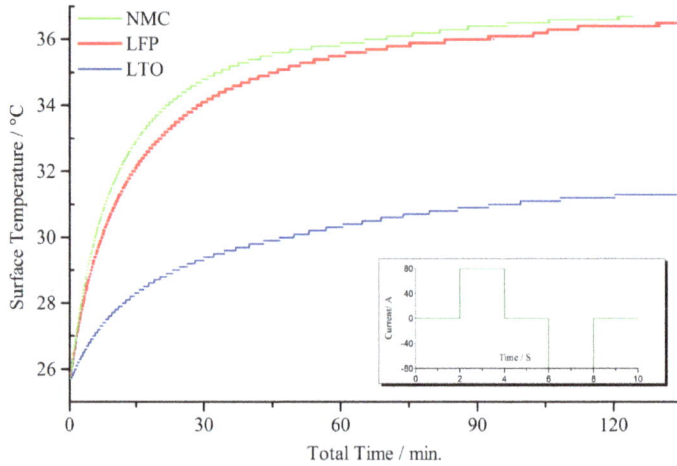

Figure 10. Thermistor based surface temperature profile of NMC, LFP, and LTO cells at 80% SoC during micro pulse cycling (~1000 cycles) at 80 A. Inset shows one single micro pulse cycle.

According to the IR images (Figure 11a–c) of NMC during micro pulse cycling, it was observed that, as the cell surface temperature proceeded towards a steady state, the spatial temperature distribution over the surface became non-uniform. In addition, the upper half portion of the cell, which is near the tabs, was comparatively hotter than the other half. It is also visible from the IR images that the hottest point is mostly located near the positive tab of the cell. Local high current density at the adjacent areas of the tab, for a very short time, can be attributed to this localization of the hottest region. However, evolution of the surface temperature contour in the case of the LFP cell showed a different pattern compared to the contour of the NMC cell (Figure 11e–f). In this case, the location of the hottest region was observed around the center region of the cell throughout the cycling. This evolution pattern is more similar to the pattern during continuous charge/discharge of the LFP cell (Figure 6). Relative spatial distribution of two phases ($LiFePO_4$/$FePO_4$) over the electrode, which changes during the charge/discharge process, might be related to this surface temperature distribution evolution [25]. In the case of the LTO cell, at the beginning of the cycling, the hottest region appeared at the upper half of the cell (the half that contains the positive tab) (Figure 11g,h). However, at a steady state, the hottest region shifted towards the lower half of the cell (Figure 11i). It is important to mention here, that, at the steady state, the highest temperature difference over the complete surface was less than 2 °C. Variation in geometrical attributes (size and position of the tabs) and material properties (LTO based anode) can be considered in order to explain this variation compared to other to cells. Further, individual

IR images contains different temperature scales in order to show the distinguishable non-uniformity of the cell surface temperature at a certain stage of operation.

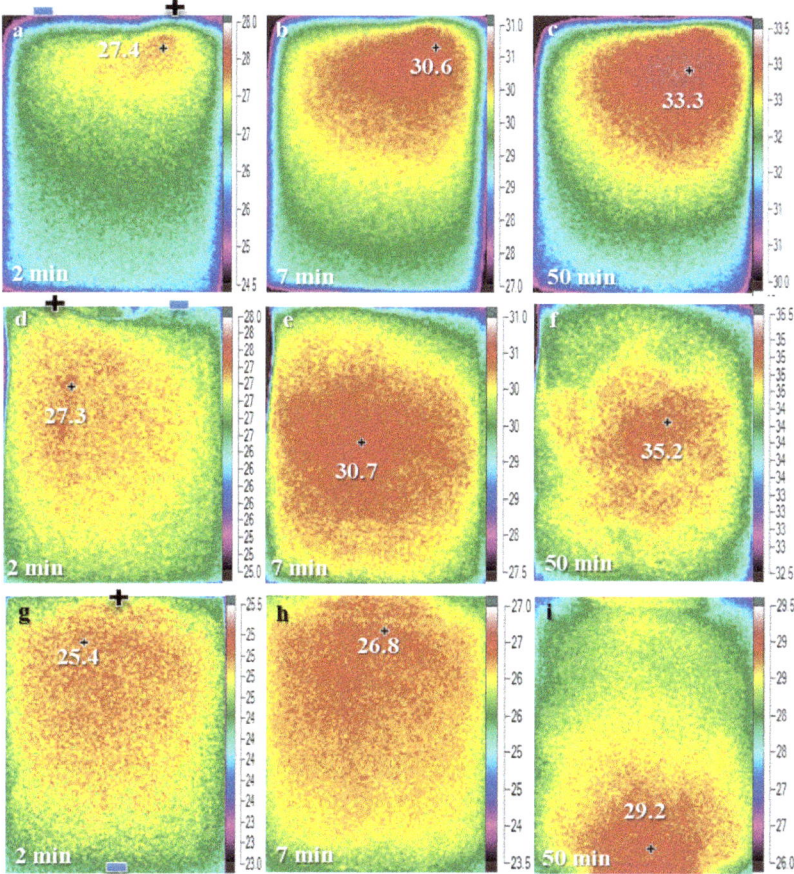

Figure 11. IR images during micro pulse cycling at 80 A of NMC, LFP and LTO cells at an 80% SoC at different points of time along with the location of respective maximum temperature. (**a**) NMC after 2 min; (**b**) NMC after 7 min; (**c**) NMC after 50 min; (**d**) LFP after 2 min; (**e**) LFP after 7 min; (**f**) LFP after 50 min; (**g**) LTO after 2 min; (**h**) LTO after 7 min; (**i**) LTO after 50 min. Different temperature scales were used here to show the distinguishable non-uniformity of surface temperature distribution.

5. Modeling Perspective

A large number of works have been published on the modeling of Li-ion batteries [8,27–33]. These models were created to predict electrical, electrochemical, thermal and mechanical behavior of the battery system, either independently or in combination (coupled). According to the literature, the scale of these models may vary from the microscopic level (molecular level quantum behavior based) to the mesoscopic level (electrochemical process and transport phenomenon based) or to the macroscopic level (based on the distribution of current and voltage on electrodes) to blackbox system level (empirical, data fit based). A comprehensive review in this regard can be found in Reference [8].

Thermal models generally describe the thermal behavior in terms of heating of the battery and are built on the energy conservation of the system as follows:

$$\rho\, C_p \frac{\partial T}{\partial t} = K\nabla^2 T + Q \tag{2}$$

where, T—temperature of the system domain, ρ—density of the system component, C_p—specific heat capacity, K—thermal conductivity and Q—volumetric heat source.

Depending on the level of the modeling, Equation (2) can be applied on the separate domain of the battery. For instance, for macroscopic level the battery can be divided in three domains: electrode, tab and casing. Thus, Equation (2) for the three domains will be:

$$\rho\, C_p \frac{\partial T}{\partial t} = K\nabla^2 T, \text{ for casing as there is no heat source} \tag{3}$$

$$\rho\, C_p \frac{\partial T}{\partial t} = K\nabla^2 T + Q, \text{ for electrode and tab domain} \tag{4}$$

At the boundaries of the domains heat transfer to the surroundings by radiation and convection,

$$Q_{transfer} = (h_{conv} + h_{rad})\,(T - T_{amb}) \tag{5}$$

where, h_{conv}—convection coefficient, h_{rad}—radiation coefficient and T_{amb}—ambient temperature.

Now, the variation between the existing modeling principal lies in the determination of the volumetric heat Q. Q can be formulated as:

$$Q = Q_{Ech} + Q_J \tag{6}$$

where, Q_{Ech} and Q_J are electrochemical heat and joule heating, respectively.

For modeling at the system level, considering the system as a blackbox, the right side of Equation (6) can be formulated as:

$$Q = \varepsilon_e T (\frac{dV_{ocv}}{dT})I + RI^2, \text{ for the electrode domain} \tag{7}$$

$$Q = \varepsilon_t R' I^2, \text{ for the tab domain} \tag{8}$$

where, ε_e, ε_t are the volume ratio factors of electrodes and tabs respectively, R is the internal resistance of the cell, R' is the resistance of the tab, V_{ocv} and I are the charge/discharge currents.

To model the non-uniformity of the heat across the area of the cell a thermal equivalent circuit model can be assumed in the different direction of the cell. The circuit generally consists of several thermal resistance and thermal capacitance network. Through electrical characterization, relative amount of heat generation in different direction can be can be modeled. For instance, in Reference [9], a first order Cauer model was considered and found in fair agreement between the modeled and real thermal behaviors. Due to the simplicity and low number of parameters, this modeling is efficient in terms of computational power. As shown previously, the LTO cell has a lesser degree of temperature non-uniformity compared to the NMC cell and the LFP cell, a thermal equivalent circuit model can be very effective in this case. However, as the model is built based on characterization data at a sets of test conditions, the model accuracy can be low in dynamic conditions, especially in the case of a cell with severe temperature non-uniformity, as the NMC and LFP cells presented above.

In order to overcome the accuracy limitations, the model can be extended to macroscopic level, taking into account the potential gradient over the surface of the electrodes [29,30,33]. In this model formulation, charge balance can be applied to a pair of electrode as:

$$\sigma_j \nabla^2 V_j + \frac{Jn}{\delta_j} = 0 [j = \text{positive (p)}, \text{negative (n)}] \tag{9}$$

where, σ_j is the electrical conductivity of the current collector σ, V_j is the potential distribution on the current collector, J is the transvers current density perpendicular to the electrode surface, n is the unit normal vector, δ_j is the thickness of the current collector and the subscript j corresponds to the positive and negative electrode. In this case, Equations (6) and (7) become as follows:

$$Q = \frac{J}{(\delta_{pe} + \delta_s + \delta_{ne})} T \frac{dV_{ocv}}{dT} + \frac{J}{(\delta_{pe} + \delta_s + \delta_{ne})} (V_{ocv} - V_p + V_n) + \frac{\sigma_j (\nabla V_j)^2}{\delta_j} \tag{10}$$

where, δ_{pe}, δ_s and δ_{ne} are the thickness of positive electrode, separator and negative electrode respectively.

The transverse current density can be expressed in terms of open circuit voltage and voltage of the electrodes [29] as:

$$J = \frac{V_{ocv} - (V_p - V_n)}{R} \tag{11}$$

541

Here, $\underline{V}_{ocv}$ and R can be determined experimentally. With this model, the thermal behavior of any pouch type cell can be modeled with very high accuracy during constant charge or discharge. However, against dynamic current profile such as micro cycle as presented previously, this model may show less accuracy. In order to overcome this limitation, the non-faradic component of the transverse current can be taken into account. As suggested by Taheri *et al.* [34], the transvers current density can be expressed as:

$$J = J_{\text{faradic}} + J_{\text{non-faradic}} = Y(V_p - V_n - U) + C_{dl}\frac{d(V_p - V_n)}{dt} \tag{12}$$

where, in the faradic component, Y and U are fitting parameters and expressed in terms of depth of discharge (DoD). Additionally, in the non-faradic component, C_{dl} is the lumped double layer capacitance.

As presented above, a macroscopic level modeling, assumption was made that electrochemical reactions are homogenous over the electrode area. At the mesoscopic level, modeling the inhomogeneity of electrochemical reaction parameters, such as charge balance in the liquid and solid phases, can be incorporated to improve the accuracy of the model [30]. For instance, in order to model the shifting of the hottest point at the end of a continuous discharge process, a mesoscopic electrochemical-thermal model can be incorporated for higher accuracy. However, this kind of modeling is computationally challenging. Moreover, the parameters needed for the model are very difficult to determine.

6. Conclusions

Battery cell surface temperature distribution under high current continuous charge and discharge up to 100 A, as well as under high current micro-pulse cycling at 80 A, were studied for three different commercial pouch type cells: NMC cathode based 20 Ah cell, LFP cathode based 14 Ah cell, and LTO anode based 5 Ah cells, by using both contact thermistor measurements and infrared thermography. Variations in temperature profiles between the different types were observed during continuous charge/discharge. However, the spatial distribution patterns of the surface temperature were observed to be similar in the case of these three types of cells. At the beginning of the continuous charge/discharge test, the hottest region appeared near the positive tab but as the process progressed towards the end, the temperature distribution became more uniform with the hottest region located at the center region of the cell's surface. Comprised of a titanate based anode, the LTO cell showed a very low temperature rise, which proved it as the best option from the thermal safety point of view. On the other hand, during micro pulse cycling, a similar trend of the thermistor based temperature profiles was observed. According to this trend, the temperature rose sharply until a common point in time and became steady without any significant rise until the end of the cycling. The location of the hottest region was observed near the positive tab throughout the micro pulse cycling in the case of the NMC cell. The pattern of the surface temperature distribution throughout the micro pulse cycling in the case of the LFP cell was similar to the pattern observed during continuous charge and discharge, showing the hottest region in the center area of the cell. While in the case of the LTO cell, although the hottest region was located at the upper half (towards positive tab) at the beginning of cycling, it shifted towards the area adjacent to the negative tab of the cell at the steady state.

Non-uniformity of the surface temperature cannot be attributed to a single independent factor. It is a result of a combined influence of electrochemical properties, such as the inhomogeneity of the reactions across the electrode, of electrical properties, such as localization of current density and SoC, and of physical properties, such as the cell size and the geometry, and locations of the tabs. In most cases those factors are interdependent. This makes it very cumbersome to point out the most influencing factor. In order to model the non-uniformity of the cell temperature, several modeling methods have been adopted by researchers. These methods have been discussed in this work. Moreover, the effectiveness and limitations of these methods were concisely discussed from the point of view of the thermal behavior of the investigated cells.

Nevertheless, as temperature is one of the major factors in battery cell ageing, non-uniformity of the temperature distribution may lead to imbalanced ageing, and, ultimately, may lead to the degradation and imbalance of cell performance. In order to avoid such undesirable behaviors, improvements needs to be made at both the cell level and pack level. At the cell level, material properties and physical properties can be improved, whereas at the pack level, improvements in cooling architecture can be incorporated.

Acknowledgments: This research has been made possible, thanks to the research project "Battle" and was funded by the Flemish Agency for Innovation by Science and Technology (IWT130019). Further, we acknowledge Flanders Make for the support to our research team.

Author Contributions: Shovon Goutam set up and conducted all the experimental work and analysis and wrote the manuscript. Jean-Marc Timmermans designed and supervised the experimental work, the analysis and writing of the manuscript. Noshin Omar, Peter Van den Bossche and Joeri Van Mierlo provided guidance and technical advice.

Conflicts of Interest: The authors declare no conflict of interest.

References

1. International Energy Agency (IEA). *Technology Roadmap: Electric and Plug-in Hybrid Electric Vehicles*; IEA: Paris, France, 2011.
2. Scrosati, B.; Garche, J. Lithium batteries: Status, prospects and future. *J. Power Sources* **2010**, *195*, 2419–2430.

3. Omar, N.; Daowd, M.; van den Bossche, P.; Hegazy, O.; Smekens, J.; Coosemans, T.; van Mierlo, J. Rechargeable Energy Storage Systems for Plug-in Hybrid Electric Vehicles—Assessment of Electrical Characteristics. *Energies* **2012**, *5*, 2952–2988.

4. Barré, A.; Deguilhem, B.; Grolleau, S.; Gérard, M.; Suard, F.; Riu, D. A review on lithium-ion battery ageing mechanisms and estimations for automotive applications. *J. Power Sources* **2013**, *241*, 680–689.

5. Omar, N.; Monem, M.A.; Firouz, Y.; Salminen, J.; Smekens, J.; Hegazy, O.; Gaulous, H.; Mulder, G.; van den Bossche, P.; Coosemans, T.; *et al.* Lithium iron phosphate based battery—Assessment of the aging parameters and development of cycle life model. *Appl. Energy* **2014**, *113*, 1575–1585.

6. Bandhauer, T.M.; Garimella, S.; Fuller, T.F. A critical review of thermal issues in lithium-ion batteries. *J. Electrochem. Soc.* **2011**, *158*.

7. Van den Bossche, P.; Omar, N.; Al Sakka, M.; Samba, A.; Gualous, H.; van Mierlo, J. The challenge of PHEV battery design and the opportunities of electrothermal modeling. In *Lithium-Ion Batteries*; Pistoia, G., Ed.; Elsevier: Amsterdam, The Netherlands, 2014; pp. 249–271.

8. Ramadesigan, V.; Northrop, P.W.C.; De, S.; Santhanagopalan, S.; Braatz, R.D.; Subramanian, V.R. Modeling and simulation of lithium-ion batteries from a systems engineering perspective. *J. Electrochem. Soc.* **2012**, *159*, R31–R45.

9. Samba, A.; Omar, N.; Gualous, H.; Firouz, Y.; van den Bossche, P.; van Mierlo, J.; Boubekeur, T.I. Development of an advanced two-dimensional thermal model for large size lithium-ion pouch cells. *Electrochim. Acta* **2014**, *117*, 246–254.

10. Strobridge, F.C.; Orvananos, B.; Croft, M.; Yu, H.-C.; Robert, R.; Liu, H.; Zhong, Z.; Connolley, T.; Drakopoulos, M.; Thornton, K.; *et al.* Mapping the inhomogeneous electrochemical reaction through porous LiFePO$_4$-electrodes in a standard coin cell battery. *Chem. Mater.* **2015**, *27*, 2374–2386.

11. Zhao, W.; Luo, G.; Wang, C.-Y. Effect of tab design on large-format Li-ion cell performance. *J. Power Sources* **2014**, *257*, 70–79.

12. Eddahech, A.; Briat, O.; Vinassa, J.-M. Thermal characterization of a high-power lithium-ion battery: Potentiometric and calorimetric measurement of entropy changes. *Energy* **2013**, *61*, 432–439.

13. Robinson, J.B.; Darr, J.A.; Eastwood, D.S.; Hinds, G.; Lee, P.D.; Shearing, P.R.; Taiwo, O.O.; Brett, D.J.L. Non-uniform temperature distribution in Li-ion batteries during discharge—A combined thermal imaging, X-ray micro-tomography and electrochemical impedance approach. *J. Power Sources* **2014**, *252*, 51–57.

14. Veth, C.; Dragicevic, D.; Merten, C. Thermal characterizations of a large-format lithium ion cell focused on high current discharges. *J. Power Sources* **2014**, *267*, 760–769.

15. Goutam, S.; Timmermans, J.M.; Omar, N.; van den Bossche, P.; van Mierlo, J.; Rodriguez, L.; Nieto, N.; Swierczynski, M. Surface temperature evolution and the location of maximum and average surface temperature of a lithium-ion pouch cell under variable load profiles. In Proceedings of the EEVC European Electric Vehicle Congress EEVC-2014, Brussels, Belgium, 3–5 December 2014.

16. Samba, A.; Omar, N.; Gualous, H.; Capron, O.; van den Bossche, P.; van Mierlo, J. Impact of tab location on large format lithium-ion pouch cell based on fully coupled tree-dimensional electrochemical-thermal modeling. *Electrochim. Acta* **2014**, *147*, 319–329.

17. Lin, C.; Xu, S.; Chang, G.; Liu, J. Experiment and simulation of a LiFePO$_4$ battery pack with a passive thermal management system using composite phase change material and graphite sheets. *J. Power Sources* **2015**, *275*, 742–749.

18. Karimi, G.; Li, X. Thermal management of lithium-ion batteries for electric vehicles. *Int. J. Energy Res.* **2013**, *37*, 13–24.

19. Mulder, G.; Omar, N.; Pauwels, S.; Meeus, M.; Leemans, F.; Verbrugge, B.; de Nijs, W.; van den Bossche, P.; Six, D.; van Mierlo, J. Comparison of commercial battery cells in relation to material properties. *Electrochim. Acta* **2013**, *87*, 473–488.

20. Takano, K.; Saito, Y.; Kanari, K.; Nozaki, K.; Kato, K.; Negishi, A.; Kato, T. Entropy change in lithium ion cells on charge and discharge. *J. Appl. Electrochem.* **2002**, *32*, 251–258.

21. Jalkanen, K.; Aho, T.; Vuorilehto, K. Entropy change effects on the thermal behavior of a LiFePO$_4$/graphite lithium-ion cell at different states of charge. *J. Power Sources* **2013**, *243*, 354–360.

22. Viswanathan, V.V.; Choi, D.; Wang, D.; Xu, W.; Towne, S.; Williford, R.E.; Zhang, J.-G.; Liu, J.; Yang, Z. Effect of entropy change of lithium intercalation in cathodes and anodes on Li-ion battery thermal management. *J. Power Sources* **2010**, *195*, 3720–3729.

23. Li, Z.; Zhang, J.; Wu, B.; Huang, J.; Nie, Z.; Sun, Y.; An, F.; Wu, N. Examining temporal and spatial variations of internal temperature in large-format laminated battery with embedded thermocouples. *J. Power Sources* **2013**, *241*, 536–553.

24. Orikasa, Y.; Maeda, T.; Koyama, Y.; Murayama, H.; Fukuda, K.; Tanida, H.; Arai, H.; Matsubara, E.; Uchimoto, Y.; Ogumi, Z. Transient phase change in two phase reaction between LiFePO$_4$ and FePO$_4$ under battery operation. *Chem. Mater.* **2013**, *25*, 1032–1039.

25. Robert, D.; Douillard, T.; Boulineau, A.; Brunetti, G.; Nowakowski, P.; Venet, D.; Bayle-Guillemaud, P.; Cayron, C. Multiscale phase mapping of LiFePO$_4$-based electrodes by transmission electron microscopy and electron forward scattering diffraction. *ACS Nano* **2013**, *7*, 10887–10894.

26. Chen, Z.; Belharouak, I.; Sun, Y.-K.; Amine, K. Titanium-based anode materials for safe lithium-ion batteries. *Adv. Funct. Mater.* **2013**, *23*, 959–969.

27. Kwon, K.H.; Shin, C.B.; Kang, T.H.; Kim, C.S. A two-dimensional modeling of a lithium-polymer battery. *J. Power Sources* **2006**, *163*, 151–157.

28. Li, J.; Cheng, Y.; Ai, L.; Jia, M.; Du, S.; Yin, B.; Woo, S.; Zhang, H. 3D simulation on the internal distributed properties of lithium-ion battery with planar tabbed configuration. *J. Power Sources* **2015**, *293*, 993–1005.

29. Wu, B.; Li, Z.; Zhang, J. Thermal design for the pouch-type large-format lithium-ion batteries: I. Thermo-electrical modeling and origins of temperature non-uniformity. *J. Electrochem. Soc.* **2014**, *162*, 181–191.

30. Guo, M.; White, R.E. A distributed thermal model for a Li-ion electrode plate pair. *J. Power Sources* **2013**, *221*, 334–344.

31. Gerver, R.E.; Meyers, J.P. Three-dimensional modeling of electrochemical performance and heat generation of lithium-ion batteries in tabbed planar configurations. *J. Electrochem. Soc.* **2011**, *158*, 835–843.

32. Yazdanpour, M.; Taheri, P.; Mansouri, A.; Bahrami, M. A distributed analytical electro-thermal model for pouch-type lithium-ion batteries. *J. Electrochem. Soc.* **2014**, *161*, 1953–1963.

33. Yi, J.; Lee, J.; Shin, C.B.; Han, T.; Park, S. Modeling of the transient behaviors of a lithium-ion battery during dynamic cycling. *J. Power Sources* **2015**, *277*, 379–386.

34. Taheri, P.; Mansouri, A.; Yazdanpour, M.; Bahrami, M. Theoretical analysis of potential and current distributions in planar electrodes of lithium-ion batteries. *Electrochim. Acta* **2014**, *133*, 197–208.

Energy Consumption Prediction for Electric Vehicles Based on Real-World Data

Cedric De Cauwer, Joeri Van Mierlo and Thierry Coosemans

Abstract: Electric vehicle (EV) energy consumption is variable and dependent on a number of external factors such as road topology, traffic, driving style, ambient temperature, *etc.* The goal of this paper is to detect and quantify correlations between the kinematic parameters of the vehicle and its energy consumption. Real-world data of EV energy consumption are used to construct the energy consumption calculation models. Based on the vehicle dynamics equation as underlying physical model, multiple linear regression is used to construct three models. Each model uses a different level of aggregation of the input parameters, allowing predictions using different types of available input parameters. One model uses aggregated values of the kinematic parameters of trips. This model allows prediction with basic, easily available input parameters such as travel distance, travel time, and temperature. The second model extends this by including detailed acceleration data. The third model uses the raw data of the kinematic parameters as input parameters to predict the energy consumption. Using detailed values of kinematic parameters for the prediction in theory increases the link between the statistical model and its underlying physical principles, but requires these parameters to be available as input in order to make predictions. The first two models show similar results. The third model shows a worse fit than the first two, but has a similar accuracy. This model has great potential for future improvement.

Reprinted from *Energies*. Cite as: De Cauwer, C.; Van Mierlo, J.; Coosemans, T. Energy Consumption Prediction for Electric Vehicles Based on Real-World Data. *Energies* **2015**, *8*, 8573–8593.

1. Introduction

The energy consumption of vehicles has been a widely assessed research topic. Until recent years, research was focused on fuel consumption of conventional vehicles, and how driving styles, ITS (Intelligent Transport System) systems, and other intelligent systems impact fuel consumption [1]. Recently, these studies have been extended to the energy use of electric vehicles (EVs). EV energy consumption studies can generally be divided according to their purpose and model calculation methodology. Studies report the development of energy models for the purpose of EV drivetrain design and optimization [2,3], assessment of the influences on the energy consumption [4–6], and global energy consumption or grid impact due to the introduction of EV or hybrid vehicles [7,8]. In some cases the energy model is used for an (all-electric) range prediction [9]. The methodology for the calculation of energy consumption either consists of creating a vehicle model that simulates electrical parameters based on kinematic and dynamic requirements (backwards simulation) [3,5–7] or by means of statistical models based on measurements of the EV consumption, either from real-world data [4,9] or test cycles [2]. Using real-world measurements has the advantage of predicting more realistic values for energy consumption, but relies on available data and statistical modeling and is often uncoupled from the vehicle dynamics and drivetrain behavior. In contrast, using a vehicle model gives you a direct link with the vehicle dynamics and drivetrain behavior, making the identification of influences of the drivetrain parameters on energy consumption more clear. Although Shankar *et al.* [4] and Neaimeh *et al.* [9] both use real-world measurements for the calculation of the energy consumption, only a limited number of external parameters are included in their models and the link with the vehicle dynamics is low. In Neaimeh *et al.* [9], the prediction was based on a simplified vehicle dynamics model and only included road inclination as an external influencing parameter, whereas Shankar *et al.* [4] use a purely statistical approach based on road type detection and its linked average consumption. This paper proposes to extend those models and fill this knowledge gap by including more external parameters while increasing the link with the vehicle dynamics. The goal here is to detect and quantify correlations between the kinematic parameters of the electric vehicle and its energy consumption using real-world data, in order to predict the real-world consumption of electric vehicles as an intermediate step towards range prediction. A model that uses real-world data compared to using cycle test values produces more realistic energy consumptions, and has added value for research topics relying on consumption data as input. Life Cycle Assessments (LCA) and Total Cost of Ownership (TCO) studies, for example, still use energy consumption based on the New European Drive Cycle (NEDC) [10–12] or other standardized cycles.

2. Energy Models

The goal is to build up EV energy consumption models based on real-world measurements. The proposed models are statistical models based on the underlying physical principles of the vehicle dynamics and kinematics. As the ultimate goal of this study is all-electric range (AER) prediction for electric vehicles, the energy consumption considered in this paper is the energy consumption on a battery-to-wheel scope as defined in De Cauwer *et al.* [13] and corresponds to the energy drawn from the battery. Therefore, energy losses in the energy supply chain prior to the battery are not considered as they do not impact the range of the EV. As such, grid losses and charging losses are not included in this model. As De Cauwer *et al.* [13] argues, it does, however, influence real-world tank-to-wheel consumption and therefore the costs associated with real-world EV use. The battery-to-wheel consumption of an electric vehicle is a function of the required mechanical energy at the wheels, determined by the kinematic parameters over a trajectory, the drivetrain efficiency, and the energy consumption of auxiliaries. The total required mechanical energy at the wheels as a function of the kinematic parameters describing vehicle movement can be expressed in the vehicle dynamics equation:

$$E_{ij} = \tfrac{1}{3600} \left[m_{ij}.g.\left(f.cos\varphi + sin\varphi \right) + 0.0386.\left(\rho.C_x.A.v_{ij} \right) + \left(m_{ij} + m_f \right).\tfrac{dv}{dt} \right] d_{ij}, \quad (1)$$

where:

E_{ij} = Mechanical energy required at the wheels to drive on a distance d_{ij} [kWh]

m_{ij} = Total vehicle mass [kg]

m_f = Fictive mass of rolling inertia [kg]

g = Gravitational acceleration $\left[m/s^2 \right]$

f = Vehicle coefficient of rolling resistance [−]

φ = Road gradient angle [°]

ρ = Air density $\left[kg/m^3 \right]$

C_x = Drag coefficient of the vehicle [−]

A = Vehicle equivalent cross section $\left[m^2 \right]$

v_{ij} = Vehicle speed between the point i and the point j [km/h]

d_{ij} = Distance driven from point i to point j [km]

Equation (1) includes five terms, each describing a contribution to the energy consumption. These terms describe, respectively, the rolling resistance, potential energy, aerodynamic losses, kinetic energy, and energy for the acceleration of rotational parts. The aerodynamic losses and rolling resistances are pure energy losses. The potential and acceleration (kinetic) energy for EVs, unlike for conventional vehicles, can partly be recovered by regenerative braking. It is the EV battery that provides, through the numerous energy conversion steps of the drivetrain, the required energy for traction. Additionally, the EV battery also provides the energy to supply the auxiliaries. Heating and air-conditioning systems can make a significant contribution to the EV energy consumption and are ambient temperature dependent.

The proposed models are statistical models that correlate these kinematic parameters over a trajectory and the measured energy consumption at the battery, based on the vehicle dynamics equation. They are application specific, because different research topics do not necessarily have the same input data available for the model, nor do they require the same level of accuracy. Three models will be proposed here: one using the aggregated trip data, a second using aggregated trip data with additional acceleration data, and a third using shorter trip segments or "micro-trips" to calculate energy consumptions. These models have a direct link with the vehicle kinematics (vehicle speed, distance, altitude, *etc.*) but do not yet allow extraction of secondary effects that impose kinematic constraints (such as traffic and weather), although this information is inherently included in the real-world data. The first energy model is a model based on macro trips, *i.e.*, uses aggregated values of the kinematic parameters that describe trips. These are the average vehicle speed, the distance driven, and the elevation. By applying multiple linear regression on the real-world measured values of the consumed energy and kinematic parameters for trips, a linear model is created based on the vehicle model presented above. The model presented in Lebeau *et al.* [14] used the vehicle dynamics equation to calculate the required mechanical energy and added a temperature- and time-dependent term to account for auxiliaries and temperature-dependent efficiencies. The difference here is that the prediction is purely statistical without any calculation based on theoretical values, and based on the underlying known relations between energy consumption and a number of independent variables presented in Equation (1). As such we must have a distance-dependent term describing rolling resistance, a speed-squared-dependent term describing aerodynamic losses, a height-dependent term describing potential energy, and a term describing auxiliary consumption. Two obvious factors influencing energy consumption that are present in the physical model, *i.e.*, acceleration and weight variation, are absent here. This approach is justified for applications where the input data for the prediction only contain aggregated values of the kinematic parameters (such as those presented in Lebeau *et al.* [14]). The weight factor cannot be extracted as this information was not in the available data. To describe the linear dependency of the auxiliary use and the energy consumption, we introduce a temperature-scaled, time-dependent term. To be able to introduce such a term into the linear regression, we reform the temperature scale to an absolute one, with 20 °C being the neutral temperature, and introduce the parameter $aux \in [0,1]$, with its value being the ratio of the time the auxiliaries are on. Now the independent variable in the regression analysis describing auxiliary consumption becomes:

$$|20 - T| * aux * t \qquad (2)$$

with:

$$T : ambient\ temperature\ in\ °C$$

$$\Delta t : time\ difference$$

$$aux : \frac{Duration\ of\ auxiliaries\ switched\ on}{Total\ duration\ of\ trip}$$

550

The resulting formula using linear regression describing energy consumption in this simplified model then becomes:

$$E = B_1 * s + B_2 * \langle v \rangle^2 * s + B_3 * |20 - T| * aux * t + B_4 * Hpos + B_5 * Hneg + \varepsilon \quad (3)$$

with:

B_i : regression coefficients
ΔE : energy
$\langle v \rangle$: average speed
Δs : distance
$|20 - T| * aux$: scaled temperature value with T in °C
Δt : time
$\Delta Hpos$: positive elevation
$\Delta Hneg$: negative elevation
ε : error term

Except for the temperature, the unit of each parameter does not matter as a linear relationship is assumed; it will only change the regression coefficients but not the result. This model will be referred to as the macro model. Using the average speed instead of the integration in the aerodynamic term $(B_2 * \langle v \rangle^2 * s)$ in Equation (3) (or sum because of the discrete nature of the data) introduces a (large) approximation because of the quadratic nature of the term. Moving towards a model that is no longer restricted by the available input data for prediction, the aerodynamic term in Equation (3) can be replaced by:

$$B_2 * \sum_{i=1}^{n} v_i^2 * \Delta s_i \quad (4)$$

with:

$$\sum_{i=1}^{n} \Delta s_i = \Delta s \quad (5)$$

where n is the number of recorded data points.

Similarly, the v_i can be used to include a term describing acceleration, so kinetic energy changes are calculated by the sum of the squared speed differences. If divided by the distance driven, this becomes a characteristic metric that is here called the constant motion factor (CMF) and describes the kinetic energy changes per unit distance:

$$CMF = \frac{\sum_{i=2}^{n} |v_i - v_{i-1}|}{\Delta s} \quad (6)$$

Over a complete trip there will be as much positive change as negative change and the contributions of decelerations and accelerations to the CMF will be virtually equal. Including Equations (5) and (6) in Equation (3) gives:

$$\Delta E = B_1 * \Delta s + B_2 * \sum_{i=1}^{n} v_i^2 * \Delta s_i + B_3 * abs(20 - T) * aux * \Delta t + B_4 * \frac{\sum_{i=2}^{n} |v_i^2 - v_{i-1}^2|}{\Delta s_i}$$

$$+ B_5 * \Delta Hpos + B_6 * \Delta Hneg + \varepsilon$$

$$= B_1 * \Delta s + B_2 * \sum_{i=1}^{n} v_i^2 * \Delta s_i + B_3 * |20 - T| * aux * \Delta t + B_4 * CMF * \Delta s$$

$$+ B_5 * \Delta Hpos + B_6 * \Delta Hneg + \varepsilon \qquad (7)$$

This model uses non-averaged data of the speed and distance parameters and sums them up over the trip before performing the linear regression on the aggregated values. The model and its coefficients are still based on the aggregated trips but not yet based on micro-trips. Therefore we will refer to this model as the *hybrid model*.

Another option is to segment a trip into pieces of equal duration (micro-trips) and use the detailed measuring points to make a prediction on the micro trip consumption using linear regression. Thereafter the consumptions of the segments are integrated over the trip to predict the total consumed energy. Unlike for the macro trips, for the micro trips the contributions of the positive and negative speed changes to the CMF will not necessarily be equal and the CMF should be split in CMF^+ and CMF^- when $v_{i+1} > v_i$ and $v_{i+1} < v_i$ respectively.

In this model, Equation (7) evolves to:

$$\Delta E = \sum_{trip} \Delta Emicro$$

$$= \sum_{segments} B_1 * \sum_{i}^{segment} \Delta s_i + B_2 * \sum_{i}^{segment} v_i^2 * \Delta s_i + B_3 |20 - T_{segment}|$$

$$* aux_{segment} * \Delta t_{segment} + B_4 * \left(\sum_{i}^{segment} CMF_i^+ \right) * \Delta s + B_5 \qquad (8)$$

$$* \left(\sum_{i}^{segment} CMF_i^- \right) * \Delta s + B_6 * \sum_{i}^{segment} \Delta Hpos_i + B_7 * \sum_{i}^{segment} \Delta Hneg_i + \varepsilon$$

This model has the potential to be the most accurate as it does not apply averaging and so more information resides in the values of the parameters. However, these detailed values of the kinematic parameters to be fed into the model for prediction are not necessarily readily available for any random trip, making prediction with this model impossible. To make this model applicable for the prediction of energy consumption of new trips, an additional correlation has to be done between characteristic values of these kinematic and physical (measured here) parameters and external factors so that their occurrence in the chosen trip can be predicted. This model will be referred to as the *micro model*.

3. Data Availability and Reliability

The available data were generated with a logger device recording the CAN bus signals of a 2012 Nissan Leaf (24 kWh Li-Ion battery, 80 kW electric motor, 1700 kg vehicle mass [15]) during a two year period of 2013 and 2014. A CAN logger registered the battery current and voltage, the state-of-charge (SoC), the GPS coordinates, and the timestamp at a 1 Hz frequency. This provided a dataset containing 23,700 km distance covered, spread over the 261 days in 2013 and 312 days in 2014 when the vehicle was driven (out of a total of 730 days registered). The vehicle was driven by up to 10 individuals on a regular basis, with no restrictions of use, and was located in the outer region of the city of Brussels. This way most types of use and road topology (urban, highway, rural, *etc.*) are represented in the data. However, because the data are generated by only one vehicle, the constructed models are drivetrain specific and potential differences in the impact of the external factors on real-world energy consumption for different EV types cannot be extracted from the data. The speed and acceleration values used are based on position coordinates; therefore an error exists on speed and especially acceleration. The acceleration often showed unrealistic acceleration peaks of over 2 m/s^2. Therefore, if the speed was monotonically increasing (positive acceleration) or decreasing (negative acceleration) over multiple measurement points, the acceleration was averaged up to a maximum of three measurement points (3 s). By increasing the duration over which the acceleration is calculated, the relative error of the acceleration decreases, but the acceleration is more averaged. The maximum of three measurement points (so 3 s) was chosen to have a balance between both. The acceleration peaks and the averaged values are illustrated in Figure 1 by plotting the acceleration values for 2014 for a selection of data points.

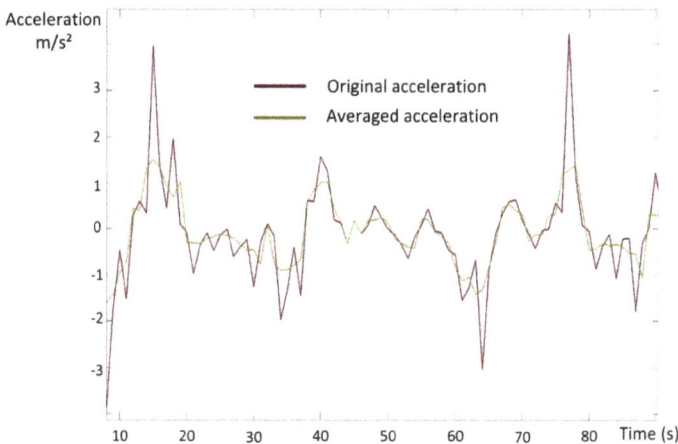

Figure 1. Acceleration and averaged acceleration for a number of data points.

553

Occurrences of loss of satellite connection caused the data to "jump", *i.e.*, divert from the 1 Hz data point registration. Because of this, the vehicle "jumps" to the next location, making the distance between the points the straight line distance, causing the distance and acceleration to be wrong. Moreover, as the energy consumption is calculated by integrating the battery voltage and current over time, decreasing the log frequency increases integration error. This effect can be considerable when driving into tunnels. Making a hard filter to eliminate all trips that recorded events of time lapses over 3 s resulted in only 15%–30% of the total available trips remaining. Investigation of these particular trips with time lapses showed they did not significantly alter the results of the models on a systematic basis. Therefore, when using the data structured under the form of trips, we decided not to use this hard filter but to eliminate individual cases (entire trips) after analysis using statistical post-processing tools: extreme values of the standard residuals (>2) and Cook's distances (>1) were individually investigated and eliminated in the case of non-reliable data. Figure 2 shows a part of a trip where time lapses are registered. It also shows the current value at those points to illustrate that data jumps at non-average current values lead to large energy errors due to the integration.

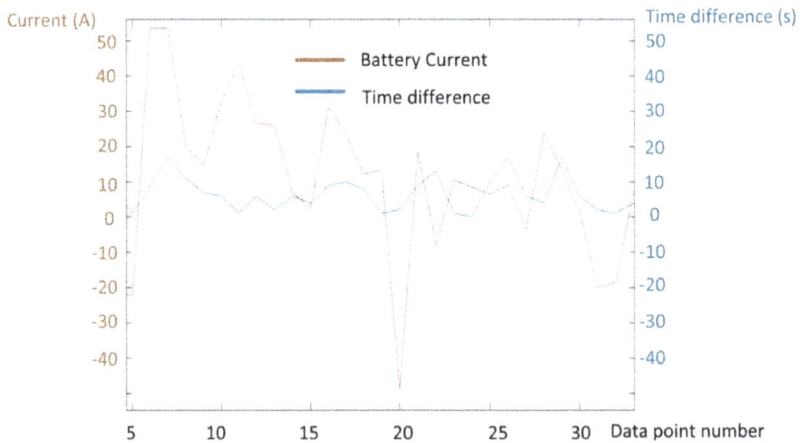

Figure 2. Example of data point jumps and the current value at those points.

An additional problem was the occasional speed spike. The trips containing unrealistic high speed values were deleted *a posteriori* after detection of outliers through the same statistical post-processing tools. After the post-processing and filtering was applied, the remaining usable data for the model construction consisted of 21,300 driven kilometers. All the remaining usable data was used for the multiple linear regression and not split up into parts for model construction and model validation. The statistical nature of the model, the vast amount of measured data used, and post-processing sensitivity analysis justify this approach. However, for further improvement and usefulness of the model, the considered vehicle fleet could be extended and further validation of the individual contributions of external factors could be performed.

4. Model Results

4.1. Macro Models

The results of the multiple linear regressions for both the *macro* model and *hybrid* model are presented in Table 1. The table gives the correlation coefficient (R^2), an overview of the regression coefficients (B_i) and its standard deviation (SE B_i) and confidence level (P), the standardized beta estimates (β_i), and the relative average contribution of each term. The standardized beta estimates give the number of standard deviations by which the prediction will change when the predictor changes one standard deviation and is, as such, an indication for the importance of a predictor. The relative average contribution of a term (expressed as a percentage) is the contribution of that term to the total energy consumption and is calculated by taking the sum over all trips of that term, multiplied by its regression coefficient and then divided by the sum of all energy consumption.

The confidence levels of the regression coefficients show that all predictors are very significant. The relative average contribution is an indication of how each predictor (term) contributes to the total energy consumption. The difference of the relative average contributions and standardized beta coefficients for the different predictors between the macro and hybrid model shows that the model is sensitive to the presence or absence of significant predictors. Introducing the CMF to include acceleration in the model has a big impact on the regression coefficients and thus the average contribution of the terms to the total consumption. The acceleration term accounts for almost 20% of the energy consumption, mostly at the cost of the rolling resistance contribution, which decreases by 25%. Introducing an acceleration term also causes minor shifts in the other terms. It is, however, logical that with the absence of an acceleration term in the macro model, the contribution of the acceleration shifts to the distance term because, on average, longer trips will have more acceleration actions. However, on individual trips, this is not the case at all because for trips with an equal distance, the amount and magnitude of acceleration actions can vary significantly depending on the road type, congestion level, and driving style. In both models a constant term also makes a contribution to the prediction. These constant terms come out of the linear regression, but were not defined in the physical model. This means part of the variability of the data is not accounted for by the predictors from the physical model and resides in this intercept term. One cause can be the non-linear auxiliary consumption during trips, which is caused by the conditions of the vehicle after a period of standstill. After a period of standstill, the vehicle first has to be brought to the requested temperature before maintaining this temperature for the rest of the trip, during which the consumption can be expected to be linearly time dependent. Although average contributions of the intercept terms are small for both models (3%–4%), the contribution per individual trip can be much higher. The average contribution is a good measure of how much each term contributes to the total energy consumption but gives no information on the distribution of contributions across the trips. There are considerable variations in the relative contribution of each term to the consumed energy across these trips. For the constant term, for example, the absolute contribution will be the same for a very short or very long trip, making the relative contribution vary strongly. For the auxiliaries, the distribution of contributions also includes trips with no auxiliary use, thus pulling the average value down. To illustrate these examples, the distributions of the contribution to the energy consumption per trip by the auxiliary and constant term resulting from the model are shown in Figures 3 and 4, respectively.

Table 1. Multiple linear regression results for the macro and hybrid model.

	Adjusted R^2	B_i	SE B_i	β_i	P	Avg. % Contribution
Model 1 macro	0.956				<0.001	
Constant		1.19E-1	2.65E-2		<0.001	4%
Rolling resistance		1.32E-1	1.84E-3	0.83	<0.001	78%
Aerodynamic		5.00E-6	4.45E-7	0.11	<0.001	6%
Auxiliaries		1.83E-1	7.35E-3	0.14	<0.001	7%
Pos. elevation		3.08E-3	3.72E-4	0.23	<0.001	26%
Neg. elevation		−2.54E-3	3.68E-4	−0.19	<0.001	−21%
Model 2 hybrid	0.968					
Constant		9.24E-2	2.33E-2		<0.001	3%
Rolling resistance		9.36E-2	2.40E-3	0.58	<0.001	55%
Acceleration		1.00E-2	7.53E-3	0.16	<0.001	18%
Aerodynamic		5.59E-6	2.35E-7	0.24	<0.001	15%
Auxiliaries		· 1.78E-1	6.28E-3	0.14	<0.001	7%
Pos. elevation		2.75E-3	3.18E-4	0.21	<0.001	23%
Neg. elevation		−2.59E-3	3.15E-4	−0.20	<0.001	−22%

R^2: correlation coefficient; B_i: regression coefficient; SE B_i: standard deviation of the regression coefficient; β_i: standardized beta estimate; P: confidence level; Avg. % contribution: the relative average contribution.

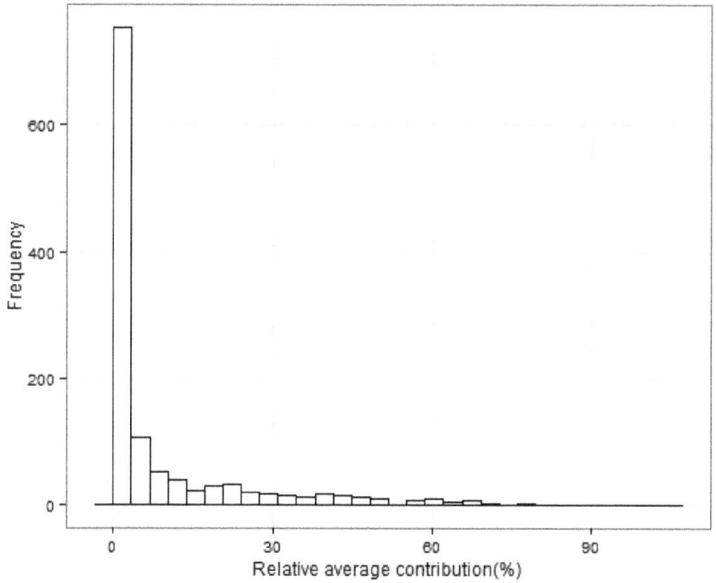

Figure 3. Distribution of the relative contribution of the auxiliary term to the energy consumption of all trips.

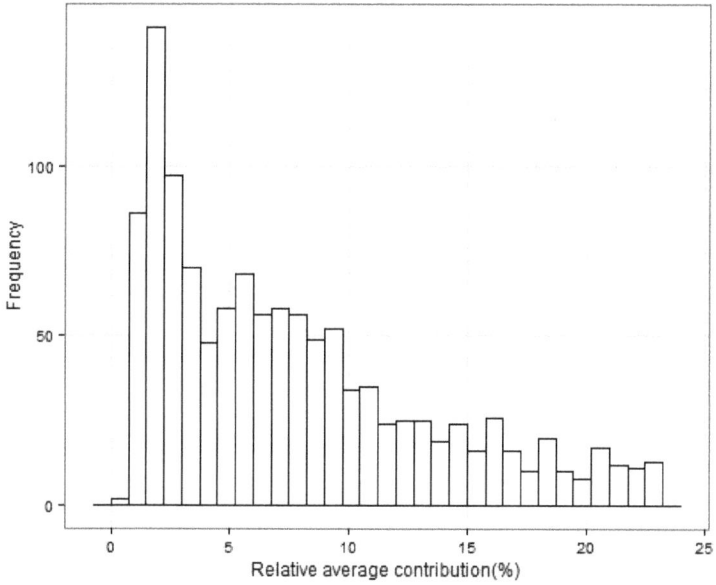

Figure 4. Distribution of the relative contribution of the constant term to the trip consumption for all trips.

557

The distribution of the contribution of the auxiliaries to the energy consumption given by the model is hyperbolic, with a spike for non-auxiliary use trips and contributions running up to about 80% (rare occasions). The trips with a prediction of a high relative contribution of the auxiliary energy consumption are trips with low average speed. Relative contributions of the auxiliary consumption for trips with an average speed of over 40 km/h do not exceed 40%. A low average speed means a low average power demand (or increased standstill periods), increasing the auxiliary consumption relative to the total consumption. This effect is most noticeable in this simple macro model, where the contribution of the accelerations is not accounted for and this driving characteristic is only represented in the speed-dependent (quadratic) aerodynamic term. This shows how a different use of the vehicle results in significantly different energy consumption. The distribution of the constant term contribution to the energy consumption prediction is a direct result of the distribution of the trip consumptions and ranges up to almost 25%.

Using R to construct the 0.99 confidence interval for the predicted value, then dividing those values by the predicted value gives the confidence interval on the prediction relative to the value itself. The results are shown in Figure 5. For trips over 5 kWh, the 0.99 confidence interval remains within 5% of the predicted value, meaning that the distributions of the regression coefficients are narrow and very significant and the average of those predictions are statistically not likely to be more than 5% off. Looking at the actual relative error on the prediction (that is, the measured value minus the predicted, divided by the measured value), errors can be quite large for trips with a low quantity of energy consumed (<2.5 kWh), although the concentration of trips remain within the 25% error margin. The relative error on the prediction for the macro model and its distribution can be seen in Figures 6 and 7, respectively. The graph in Figure 6 also shows that the average relative error decreases for trips with a higher quantity of energy used and stays within the 25% error band for trips higher than 2.5 kWh consumption (except for a few negative errors). This band narrows further with increased consumption.

The plot of relative errors of the hybrid model has a very similar shape to the macro model, but with lower errors and an error band that narrows down more quickly with increased energy consumption trips, as can be seen in Figure 8.

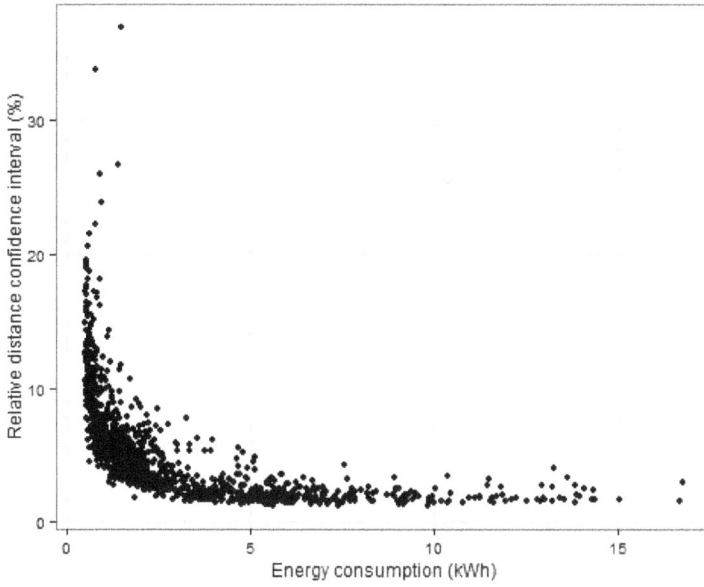

Figure 5. Confidence intervals for the predicted values of the macro model, expressed in relative distance (%) to the predicted value.

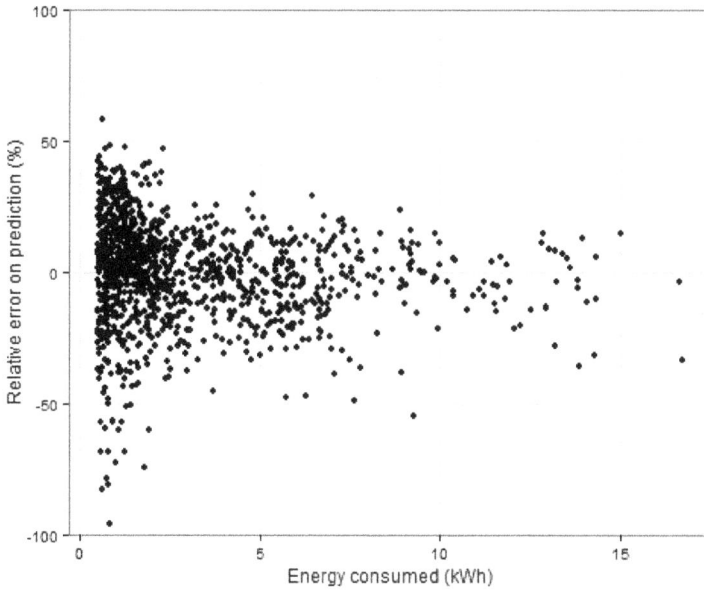

Figure 6. Relative error prediction (%) as a function of the consumed energy (kWh) in the trip for the macro model.

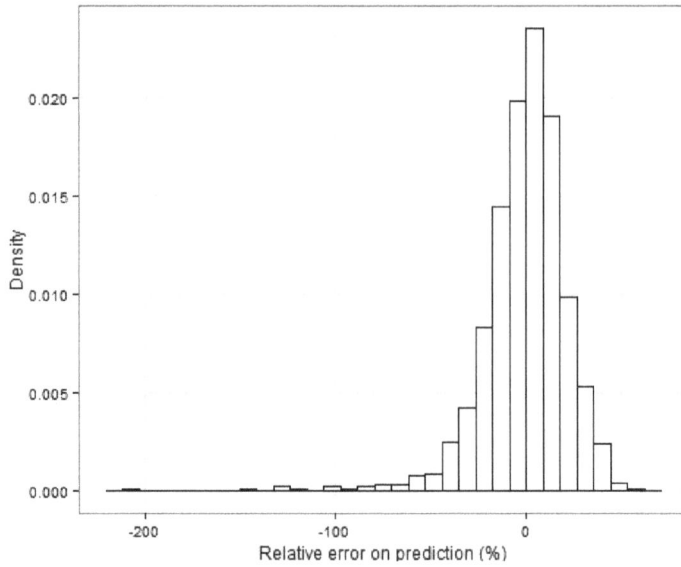

Figure 7. Distribution of the relative errors (%) for the macro model.

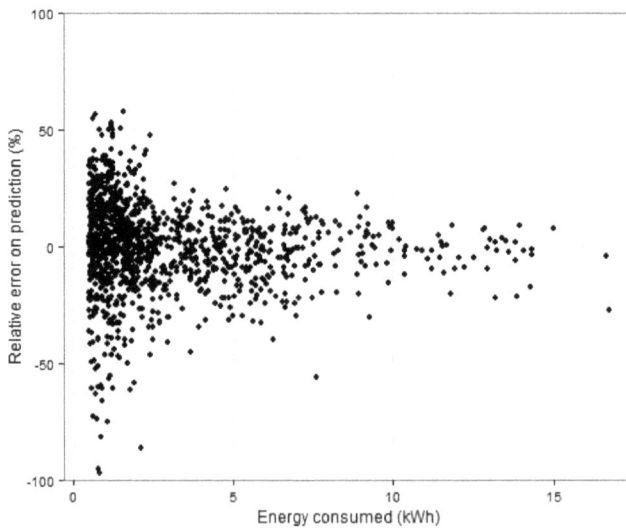

Figure 8. Relative error on the prediction (%) as a function of the consumed energy (kWh) in the trip for the hybrid model.

4.2. Micro Model

The results of the hybrid model show that moving towards more detailed values of the predictors shifts the regression coefficients and results in a better correlation.

A next step can be dividing the trips into smaller segments to build the regression model, and then aggregating the predictions of the energies for the segments to obtain the energy consumption of a considered trip. Using this approach, detailed values of the kinematic parameters, instead of averaged or aggregated values, are required to be able to calculate the predictor values over segments, making this model unusable for cases where these data are not available. Yet, if it is possible to link characteristic values of these predictors to known parameters, such as congestion levels and road type, this approach has great potential for improving the above model. Following this logic, these segments need to have sufficiently significant values for the predictors but avoid over aggregation. Therefore trips have been segmented into blocks with an equal duration of 2 min, 5 min and 10 min for experiments with the magnitude of the blocks, starting with 2 min segments. The results of the multiple linear regression using 2 min micro-trips are listed in Table 2.

Table 2. Results of the multiple linear regression for the micro-trip model.

	Adjusted R^2	B_i	SE B_i	β_i	P	Avg. % Contribution
Model 3 Micro	0.89					
Constant		3.38E-2	1.49E-3		<0.001	18%
Rolling resistance		6.43E-2	1.80 E-3	0.35	<0.001	35%
Acceleration (pos)		3.14E-3	3.81E-3	0.52	<0.001	59%
Acceleration (neg)		1.98E-3	3.87E-3	0.33	<0.001	-136%
Aerodynamic		7.85E-6	1.58 E-7	0.41	<0.001	17%
Auxiliaries		1.94E-2	3.44E-3	0.22	<0.001	9%
Pos. elevation		1.05E-3	4.01E-5	0.12	<0.001	11%
Neg. elevation		−1.16E-3	3.99E-5	−0.13	<0.001	-12%

Again, all predictors are very significant. The relative average contributions of the auxiliaries' term, aerodynamic term, and acceleration term are very similar to the hybrid model, but other contributions are different. The 2 min segment model showed the following problems:

1. A large contribution of the constant (error) term (20% on average);
2. A bigger negative contribution of the negative elevation than the positive contribution of the positive elevation, which does not correspond to the physics;
3. Low correlation coefficient (around 0.89).

This resulted in a process of detailed investigation to find the reasons for these problems and how to solve them. The problems were not caused by an insufficient number of data points per segment, as increasing the segment duration (and thus the level of aggregation) did not resolve this problem. One way of solving the problem is to combine a dataset of 2 min segments with a dataset of 10 min segments. This led to results that correspond to the physical principles of the model, yet are statistically unsound because of the artificial distribution of energy consumptions with the two levels of duration. Therefore the two min segment seems the best choice and gave the best results. However, the problems listed above demonstrate that a lot of the variability in the data is not accounted for by the model. The suspected reason for this is that because of the low predictor values, the errors in the measurements (such as integration error, assumptions in Equation (2), *etc.*) become more significant and influence the linear regression. The relative error on the prediction of the micro-trips (segments) and its distribution are depicted in Figures 9 and 10, respectively. The figures show that large error exists for the prediction segment energy consumption, but the bulk of the error lies within the 25% error band. The existence of large error for the segment predictions is consistent with the observation above that less variability is accounted for in the micro model. Combining the micro segment predictions, the energy consumption for the trips can be calculated. The relative error on that prediction is given in Figure 11. Despite the flagrant problems with the model listed above, it shows comparable accuracy to the macro and hybrid model for the prediction of energy consumption of complete trips.

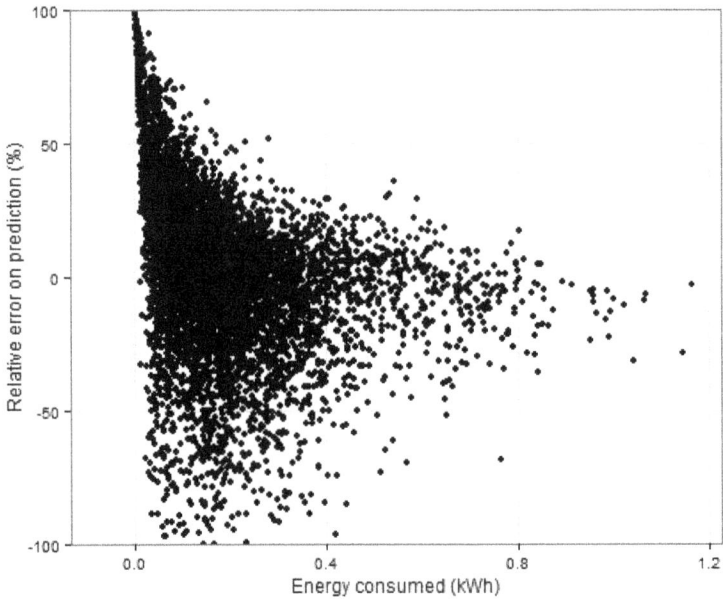

Figure 9. Relative errors on the micro-trip energy consumption predictions using the micro model.

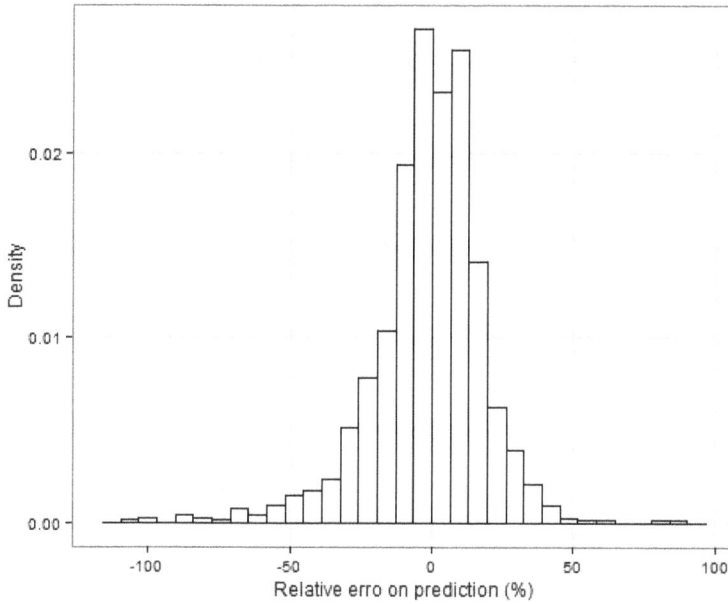

Figure 10. Distribution of the relative errors on the micro-trip energy consumption predictions using the micro model.

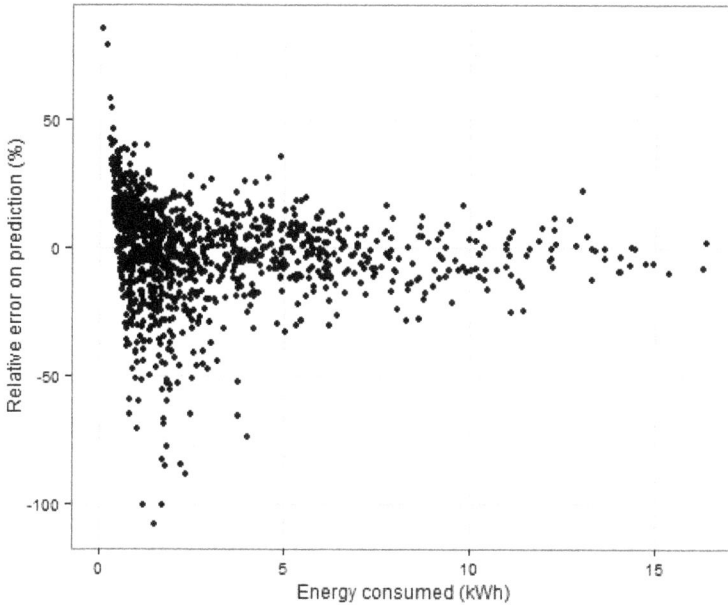

Figure 11. Relative error in the trips' energy consumption predictions using the micro model.

Comparing Figure 11 with Figures 6 and 8 demonstrates that the micro model accuracy is similar to the accuracy of the macro and hybrid model, with errors drastically decreasing when the quantity of consumed energy is above 2.5 kWh and is within the 25% error band for most of the trips.

Another approach would be to split up the energy consumption into positive (tractive power) and negative (regenerative braking) power, and construct a linear model for the regenerative energy with the negative acceleration and elevation. Using this approach, however, the negative elevation turns out not to be a significant parameter. This is not illogical, as gentle elevation differences do impact the positive power (tractive effort required) but do not necessarily result in negative power, as is further discussed in Section 5.

5. The Regenerative Braking Influencing Factors

Regenerative braking for EVs is an important part of the EV drivetrain efficiency [16] and regenerative braking strategies are a topic of research [16–18]. Figure 12 depicts the distribution of the relative amount of regenerated energy, defined as the ratio of the total amount of energy returned to the battery to the total amount of energy extracted from the battery. Figure 12 shows that while the average percentage of regenerated energy is about 15%, regenerated percentages can go up to 40% in extreme cases.

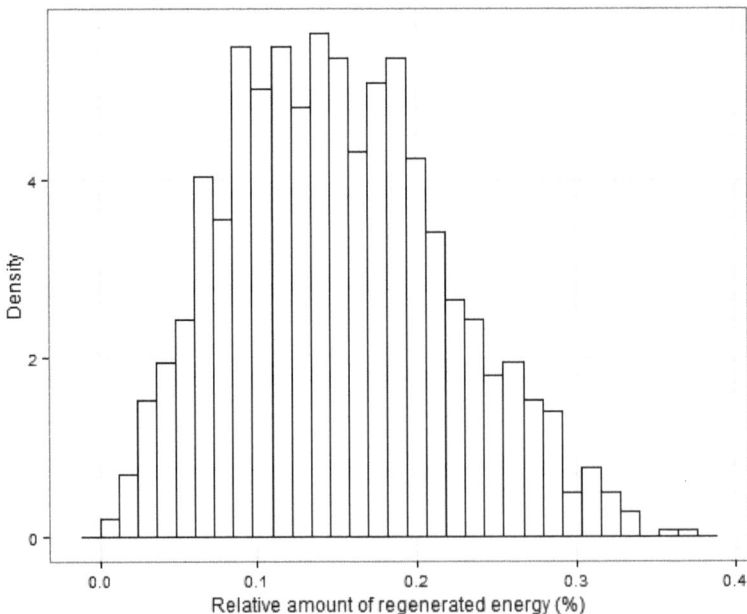

Figure 12. Distribution of regenerated energy percentage for all trips.

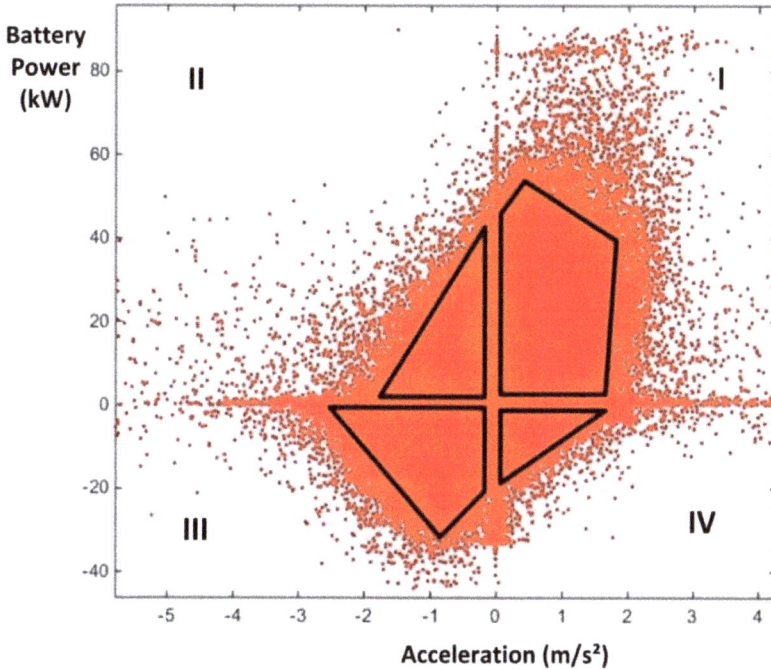

Figure 13. Four quadrant operation data points for the 2013 data.

The regenerative energy percentage is heavily dependent on the number of braking actions performed during a trip, which depends on road topology, traffic situation, and driving style, in combination with the regenerative braking strategy and drivetrain efficiency. This is shown in Mammosser *et al.* [17], where efficiency maps for regenerative braking are generated as a function of the braking force and vehicle speed using a vehicle simulation tool. In the second and third models, part of the use of the vehicle (braking actions) is represented in the CMF$^+$ and CMF$^-$, but these parameters do not include all effects. Regenerative braking only occurs within a torque and power window and driving style can have a further effect on efficient regenerative braking. Figure 13 depicts all the 2013 data points in an *acceleration–battery power* plane and illustrates the effect of driving style on regenerative braking. From Figure 13 it is observed that the regenerative power does not exceed 40 kW and has its largest value at a specific deceleration (at about -1 m/s^2 here, but for more information on the accuracy of this value see Section 3 (Data Availability and Reliability)). This highlights the non-linear principle of the regenerative braking energy and should be thoroughly investigated to further improve the model.

The battery power-vehicle acceleration plane can be divided into four quadrants. Quadrant one consists of actions using traction power for acceleration (so positive battery power and acceleration) and quadrant three for regenerative braking (negative power and acceleration). Less obvious are the actions residing in quadrant 2 and 4. Quadrant 2 has use of traction power but decreasing speed (negative acceleration), meaning traction power is still being used but not sufficiently to overcome resistive forces (for example, slowing down while climbing a hill). Quadrant 4 consists of actions where there is positive acceleration yet power is regenerated. This seems a very artificial condition, but the easiest example of a situation that would allow that to happen is driving steeply downhill. For ease of view, the bulk of the data in each of these quadrants are marked with a black cross section. Figure 13 seems to have a great number of outliers or extreme values, but since more than 1 million data points are depicted, this is not abnormal. Although the arguments above show why further analysis is required to better model regenerative braking and driving style, the CMFs can be used to categorize the use of the vehicle and are important parameters in the linear regression models above. Figures 14 and 15 depict the distribution of the CMFs for macro-trips and micro-trips, respectively. Both show similar distributions and show that occurrences of CMFS greater than 1 are very rare and greater than 2 non-existent (macro trips did show CMFs greater than 2 but from statistical analysis these outliers were investigated and deleted from the sample because of the corrupted data). These CMFs are calculated from the speed in km/h and distance in km. Having a value varying between 0 and 1 is convenient for the characterization, as the CMF basically indicates the amount of kinetic energy change per km, which will vary according to the road topology/traffic situation and driving style.

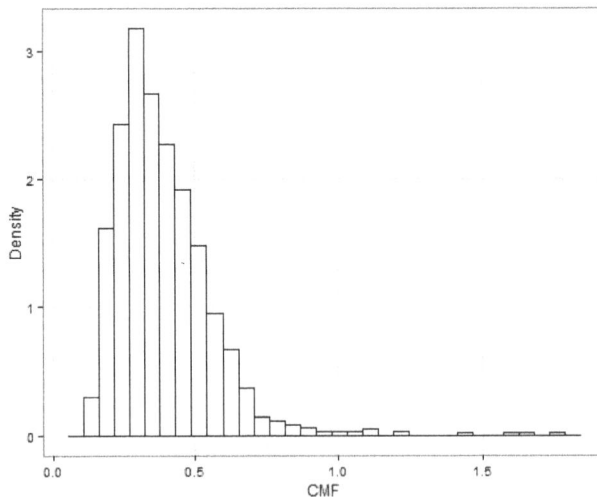

Figure 14. Histogram of the constant motion factor for the macro trips.

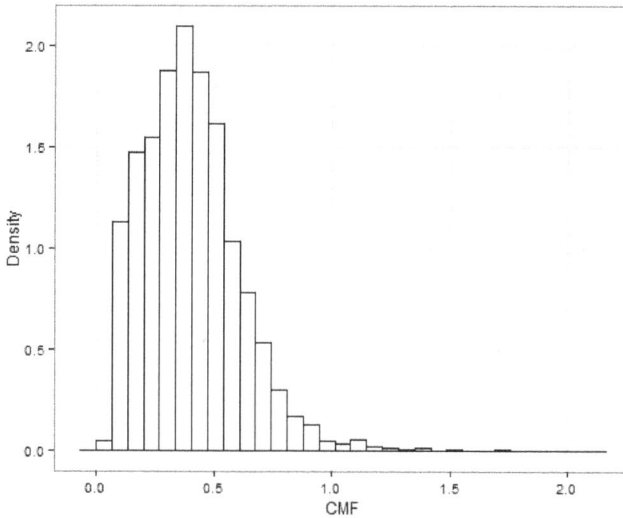

Figure 15. Histogram of the constant motion factor for the micro-trips.

Another parameter influencing regenerative braking is the weight of the vehicle. Increased mass results in an increased kinetic energy per unit speed. Moreover, this can shift the regenerative braking window. To investigate the influence of increased mass on vehicle consumption, during a three-month test phase in 2015 the vehicle was loaded with 200 kg. We expected to see shifted regression coefficients for the acceleration and elevation terms, as well as increased average consumption. This was not, in fact, observed, but a longer monitoring period might be required for the energy consumption to be significantly different.

6. Conclusions

Using the underlying physical relationships expressed in the vehicle dynamics equation, three models for EV energy consumption prediction have been constructed by applying the statistical method of multiple linear regression to real-world trip and energy consumption data for an EV. Because the data set consists of one vehicle's data only, the constructed models are drivetrain specific and potential differences in the impact of the external factors on real-world energy consumption for different EV types are not included in these models. Each model uses a different level of aggregation for the input parameters (predictors), which allows for prediction according to the available input data. One model uses kinematic parameters aggregated over trips; one model extended this model by including detailed acceleration parameters over the trip; and a third model uses detailed values of the kinematic parameters to predict the energy consumption over micro-trips. Multiple linear regression with a high correlation coefficient proved to be a good technique for predicting the energy consumption using aggregated trip data by accounting for a large portion of the variability residing in the data. Introducing more detailed trip data in the second model shifted the correlation coefficients and showed the importance of including a term representing the acceleration. From the results it is clear that moving towards a model with a higher level of detail has the potential to create a more accurate model. Both the first and second model had the concentration of the prediction errors within the 25% and stayed within that level for trips with energy consumption over a few kWhs. Using multiple linear regression on micro-trips in the third model resulted in regression coefficients that do not correspond to the underlying physical principles. Moreover, to make this model applicable, this detailed level of input has to be extracted and predicted from road, environmental, and traffic situations. Despite these drawbacks for the third model, in addition to a lower correlation coefficient and high constant term (so there is less variability accounted for), the prediction of the consumption over a complete trip had similar accuracy to the other two models, which indicates the potential of this method. So, further study of statistical outliers, mapping the non-linear effects of regenerative braking, and linking traffic situations and driving style with characteristic parameters such as the CMF have the potential to contribute to a model that is both accurate and extractable from available and statistical road data.

Acknowledgments: The authors would like to acknowledge the Agency for Innovation by Science and Technology in Flanders (IWT) as the funder for the PhD grant of Cedric De Cauwer, and Flanders Make for the support to our team. The authors would also like to acknowledge the EVA, iMOVE and EVTeclab platforms of the Flanders' Living labs for providing the required framework to perform the research for this paper.

Author Contributions: Cedric De Cauwer contributed in all parts of the process of this publication: collecting the data, processing the data, constructing the model, statistical analysis, interpreting the results, and writing the paper. Joeri van Mierlo contributed in the construction of the model, the interpretation of the results, and reviewing the paper. Thierry Coosemans contributed in the construction of the model, the interpretation of the results, and reviewing the paper.

Conflicts of Interest: The authors declare no conflict of interest.

References

1. Fotouhi, A.; Yusof, R.; Rahmani, R.; Mekhilef, S.; Shateri, N. A review on the applications of driving data and traffic information for vehicles' energy conservation. *Renew. Sustain. Energy Rev.* **2014**, *37*, 822–833.
2. Wager, G.; McHenry, M.P.; Whale, J.; Bräunl, T. Testing energy efficiency and driving range of electric vehicles in relation to gear selection. *Renew. Energy* **2014**, *62*, 303–312.
3. Dib, W.; Chasse, A.; Moulin, P.; Sciarretta, A.; Corde, G. Optimal energy management for an electric vehicle in eco-driving applications. *Control. Eng. Pract.* **2014**, *29*, 299–307.

4. Shankar, R.; Marco, J. Method for estimating the energy consumption of electric vehicles and plug-in hybrid electric vehicles under real-world driving conditions. *Intell. Transp. Syst. IET* **2013**, *7*, 138–150.

5. Yao, E.; Yang, Z.; Song, Y.; Zuo, T. Comparison of electric vehicle's energy consumption factors for different road types. *Discret. Dyn. Nat. Soc.* **2013**, *2013*, 328757:1–328757:7.

6. Badin, F.; le Berr, F.; Briki, H.; Petit, M.; Magand, S.; Condemine, E. Evaluation of EVs energy consumption influencing factors. In Proceedings of the 2013 World Electric Vehicle Symposium and Exhibition (EVS27), Barcelona, Spain, 17–20 November 2013; pp. 1–12.

7. Kambly, K.R.; Bradley, T.H. Estimating the HVAC energy consumption of plug-in electric vehicles. *J. Power Sources* **2014**, *259*, 117–124.

8. Yuksel, T.; Michalek, J.J. Effects of Regional Temperature on Electric Vehicle Efficiency, Range, and Emissions in the United States. *Environ. Sci. Technol.* **2015**, *49*, 3974–3980.

9. Neaimeh, M.; Hill, G.A.; Hübner, Y.; Blythe, P.T. Routing systems to extend the driving range of electric vehicles. *Intell. Transp. Syst. IET* **2013**, *7*, 327–336.

10. Messagie, M.; Boureima, F.S.; Coosemans, T.; Macharis, C.; Van Mierlo, J. A range-based vehicle life cycle assessment incorporating variability in the environmental assessment of different vehicle technologies and fuels. *Energies* **2014**, *7*, 1467–1482.

11. Messagie, M.; Lebeau, K.; Coosemans, T.; Macharis, C.; van Mierlo, J. Environmental and financial evaluation of passenger vehicle technologies in Belgium. *Sustainability* **2013**, *5*, 5020–5033.

12. Nordelöf, A.; Messagie, M.; Tillman, A.M.; Söderman, M.L.; van Mierlo, J. Environmental impacts of hybrid, plug-in hybrid, and battery electric vehicles-what can we learn from life cycle assessment? *Int. J. Life Cycle Assess.* **2014**, *19*, 1866–1890.

13. de Cauwer, C.; Maarten, M.; Coosemans, T.; van Mierlo, J.; Heyvaert, S. Electric vehicle use and energy consumption based on real-world electric vehicle fleet trip and charge data and its impact on existing EV research models. In Proceedings of the International Electric Vehicle Symposium and Exhibition, Kintex, Korea, 3–6 May 2015; pp. 1–11.

14. Lebeau, P.; de Cauwer, C.; van Mierlo, J.; Macharis, C.; Verbeke, W.; Coosemans, T. Conventional, hybrid, or electric vehicles: Which technology for an urban distribution centre? *Sci. World J.* **2015**, *2015*, 302867:1–302867:11.

15. Lohse-Busch, H. *Advanced Powertrain Research Facility, Nissan Leaf Testing and Analysis*; U.S. Department of Energy: Washington, DC, USA, 2012.

16. Lv, C.; Zhang, J.; Li, Y.; Yuan, Y. Mechanism analysis and evaluation methodology of regenerative braking contribution to energy efficiency improvement of electrified vehicles. *Energy Convers. Manag.* **2015**, *92*, 469–482.

17. Mammosser, D.; Boisvert, M.; Micheau, P. Designing a set of efficient regenerative braking strategies with a performance index tool. *Proc. Inst. Mech. Eng. Part D J. Automob. Eng.* **2014**, *228*, 1505–1515.

18. Zou, Z.; Cao, J.; Cao, B.; Chen, W. Evaluation strategy of regenerative braking energy for supercapacitor vehicle. *ISA Trans.* **2014**, *55*, 234–240.

MDPI AG

St. Alban-Anlage 66

4052 Basel, Switzerland

Tel. +41 61 683 77 34

Fax +41 61 302 89 18

http://www.mdpi.com

Energies Editorial Office

E-mail: energies@mdpi.com

http://www.mdpi.com/journal/energies

www.ingramcontent.com/pod-product-compliance
Lightning Source LLC
Chambersburg PA
CBHW051926190326
41458CB00026B/6426